SUMMARY OF TOOLS FOR CONTINUOUS IMPROVEMENT

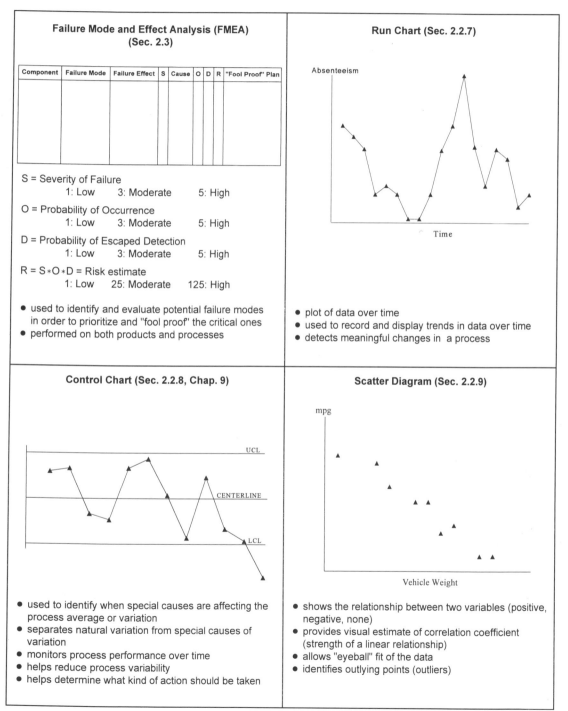

Failure Mode and Effect Analysis (FMEA) (Sec. 2.3)

Component	Failure Mode	Failure Effect	S	Cause	O	D	R	"Fool Proof" Plan

S = Severity of Failure
    1: Low   3: Moderate   5: High

O = Probability of Occurrence
    1: Low   3: Moderate   5: High

D = Probability of Escaped Detection
    1: Low   3: Moderate   5: High

$R = S * O * D$ = Risk estimate
    1: Low   25: Moderate   125: High

- used to identify and evaluate potential failure modes in order to prioritize and "fool proof" the critical ones
- performed on both products and processes

Run Chart (Sec. 2.2.7)

Absenteeism

Time

- plot of data over time
- used to record and display trends in data over time
- detects meaningful changes in a process

Control Chart (Sec. 2.2.8, Chap. 9)

UCL

CENTERLINE

LCL

- used to identify when special causes are affecting the process average or variation
- separates natural variation from special causes of variation
- monitors process performance over time
- helps reduce process variability
- helps determine what kind of action should be taken

Scatter Diagram (Sec. 2.2.9)

mpg

Vehicle Weight

- shows the relationship between two variables (positive, negative, none)
- provides visual estimate of correlation coefficient (strength of a linear relationship)
- allows "eyeball" fit of the data
- identifies outlying points (outliers)

(continued on inside of back cover)

BASIC STATISTICS
TOOLS FOR CONTINUOUS IMPROVEMENT

ABOUT THE BOOK

The primary motivation for this book is to educate the readers (students, managers, engineers, researchers, analysts, practitioners, and statisticians) in the application of statistical tools to achieve continuous improvement in how they do business.

ABOUT THE AUTHORS

Mark J. Kiemele, co-founder and senior partner of Air Academy Associates, has more than 25 years of teaching and consulting experience. Having trained more than 10,000 scientists, engineers, managers, trainers, practitioners, and college students from more than 20 countries, he is world-renowned for his approach to teaching statistical methods to nonstatisticians. His support is requested by an impressive list of clients, including Sony, Microsoft, General Electric, Raytheon, AlliedSignal, Lockheed Martin, EG&G, Corning, Data General, Coilcraft, Chevron, Walker Parking, and Abbott Laboratories. He earned a B.S. and M.S. in Mathematics from North Dakota State University and a Ph.D. in Computer Science from Texas A&M University. During his time in the U.S. Air Force, Dr. Kiemele supported the design, development and testing of various weapon systems, including the Maverick and Cruise Missile systems, and was a professor at the USAF Academy. In addition to many published technical papers, he has co-authored the books *Knowledge Based Management (KBM)*, and the AT&T Bell Labs book entitled *Network Modeling, Simulation, and Analysis*. He has also edited the text *Understanding Industrial Designed Experiments*.

Stephen R. Schmidt, founder of Air Academy Associates, a multi-million dollar international consulting business, has more than 20 years of teaching and consulting experience in process improvement methodology. His "Keep It Simple Statistically" (KISS) approach has gained him widespread popularity and an impressive list of clients: General Electric, Sony, Motorola, Boeing, Texas Instruments, Lockheed Martin, AlliedSignal, Levi Strauss, Abbott Labs, Citicorp, Ford Motor Company, Dannon, National Semiconductor, plus many others. He has served as an adjunct faculty member to the Motorola Six Sigma Research Institute where he trained the first wave of Black Belts at Motorola and Texas Instruments and he provides management training and serves as a principal instructor for the Accelerated Six Sigma program at The University of Texas at Austin. He has a B.S. in Math from the USAF Academy, an M.S. in Operations Research from the University of Texas, and a Ph.D. in Applied Statistics from the University of Northern Colorado. He served for 20 years in the U.S. Air Force as an instructor pilot and tenured professor at the USAF Academy. Dr. Schmidt also co-authored three recent texts: *Understanding Industrial Designed Experiments; Knowledge Based Management (KBM)* and *Total Quality: A Textbook of Strategic Quality, Leadership and Planning*.

Ronald J. Berdine is an experienced facilitator in the areas of problem solving, metric development, and implementation of process improvement strategies. Dr. Berdine retired after twenty years in the U.S. Air Force where he served as an Electronic Warfare Officer, Navigator, Associate Professor and Head of the Department of Mathematical Sciences at the USAF Academy. He has taught and consulted for AlliedSignal, Atmel, Sony, Lockheed Martin, Bombardier, General Electric, Motorola, Abbott Labs, Lenox, AB Dick and numerous Department of Defense Agencies. His extensive consulting background involves industrial and service applications of DOE, SPC, reliability, and management techniques. He is co-author of the text *Knowledge Based Management (KBM)*. He received his B.S. in Mathematics from Iowa State University, his M.S. in Operations Research from Stanford University, and his Ph.D. in Statistics from Texas A&M University.

BASIC STATISTICS

Tools for Continuous Improvement

Fourth Edition

Mark J. Kiemele   **Stephen R. Schmidt**

Ronald J. Berdine

Colorado Springs, Colorado

Library of Congress Catalog Card Number: 97-070783

ISBN 1-880156-06-7

Printed in the United States of America

26 25 24   -17 16 15

Cover design: Bernard Sandoval, Sandia, Colorado Springs, CO
Production/Graphics: Beatriz Orozco, Air Academy Press, Colorado Springs, CO

The authors recognize that perfection is unattainable without continuous improvement.
Therefore, we solicit comments as to how to improve this text.   To relay your comments or
to obtain further information, contact:

AIR ACADEMY PRESS & ASSOCIATES, LLC
1650 Telstar Drive, Suite 110
Colorado Springs, CO 80920
Phone: (719)531-0777 ◆ FAX: (719) 531-0778
email: aaa@airacad.com
website: www.airacad.com

Preface

According to Francis Bacon, an English philosopher and statesman and father of the scientific method, "Knowledge is Power." Knowledge is not based on opinion; rather, knowledge is derived from facts and data. In order to efficiently collect and effectively analyze data to extract the maximum knowledge available, society must rely on statistical techniques. For years these techniques have been taught at a level that turned many people more "off" than "on" as to the power of the techniques. This book is our attempt to revolutionize the presentation of basic statistical techniques by incorporating our well known Keep It Simple Statistically (KISS) approach.

Today's international marketplace is engulfed with stiff competition and necessitates a change in how we do business. Customers for any product or service are more discerning than ever before and demand the highest quality possible at the most competitive prices. To remain competitive in the 1990's and on into the 21st century, U.S. businesses must learn from their competition. It is well documented that the Japanese have utilized statistical tools to design and produce high quality goods and services at low cost. A nationwide culture change has been underway in the form of the total quality movement (see Chapter 1). To experience continuous quality improvement in the long run, companies will need to utilize the power of statistical techniques. According to Mr. Craig Barrett, Senior Vice President of Intel (certainly one of America's most successful companies), "Statistics is the key to industrial competitiveness." The Accreditation Board for Engineering and Technology (ABET) is also advocating the blending of statistical tools into existing engineering curriculums [Fong, Jeffery T., "Engineers' Statistical Literacy is Key to U.S. Competitiveness," *ASME News*, Vol. 9, No. 3, Oct 1989].

Since the problem has not been a lack of statistical tools but rather the way they have been presented, this fourth edition has been written to provide an understandable approach to the presentation of a wide ranging scope of statistical tools, including the techniques necessary to comply with ISO-9000, QS-9000, D1-9000, process characterization and process validation. Although the subject discussed in this text is basic statistics, keep in mind that the real objective is to use simple statistical techniques to transform data into the knowledge needed to make good decisions.

This fourth edition is the culmination of many contributions, suggestions and constructive criticisms from our many devoted industrial and academic readers. The following major additions and improvements have been implemented in the fourth edition, making it a much more complete and comprehensive treatment than the previous edition.

(1)   Increased emphasis on the use of statistical tools for gaining knowledge about processes, products, people, and organizations.

(2)   New or revised sections on:

    (a)   Knowledge Based Management

    (b)   Developing Valid Metrics and Scorecards

    (c)   Success Stories

    (d)   Affinity Diagrams

    (e)   Fault Tree Analysis

    (f)   Failure Mode and Effect Analysis

    (g)   Box Plots

    (h)   Discrete Uniform Distribution

    (i)   Continuous Uniform Distribution

    (j)   Residual Analysis

    (k)   Three-Level Designs for Non-Linear Models

    (l)   Three-Level Screening Designs

    (m)   Selecting a Design and a Sample Size

    (n)   Understanding Variation

    (o)   Process Control vs Process Capability

    (p)   Control Chart Philosophy and Interpretation

    (q)   Reliability Confidence Limits for the Exponential Failure Model

    (r)   Quality Function Deployment

    (s)   Control Chart Background

(3)   Eleven new case studies:

    (a)   A Generalized Transplantation Process for Patients with End Stage Renal Disease

    (b)   Relationship Between Spousal Abuse and Spousal Murder

    (c)   Historical Data Analysis of a Plating Process

    (d)   Reducing Radar Cross Sections

    (e)   Process Development for Bonding Titanium to Cobalt Chrome

    (f)   Process Control and Capability in the Automotive Industry

    (g)   Pesticide Mix Interaction Effect

    (h)   Ice Cream, Engineering, and Statistics

    (i)   Developing a Risk Management Model

    (j)   Lemon/Lime Juicer Project

    (k)   Statistical Analysis of a Randomization Algorithm

(4)   An expanded treatment of designed experiments to include 3-level modeling and screening designs, as well as a simplified approach to selecting a design and a sample size.

(5)   A completely rewritten Chapter 1 (Why Statistics) emphasizing the use of statistical tools as a means to an end rather than an end in itself.

(6)   All 30 case studies have been put in their own chapter, Chapter 12. There are approximately 200 pages of case studies.

(7)   A completely rewritten chapter on Statistical Process Control (Chapter 9), blending more service oriented examples into the applications.

(8)   A simplified treatment of what Six Sigma quality means, as well as a comparison and contrast of the following capability measures: dpm, σ_{level}, C_{pk}, and C_p.

(9)   A detailed treatment, with graphical illustrations, of out-of-control symptoms.

(10)   A new, more powerful software package, SPC KISS Student Version, accompanies the text.

This text is recommended for use at the college level and within industry to educate the readers (students, managers, engineers, scientists, researchers, analysts, practitioners, and statisticians) in the application of statistical tools to achieve continuous improvement in how they do business. We designed this text specifically for those who are taking a first course in statistics, as well as for those who may have already taken a statistics course (perhaps some time ago) but need a refresher. This fourth edition could be used in a two-semester sequence of courses in statistical techniques for continuous process improvement. It should serve well as a reference handbook for any practitioner using statistical tools for process improvement. Calculus is not a prerequisite for studying the material in this book, although the astute student who has had calculus should be able to see where calculus could have been used to obtain some of the results presented. There are no proofs in this book, and the emphasis is on applications, examples, and the view of practitioners.

The text can be partitioned into five major categories:

Basic Tools:	Chapters 1-4
Intermediate Tools:	Chapters 5-7
Advanced Experimentation, Modeling, and Process Analysis:	Chapters 8-9
Advanced Topics:	Chapters 10-11
Case Studies:	Chapter 12

Chapter 1 is a motivational chapter to show the reader why statistics is needed and where and how statistical tools have been used successfully. An important feature of this chapter is to show how processes, measurement, metrics, and quality improvement are related to the fundamental principal of gaining the right kinds of knowledge. Chapter 2 introduces the basic terminology and tools of descriptive statistics, emphasizing the use of graphical and measurement tools to transform data into information. Measures of central tendency and dispersion are covered for both individually listed data points and grouped data. Chapters 3 and 4 present a minimal, yet necessary, amount of information on probability and probability distributions.

The distributions are presented because: (1) they can be used directly to solve problems and (2) they form the basis for other process improvement tools such as control charts.

Chapter 5, on sampling distributions and estimation, provides a link to inferential statistics, whereby sample data can be used to infer something about the population from which the sample data was extracted. Confidence intervals and sample size determination for both infinite and finite populations are presented in this chapter. The intermediate tools of hypothesis testing and regression are covered in Chapters 6 and 7. Rules of Thumb for detecting shifts in average and standard deviation are presented to provide simplified alternatives for the t-test and F-test, respectively.

Chapter 8 provides detailed coverage of the topic of designed experiments. A blending of the classical and Taguchi approaches is presented. Ready-to-use experimental design templates are provided, along with Rules of Thumb for determining which design to use and how many replicates to perform. Models for the process average and process standard deviation are developed and interpreted. Screening and modeling designs for both 2 and 3-level designs are covered. Statistical Process Control (SPC) is the topic of Chapter 9, which has been rewritten to provide a more holistic view of understanding variation. Process control and capability are compared and contrasted, and out-of-control symptoms are presented. The concept of Six Sigma is addressed and capability measures such as dpm, σ_{level}, C_{pk}, and C_p are defined and illustrated.

Chapters 10 and 11 provide brief introductions to the topics of reliability and quality function deployment, respectively. Reliability is presented as an application of some of the tools presented earlier; and quality function deployment (QFD) is a quality improvement activity designed to effectively and efficiently translate customer requirements into those key quality characteristics that must be measured and for which tolerances must be established. Chapter 11 has been completely rewritten to provide a simpler and more straightforward approach to the topic of QFD.

Chapter 12 provides 30 case studies and approximately 200 pages of real applications of the statistical tools presented in the text. These case studies span a wide variety of applications, both in service and manufacturing. In some cases, the actual data from the case study is not presented due to the proprietary nature of the application.

This edition provides the reader with a powerful statistical package called SPC KISS (Student Version). When installed, SPC KISS runs with MS Excel (Version 5.0 and later) as a new main menu option on the MS Excel menu bar. The user's guide for SPC KISS is provided in Appendix L.

H. G. Wells, the noted English author, stated almost a century ago, "Statistical thinking will one day be as necessary for efficient citizenship as the ability to read and write." The authors believe that day has come. This text offers the interested reader an opportunity to improve the quality of his or her statistical literacy and thus become a better problem solver and decision maker.

This text was written based on how we perceived the "voice of the customer." If you, our customer, have any suggestions or criticisms, please write or call Air Academy Press. Since continuous improvement is our goal, we will be implementing changes and additions as the need arises. Thank you for purchasing our text — we wish you the best at implementing these techniques.

Mark J. Kiemele   Stephen R. Schmidt   Ronald J. Berdine

April 15, 1997

Acknowledgments

Teamwork is a critical ingredient to the success of any industry and it is a major part of today's quality movement. If it were not for teamwork, this text would not exist. Although our names are listed as the authors, we wish to thank all those who were part of the team that produced this text. The following is no doubt only a partial list of those who have contributed.

Jonathon Andell

Barbara Bicknell

Mike Bishop

Carl Bodenschatz

Tuiren Bratina

James Brickell

Tom Cheek

Rai Chowdhary

Susan Darby

Ron Duck

Bob Frost

Jan Gaudin

Bill Gaught

Jay Gould

Mikel Harry

Charles Hendrix

Suellen Hill

Pete Hofmann

Bert Hopkins

Peter Jessup

Carol Kiemele

Kenneth Knox

James Kogler

K. D. Lam

Dan Litwhiler

Dennis Mahoney

Kim Mayfield

Philip Mayfield

Al Memmolo

Ken Mooney

W. T. Motley

Terry Newton

Cher Nicholas

Jim Norton

Beatriz Orozco

Lee Pollock

Lisa Reagan

James Riggs

Ed Robertson

Kaye de Ruiz

Jim Rutledge

Paul Ruud

Vicki Schmidt

Wen Shih

Augustine Smith

Gary Steele

John Stinson

Heather Tavel

John Tomick

George Trudeau

Jim Watters

Glen Wiggy

Buddy Wood

Vern Yetzer

Special thanks to our families who stood by us and encouraged us despite the demands of putting a text such as this together. And most importantly, thanks to God for giving us the talents, wisdom and strength to make this book a reality.

Contents

Case Studies

Applicability Matrix

Case Studies

Chapter	1	2	3	4	5	6	7	8	9	10	11	12	13	14	15	16	17	18	19	20	21	22	23	24	25	26	27	28	29	30
1	*																													
2	*	*	*	*	*	*	*						*		*													*	*	
3								*	*	*	*	*													*					
4									*	*	*	*													*					*
5																*														*
6																*														
7													*																	
8							*							*	*	*	*	*	*					*						
9																				*	*	*	*	*	*					
10																														
11																										*	*	*	*	

The shaded boxes indicate Case Studies for Transactional Processes.

Chapter 1

Why Statistics

1.1 Applications of Statistics[1]

Through the use of advanced electronic communication and computer technologies, today's society is inundated with vast amounts of data which is typically stored in high speed computers. In its raw form (e.g., long lists of numbers, names, places), this data is of little value. However, when manipulated with statistical tools, the data can be transformed into valuable information (numeric and/or graphic). This knowledge is vital for drawing conclusions and making decisions. Since statistical tools are required to gain useful information or knowledge from data, the Internet Age of today is rapidly generating the statistical age of tomorrow. Interestingly enough, a well-known English author predicted this trend some time ago.

> *"Statistical thinking will one day be as necessary for efficient citizenship as the ability to read and write."*
>
> **H.G. Wells**
> (ca 1925)

The world-wide spread of the quality movement begun by W. Edwards Deming, Joseph M. Juran and many others is also an indisputable realization that the statistical age is upon us. Consider the following quotes.

---

[1]Much of this section has been taken with permission from D. Hinders, *Focus on Statistics*, the Woodrow Wilson National Fellowship Foundation, P.O. Box 642, Princeton, NJ 08542.

> *"Sound understanding of statistical control is essential to management, engineering, manufacturing, purchase of materials, and service."*
>
> **W. Edwards Deming**
> (ca 1955)

> *"Product and service quality requires managerial, technological, and statistical concepts throughout all the major functions in an organization."*
>
> **Joseph M. Juran**
> (ca 1960)

Far too often the meetings and conversations of today tend to focus on opinions, intuition, emotions, and other non-numeric or subjective sources of information, to which a noted modern-day columnist has responded.

> *"Not knowing the difference between opinion and fact makes it difficult to make good decisions."*
>
> **Marilyn Vos Savant**
> (ca 1995)

Although there is value in non-numeric information, continuous improvement in knowledge requires the inclusion of numeric sources as well. Consider the following question:

Should the wearing of seat belts be compulsory by law?

Without numeric information, the public really lacks sufficient knowledge to answer this question. Extracting data from police reports and summarizing it through the use of basic statistics enables the public to formulate an answer. Table 1.1 summarizes casualty data recorded prior to when seat belts were made mandatory. After reviewing the data in Table 1.1 what would you conclude?

Casualties from Automobile Accidents (Drivers & front seat passengers in cars and light vans)				
Safety Belt	**Killed**	**Serious**	**Slight**	**Total**
Worn	4	72	447	523
Not Worn	62	612	2543	3217
Total	66	684	2990	3740

Table 1.1 Data on the Use of Safety Belts

Statistics is playing an increasingly important role in nearly all aspects of society. Unfortunately, the lack of a clear understanding of statistics sometimes results in inaccurate information. One very good source for helping the reader understand the ways in which statistics can be misused is *How To Lie With Statistics* by Darrell Huff. It is short, humorous, and full of clever examples of ways to dupe your neighbor using "good" statistical techniques. It is somewhat out of date (e.g., "The average Yale graduate makes $13,000 a year.") but the principles discussed are timeless. Many of the misuses of statistics result from a poor understanding of the techniques. When presented in a Keep It Simple Statistically (KISS) manner, the power of statistics is within the grasp of almost everyone. Society as a whole could benefit substantially from a working knowledge in the correct application of the basic, yet powerful, statistical tools presented in this text. Some examples reflecting the broad applications of statistics and its uses/abuses are as follows:

ADVERTISING

The reference for the following example is *Consumer Reports*, August, 1971. *Consumer Reports*, by the way, is another excellent and continuing source of examples of the uses of statistics in society.

In the late 1960's and early 1970's, an Excedrin commercial starring David Janssen (of "Fugitive" fame) ran on television. He seriously proclaimed, "Tests conducted at a famous hospital have proven that two Excedrin contain twice as much pain reliever as four of the best selling aspirin."

The implication here, of course, was that, if we have a headache, we can use less Excedrin (less cost??) than if we were to use aspirin. What Mr. Janssen, in his enthusiasm for Excedrin, failed to tell the viewers is that the tests (conducted in a Philadelphia hospital) were not conducted on headache pain but on post-partum pain. New mothers who complained of pain were given either Excedrin, aspirin, or a placebo. This example is what *How To Lie With Statistics* would describe as follows: when we can't prove what we want to prove, prove something else instead and pretend they are the same thing. There were some other strange things in the data and at least one network, NBC, recognized the fallacious nature of the Excedrin commercials and refused to run them.

OPINION POLLING

We are inundated with polls in America. When an election nears, it is almost impossible to pick up a newspaper and not read about one poll or another. We must understand that polls are subject to experimental error under the best of conditions and that, improperly done, they may simply be wrong. Gallup was roundly criticized for miscalling the 1948 Truman-Dewey presidential race. He had Dewey on top in a very close race and looked bad when the final figures gave the election to Truman. The primary reason for Gallup's inaccurate prediction was his use of a non-representative sample from which he drew erroneous conclusions.

More recently, the polls erred badly in calling the 1980 presidential race between Carter and Reagan. The consensus of the major polls was that the race was probably too close to call, although a shift toward Reagan was being picked up. Much soul-searching occurred among the major pollsters and many press releases were issued explaining away the disaster when Reagan swept the country. Interestingly enough, the private pollsters for both Carter and Reagan knew very clearly before the election that a landslide was in the making. These private pollsters had a lot more money and were able to run polls almost daily prior to the election.

Remember that a poll is accurate only at the instant it is taken. In 1980, there was apparently a major shift to Reagan very close to the election (hastened by a downturn in the fortunes of the hostages in Iran the weekend before the election), and the major pollsters simply were not able to poll close enough to the election to detect the shift. They were not necessarily wrong in the way they were calling the race; they were just calling it too early.

Sometimes things are done incorrectly. In 1936, *Literary Digest* magazine sampled some 10 million voting age persons for their preferences in the Landon-Roosevelt presidential race. Almost 2.4 million of these persons returned the questionnaire indicating their preferences for president. Since the major pollsters now use samples in the 1000-1500 voter range, that sample size should have been more than adequate to make a correct inference about the entire voting population of the United States in 1936. Based on the survey results, the now defunct *Literary Digest* predicted a massive Landon victory. They were wrong by 19 percentage points!

Two explanations have been advanced for this disaster, which some have argued led to the demise of the *Digest*. The first, and most commonly advanced, is that the *Digest* chose its 10 million person sample mostly from lists of automobile and telephone owners. In 1936, this represented an economically more advantaged group that was, hence, more Republican. The argument here is that the 2.3 million returned questionnaires may have been random for the universe from which they were selected but, because the universe was overwhelmingly Republican, the poll basically found that most Republicans were probably going to vote Republican.

The other, and more statistically satisfying, explanation is that a 23% return rate on a survey is likely to be biased in some way. There is simply no way to generalize to a large population when only 23% of a sample, regardless of its size, elects to return the questionnaire. There are ways of dealing with small sample returns, but they were not employed by the *Literary Digest*. Clearly, under this alternative explanation, the 23% that did choose to return their questionnaires were overwhelmingly Republican and, hence, the prediction that Landon would win.

In fairness to the major pollsters, on whom much abuse was heaped because of the 1948 and 1980 elections, we should mention that Gallup, Roper, and Crossley, using more scientific sampling techniques, all called the 1936 race accurately in favor of Roosevelt.

SPORTS

Introductory concepts in statistics can be taught more effectively if the data serve to motivate students. Many students have some interest in sports, and more raw data abounds in sports than in any other area of modern American society. One only has to read the daily paper to find a wealth of examples of sports statistics. One especially rich source for data is the *Baseball Abstract* published annually by Bill James. The interested reader is referred to this and other sports almanacs for a myriad of sports data.

DRAFT LOTTERY[2]

A curious example of a non-random drawing occurred during the first draft lottery held on December 1, 1969. The idea was that those men who would turn 18 in 1970 would be ranked in terms of their eligibility for the draft. Because we were deeply involved in Vietnam at the time, one's location on the list could be viewed as a life or death matter. The idea, basically, was that the 366 possible birthdays would be written on slips of paper, put into a hat, and drawn in random order. The first birthday drawn would be number 1 to be drafted, the second birthday would be number 2, etc. In order to be fair, the intent was that the likelihood of being drafted high or low was even, and everyone was to have an equal chance of being drafted or not being drafted.

The idea was sound except that something went wrong. When the data from the lottery were analyzed, those who were born in the final six months of the year were found to be much more likely to have a low draft number than those born in the first six months of the year. That is, your chances of being drafted were much greater if you were born in July-December than if you were born in January-June. The highest average draft number was 225.8 in March, and the numbers trailed down to only 121.5 in December.

---

[2]The reference for this information is Mosteller, et. al., Statistics By Example: Finding Models, Addison-Wesley Co., Menlo Park, CA, 1973. This is the fourth of four books in that *Statistics By Example* series, all of which contain a wealth of good examples of the use of statistics in society.

Statistical analysis of this event indicated that the pattern was unlikely to have occurred by chance. The reason was pretty obvious. The capsules containing the slips of papers were put into a box month by month, putting January in first, then February, etc. Then the box was shaken a few times and the contents poured into a large fishbowl, but they were not stirred. Apparently this resulted in December's being mostly in the bottom layer of the fishbowl, November's being mostly in the next layer, etc., until the top layer was mostly January. The "random" drawing then tended to select capsules from the bottom so that the later months had the low numbers and the earlier months the high draft numbers. The process was corrected in 1971 to produce a more random drawing.

MEDICINE[3]

Statistics has one of its most valuable uses in the medical field because the payoffs for successful research are so widespread. By the same token, bad research fails to find cures and may produce "cures" that really aren't, thus giving false hope to the afflicted.

One of the largest medical research projects ever undertaken was the testing of a vaccine for polio. In the 1950's, everyone was terrified of polio. Large, tube-like devices called iron lungs were home for many seriously affected polio victims. Even today, many adults retain the withered limb obtained from a bout with childhood polio. Many adults over the age of 50 well remember the fear of community swimming pools because of a polio scare. Polio was a strange disease because it tended to be more prevalent in well-to-do areas. It would be epidemic in one community and barely touch a neighboring community, or it would be epidemic in a community one year and barely appear the next. These characteristics of polio made it difficult to design a field study for a vaccine because there seemed to be large numbers of confounding variables. Even more confounding was the fact that polio, even when epidemic, simply didn't strike very many people. It might strike only 50 out of every 100,000 or so. Thus, any field test would have to

---

[3]This example of the uses of statistics in medical and scientific research, as well as many others, is contained in Tanur, Mosteller (Eds.), ***Statistics: A Guide To The Unknown***, Holden-Day, Inc., San Francisco, CA, 1972.

involve more than a million people in order to have an adequate sample size. The field test on the Salk vaccine involved 1.83 million people and was a double-blind experiment.   That is, neither the persons taking the vaccine nor the persons administering it were aware of whether they were dealing with the vaccine or a placebo. The results of the field test were convincing enough that the Salk vaccine was approved as an anti-polio treatment. The Salk vaccine was soon replaced by another more effective vaccine, but the massive field test, statistically analyzed, was a major step in the successful battle against polio.

CRYPTOLOGY

At first glance, the reader might consider this application of statistics to be somewhat odd. While cryptology is considered to be the art and science of encrypting and decrypting messages, statistics is involved in the organization and analysis of data. How are the two related? Statistics can be used to study the frequencies of symbols in an encrypted message (e.g., a cryptogram) and relate these frequencies to the frequencies of letters that occur in normal text of the English language. For example, suppose the cryptanalyst (a cryptologist who specializes in decrypting messages) wishes to decrypt the following secret message:

WKH GHFODUDWLRQ RI LQGHSHQGHQFH

First, by studying ordinary English text, the cryptanalyst is able to develop a table of frequencies (i.e., a frequency distribution) for the various letters in the alphabet. This table might appear as follows (from most frequently occurring letters to the least):

E:	13.6%	L:	3.4%	W:	1.5%
T:	9.2%	H:	3.1%	V:	1.4%
N:	7.9%	C:	2.9%	B:	1.1%
I:	7.4%	F:	2.8%	X:	0.8%
O:	7.3%	P:	2.8%	K:	0.6%
A:	7.2%	U:	2.5%	J:	0.4%
R:	7.1%	M:	2.3%	Z:	0.3%
S:	6.0%	Y:	1.8%	Q:	0.2%
D:	4.8%	G:	1.6%		

The cryptanalyst can then reason that the most frequently occurring symbols in the cryptogram might correspond to the most frequently occurring letters in the table. For the previous cryptogram, in which 'H' is the most frequently occurring symbol, it might be a good guess that 'H' in the encrypted message corresponds to the letter 'E' in ordinary English. This may or may not be correct, but it certainly is more likely that the 'H' stands for an 'E' rather than for a 'Q' or 'Z.'

Using the frequency rationale just presented, try your hand at decrypting the cryptogram. The spaces between words have been preserved to make life easier for you. Reference Problem 1.4, the answer to which is given in Appendix M: Answers to Selected Problems. If you correctly decrypted the message, consider yourself a worthy contestant for "The Wheel of Fortune."

LAW

There are many instances of the use of probability in legal settings. Some involve examining juries to see if the jury fairly represents the community from which it is drawn. For example, there is a classic case, Dr. Spock's anti-war activity trial, in which it was found that women were underrepresented (see Tanur,

Statistics, A Guide To The Unknown, pp. 139-149). Another more recent case in Richmond, California, determined that the jury underrepresented the percentage of blacks in the community and that this was a condition that needed to be changed in the interest of a just legal system.

There are many applications of the use of probability and statistics in law. A good source having a wide variety of examples is [Degr 86]. An interesting case involved a purse snatching incident in the Los Angeles suburb of San Pedro. In 1964, a Mrs. Juanita Brooks, 71, was knocked down on the street and her purse containing $35 was stolen. Mrs. Brooks was unable to positively identify the defendants in the case, Michael and Janet Collins, and it appeared at first that they would be acquitted of the charges against them.

The district attorney figured otherwise, however. It seems that a couple bearing certain characteristics had been seen running from the alley in which the mugging occurred, and these characteristics matched the Collins' very closely. The D.A. brought in as an "expert" witness a mathematics professor from a local college who presented the following probabilities:

(1)	Man with beard	....	1/10
(2)	Blond Woman	....	1/4
(3)	Yellow car	....	1/10
(4)	Woman with ponytail	....	1/10
(5)	Man with mustache	....	1/3
(6)	Interracial couple in car	....	1/1000

Michael Collins was a black man with a mustache and a beard. Janet Collins was a blond white woman with a ponytail. They were married and owned a yellow car. The mathematics professor argued that the correct way to determine the probability that a given couple simultaneously possessed all six characteristics was to multiply the individual probabilities together. Hence, he reasoned, there was only one chance in 12,000,000 of finding such a couple. Since such a couple demonstrably did exist — the Collins were, after all, on trial — and since there was such a low probability of turning up such a couple by chance alone, it was likely, argued the D.A., that the Collins were, in fact, the guilty couple.

This was enough for the jury. Convinced by the "mathematics" they had seen, they duly convicted the Collins of the crime. It was, of course, appealed and eventually overturned because of some interesting statistical analyses we present in Case Study 9: Probability and the Law. Please refer to this case study for the details.

INDUSTRY

Not long ago, manufacturing industries in North America provided the majority of jobs for the working population. This is now no longer the case. Far more people work in information related jobs and/or the service industries than in manufacturing. Furthermore, all industries, be they service or manufacturing, are now required to be more efficient in order to compete worldwide. This shift toward increased efficiency, and toward "informed industrial decision making," requires workers to play an integral part in the collection of needed information. To do so, workers require a much more sophisticated knowledge of statistics than they had before.

Throughout this text, the reader will find that to remain competitive in today's industrial environment requires the use of basic statistical tools for continuous improvement in quality. Global competitiveness and fixed pricing in health care and government contracts is also forcing the non-manufacturing industry to make quality improvements and reduce waste. As a result we must all be open to new concepts concerning quality and the use of statistics.

Consider the following erroneous perception of quality in American industry, as illustrated in Figure 1.1. By building parts to specification (to blueprint) we imply that parts anywhere within specification are of equal quality. This is simply not true. Consider the diameter of a shaft built at the upper specification limit which is mated with a sleeve built at the lower specification limit. It should be obvious that performance and reliability of these parts will not be the same as if they were both built at the designed target values.

Today's consumers of goods and services are demanding a new philosophy as indicated in Figure 1.2. Based on this new philosophy, producers of goods and services must focus on target values for critical dimensions. To achieve the ultimate in quality (all critical dimensions are produced exactly the same at their properly designed targets) requires the use of basic statistical tools. The next section provides an in-depth look at the quality movement which has spread across America.

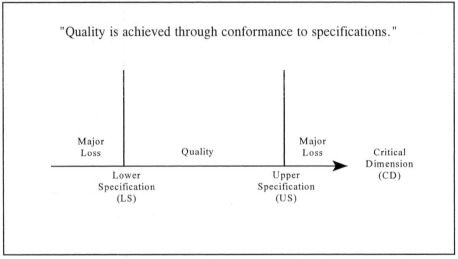

Figure 1.1 Erroneous Perception of Quality through Specs

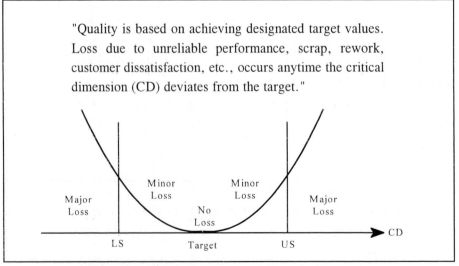

Figure 1.2 Quality Based on Achieving Target Value

1.2 The Quality Movement

America's industrial strength during the decade following World War II was the envy of all nations. Our market was the world. The demand for our products outpaced our ability to produce. Jobs were secure. Corporate profits soared. And in America we enjoyed what was perceived by the rest of the world to be the "good life." But it is difficult to manage success. American industry failed to have a clear vision for the future and fell behind to a country (Japan) willing to seek new methods in gaining knowledge. Evidence of this decline is shown in the following [Graz 89]:

■ The United States ranks last among the industrialized nations in the rate of productivity growth (as of January 1990).

■ Between 1970 and 1988, our share of the USA's consumer electronics market fell from 100% to under 5%.

■ In Japan it takes an auto worker 11 hours to build a car; in the United States it takes 31 hours (1989).

Having slipped from its pedestal of dominating world markets, American industry began to examine the successes of the Japanese. What they found were the philosophies of Deming, Juran, Ishikawa, and others who revolutionized how we do business. This revolution included the use of statistical techniques. For more information on these philosophies and the quality movement, please refer to *Knowledge Based Management* [Schm 96].

The 1980's and early 1990's were still difficult times for many industries. However, today we see that many companies have successfully fought back. It is important to note that they did not accomplish this success by only working harder. Rather, success in many cases came from the use of Keep It Simple Statistically (KISS) techniques to gain knowledge quickly in order to generate continuous quality improvement — in other words to make products and services better, faster, and at lower cost. This competitive advantage has positively impacted the bottom line of many companies.

Consider the following statements taken from ***Knowledge Based Management*** [Schm 96].

> *"For us quality is not a cost, it is a savings ($1.4 billion)."*
>
> Richard C. Buetow
> Motorola Senior Vice President
> and Director of Quality

> *"KISS techniques helped us fuse titanium and cobalt chrome for the first time in the world. Over the last 7 years this knowledge has been worth approximately $400 million."*
>
> Rai Chowdhary
> Intermedics Orthopedic Engineer

> *"Quality improvement is the key to our staggering increase of more than 60% in quarterly earnings."*
>
> Ray Strata
> Analog Devices, CEO

Another major success story is the comeback of our semiconductor industry. Here is what a Senior Vice President from Intel said at a 1989 ABET (Accreditation Board for Engineering and Technology) conference:

> *"Statistical literacy is the key to our industrial competitiveness."*
>
> Craig Barrett
> Intel Senior Vice President

The point is this: statistical techniques are needed to acquire the right kind of knowledge needed for success, no matter how success may be defined. The next section addresses a strategy for gaining that knowledge.

1.3 Knowledge Based Management (KBM)

The quality improvement philosophies of Deming, Juran, and Ishikawa have successfully captured a significant following. Yet many see these philosophies as emphasizing the "what" of quality improvement, leaving them struggling with "how" to make it happen. A closer look at these philosophies reveals a common but unwritten theme: that enhanced process and product/service knowledge is needed for improved quality. Deming used the term "profound knowledge." The basic building block for improving processes is knowledge about our processes, products, people, and organization. Knowledge Based Management (KBM) is an extension of the works of Deming, Juran, and Ishikawa with a focus on the "hows." KBM is a strategy for obtaining the right knowledge at the right time by the right people. Three major ingredients form the strategy of Knowledge Based Management (KBM). These are shown in Figure 1.3 as Questions Managers Need to Answer, Questions Managers Need to Ask, and Tools and Techniques to Answer Questions and Improve the Scorecard.

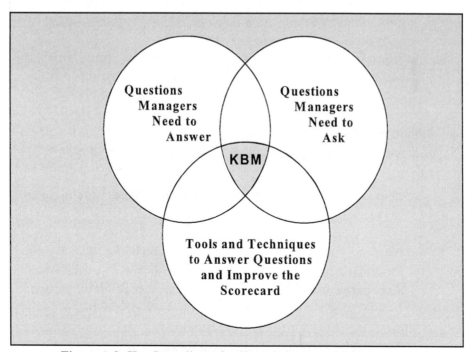

Figure 1.3 Key Ingredients for Knowledge Based Management

The reader interested in learning more about the questions managers need to answer and how the philosophy of KBM ties the questions and tools together is referred to the text ***Knowledge Based Management (KBM)*** [Schm 96]. Getting the right knowledge requires asking the right questions, and the questions managers need to ask should provide the impetus for gaining the knowledge needed to improve processes and quality. We list this set of questions in Table 1.2 for the purpose of providing the reader a reference point for why we need the tools presented in this text. A quick glance to the inside front and back covers of this book will give you an overview of the statistical tools that are needed to answer the questions in Table 1.2.

**QUALITY IMPROVEMENT ORIENTED QUESTIONS
MANAGERS NEED TO ASK THEIR PEOPLE!**

1. What processes (activities) are you responsible for? Who is the owner of these processes? Who are the team members? How well does the team work together?

2. Which processes have the highest priority for improvement? How did you come to this conclusion? Where is the data that supports this conclusion?

For those processes to be improved,

3. How is the process performed?

4. What are your process performance measures? Why? How accurate and precise is your measurement system?

5. What are the customer driven specifications for all of your performance measures? How good or bad is the current performance? Show me the data. What are the improvement goals for the process?

(Continued)

Table 1.2 Quality Improvement Oriented Questions
Managers Need To Ask Their People!

QUALITY IMPROVEMENT ORIENTED QUESTIONS
MANAGERS NEED TO ASK THEIR PEOPLE!

6. What are all the sources of variability in the process? Show me what they are.

7. Which sources of variability do you control? How do you control them and what is your method of documentation?

8. Are any of the sources of variability supplier-dependent? If so, what are they, who is the supplier, and what is being done about it?

9. What are the key variables that affect the average and variation of the measures of performance? How do you know this? Show me the data.

10. What are the relationships between the measures of performance and the key input variables? Do any key variables interact? How do you know for sure? Show me the data.

11. What setting for the key variables will optimize the measures of performance? How do you know this? Show me the data.

12. For the optimal settings of the key variables, what kind of variability exists in the performance measures? How do you know? Show me the data.

13. How much improvement has the process shown in the past 6 months? How do you know this? Show me the data.

14. How much time and/or money have your efforts saved or generated for the company? How did you document all of your efforts? Show me the data.

Table 1.2 Quality Improvement Oriented Questions
Managers Need To Ask Their People!

At this point it is important that the reader understand the context of the use of statistical tools, namely, that they are not an end in themselves but rather a means to a much grander purpose. This purpose is shown in Figure 1.4 where much broader concepts associated with quality of life are shown as we ascend the staircase. Unfortunately, the thrust for quality improvement that many companies ignited in the 1980's has fizzled in the 1990's. One of the major reasons is their failure to recognize that the proper knowledge about a process is a necessary step prior to quality improvement and any subsequent higher steps. Without the stable foundation stone of knowledge, we can easily stumble on our journey to process improvement and beyond.

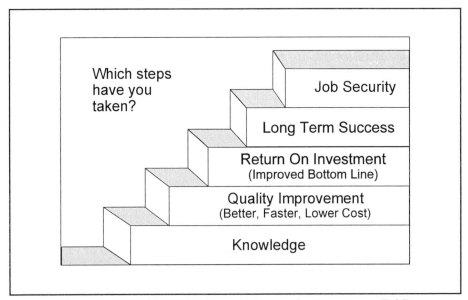

Figure 1.4 Critical KBM Steps to Return On Investment (ROI)
and Beyond

The serious student will undertake the study of the tools and techniques presented in this text with the end goal always in sight. That is, statistics are needed to answer the questions. The answers ultimately provide people with the knowledge needed to become better decision makers and successfully climb the staircase to a better quality of life. The next section addresses the context or scenario in which statistics play such an important role.

1.4 Processes, Measurement, and Process Improvement Tools

Whether we are involved in the service or manufacturing industry we come face to face with some type of process each day. In a general sense we will define a process as a blending of inputs to achieve some desired output. Figure 1.5 shows a general process diagram, also known as an Input-Process-Output (IPO) diagram. The outputs generally fall into one of three possible categories: performing a service, producing a product, or completing a task. The inputs typically are aligned under such categories as people, material, equipment, policies, procedures, methods, environment, etc.

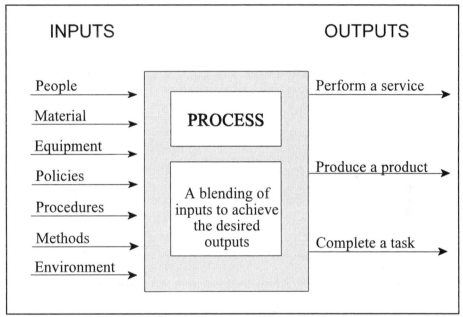

Figure 1.5 General Diagram of a Process

For a specific process, the practitioner should begin to think of an output as a *measure of performance*. The choice of outputs or performance measures should be based on how well the process performs with respect to customer (internal or external) requirements. Most performance measures will have something to do with cost, time, defect or error rate, or some other critical quality measure that is associated with fitness of use for the customer. It is important to note that these outputs or performance measures must be metrics, for if we cannot measure a process we will not know if or when a process is improving. It is difficult to improve anything if we cannot measure it. We don't know what we don't know, and we won't know unless we measure it.   Examples of specific measurable outputs for both manufacturing and transactional types of processes can be seen in the IPO diagrams shown in Figures 1.6 through 1.19. The processes shown in these figures address the following applications:

> Billing Process
> Machining Process
> Composite Material Process
> Mail Sorting Process
> Training Process
> The Process of Living
> Software Coding Process
> Purchasing Process
> Accounts Receivable Process
> Surgical Procedure
> Help Desk Call Center Process
> Mortgage Loan Administration Process
> Hiring Process
> Contract Preparation Process

By no means do the outputs shown in these diagrams reflect all of the critical performance measures that could be considered for these processes. They do show, however, that any process, whether it is a manufacturing or non-manufacturing process, can be described by inputs and outputs, where the outputs can be defined in terms of critical performance measures or metrics.   How well we measure these outputs is crucial to determining if supposed improvements have been made, as well as to making future improvements. For discussions on measurement system capability, see Case Study 7: A Simple Measurement System Study and Section 9.8:   Gage Capability Analysis.

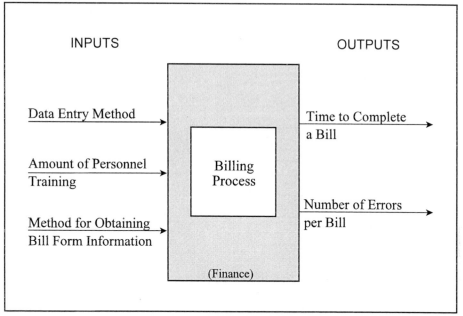

Figure 1.6 Billing Process Diagram

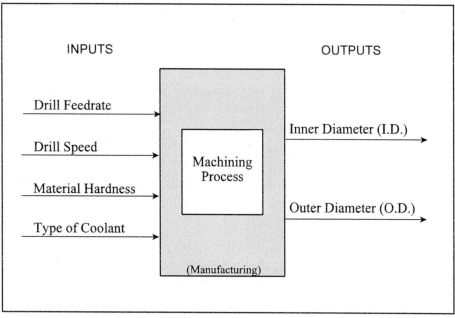

Figure 1.7 Machining Process Diagram

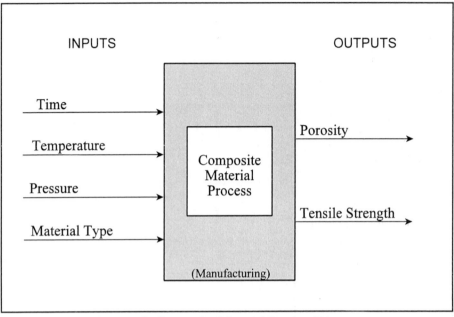

Figure 1.8 Composite Material Process Diagram

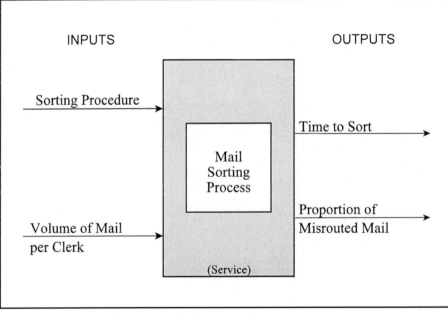

Figure 1.9 Mail Sorting Process Diagram

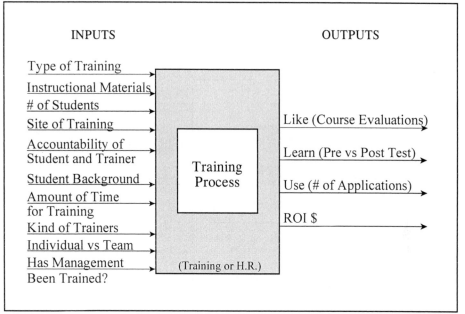

Figure 1.10 Training Process Diagram

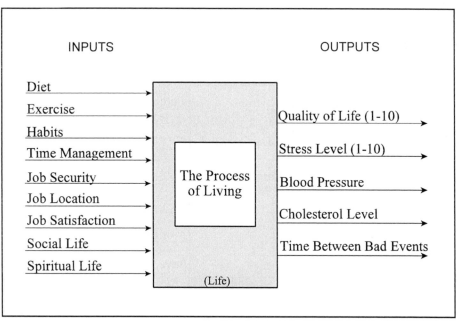

Figure 1.11 Living Process Diagram

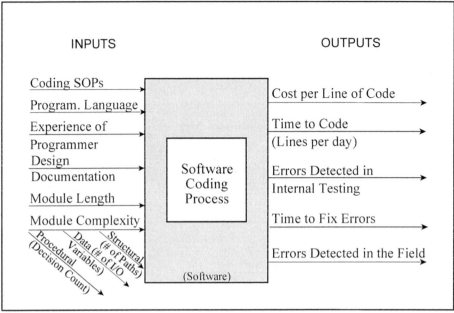

Figure 1.12 Software Coding Process Diagram

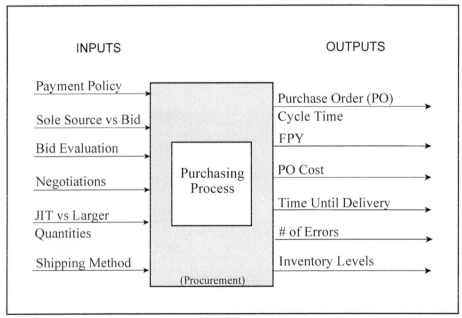

Figure 1.13 Purchasing Process Diagram

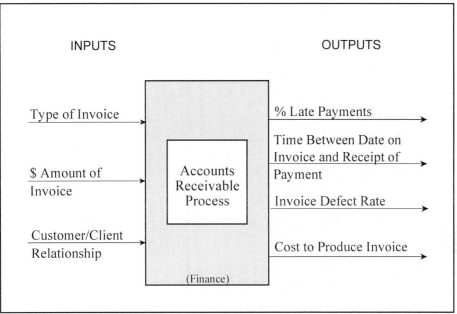

Figure 1.14 Accounts Receivable Process Diagram

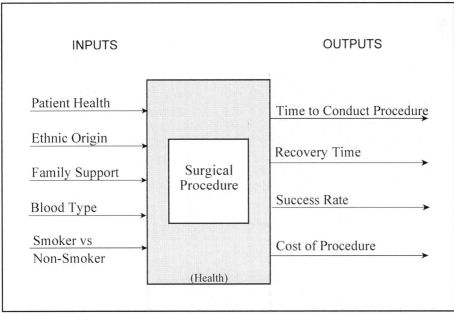

Figure 1.15 Surgical Procedure Diagram

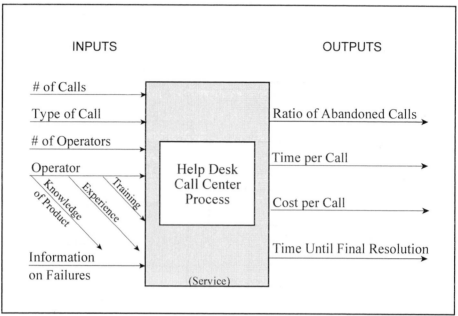

Figure 1.16 Help Desk Call Center Process Diagram

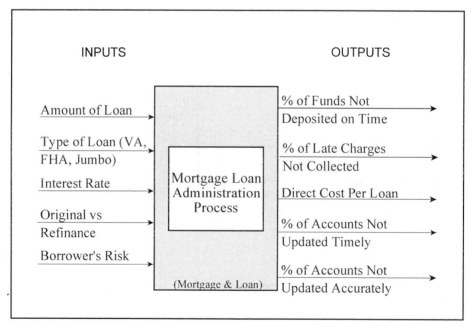

Figure 1.17 Mortgage Loan Administration Process Diagram

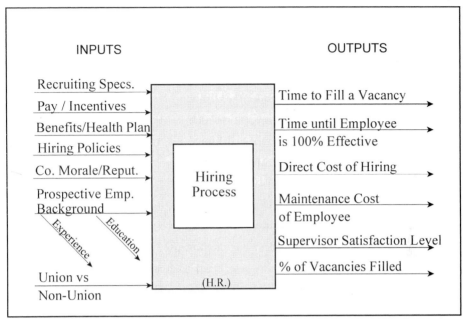

Figure 1.18 Hiring Process Diagram

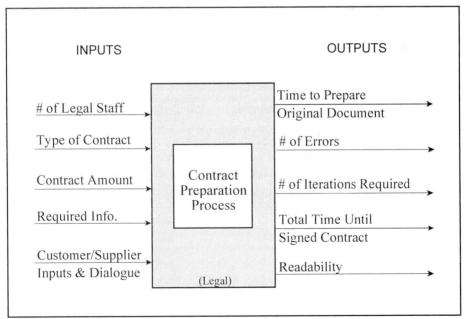

Figure 1.19 Contract Preparation Process Diagram

A process will also have inputs (in some cases there are hundreds of inputs). Inputs should be thought of as **sources of variation**. That is, the inputs are the variables or sources from which variation in the performance measure originates. To develop a complete list of inputs, the crossfunctional team must use some sort of brainstorming technique together with a tool such as the cause and effect (or fishbone) diagram described in Chapter 2. Each of the inputs (potential causes or sources of variability) on the fishbone diagram should be classified as either a C, N, or X variable, as defined below:

C = those inputs which must be held constant and require standard operating procedures to insure consistency. Consider the following examples: the method used to enter information on a billing form, the method used to load material in a milling or drilling process, the method used to lay up a composite material prior to placement in the autoclave (a pressure cooking type oven).

N = those inputs which are noise or uncontrolled variables and cannot be cheaply or easily held constant. Examples are room temperature or humidity.

X = those inputs considered to be key process (or experimental) variables to be tested in order to determine what effect each has on the outputs and what their optimal settings should be to achieve customer-desired performance.

This partitioning of the input variables is discussed in more depth in Section 2.2.2, Cause and Effect Diagrams with CNX.

To fully understand our processes and make continuous improvements requires the gathering, displaying and analysis of process data. Motorola has captured the essence of this statement by paraphrasing a century-old quote from Lord Kelvin, noted British physicist, as follows:

- *"If what we know about our processes can't be expressed in numbers, we don't know much about them.*

- *If we don't know much about them, we can't control them.*

- *If we can't control them, we can't compete."*

Clearly, objective data from our processes is a necessary ingredient for process improvement. However, before we collect data we must take steps to insure the accuracy and validity of the data. Table 1.3 indicates what we must consider before collecting any process data.

The following is a list of comments on each of the steps in Table 1.3.

1. If management is not willing to respond to and act upon good data, it is of little value to collect it prior to a change in management attitudes. Managers are also responsible for motivating employees and insuring that they have the proper training to do their jobs right. Finally, collecting data without a long term plan to document all continuous improvement efforts is unconscionable. However, the authors in their many years of experience find few companies that truly document all process improvement efforts.

2. The team should know the specific inputs and outputs of the process. Constructing an IPO diagram should help in this regard. It will also help identify suppliers and customers of the specific process. A process flow diagram describing each step of the process should be accomplished and used to remove non-value added steps, identify problem areas, and estimate overall cycle times and defect rates. Process flow diagrams are discussed in detail in Chapter 2.

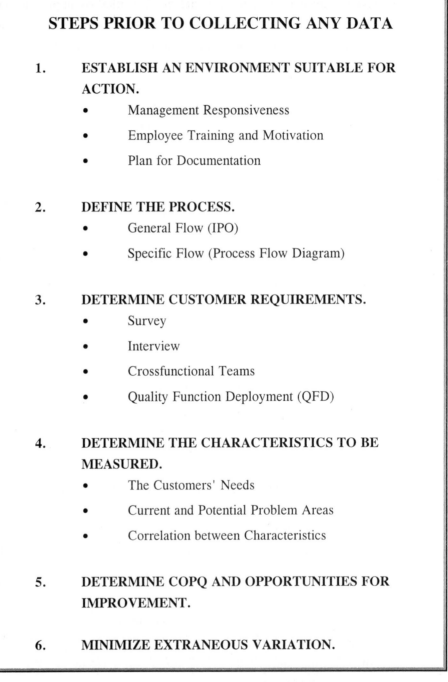

STEPS PRIOR TO COLLECTING ANY DATA

1.   **ESTABLISH AN ENVIRONMENT SUITABLE FOR ACTION.**
   - Management Responsiveness
   - Employee Training and Motivation
   - Plan for Documentation

2.   **DEFINE THE PROCESS.**
   - General Flow (IPO)
   - Specific Flow (Process Flow Diagram)

3.   **DETERMINE CUSTOMER REQUIREMENTS.**
   - Survey
   - Interview
   - Crossfunctional Teams
   - Quality Function Deployment (QFD)

4.   **DETERMINE THE CHARACTERISTICS TO BE MEASURED.**
   - The Customers' Needs
   - Current and Potential Problem Areas
   - Correlation between Characteristics

5.   **DETERMINE COPQ AND OPPORTUNITIES FOR IMPROVEMENT.**

6.   **MINIMIZE EXTRANEOUS VARIATION.**

Table 1.3 Steps Prior to Collecting Data

3.   Somehow we must understand our customers' needs so that we measure the right process outputs. Quality Function Deployment (QFD) is a scientific tool (discussed in Chapter 11) to help us understand *who* our customers are, *what* our customers need, and *how* we can measure our performance in satisfying customer needs. Survey data, along with interviews and the use of crossfunctional teams, can also aid in understanding customer requirements.

4.   It is necessary to manage and measure the same thing. Ascertaining the customers' needs and being aware of the current and potential problem areas can help in determining the quality characteristics to be measured. It may not be necessary to measure all of the quality characteristics of interest, especially if a high correlation between quality characteristics exists. Correlation is discussed in Chapter 2.

5.   Determining the Cost of Poor Quality (COPQ) is essential for: i) monitoring the process to determine where we want to make improvements and ii) evaluating how much improvement we have made by finding the difference between the before and after quality improvement COPQ values. Often the price tag associated with collecting and analyzing data can be high. This can discourage many knowledge gaining efforts. However, knowing our COPQ is analogous to knowing the cost of doing nothing. A simple comparison of COPQ to the price of gaining knowledge can assist us in determining the potential value associated with our future efforts.

6.   To minimize extraneous variation prior to collecting data, we must spend sufficient time on the cause and effect diagram, including the development of standard operating procedures to hold most of our process inputs constant. For example, we cannot tolerate large differences in operator performance. If necessary, process data should be collected on each operator and a non-threatening meeting held to discuss how to achieve consistency. In other words, show them the problem and solicit their help in finding a solution.

METHOD	RETURN ON INVESTMENT	COMMENTS
- "I think" and "I feel"	Low	We don't discourage thoughts, feelings and preconceived notions. They must, however, be supported with data.
- Sort Good from Bad (Inspection)	Low	The technique does not improve a process. It just minimizes the proportion of bad product sold to the customer. Bad product is still being produced.
- Basic Statistical Tools - Historical Data Analysis - Control Charts (SPC)	Medium	If we insure that we have good data, these tools turn data into information. See Chapters 2, 3, 4, 5, 6, 7 and 9.
- Hypothesis Testing - Regression Modeling - Design of Experiments (DOE) - Robust Designs	High	These tools are very powerful for maximizing process knowledge with minimum resources. See Chapters 6, 7, and 8.
- Concurrent Engineering - Quality Function Deployment (QFD)	High	These tools ensure that we are working on the right problems from a customer standpoint and that we do so crossfunctionally. See Chapter 11.
- All of the above plus teamwork and documentation	Maximum	The only way to maximize success in our quality improvement efforts is to use all the tools, each at the right time.

Table 1.4 Evolution of Process Improvement Tools

Table 1.4 depicts how some companies have evolved with regard to the use of process improvement tools. This text provides a working knowledge of each of these tools. Our hope is that you implement the tools and become extremely successful in your process improvement efforts.

1.5 Developing Valid Metrics and Scorecards

"You think you've had trouble coming to grips with metrics. Consider the case of a befuddled Australian woman after her country made the switch (to the metric system). *'Since eggs went metric,'* she complained to her government, *'they have been pale in color and lacking in freshness.   This clearly shows that chickens cannot adjust to laying different size eggs. We tamper with nature at our peril.'"*

John Flinn
San Francisco Examiner

The authors' experience has been that the term "metrics" means different things to different folks. Furthermore, we have found that practitioners sometimes struggle with what it is they should be measuring. In this section we formally define the term "metric." Then we will build the framework or context in which metrics are used and show a technique (i.e., a process) by which valid metrics can be identified.

From the American Heritage Dictionary, we have:

metric (noun):   A standard of measurement

metrical (adj):   Of or pertaining to measurement

-metrics (suff): The application of statistics and mathematical analysis to a specified field of study (e.g., biometrics, econometrics)

Clearly, metrics are related to measurement. We now define the term "metric" in the context of process improvement.

> A *metric* is an objective indicator or measure which facilitates process improvement.

A "metric" is really the name of a variable or performance measure that we will track to help us improve and know when we are improving a process. A metric is a verbal or written set of words. It is not a number. Examples of metrics are: "market share," "number of defective parts," "reliability of a system," "net sales," and "time to process engineering change proposals." We will track or quantify a metric via numerical measures like statistics. For example, the mean (or average) and standard deviation are statistical measures discussed in Chapter 2 that we may use to help quantify a metric. Suppose "gas mileage" is the metric of interest and we collect data on (or track) this metric. After collecting 20 gas mileage data points (one from each of 20 fills of the tank), where the unit of measure is miles per gallon (mpg), suppose we find the average to be 25 mpg and the standard deviation to be 2 mpg. We have just defined "gas mileage" as one metric or performance measure of the process of driving a vehicle. Certainly there are others. The average (25 mpg) and standard deviation (2 mpg) are statistics or numerical values that describe the metric called "gas mileage."

Over the years, we have asked our students and clients why we need metrics. A list of their brainstormed responses is shown in Table 1.5. Although not exhaustive, this list gives a fairly comprehensive look at the practicality of metrics.

Developing valid metrics is not always as easy as it sounds, especially in the service industry. In manufacturing, some of the metrics are quite obvious. If we are to drill holes of a specified diameter, the metric is easy: "diameter of hole." If we track the diameter of the holes we drill, we will be able to tell how well the process of drilling holes is performing. In service, life is not always so pleasant. For example, consider the architect or engineer who is designing a structure of some sort. The typical concern is: "What do I measure? Every design is different.

WHY WE NEED METRICS

- Perception and intuition are not always reality

- To gather the facts for good decision making

- Paradigms can limit our thought process

- To identify/verify problem areas/bottlenecks

- To understand our processes better

- To characterize our processes (to know how inputs and outputs are related)

- To validate our processes (are they performing within the requirements/specs)

- To evaluate customer satisfaction

- To document our processes and communicate about them

- To baseline a process

- To see if our processes are improving

- To determine if a process is stable or predictable and how much variation is inherent in the process

- "In God we trust; all others bring data."

Table 1.5 Students' Responses to Why We Need Metrics

Every customer is different and I do different things for each of my customers."
In essence, there are applications both in service and manufacturing where it may not be obvious what the key metrics are. The following is presented to help practitioners (hopefully working in crossfunctional teams) develop and identify key metrics.

Before presenting a step-by-step process for identifying metrics, we consider the framework or environment in which this is to be accomplished. The key metrics in any organization will be linked to the mission, vision, and goals of the organization. We briefly address each of these areas.

Mission

The mission statement embraces the *"whats"* of an organization and answers the following questions:

1. What do we produce and/or what service do we provide? The mission implicitly defines key or critical processes which, if not performed well, will spell the demise of the organization.

2. What characteristics of this product/service make it a valuable commodity to our customers? These customers may be internal or external to the company.

Vision

Vision embodies *"where"* we are going and answers the following questions:

1. How do we define "success" as a company or organization?

2. How are we going to ensure that we will be successful in the future?

3. What direction will we take in order to lead the industry?

Goals

Goals follow naturally from the vision and mission statements, and often many goals come to mind when the vision and mission statements are adopted. Suppose part of our vision is to be the world's leading manufacturer of brake linings for automobiles. A goal might be to increase our market share from 10% to 50% over the next ten years. Goals like this are sometimes separated into long term and short term goals. Long term goals are goals that would force us to look relatively far into the future. They may be possible to attain but we realize that they are on the limit of attainability. Short term goals provide specific milestones for meeting the long term goal. For example, to realize a five-fold increase in market share over ten years, we might strive for a 20% increase each year (e.g., from 10% to 12% the first year). Note that the metric here is "market share."

Again, metrics are those objective indicators and measures which allow us to track progress toward our goals. While "market share" is a nice metric at the corporate level, the rank and file members of an organization will certainly not be tracking "market share" on a daily basis. That is why we need to find metrics at the individual process level that will contribute to measuring progress toward both the corporate goals and the derived goals at the process level. Since process improvement means doing things better, faster, and at lower cost, and since process improvement is certainly an overall goal, almost every goal at any level of an organization will be related in some way to either ***performance, schedule***, or ***cost***. Thus, we use these three categories as the major process improvement criteria into which any goal or metric will fall. We must also identify the key or critical processes which relate directly to the mission and which keep us in business. Then for each critical process/process improvement criterion, we search for those metrics that measure progress. Table 1.6 shows the Critical Processes/Process Improvement Criteria Matrix, also known as a "scorecard."

We now summarize the steps in searching for and identifying key metrics to be placed in the scorecard.

Step 1.   Identify process improvement criteria that will lead the organization toward its defined goals. The three identified above, performance, schedule, and cost, apply to almost any organization. There may be others, such as safety, ergonomics, or environment. The vision statement should give guidance as to what issues are important in determining the criteria that define corporate success.

Step 2.   Identify critical or important processes that are essential for the success of the company. Here the mission statement is the key to defining critical processes.

Step 3.   Build a scorecard as shown in Table 1.6.

		VISION		
		Process Improvement Criteria		
Critical Processes	Definition	Performance	Schedule	Cost

Staff
   Accounting
   Marketing
   Human Resources
   Training				

Operations
   Manufacturing
   Service				

Design				
Research and Development				

(Left vertical banner: **M I S S I O N**)

Table 1.6 Critical Processes / Process Improvement Criteria Matrix
or Scorecard

Step 4.   Fill in the definition portion of the scorecard by defining each process in detail. Use a process flow diagram (see Chapter 2) to do this.

Step 5.   If direct metrics are available, use them. If not, brainstorm potential factors that affect the process improvement criteria. Use a cause and effect diagram (see Chapter 2) to do this.

Step 6.   Using the process flow diagram and completed cause and effect diagram, select applicable measures to track.

Table 1.7 is an example of a fairly balanced scorecard. Note that Steps 4-6 make reference to tools we have not yet covered, indicating the tools can be used to develop the metrics upon which the tools themselves are applied.

There are a few cautions in using this matrix. We don't have to measure everything. Remember, it is better to measure a few things well than to spend all of our time collecting data on metrics that may do very little in aiding process improvement. On the other hand, we should not hesitate to measure those variables that appear to be important and which show up in multiple cause and effect diagrams. Some practitioners hesitate to measure anything because they feel they might be measuring the wrong thing. Sometimes measuring the wrong things leads to measuring the right things. Measurement is much like the dilemma of "purchasing a computer." That is, we can wait for the price of a computer to become low enough to buy it. Since computer prices continue to decline, we run the risk of never owning one because next month's price will no doubt be lower than this month's. Likewise with measurement, we may wait forever to find and measure the "right thing" and end up measuring nothing.

There is a fine line between too many metrics and too few metrics. This line can be clarified by validating our metrics. That is, if we are measuring something but never use it to help make decisions or improve processes, then it should be removed from the active list of metrics. Like clothes in a closet, metrics must be evaluated for usefulness. If no longer needed, they must be removed from the inventory. Also, there may be some areas where it costs more to measure something than it is worth. If we want to measure customer satisfaction on a weekly basis, it may indeed be too costly. However, techniques such as sampling may relieve problems like this.

Finally, we want to establish accountability for each metric. Each metric should have a person who will collect data and report on it on a regular basis. Feel free to revise metrics when necessary but remember that as metrics are changing, it may become more difficult, if not impossible, to compare them to previous metrics. For example, if a questionnaire is used to measure customer satisfaction and we decide to change the questions the next time we collect data, we may not be able to compare results from one questionnaire to another.

Critical Processes	Definition	Process Improvement Criteria			
		Performance	Schedule	Cost	
Accounting		Billing Errors	Invoice Cycle Time	Annual Budget Software Costs	
Marketing		New Sales	Order Time	Expense Reports Advertising Costs	
Human Resources		Production Rate Suggestion Rate Turnover Rate	Training Time	Absenteeism Benefit Costs	
Training		Return on Investment from Training	Training Time	Per Student Cost	
Operations Manufacturing		Process Capability (C_{pk}) Customer Surveys Defect Rate	Cycle Time Down Time Setup Time	Production Costs Warranty Costs COPQ Scrap Rework	
Research and Development		Design Changes	Development Time	Expected Life Cycle Costs	

VISION

M I S S I O N

Table 1.7 A Scorecard Filled with Metrics

1.6 Malcolm Baldrige National Quality Award

As previously discussed, the use of statistical tools to gain knowledge will greatly improve the success of any company or organization. The U.S. government has recognized the power of statistics and its role in quality improvement. Thus, to further the use of these techniques, our government has developed a national quality award.

> *"The improvement of quality in products and the improvement of quality in service — these are national priorities as never before ... all American firms benefit by having a standard of excellence to match and perhaps, one day to surpass. There can be no higher standard of quality management than those provided by the winners of the Malcolm Baldrige National Award."*

George Bush

U.S. President (1988-1992)

The Malcolm Baldrige National Quality Award was created by Public Law 100-107 and signed into law on August 20, 1987. The Award Program, responsive to the purposes of Public Law 100-107, led to the creation of a new public-private partnership. Principal support for the program comes from the Foundation for the Malcolm Baldrige National Quality Award, established in 1988.

The Award is named for Malcolm Baldrige, who served as Secretary of Commerce from 1981 until his tragic death in a rodeo accident in 1987. His managerial excellence contributed to long term improvement in efficiency and effectiveness of government.

The award is managed by the US Department of Commerce, National Institute of Standards and Technology, Gaithersburg, MD 20899. The Findings and Purposes Section of Public Law 100-107 states that:

1. The leadership of the United States in product and process quality has been challenged strongly (and sometimes successfully) by foreign competition, and our Nation's productivity growth has improved less than our competitors' over the last two decades.

2. American business and industry are beginning to understand that poor quality costs companies as much as 20 percent of sales revenues nationally, and that improved quality of goods and services goes hand in hand with improved productivity, lower costs, and increased profitability.

3. Strategic planning for quality and quality improvement programs, through a commitment to excellence in manufacturing and services, are becoming more and more essential to the well-being of our Nation's economy and our ability to compete effectively in the global marketplace.

4. Improved management understanding of the factory floor, worker involvement in quality, and greater emphasis on statistical process control can lead to dramatic improvements in the cost and quality of manufactured products.

5. The concept of quality improvement is directly applicable to small companies as well as large, to service industries as well as manufacturing, and to the public sectors as well as private enterprise.

6. In order to be successful, quality improvement programs must be management-led and customer-oriented and this may require fundamental changes in the way companies and agencies do business.

7. Several major industrial nations have successfully coupled rigorous private sector quality audits with national awards giving special recognition to those enterprises the audits identify as the very best.

8. A national quality award program of this kind in the United States would help improve quality and productivity by:

A. helping to stimulate American companies to improve quality and productivity for the pride of recognition while obtaining a competitive edge through increased profits;

B. recognizing the achievements of those companies which improve the quality of their goods and services and provide an example to others;

C. establishing guidelines and criteria that can be used by business, industrial, governmental, and other organizations in evaluating their own quality improvement efforts; and

D. providing specific guidance for other American organizations that wish to learn how to manage for high quality by making available detailed information on how winning organizations were able to change their cultures and achieve eminence.

MALCOLM BALDRIGE NATIONAL QUALITY AWARD WINNERS

Company Name	Year (Award Area)
ADAC Laboratories	1996 (Manufacturing)
Ames Rubber Corporation	1993 (Small Business)
Armstrong World Industries	1995 (Manufacturing)
AT&T Consumer Communication Services	1994 (Service)
AT&T Network Systems Group	1992 (Manufacturing)
AT&T Universal Card Services	1992 (Service)
BI	1999 (Service)
Boeing Airlift and Tanker	1998 (Manufacturing)
Cadillac Motor Car Company	1990 (Manufacturing)
Corning, Inc. Telecommunications Products Division	1995 (Manufacturing)
Custom Research	1996 (Small Business)

Company Name	Year (Award Area)
DANA Commercial Credit Corporation	1996 (Service)
Eastman Chemical Company	1993 (Manufacturing)
Federal Express Corporation	1990 (Service)
Globe Metallurgical, Inc.	1988 (Small Business)
Granite Rock Company	1992 (Small Business)
GTE Directories Corporation	1994 (Service)
IBM Rochester	1990 (Manufacturing)
Marlow Industries	1991 (Small Business)
Merrill Lynch Credit Corp.	1997 (Service)
Milliken & Company	1989 (Manufacturing)
Motorola, Inc.	1988 (Manufacturing)
Solar Turbines	1998 (Manufacturing)
Solectron Corporation	1991 (Manufacturing)
Solectron Corporation	1999 (Manufacturing)
STMicroelectronics	1999 (Small Bus./Manuf.)
Sunny Fresh Foods	1997 (Manufacturing)
Texas Instruments, Inc. (Defense Systems & Electronics Group)	1992 (Manufacturing)
Texas Nameplate	1998 (Small Business)
The Ritz-Carlton Hotel Company	1992 (Service)
The Ritz-Carlton Hotel Company	1999 (Service)
3M Dental Products Division	1997 (Manufacturing)
Trident Precision Manufacturing	1996 (Small Business)
Wainwright Industries, Inc.	1994 (Small Business)
Wallace Company	1990 (Small Business)
Westinghouse Electric Corporation (Commercial Nuclear Fuel Division)	1988 (Manufacturing)
Xerox Business Services	1997 (Service)
Xerox Corporation (Business Products and Systems)	1989 (Manufacturing)
Zytec Corporation	1991 (Manufacturing)

1.7 SIX SIGMA FOR MANUFACTURING AND NON-MANUFACTURING PROCESSES

Six Sigma is a quality improvement and business strategy that began in the 1980's at Motorola. Emphasis is on reducing defects to less than 4 per million, reducing cycle time with aggressive goals such as 30-50% reduction per year, and reducing costs to dramatically impact the bottom line. The statistical and problem solving tools are similar to other modern day quality improvement strategies. However, Six Sigma stresses the *application* of these tools in a methodical and systematic fashion to gain knowledge that leads to breakthrough improvements with dramatic, measurable impact on the bottom line. The secret ingredient that really makes Six Sigma work is the infrastructure that is built within the organization. It is this infrastructure that motivates and produces a Six Sigma culture or "thought process" throughout the entire organization. The power of a Six Sigma approach is best described by proven return-on-investment (ROI) as shown next from Motorola, AlliedSignal, and General Electric (GE).

Motorola ROI

1987-1994

- Reduced in-process defect levels by a factor of 200.
- Reduced manufacturing costs by $1.4 billion.
- Increased employee production on a dollar basis by 126%.
- Increased stockholders share value fourfold.

AlliedSignal ROI

1992-1996

- $1.4 Billion cost reduction.
- 14% growth per quarter.
- 520% price/share growth.
- Reduced new product introduction time by 16%.
- 24% bill/cycle reduction.

General Electric ROI

1995-1998

- Company wide savings of over $1 Billion.
- Estimated annual savings to be $6.6 Billion by the year 2000.

Based on the number of articles written the last two years about GE and its CEO, Jack Welch, GE has now become the standard bearer for how Six Sigma is implemented to successfully drive positive bottom line impact along with recognized "World Class" status. Other highly respected and successful companies such as SONY are benchmarking off of GE and implementing a similar strategy.

The companies mentioned thus far are certainly well known for their engineering and manufacturing excellence. What is not as well known is their view of the importance of Six Sigma in non-manufacturing or transactional areas. Bob Galvin, former President and CEO of Motorola, has stated that the lack of initial Six Sigma emphasis in the non-manufacturing areas was a mistake that cost Motorola at least $5 Billion over a 4-year period. It is common these days to hear comments like, "Yes, Company X has a great product, but they sure are a pain to do business with!" Consequently, Jack Welch is mandating Six Sigma in all aspects of his business, most recently in sales and other transactional (non-manufacturing) processes. Unfortunately, the typical response from non-manufacturing employees has been, "*We're different. Six Sigma makes sense for manufacturing but does not apply to us!*" This is simply an excuse in order to avoid being held to the same accountability standards as manufacturing.

The point to be made here is that any process can be represented as a set of inputs which, when used together, generates a corresponding set of outputs. An abbreviated pharmaceutical tablet manufacturing process might appear as shown below:

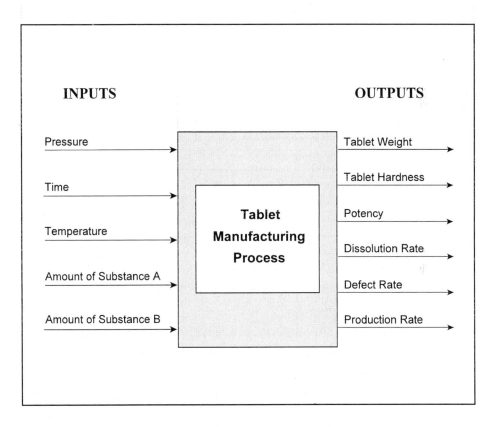

Transactional organizations simply are not accustomed to looking at their processes in this manner and thus will struggle a little in developing a similar abbreviated diagnosis of a transactional process. An Input-Process-Output (IPO) diagram for a sales process is shown next:

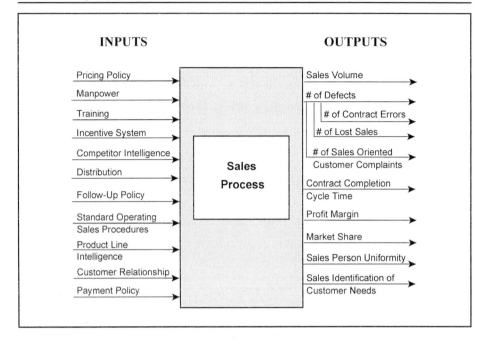

Thus, a process is a process, regardless of the type of organization or function. All processes have inputs and outputs. All processes have customers and suppliers, and all processes exhibit variation. Since the purpose of Six Sigma is to gain breakthrough knowledge on how to improve processes to do things **Better**, **Faster**, and at **Lower Cost**, it applies to everyone. Furthermore, since processes such as sales have historically relied less on scientific methods than engineering and manufacturing, the need for Six Sigma (i.e., a structured and systematic methodology) is even stronger here. This has been and is being recognized by World Class CEO's such as Bob Galvin, Larry Bossidy, Dan Burnham, Nobuyuki Idei, and Jack Welch.

The method to implement Six Sigma for non-manufacturing processes is simple: the same way we implement it for engineering and manufacturing processes at Motorola, Texas Instruments, GE, Lockheed Martin, Corning, Sony, etc., with only slight modifications. These modifications are typically confined to the type and depth of statistical tools that need to be included in the training. Obviously, the slant on applications must also be directed toward the non-manufacturing processes.

A specific strategy for Six Sigma manufacturing and non-manufacturing processes would look similar to what is shown below:

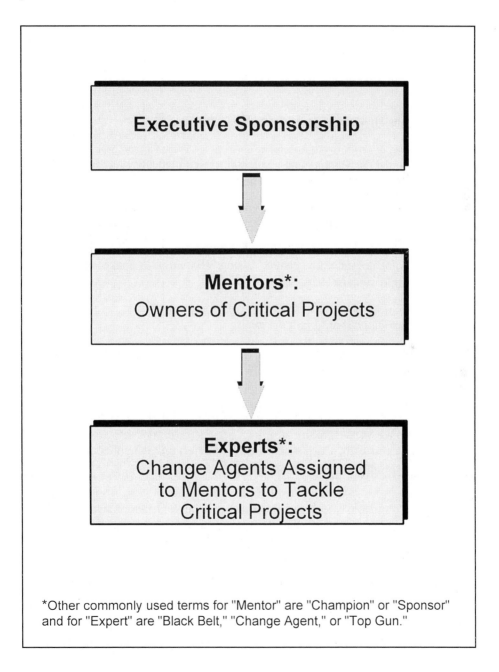

*Other commonly used terms for "Mentor" are "Champion" or "Sponsor" and for "Expert" are "Black Belt," "Change Agent," or "Top Gun."

The executives must have a total commitment to the implementation of Six Sigma and accomplish the following:

1.   Establish a Six Sigma Leadership Team.

2.   Identify key business issues.

3.   Assign Mentors to each key business issue.

4.   Assist the Mentors and Leadership Team in identifying critical projects that are tied to the key business issues.

5.   Assist the Mentors and Leadership Team in selecting Expert candidates.

6.   Allocate time for change agents (Experts) to make breakthrough improvements.

7.   Set aggressive Six Sigma goals.

8.   Incorporate Six Sigma performance into the reward system.

9.   Direct finance to validate all Six Sigma ROI.

10.   Evaluate the corporate culture to determine if intellectual capital is being infused into the company.

11.   Continuously evaluate the Six Sigma implementation and deployment process and make changes if necessary.

The roles of the Leadership Team, Mentors, and Experts are defined next.

COMPARISON OF ROLES

	EXPERT	MENTOR	LEADERSHIP TEAM
PROFILE	• technically oriented • respected by peers and management • master of basic and advanced tools	• senior manager • respected leader and mentor of business issues • strong proponent of Six Sigma who asks the right questions	• highly visible in company • trained in Six Sigma • respected leaders and mentors for Experts
ROLE	• leads strategic, high impact process improvement projects • change agent • trains and coaches on tools and analysis • teaches and mentors cross-functional team members • full-time project leader • converts gains into $	• selects projects and Experts • provides resources and strong leadership for projects • inspires a shared vision • establishes plan and creates infrastructure • develops metrics • converts gains into $	• develop a training Master Plan to implement Six Sigma • schedule training • select projects and Experts • determine certification requirements and certify Experts • develop an Expert network to enhance communication • review and improve the Six Sigma process
TRAINING	• three to four 1-week sessions with three to six weeks in between to apply • project review in every session	• 4 days of Mentor training • Six Sigma development and implementation plan	• one day of Basics of Six Sigma training or 4 days of Mentor training
NUMBERS	• 1 per 50 employees (2 %)	• 1 per business group or major working site	• 4 - 6 member team

The overall approach to obtaining the **right kind of knowledge** is focused on finding the **answers to the 14 questions** shown next. These questions, which are partitioned into a strategy (**P**rioritize, **C**haracterize, **O**ptimize, and **R**ealize (PCOR)), form the Six Sigma Project Master Strategy.

SIX SIGMA PROJECT MASTER STRATEGY

Prioritize	Characterize

Prioritize

1. What processes (activities) are you responsible for? Who is the owner of these processes? Who are the team members? How well does the team work together?

2. Which processes have the highest priority for improvement? How did you come to this conclusion? Where is the data that supports this conclusion?

Characterize

3. How is the process performed?

4. What are the process performance measures? Why? How accurate and precise is the measurement system?

5. What are the customer driven specifications for all of the performance measures? How good or bad is the current performance? Show me the data. What are the improvement goals for the process?

6. What are all the sources of variability in the process? Show me what they are.

7. Which sources of variability do you control? How do you control them and what is your method of documentation?

8. Are any sources of variability supplier-dependent? If so, what are they, who is the supplier and what is being done about it?

9. What are the key variables that affect the average and variation of the measures of performance? How do you know this? Show me the data.

10. What are the relationships between the measures of performance and the key input variables? Do any key variables interact? How do you know for sure? Show me the data.

SIX SIGMA PROJECT MASTER STRATEGY (Cont.)

Optimize	Realize
11. What setting for the key variables will optimize the measures of performance? How do you know this? Show me the data. 12. For the optimal settings of the key variables, what kind of variability exists in the performance measures? How do you know? Show me the data.	13. How much improvement has the process shown in the last 6 months? How do you know this? Show me the data. 14. How much time and/or money have your efforts saved or generated for the company? How did you document all of your efforts? Show me the data.

    The Six Sigma tools and methodology must be taught to Mentors, Experts and other managers at a level they can grasp and feel confident to apply. A proven instructional approach developed by Air Academy Associates is shown below:

    A **Keep-It-Simple-Statistically** (**KISS**) approach is used, with the intention to avoid statistical complexity. Statistics is not presented as an "end", but rather the means to gaining knowledge for making good decisions which are critical for success. There are a variety of Six Sigma tools and techniques, and we will use the

"Present/Practice/Apply/Review" instructional strategy. That is, we will *present* a tool or method, give you a chance to *practice* that tool in class, then have you *apply* that tool to your project, and finally have you *review* the results of the application to your project. A final report will be written to document your success story and its impact to the company's bottom line.

Another critical piece to a successful Six Sigma experience is the reward structure. Recall that many companies struggled to engage the entire organization in implementing TQM. To overcome this problem, Jack Welch has made the following statements:

1.   To get promoted you must be Six Sigma trained.

2.   Forty percent of top management bonuses are tied to Six Sigma goals.

3.   Stock options are tied to Six Sigma performance.

As you can imagine, General Electric has very few problems engaging the entire organization in its Six Sigma initiative.

Thus, the modern day Six Sigma movement has fully embraced a Knowledge Based Management approach. Numerous companies, such as General Electric, Raytheon, and Sony, are demonstrating that this approach has a high return on investment. This new and improved Six Sigma business strategy is much more powerful than the original Six Sigma developed at Motorola. For more information on the power of Six Sigma, see General Electric's 1997 and 1998 Annual Reports to its share holders, employees, and customers. The following dual Input Process Output (IPO) diagram, a Six Sigma tool, summarizes the essence and power of a Six Sigma business strategy.

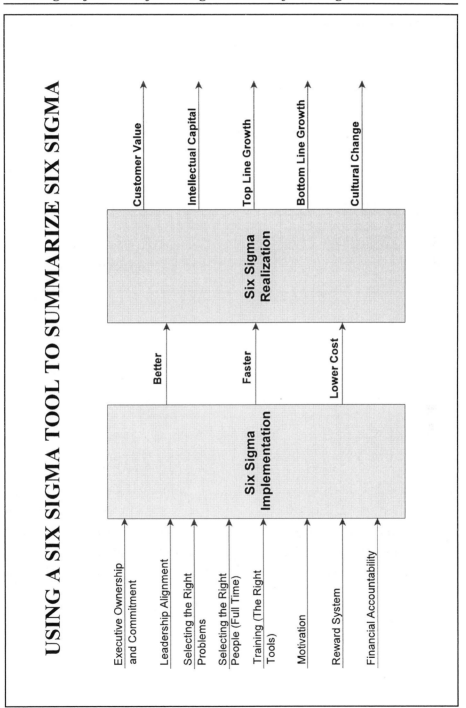

USING A SIX SIGMA TOOL TO SUMMARIZE SIX SIGMA

1.8 Success Stories

The primary purpose of this chapter is to emphasize the power of statistical tools for obtaining knowledge that leads to good decisions. These decisions should generate better, faster, and lower cost products, processes, and service. We close this chapter with several success stories that resulted in an improved bottom line. As seen here, a company need not necessarily be a Baldrige Award winner to generate successes.

Company	Benefits
Ford Motor Company [Afrm 89]	Increased market share and profits; 65% reduction in customer reported defects; 35% increase in customer satisfaction
Xerox [Afrm 89]	Manufacturing costs down 20% (1982-1986); cycle time reduced by 60%; revenue produced per employee up 20%
Westinghouse Electric Corp. (Commercial Nuclear Fuel Division) [Afrm 89]	Increased manufacturing by over 37%, thus reducing scrap, rework, and manufacturing cycle time
ITT [Afrm 89]	$35,000,000 cost savings per year (1983-1987)
Hewlett Packard (Yokohama) [Afrm 89]	Profit up 244% from 1977-1984; hardware failure rates down 79%; manufacturing costs down 42%; productivity up 120%; and market share up 19%

Company	Benefits
Boeing Aerospace Co. [Wats 90]	For the Initial Upper Stage Program: billing errors reduced to 0, cycle time reduced from 20 days to 3, technical order processing streamlined — saving $875 and 3.75 man-hours per T.O.; for the AWACS contract: billing delinquencies reduced by 50%; Overall Savings: $1.5 million per year
Pittron Steel Foundry [Wats 90]	Sales increased by 400%; profits up 30%; productivity up 64%
U.S. Navy [Wats 90]	F-14 Overhaul Program: cut average cost from $1.6 million per aircraft in 1986 to $1.2 million in 1989; Cherry Point: aircraft failure rates reduced by 90% between 1987 and 1988; Overhaul of USS Saratoga: expected to save $10 million and 22,000 man-days; Norfolk Naval Shipyard: reduced rejection rate in electronic connectors from 55% to 6%
Martin Marietta [Wats 90]	Titan IV: "boattail" assembly cycle time reduced from 2500 to 1200 hours; wire crimping process operations reduced from 11 to 1 with savings of over $18,000 per month

Company	Benefits
P.I.E. Trucking [Afrm 89]	Freight bill accuracy errors down from 10% to 1%; productivity up 30%; all auditing and inspection costs eliminated
Internal Revenue Service (IRS) [Burs 88]	Processing errors reduced from 30,000 (1986) to 3,000 (1987)
U.S. Forest Service [Burs 88]	Experimental forest unit costs reduced 15%, saving $175,000
Social Security Administration [Burs 88]	Claim processing reduced to 73.9 days in 1987 from 81 days in 1986
Department of Housing and Urban Development [Burs 88]	Average time to process a loan for property improvement and purchase of manufactured housing was 85 days (1985), 29 days (1986), and 22 days (1987)
Department of Education [Burs 88]	Reduced total processing time for determining the eligibility and certification of higher education institutions by 20% (1986)
IBM [Burs 88]	Reduced the number of software errors in complex NASA programs to near zero and lowered reconfigure time of flight software by 50%

Company	Benefits
U.S. Air Force: Electronic Systems Center [Hopk 93]	A Pareto analysis of the deviations/waivers to the Joint Tactical Information Distribution System (JTIDS) led to $175,000 savings in work avoidance
U.S. Air Force: Propulsion Systems Program Office [Lead 90]	40% reduction in report processing time; average paperwork turnaround time dropped from 21.6 days to 6.3 days; 43% improvement in Engineering Change Proposal sub-processes; 95% reduction in monthly management review process
U.S. Air Force: Ballistic Missile Organization [Lead 90]	Peacekeeper Program: $3,000,000 cost avoidance through the use of tools such as statistical predictions of equipment failure to authorize unscheduled maintenance
U.S. Air Force: Sacramento Air Logistics Center [Lead 90]	F-111 remanufacture of horizontal stabilizer and attack radar process changes saved nearly $1,000,000; Mean Time Between Failures (MTBF) for A-10 actuator cylinder system extended from 1,840 to 4,000 hours
Motorola [Moto 88]	Cellular telephone: 30 to 1 reduction in factory cycle time; 4 to 1 reduction in defects per unit; reduced part counts from 1,378 to 523; 10 to 1 improvement in reliability; In Communications, Semi-conductor and General Systems Groups, sales up by 23% to 38% and orders up 21% to 51%

Company	Benefits
Massachusetts General Hospital [Berw 91]	Improved the billing process by reducing monthly defects by 52%, resulting in a projected $189,000 savings per year
Intermedics [Schm 96]	Rai Chowdhary began a 4-month quest for quality improvement on the coating of titanium on a cobalt-chrome substrate. The solution to this problem defied all previous medical product expert opinions and approximately 10 years of research from some of the nation's leading material scientists. His efforts in using KISS techniques generated the knowledge to patent the process and save his company from ditching a product which subsequently has generated $60 million of revenue annually
Parkview Medical Center, Pueblo, CO [Schm 96]	Improved process of producing good cultures from 35% to 95%; reduced cycle time in accounts receivable from 82 days to 40 days, saving more than $4 million in a 3-year period
Friendly's Ice Cream [Park 96]	SPC techniques saved ½ million dollars annually by reducing variability in filling and packing ice cream containers
Teradyne and Analog Devices [Stat 95]	Increased quarterly earnings by more than 60% in highly competitive micro-electronics industry

The remainder of this book explains how to use statistical methods as tools to achieve continuous improvement and develop success stories such as those presented in this section.

1.9 Problems

1.1   List 5 different business areas where statistics can be used.

1.2   What is meant by "continuous improvement" and why is it important?

1.3   Where does "statistics" fit into the area of "continuous improvement?"

1.4   Decrypt the encoded message on page 1-8 using the frequency analysis mentioned there.

1.5   What is the difference between the old and the new philosophies of quality?

1.6   Describe the role of management in an environment conducive to process improvement.

1.7   What is the essence of Knowledge Based Management (KBM)?

1.8   Define an application within your area of endeavor where statistics could provide improved knowledge and help make you a better decision maker.

1.9   What is common among the philosophies of Deming, Juran, and Ishikawa?

1.10   Describe your interpretation of cost of poor quality (COPQ). List three areas where you have personally witnessed COPQ.

1.11   Build an Input-Process-Output (IPO) diagram for some process with which you work.

1.12   What is meant by the term "valid metric?" What is one thing you measure that you would consider to be a "valid metric" and why?

1.13   What is the Malcolm Baldrige National Quality Award? How old is it, why was it established, and who are some of the winners?

1.14   Analyze the data in Table 1.1 and draw a defensible conclusion.

1.15   What is Six Sigma and how is it different from the quality initiatives of the past?

Chapter 2

Graphical and Measurement Tools

2.1 Probability vs Statistics

The terms "probability" and "statistics" mean different things to different people. From the statements "the odds of Florida State winning the national collegiate football championship are 2 to 1" or "Tony Gwynn's league-leading batting average is .337" to "Bill Clinton will win the election with 44% of the popular vote with a margin of error of $\pm 3\%$," probability and statistics are used daily in a wide variety of settings. Furthermore, probability and statistics influence the production or manufacture of almost every product that individuals use or come in contact with every day. For example, statistical control charts are used to monitor the variability in processes, whether it be the manufacturing of automobile tires or the process of producing purchase orders. Process variability affects the quality of life experienced daily whether we realize it or not. For some, statistics are worse than "damned lies," while to others, statistics are needed to monitor and improve processes. Thus, no matter how probability and statistics are perceived, the fact of the matter is that they do affect our lifestyle.

While music is considered by many to be the universal language, statistics makes a strong case for being the universal language of the sciences. Statistics is the art and science of collecting, classifying, presenting, interpreting, and analyzing numerical data, as well as making conclusions about the system from which the data was obtained. The field of statistics can be broken into two major areas: descriptive statistics and inferential statistics.

Descriptive statistics is the branch of statistics with which most people are familiar. It characterizes and summarizes the most prominent features of a given set of data. Some of the more commonly used descriptive measures include means, medians, standard deviations, percentiles, graphs, tables, and charts. The reader no doubt encounters some of these on almost a daily basis, regardless of the discipline in which he or she is involved. Descriptive methods can be used to describe the elements of a population as a whole (i.e., a census which represents all the possible data from a process), or they can be used to describe data that represent just a sample of elements from the entire population. This chapter addresses tools within the realm of descriptive statistics.

Inferential statistics, on the other hand, is the branch of statistics that deals with drawing conclusions about a population based on information obtained from a sample drawn from that population. While descriptive statistics has been practiced for centuries, inferential statistics is a relatively new phenomenon having its roots in the twentieth century. How is it that the Gallup, Harris, and major network polls can so accurately predict the outcomes of major elections prior to the actual election itself? How can one tell that a manufacturing process is out of control before an entire lot of defective items is produced? How can the Food and Drug Administration be confident that a certain vaccine will be effective in the population as a whole? The answer to each of these questions lies in the ability to accurately infer something about a population when information from only a sample is known.

Probability is the bridge between descriptive statistics and inferential statistics. Reference Figure 2.1. Probability is what separates witchcraft from highly reliable predictions. Although it might seem that practical statistical methods could be presented without a basic knowledge of probability, the understanding of why inferential statistical processes produce reliable estimates is clearly dependent upon at least a basic knowledge of some of the underlying principles of probability. Thus, the order of presentation in this text proceeds from descriptive statistics to probability and then to inferential statistics.

Figure 2.1 also shows that probability is a link between a population and a sample. When certain properties of a population are known or can be assumed, the probability principles involved allow one to deduce expected properties of a sample. Alternatively, probability together with sample data allows one to evaluate the amount of uncertainty when making conclusions (or inferences) about a population. Hence, probability is at the heart of both the deductive and inferential processes involved in statistical analyses.

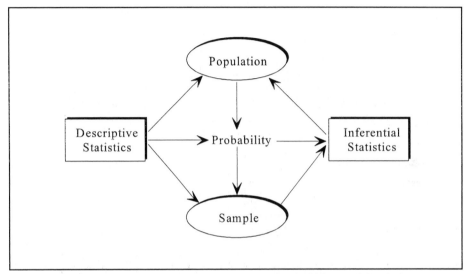

Figure 2.1 The Relationship Between Probability and Statistics

2.2 Graphical Tools in Statistics

The concise and accurate representation of data to produce information is both an art and a science. The importance of using simple graphs to represent and summarize data is reflected in Figure 2.2 which is taken from a survey of case studies reported by the Japanese Union of Scientists and Engineers (JUSE). This text will cover in detail the top 15 items of Figure 2.2 plus many other topics.

Just as pictures summarize details into information, graphical methods can be an immense aid in the process of gleaning information from raw data. This section is designed to introduce the reader to several commonly used graphical techniques in data analysis. While each of these techniques can be used in a stand-alone mode, their use is more often seen in conjunction with other statistical methods. Many statistical methods will be detailed in subsequent chapters and they will presume a working knowledge of these tools, so it is important that the reader understand the terminology and some possible uses of these graphical tools prior to addressing methodology.

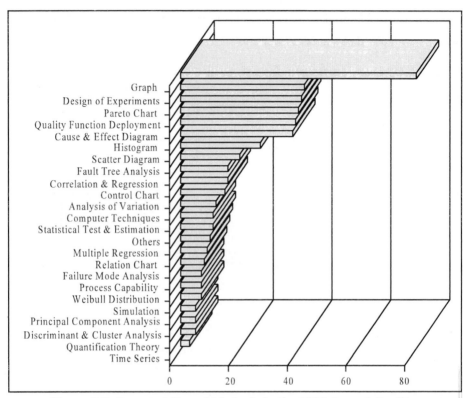

The chart lists the following methods (top to bottom): Graph, Design of Experiments, Pareto Chart, Quality Function Deployment, Cause & Effect Diagram, Histogram, Scatter Diagram, Fault Tree Analysis, Correlation & Regression, Control Chart, Analysis of Variation, Computer Techniques, Statistical Test & Estimation, Others, Multiple Regression, Relation Chart, Failure Mode Analysis, Process Capability, Weibull Distribution, Simulation, Principal Component Analysis, Discriminant & Cluster Analysis, Quantification Theory, Time Series. Horizontal axis marked 0, 20, 40, 60, 80.

Figure 2.2 Reported Use of Methods in the 1987 Annual Quality
Circle Conference (JUSE)

2.2.1 Process Flow Diagram

Before collecting data, the practitioner must be well-versed in the process from which the data will come. Deciding when and where to collect data can be as important as the data itself, and a process flow diagram (or flow chart) can help in defining the system or process under consideration. A process flow diagram is a visual representation of all the major steps and decision points in a process. It helps us understand the process better, identify critical or problem areas, and identify where improvements can be made. A team can use a flow diagram to i) evaluate the process performance (cost and cycle time) at each step, ii) determine where process bottlenecks occur, iii) identify non-value added activities which should be deleted or modified, and iv) identify new process steps to be added to improve overall performance.

The construction of a process flow diagram can be accomplished using the five standard symbols shown in Figure 2.3.

This Symbol...	*Represents...*	*Some Examples are...*
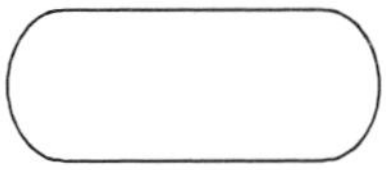 ⬭	Start/Stop	Receive Trouble Report Machine Operable
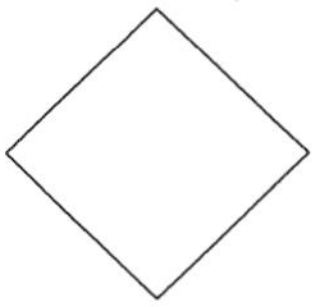 ◇	Decision Point	Approve/Disapprove Accept/Reject Yes/No Pass/Fail
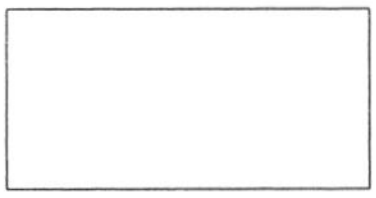 ▭	Activity	Drop Off Travel Voucher Open Access Panel
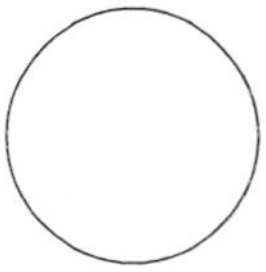 ○	Connector (to another page or part of the diagram)	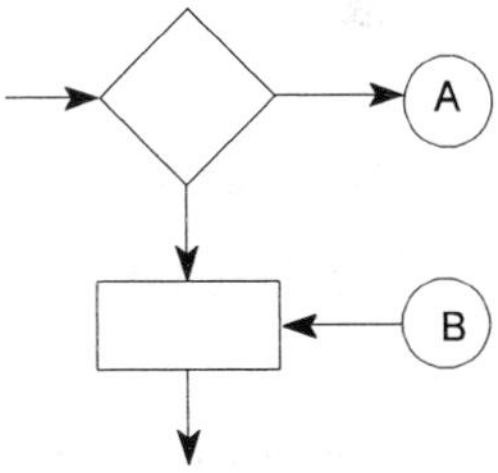
 →	Represents direction of flow	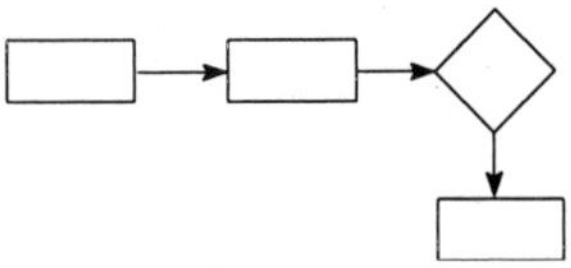

Figure 2.3 Process Flow Diagram Symbols

Almost every process or system that needs to be studied can be represented by a flow chart. Further, the level of detail that is needed can be easily adjusted by replacing a process step at one level of detail, for example, with a more detailed chart that is representative of a lower level of abstraction. Hence, the flow chart is a graphical tool that allows the practitioner the ability to describe the steps in a process in as much detail as necessary. Obviously the box labeled "Consult Manual" in Figure 2.4 could have its own flow chart which conceivably could be many times as large as Figure 2.4 itself. Flow charts should be carefully evaluated to identify non-value added steps to be removed from a process and to identify areas that could be addressed to improve quality. An example of a non-value added step in Figure 2.4 is the box labeled "Plug Power Cord In." In a "quality" process, this is a step that should not have to be accomplished over and over again.

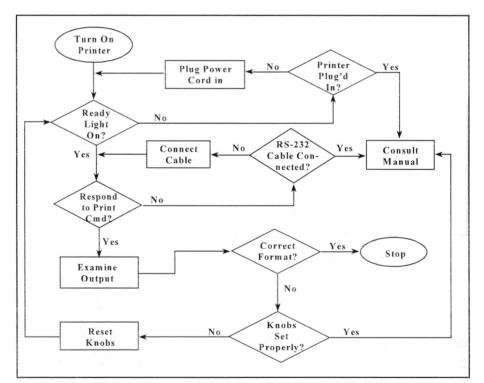

Figure 2.4 Flow Chart: Operation of a Computer Printer

Consider another example from a hospital appointment process. Figure 2.5 depicts the process as it actually existed before improvement was made.

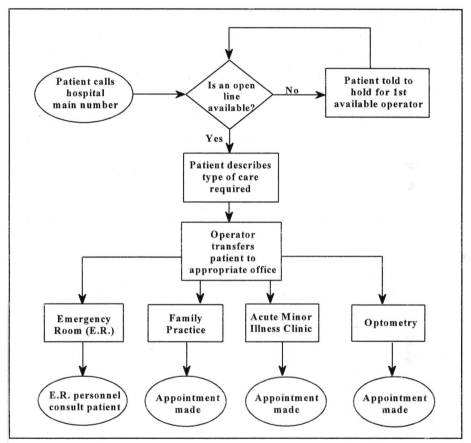

Figure 2.5 Hospital Appointment Process Flow

A hospital survey on customer satisfaction revealed that the biggest customer complaint was holding time for a hospital operator (see Sections 2.2.2 and 2.2.4). A quality improvement team was formed. They sampled the incoming calls and found that the average waiting time to get an operator was 3.4 minutes. The team suggested a new automated phone answering system with push button routing. The new process flow appears in Figure 2.6.

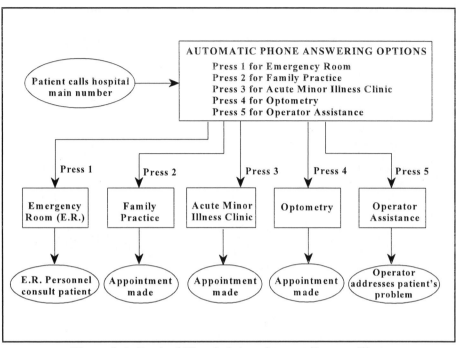

Figure 2.6 Revised Hospital Appointment Process Flow

The new process flow diagram was evaluated through a second sample of incoming call transfers. The end result was a decrease in patient waiting time from 3.4 minutes to 0.2 minute for transfer to the appropriate office. Further study will investigate individual office waiting times. For additional examples on process flow diagrams, see Chapter 12: Case Studies 1, 2, and 3.

2.2.2 Cause and Effect Diagram with CNX

The overall objective of quality control is to improve quality. That means causes of poor quality must be identified and corrected. Furthermore, the dominant causes of defects (non-conformances) need to be isolated and subsequently removed. A very useful graphical tool that can be used to identify, display, and examine possible causes of any observed condition is the cause and effect diagram. This tool is also known as an Ishikawa diagram, after Dr. K. Ishikawa of the University of Tokyo who first formalized its use in the mid 1940s [Ishi 87]. Another name for this tool, a fishbone diagram, is obvious from its structure. Reference Figure 2.7.

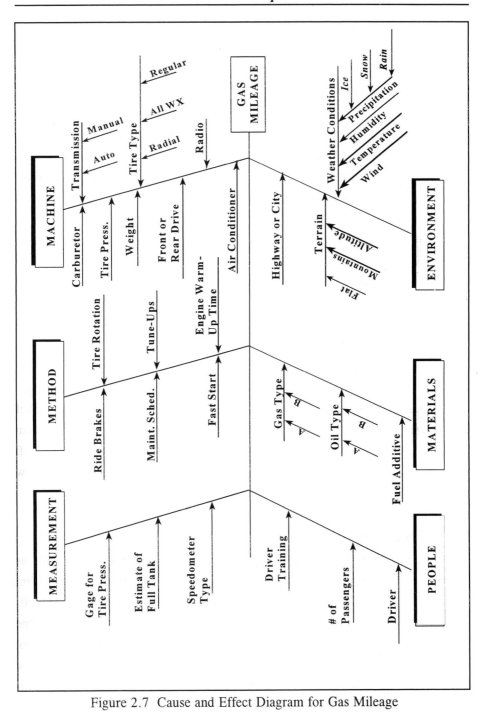

Figure 2.7 Cause and Effect Diagram for Gas Mileage

Steps in constructing a cause and effect diagram can be illustrated by way of this simple example.

1)   Identify the quality characteristic or performance measure for which a cause and effect relationship is to be established. Let us suppose that suspected poor gas mileage for a motor pool of vehicles is the performance measure to be investigated. This output or response measure is written in a box to the right of the major horizontal axis, as shown in Figure 2.7, and forms the "head" of the fish.

2)   Using structured brainstorming and the experience of knowledgeable people, generate several major classes of variables that can affect gas mileage. The classes identified for this example are measurement, method, machine, people, materials, and environment. Place these categories in boxes that are shown to feed into the major horizontal axis of the diagram, as depicted in Figure 2.7. Clearly, these six categories are applicable to most processes. However, one is not restricted to only these six. Other major categories should be used if applicable. The reason the "backbone" of the fish is broken into several major categories is to assist us in focusing on subsets of variables instead of trying to brainstorm all possible variables at one time.

3)   For each of these major categories, identify the variables or causes that fall within that category and insert these into the diagram via horizontal lines terminating at the appropriate diagonal lines emanating from the major category names. For example, gas type and oil type are variables that fit under the category of materials. See Figure 2.7. Further breakout of the causes may be possible, such as "precipitation" being a legitimate input to the branch labeled "weather conditions."

4)   Review each step in the associated process flow diagram to insure that the brainstorming session has exhausted all possible variables.

The cause and effect diagram is a living document which, along with the process flow diagram, should be dated and included in a process notebook. The final diagram should ideally display every variable known to mankind that could possibly affect the performance measure (head of the fish). As new variables are discovered, they should be added to the fishbone diagram. Once the fishbone has been completely filled in to the satisfaction of the team constructing it, the team should partition the variables in the backbone of the fish. That is, each variable or entry in the backbone should be classified (and thus labeled) as either a C, N, or X variable. The definitions of C, N, and X are as follows.

C = those variables which must be held constant and require standard operating procedures to insure consistency. Consider the following examples: the method used to enter information on a billing form, the method used to load material in a milling or drilling process, the autoclave temperature setting.

N = those variables which are noise or uncontrolled variables and cannot be cheaply or easily held constant. Examples are room temperature or humidity.

X = those variables considered to be key process (or experimental) variables to be tested in order to determine what effect each has on the outputs and what their optimal settings should be to achieve customer-desired performance.

A partitioning of the variables in Figure 2.7 is shown in Figure 2.8. For each variable labeled with a "C" there should be a standard operating procedure (SOP) written in the process notebook which details how that variable will be controlled and held as constant as possible. The SOP is the mechanism by which a "C" variable is held constant. For example, Tire Rotation is a "C" variable.

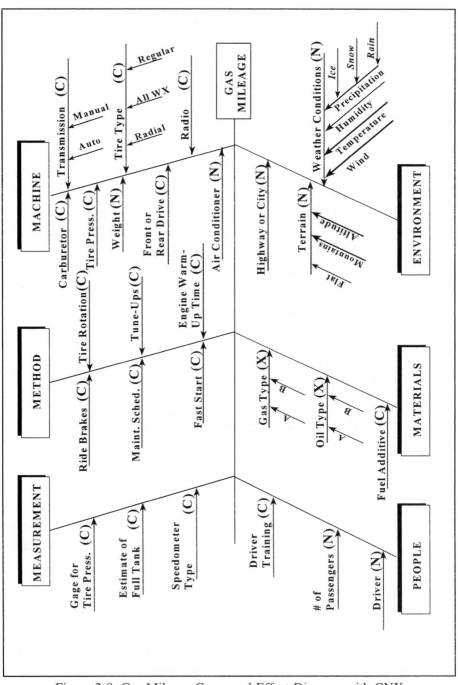

Figure 2.8 Gas Mileage Cause and Effect Diagram with CNX

The SOP should specify how often (i.e., after how many miles) and exactly how the tires are to be rotated. Operators must then be trained and motivated to follow the SOP. The perceptive reader realizes that variability in the performance measure (output) is to a great degree a reflection of the variability occurring in the input variables. Thus, the more input variables we can control, the better we can control the performance measure.

Unfortunately, there are variables that are extremely difficult to hold constant. These are noise or "N" variables. In Figure 2.8, Terrain and Weather Conditions are examples of noise or "N" variables. Ultimately, we would like to make our performance measure "robust" to these noise variables.

Gas Type and Oil Type are examples of experimental or "X" variables. When a variable is labeled with an "X", it tells the reader that we are currently experimenting or testing the effect of that variable on the performance measure, gas mileage. After the experiment, a variable like Gas Type might become a "C" variable (e.g., we will always use 87 octane gas of a particular brand).

The authors have found the technique of partitioning the variables in a fishbone diagram to be valuable from a variance reduction and return on investment standpoint. This should be a major first step in any process improvement effort. While many companies and organizations have used the fishbone, very few have gone the extra step of examining the variables in the fishbone and classifying each with either a C, N, or X and then *acting* on that information.

As a second example, see the cause and effect diagram (Figure 2.9) developed by a hospital team concerned about patient dissatisfaction. A well-constructed cause and effect diagram can help the practitioner in the analysis of a particular problem as well as in the problem identification phase. Such a diagram is useful in determining what data should be collected and where the data might best be obtained. The fishbone in Figure 2.9 was used to help construct a questionnaire sent to their customers. The cause and effect diagram is applicable to a wide variety of applications. A good exercise for any student is to construct a fishbone diagram with quality of life as the output. Consider the major bones of the fishbone to be job, family, finances, personal health, social life, and spiritual. Brainstorm all possible variables in each of these categories and be sure to label the variables as C, N, or X.

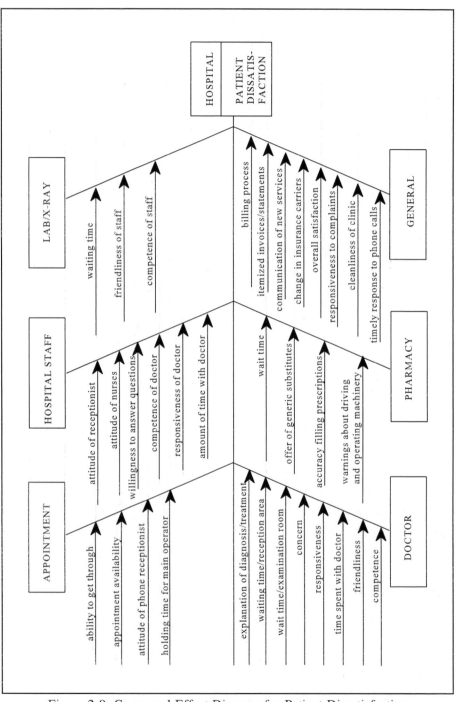

Figure 2.9 Cause and Effect Diagram for Patient Dissatisfaction

2.2.3 Affinity Diagram

Unlike the cause and effect diagram where we move from the general categories to the more specific, an affinity diagram clusters related items into a more general group. An affinity diagram is used to organize verbal information into a visual pattern. It starts with specific ideas and helps us work toward broader categories, each of which contains several of the more specific ideas. Affinity diagrams are usually generated from a group problem solving activity that is initiated by writing the problem or issue on a board or flip chart for the group to see. Using brainstorming the group tries to identify all facets of the problem and records each specific facet on a separate sticky note or index card. Then the sticky notes are clustered into major categories. This can be done by asking "What ideas are similar?" and "Is this idea related to any of the others?" For each group or category we create a separate card or title that is a more generic descriptor of that group. The resulting affinity diagram provides a hierarchical structure that will give valuable insight into the problem.

Affinity diagrams are particularly useful when analyzing written comments obtained from a survey question. The comments may vary in many ways, but what are the major themes running throughout all of the comments? Recently we asked (via a survey question) the students in our classes what they perceived to be the barriers to process improvement in their organizations. As expected, the responses were varied and numerous. Table 2.1 shows some of the individual comments we received. By "affinitizing" these individual comments, we were able to see some major categories appear. The affinity diagram in Table 2.2 shows these categories which are clearly more tangible areas for directing an improvement program than are the individual comments themselves.

Let's examine a couple of closely related entries in Table 2.1 and track them into the affinity diagram (Table 2.2). Note that both "Promotes fire-fighting" and "Too busy fire-fighting" are related to fire-fighting. However, there is not a category labeled "fire-fighting" in the affinity diagram because "Promotes fire-fighting" is more aptly associated with the Reward System and "Too busy fire-fighting" is closely related to Time.

- Communicate to managers with emotion instead of facts, data and dollars

- Continued emphasis on old ways of doing business

- Critical and/or negative attitudes

- Culture not right for sharing information

- Don't track the right metrics

- Duplicated effort

- Fear of failure using new methods

- Improper time management

- Improper allocation of resources

- Inability to think and use common sense

- Insufficient internal experts to assist others

- Insufficient manpower

- Limited resources

- Management last to be trained vs first

- Need for follow-up assistance

- Not enough people properly trained

- Not willing or knowledgeable to ask the right questions

- People not held accountable

- People (groups) compete against each other vs helping

- Poor documentation

- Poor knowledge of customer needs

- Promotes fire-fighting

- Reluctant to support new methods

- Resistance to change

- Suppliers need to be trained

- Too many layers

- Too busy reorganizing

- Too many unproductive meetings

- Too many arguments without facts and data

- Too busy fire-fighting

- Tools for properly communicating process information not known or used

- Unclear direction

- Unqualified trainers

- Unwilling to share what you know with others

- Wants immediate results vs substantial growth in knowledge through use of scientific method

Table 2.1 Alphabetized List of Comments from Survey Question
on Barriers to Process Improvement

Management

- Unclear direction
- Not willing or knowledgeable to ask the right questions
- Too many layers
- Wants immediate results versus substantial growth in knowledge through use of scientific method
- Don't track the right metrics
- Continued emphasis on old ways of doing business
- Reluctant to support new methods

Time

- Too busy fire-fighting
- Improper time management
- Duplicated effort
- Too many arguments without facts and data
- Too many unproductive meetings
- Too busy reorganizing

Communications

- Poor documentation
- Culture not right for sharing information
- Poor knowledge of customer needs
- Communicate to managers with emotion instead of facts, data and dollars
- Tools for properly communicating process information not known or used

Training

- Not enough people properly trained
- Inability to think and use common sense
- Management last to be trained versus first
- Need for follow-up assistance
- Suppliers need to be trained
- Unqualified trainers

Resources

- Limited resources
- Insufficient manpower
- Insufficient internal experts to assist others
- Improper allocation of resources

Reward System

- Promotes fire-fighting
- People not held accountable
- People (groups) compete against each other versus helping

Attitude and Motivation

- Critical and/or negative attitudes
- Resistance to change
- Fear of failure using new methods
- Unwilling to share what you know with others

Table 2.2 Affinity Diagram Example for
Barriers to Process Improvement

2.2.4 Pareto Diagram

Once all the possible factors of product quality have been identified, an effort can be undertaken to determine the dominant causes affecting the quality of the product. A Pareto diagram is a useful tool that illustrates the predominance of varying causes of poor quality by charting the causes in descending order of frequency or magnitude from left to right. In addition, this special kind of vertical bar chart can help the practitioner determine which problem should be attacked first. Pareto diagrams are named after the Italian economist Vilfredo Pareto, but J.M. Juran popularized their use by applying them to industrial problems [Ryan 89].

Pareto diagrams have become a valuable tool in industry because they identify the main sources of defects. Consider the process of fabricating integrated circuit (IC) boards. After the IC boards are fabricated certain tests are made to determine if the boards meet the shipping specifications. Of all the IC boards determined not to meet the quality specifications, it was found that the following defect types contributed to rejecting the final product: soldering, etching, molding, cracking, and a last category that is labeled "other." A Pareto diagram for this defect data is shown in Figure 2.10.

While the left vertical axis of Figure 2.10 shows the actual number of defects, the right vertical axis represents the cumulative percent frequency of defects and is a convenient scale from which to read the line graph. The line graph connects the points which represent the cumulative (from left to right) heights of the bars. The use of both a frequency axis and a percent axis conveys more information to the decision maker than the use of just a frequency axis or percent axis alone.

Figure 2.10 clearly indicates that about 65% of the defects arise from faulty soldering or etching and that perhaps management may want to concentrate its defect removal efforts in these areas. However, a few words of caution are in order. While faulty soldering is the leading culprit as far as number of defects is concerned, it may not be the leading source for loss of revenue. In fact, etching defects require a greater amount of rework than do soldering defects and, as such, represent a greater amount of monetary loss. Hence, a Pareto diagram showing the amount of lost revenue may appear as in Figure 2.11 and might be more applicable than Figure 2.10.

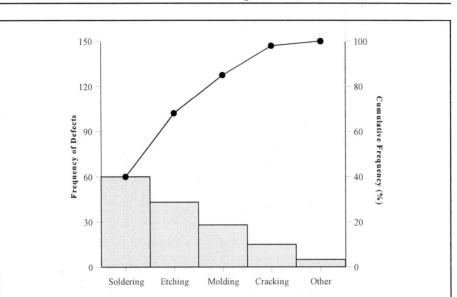

Figure 2.10 Pareto Diagram of IC Board Defects

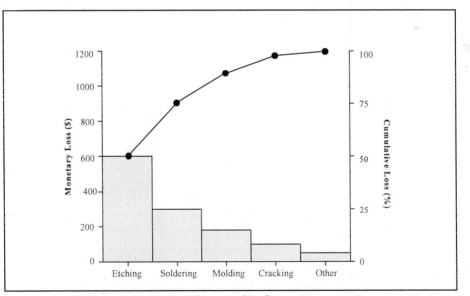

Figure 2.11 Pareto Diagram of Defect vs Monetary Loss

A Pareto diagram such as that in Figure 2.11, which shows that two kinds of defects account for 75 percent of the losses, is clearly a useful tool. While a Pareto diagram can be thought of as an extension of a cause and effect diagram, the order of presentation here is not meant to imply that a Pareto diagram must be used *after* a cause and effect diagram. On the contrary, a Pareto diagram such as shown on the preceding page may indeed instigate a brainstorming session designed to produce a cause and effect diagram with faulty etching being the quality characteristic of concern. Another important aspect of Pareto diagrams that the reader should have grasped by now is that Pareto diagrams are no better than the data they summarize. The ability to classify defects correctly is certainly a prerequisite for constructing an accurate and useful Pareto diagram.

A second example using the Pareto diagram is taken from an Environmental Protection Agency (EPA) study of particulate emissions for various hearth appliances. Particulates are fine particles which can easily be inhaled. They inhibit lung clearing mechanisms, irritate the respiratory system, may lead to increased cancer rates, and contribute to the visible "brown cloud." The results are displayed in Figure 2.12. This diagram indicates why many large cities are banning the use of wood burning stoves during peak pollution periods.

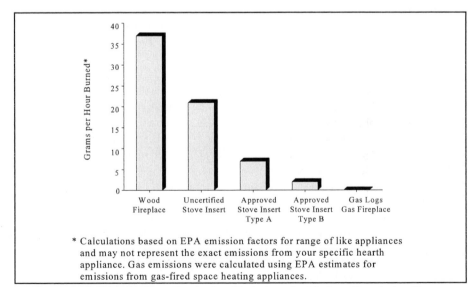

Figure 2.12 EPA Study on Particulate Emissions for Various Hearth Appliances

As a final example of the Pareto diagram consider the hospital cause and effect diagram from Figure 2.9. After the diagram was built the hospital examined one year's worth of historical data from the hospital affairs office. The major causes of complaints are listed in the Pareto diagram described in Figure 2.13. This information led to the quality improvement effort previously discussed in reference to Figures 2.5 and 2.6. Thus, Pareto diagrams should provide insight on major problem areas and give a quality improvement team an idea where to start.

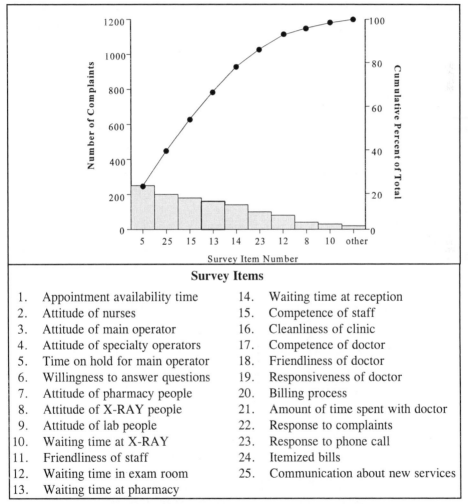

Survey Items

1. Appointment availability time	14. Waiting time at reception
2. Attitude of nurses	15. Competence of staff
3. Attitude of main operator	16. Cleanliness of clinic
4. Attitude of specialty operators	17. Competence of doctor
5. Time on hold for main operator	18. Friendliness of doctor
6. Willingness to answer questions	19. Responsiveness of doctor
7. Attitude of pharmacy people	20. Billing process
8. Attitude of X-RAY people	21. Amount of time spent with doctor
9. Attitude of lab people	22. Response to complaints
10. Waiting time at X-RAY	23. Response to phone call
11. Friendliness of staff	24. Itemized bills
12. Waiting time in exam room	25. Communication about new services
13. Waiting time at pharmacy	

Figure 2.13 Pareto Diagram of Number of Complaints

from a Hospital Survey

2.2.5 Histogram

While the Pareto diagram is a vertical bar chart that deals with frequencies of attribute data (e.g., type of defect, kind of injury, problem areas), the histogram is a bar graph that depicts frequencies of numerical data. Furthermore, whereas a Pareto diagram has its vertical bars in descending heights from left to right, the bars or rectangles in a histogram are not necessarily in descending order of height. In fact, only under special circumstances will the bars in a histogram appear to be in descending order of height. The purpose of a histogram is to provide a pictorial summary of a set of data values. That is, it provides a picture of the distribution.

Consider the data set of 50 values that is given in Table 2.3. This data was obtained from 50 different motorists and gives the gas mileage (mpg) for the auto each was driving.

18	16	30	29	28	21	17	41	8	17
32	26	16	24	27	17	17	33	19	18
31	27	23	38	33	14	13	26	11	28
21	19	25	22	17	12	21	21	25	26
23	20	22	19	21	14	45	15	24	34

Table 2.3 Gas Mileage (mpg) of 50 Vehicles

Certainly the 50 values by themselves as shown in the table do not provide a feel for what is really there. A histogram such as that shown in Figure 2.14 gives a much better synopsis of the gas mileage data. This diagram shows how the gas mileage data is distributed, namely, the center of the data set (about 24 mpg), how the data varies, and any symmetry (or lack thereof) about the center point. Note that the horizontal axis consists of classes of *numerical intervals* (in this case, gas mileage intervals). This is the distinguishing feature of a histogram and is what differentiates it from a bar chart for attribute data (e.g., the Pareto diagram). The height of any bar tells how many vehicles were observed to be in the interval for that bar.

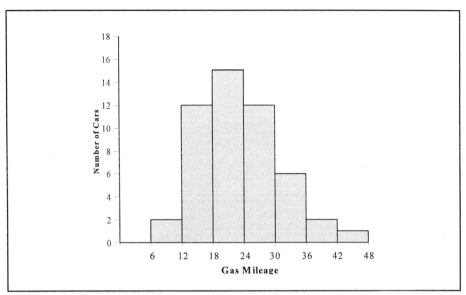

Figure 2.14 Histogram of Gas Mileage Data (k = 7)

The exact appearance of any histogram is dependent on the chosen classes for the horizontal axis. Although there are no precise rules one must follow to construct a histogram, there are certain guidelines that can be used to aid one in doing so. Assuming the raw data is given as shown in Table 2.3, the following approach will be helpful.

Guidelines for Constructing a Histogram

1.   Determine the number of data points in the data set. Call this number n (n=50 for the data in Table 2.3).

2.   Determine the range, R, of the values in the data set. $R = X_{max} - X_{min}$, where X_{max} is the largest value and X_{min} is the smallest value in the data set. In Table 2.3, $R = X_{max} - X_{min} = 45 - 8 = 37$.

3.   Determine the number of classes into which one desires to partition the data. This is perhaps the part of the process for which there are no set rules; however there are some rules of thumb that can be used.

(a)   The logarithm (base 2) rule. Let $k = \lceil \log_2 n \rceil + 1$, where $\lceil X \rceil$ means the smallest integer greater than or equal to X. $\lceil \log_2 n \rceil$ is really the smallest integer power of 2 that results in an exponentiation that meets or exceeds n. One can compute $\log_2 n$ on a calculator by (log n / log 2). For n = 50, the number of classes $k = \lceil \log_2 50 \rceil + 1 = \lceil 5.64 \rceil + 1 = 6 + 1 = 7$. Figure 2.14 displays k = 7 classes.

(b)   The square root rule. Another rule of thumb is to let $k = \lceil \sqrt{n} \rceil$. In the gas mileage example, $k = \lceil \sqrt{50} \rceil = \lceil 7.071 \rceil = 8$. Figure 2.15 is a histogram of the same gas mileage data, only with k = 8 classes. This rule tends to produce large k values whenever n > 100, so caution is in order when n > 100.

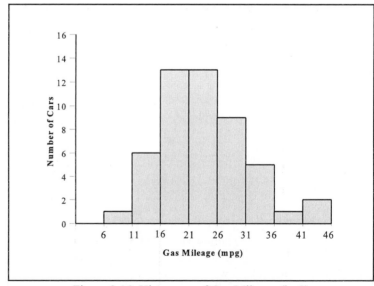

Figure 2.15 Histogram of Gas Mileage (k=8)

(c)   A combination of the two rules in (a) and (b) above will give a fairly good estimate of how many classes should be used.

(d)   The following table [Goal 88] gives a range of classes that can be considered.

Number of Data Points (n)	Number of Classes (k)
Under 50	5 - 7
50 - 100	6 - 10
100 - 250	7 - 12
Over 250	10 - 20

4.   Determine the class width, w, by dividing the range (R) by the number of classes (k) and rounding up.   From the gas mileage data, $w = R/k = 37/7 = 5.286$. Figure 2.14 shows class widths of size $w = 6$. The "rounding up" process is indeed a bit fuzzy, but rightfully so because it gives the analyst some flexibility in choosing the class widths to suit the data at hand. The authors could have chosen 5.3 or 5.5 but decided to use $w = 6$ because of simplicity and convenience and because the original data were all integers. One further note on class widths is in order. The authors encourage the use of equal class widths whenever possible. A histogram is much easier to construct and interpret if equal class interval widths are used. For example, it is much easier to compare the areas of two rectangles of equal width than it is to compare the areas of two rectangles that have different widths.

5.   After determining the number of classes (k) and the widths of each class interval (w) as shown above, one must ensure that the entire width of k class intervals, kw, will span or cover the range R. The only task remaining then is to anchor these k classes, each of width w, on the horizontal axis so that the first class on the left encompasses or contains

X_{min} and the last class contains X_{max}. This can be done in a variety of ways. In the gas mileage example, with k = 7 and w = 6, kw = (7)(6) = 42 and 42 > 37 = R. The authors elected to start the first class at 6 because of the familiarity of the sequence 6, 12, 18, ..., 48. However, starting at 7 and generating the interval boundary sequence of 7, 13, 19, 25, ..., 49 would have also been possible and correct. If either the first or last class is empty, then either the class width, the number of classes, or the anchoring of the classes should be changed.

6.   The final check on the classes should be to make sure that the classes satisfy the following two criteria:

(a)   The classes are **exhaustive**. That is, every data point can be inserted into a class. This means that the classes span or exhaust the observed range of data. This will happen if the above guidelines are followed.

(b)   The classes are **mutually exclusive**. That is, the classes do not overlap. The usual procedure is to assume that each class interval is closed on the left and open on the right. This means that the classes for the data in Figure 2.14 can be written as follows:

	Class	Equivalent Interval Notation
1.	From 6 up to but not including 12	[6,12)
2.	From 12 up to but not including 18	[12,18)
3.	From 18 up to but not including 24	[18,24)
4.	From 24 up to but not including 30	[24,30)
5.	From 30 up to but not including 36	[30,36)
6.	From 36 up to but not including 42	[36,42)
7.	From 42 up to but not including 48	[42,48)

Into which class should the value 18 mpg be inserted? It belongs in the third class, not the second class. If the classes are **exhaustive** and **mutually exclusive**, then each data point will fall into *one and only one* class, an obviously desirable property of any histogram.

Histograms can also be used to examine historical data for accuracy and/or process changes. For example, consider Figure 2.16a where the histogram has a fairly symmetric shape with no obvious outliers (extreme data values) which might be due to an error or a process change. The histogram in Figure 2.16b, however, indicates that the far right data is not a part of the symmetric shaped histogram data. Thus, we should investigate the far right data as to whether it is erroneous or something in the process changed when that data was collected. See the case studies for more detailed histogram examples.

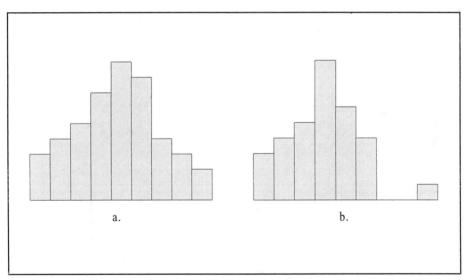

a.   b.

Figure 2.16 Use of Histograms to Evaluate Possible Data Inaccuracies
or Process Changes

2.2.6 Stem and Leaf Display

While the purpose of a histogram is to summarize the data set and give the analyst some kind of an idea about the underlying population or distribution from which the data was obtained, there is no question that some of the details in the data are not apparent in a histogram. For example, the gas mileage histogram in Figure 2.14 has a class interval of [18, 24) with an associated frequency of 15. By examining only the histogram, one cannot determine where within that interval the 15 values fall. Are all 15 values equal to 22 mpg? How many of the values are 20 mpg? It is impossible to tell. If one desires to retain this kind of detail and still get a feel for the underlying distribution, a stem and leaf display is a graphical technique that works well.

Using the gas mileage data in Table 2.3, a stem and leaf display might appear as in Table 2.4.

Stem	Leaf
0	8
1	1 2 3 4 4 5 6 6 7 7 7 7 7 8 8 9 9 9
2	0 1 1 1 1 1 2 2 3 3 4 4 5 5 6 6 6 7 7 8 8 9
3	0 1 2 3 3 4 8
4	1 5

Table 2.4   Stem and Leaf Display for Gas Mileage Data

There are many ways to create a stem and leaf display. Since the gas mileage data was all integer data with one or (mostly) two digits, the choice for the stem was easy: the "tens" digit. The "units" place then constituted the elements that make up the various leaves. Table 2.4 shows five stems or classes, and the distribution of the elements in each class is apparent. Thus, the stem and leaf display shows individual elements in each class while still retaining the feel of a histogram, even if it is a tilted histogram in the sense that the class "heights" protrude sideways. Obviously, the data must be ordered or sorted first if one is to put the data in a stem and leaf display. Case Study 6: Using a Stem and Leaf Plot to Select A Long Distance Calling Card Carrier provides an excellent example of the use of a stem and leaf display.

2.2.7 Run Chart

If data is being collected over time, a run chart is a very simple-to-construct graphical tool that can be used to record and display trends in data over time. The purpose of a run chart is to monitor a system or a process and to detect meaningful changes in the process that may take place over time. If a meaningful change (for the good or bad) can be detected, then further action may be required to effect a change in the process. Examples of the uses of run charts are plentiful. Typical applications include charting the temperature of a hospital patient every four hours; logging the amount of downtime of a computer system per week; recording the number (or percent) of required reworks of printed circuit boards per day; and tracking the productivity of a system or process, whether it is the number of integrated circuit boards accepted on the first inspection or the number of typographical errors produced by a secretarial staff. Run charts provide a simple mechanism to track a process over time.

For example, the Department of Mathematical Sciences at the USAF Academy monitors the percent of instructor substitutions on a weekly basis. Historically, a 4 percent average substitution rate has been recorded and is considered to be acceptable. The run chart shown in Figure 2.17 displays the percent substitution rate for 18 consecutive weeks in a semester. Despite the spike in the chart around week 12, this chart does not reveal any adverse trends or shifts in the average. The spike is due to a special cause affecting the process — a math conference attended by several teachers. Two very common tests that can be made on a run chart to detect meaningful systemic changes are as follows:

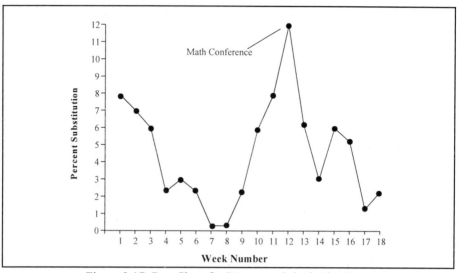

Figure 2.17 Run Chart for Instructor Substitution Rate

(a)   Since it is expected that there would be approximately the same number of points above the "average" line as there are below it, a good rule of thumb is that if there is a "run" of 7 consecutive points on one side of the average, something significant may be happening and it would probably be a good idea to call "time out."

(b)   A second test is to see whether a run of 7 or more intervals is steadily increasing or decreasing without reversals in direction. As in (a), such a pattern is not likely to occur by chance, thereby indicating something needs to be investigated.

Neither of these patterns is seen in Figure 2.17, so there was no apparent reason to investigate the "substitution" process. One caution in reading a run chart is appropriate. Looking at every variation in the data as being critical could cause undue concern and divert attention away from truly significant changes that may be occurring in the system.

Often we use a run chart to reflect performance over time. In other words, it can serve as a "scorecard" to show whether we are really making continuous improvement or just talking about it. Figure 2.18 shows a run chart for the number of software discrepancies per modified line of code detected by a customer of a software production company. The different dots over time represent data for different revisions of some major piece of source code.

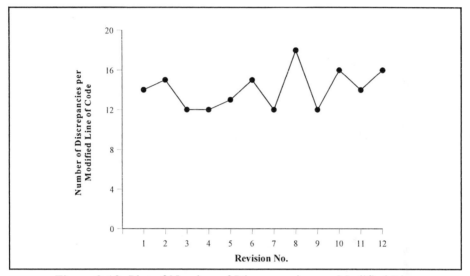

Figure 2.18 Plot of Number of Discrepancies per Modified Line
of Code Detected by Customer

Based on the data presented in Figure 2.18, the software producer could talk about continuous improvement but would have a difficult time convincing the customer it is nothing more than continuous perspiration. It should be very obvious that every manager claiming to be implementing continuous process improvement should find a way to develop a "scorecard" that reflects the real process performance. Any "scorecard" that doesn't include good data tracked over time is usually a bunch of hot air.

Sometimes it is useful to combine a run chart with a stem and leaf display or histogram so that the distribution of values can be viewed at the same time that the data over time is displayed. Consider the process of producing cylindrical steel rods. If the rods are produced in batches (or lots) and we sample 5 rods from each lot, a combined run chart with histogram for the average diameter of the sampled rods from 25 consecutive lots might appear as in Figure 2.19. Looking only at the histogram, the engineer would have no reason to suspect process problems. However, the run chart depicts a cycling of data which might be investigated as to its cause.

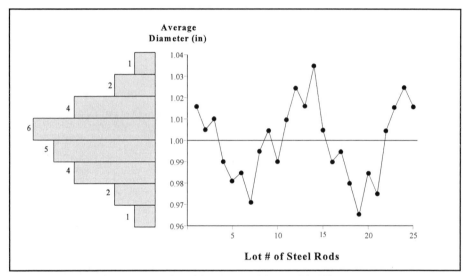

Figure 2.19 Combined Run Chart with Histogram

2.2.8 Control Chart

The run chart presented in the previous section provides a quick look at data over time. However, run charts do not provide information as to whether extreme points, trends, or cycles are statistically significant.

A control chart is a special kind of run chart which includes statistically determined upper and lower control limits as well as a center line. We can use it to detect significant trends, cycles, and outlying points (outliers). Figure 2.20 is a control chart that tracks the amount of sodium chloride (NaCl) or salt contained in a leading deodorant soap. Later we will address the subject of how the upper and lower control limits are calculated. For now, however, we emphasize that the upper and lower control limits are *not* specification limits; they are computed from data obtained from the process, and they are usually positioned three standard deviations above and below the centerline. A detailed presentation of different kinds of control charts, their uses, and how the control limits are computed is given in Chapter 9, Statistical Process Control (SPC). This section is provided as a very brief introduction to what a control chart is, what it looks like, and what it is used for.

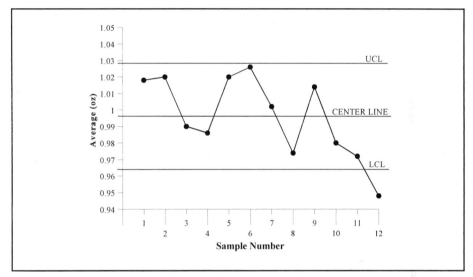

Figure 2.20 Control Chart for Salt Content in Soap

Consider the process of producing bars of soap as shown in Figure 2.20. The quality of the soap is directly related to the amount of salt (NaCl) contained in it. It is important that the manufacturer control the salt content if a quality product is to be produced. Like all processes, the process of making bars of soap has variation associated with it. Many different factors enter into a production process and a change in any factor could cause a change in the final product. Differences among machines, suppliers, incoming raw materials, and workers, among other factors, can influence and produce variability in the end product. Ultimately, it is this variability in the end product that must be controlled if the manufacturer wishes to avoid lost production, poor quality, and eventually the loss of customers.

A control chart is a graphical device that is commonly used in both service and manufacturing industries to monitor process variation and to indicate when a process is "out of control." Clearly, we would like to detect an "out of control" process as quickly as possible but, at the same time, have as few "false alarms" as possible. Statistical principles imbedded in the graphical vehicle called a control chart allow this desired situation to become reality. The control chart in Figure 2.20 plots the average amount of salt content (in ounces) from samples of size 5 taken every hour over a 12-hour production period. Table 2.5 summarizes the sample data that was collected and subsequently plotted in Figure 2.20. The average line (or centerline), the upper control limit (UCL), and the lower control limit (LCL) in this figure were determined from previously collected data (i.e., prior to this 12-hour period) when the process was actually in control. As will be seen in Chapter 9, the control limits are not always determined "a priori," or beforehand. However, like all control charts, the primary purpose of Figure 2.20 is to aid the reader of the chart to distinguish between the variability that can normally be expected in the process and the variability that could result from special or unexpected causes.

Sample	Amount of Salt (oz)					Average
1	1.02	0.99	1.01	1.03	1.04	1.018
2	1.03	1.04	1.00	1.01	1.02	1.020
3	0.97	1.04	1.01	0.97	0.96	0.990
4	0.98	0.99	1.01	1.00	0.95	0.986
5	1.05	1.03	1.02	0.99	1.01	1.020
6	0.99	1.03	1.02	1.04	1.05	1.026
7	1.01	0.98	0.99	1.02	1.01	1.002
8	0.96	0.94	0.99	1.01	0.97	0.974
9	0.99	1.03	1.02	1.01	1.02	1.014
10	0.98	0.99	0.96	0.99	0.98	0.980
11	0.96	0.99	0.95	0.97	0.99	0.972
12	0.95	0.94	0.92	0.98	0.95	0.948

Table 2.5 Sample Salt Data that Produced Figure 2.20

Points that lie within the upper and lower limits can usually be associated with "normal" variability, whereas points lying outside of the limits have a good chance of being associated with some specific cause extraneous to the "normal" sources of variation. However, even if all of the plotted points fall within the limits, certain trends can indicate that a possible adjustment in the process may be warranted. Included among these trends are the following:

(1) Seven consecutive points all on the same side of the centerline.

(2) Seven consecutive intervals all moving up or all moving down.

(3) Fourteen consecutive points that alternately move up and down.

Each of the trends given here is highly unlikely to occur if the variation is truly random. None of these trends is evident in Figure 2.20, but the lower control limit is violated by sample 12. In fact, the last few samples in Figure 2.20 suggest that an extraneous source of variation exists and that a closer look at what might have affected the process between hours 9 and 12 is justified.

The control chart in Figure 2.20 is called an $\bar{x}$-chart (read "x bar chart") because the average (or $\bar{x}$) of five measurements is the value that is plotted for each sample. Other kinds of control charts exist and are discussed in Chapter 9.

2.2.9 Scatter Diagram

The graphical tools presented thus far in Section 2.2 have dealt with data values for a single variable. For example, the gas mileage data was just that — each value represented the gas mileage of a particular vehicle. Oftentimes it is desirable to display data that is associated with more than one variable. The most common of these situations is **bivariate** data, that is, data associated with two variables. In this case the data is no longer seen as a set of individual values as shown in Table 2.3 but is instead a set of ordered pairs. The first value of each ordered pair is associated with one variable while the second value is associated with the second variable. Suppose that for the gas mileages given in Table 2.3 we also knew the weight in pounds of the vehicle associated with that mileage. Table 2.6 repeats 10 of the values from Table 2.3 along with the weight of the vehicle associated with that gas mileage.

Observation	Weight (lbs)	Mileage (mpg)
1	3000	18
2	2800	21
3	2100	32
4	2900	17
5	2400	31
6	3300	14
7	2700	21
8	3500	12
9	2500	23
10	3200	14

Table 2.6 Bivariate Data (Vehicle Weight vs Gas Mileage)

A scatter diagram is a graphical display of a set of points or ordered pairs. The scatter diagram for the 10 observations of vehicle weight and gas mileage is shown in Figure 2.21. A scatter diagram is also called a scattergram or scatter plot.

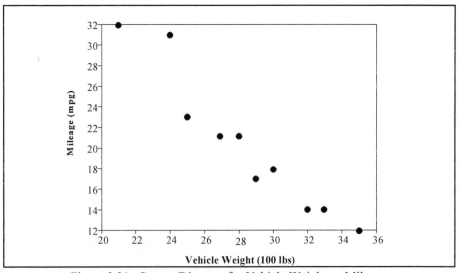

Figure 2.21   Scatter Diagram for Vehicle Weight vs Mileage

The purpose of a scatter diagram is to visually study the relationship between two variables. Figure 2.21 clearly exhibits an inverse relationship between the variables "weight of vehicle" and "gas mileage," for it seems as if increased vehicle weight is associated with lower gas mileage. Statistical methods presented later will allow for a more precise characterization of the relationship of the two variables, but a scatter diagram is a first step in visualizing and quantifying that relationship. Intuition and the laws of physics certainly give credence to the inverse relation shown in Figure 2.21. However, in general it is wise for the practitioner to abstain from making cause and effect statements based solely on the appearance of a scatter diagram. The next example illustrates this "caution."

Consider the data in Table 2.7 which shows the bivariate data of "the number of points scored" vs "the number of fouls committed" for each of 12 members of a basketball team over an entire season. The scatter diagram of this data shown in Figure 2.22 gives the viewer the feeling that those players who fouled the most also scored the most. Any coach who would read a cause and effect relationship into this data is destined for a short-lived contract, however. Certainly "committing more fouls" is *not* the cause of "scoring more points." Perhaps "amount of playing time" or some other variable not even shown here may have a significant impact on both "fouls committed" and "points scored." Hence, caution is in order when deducing any cause and effect relationship between the variables displayed in a scatter diagram.

Case	Fouls	Points
1	3	4
2	8	12
3	13	19
4	18	41
5	31	52
6	42	78
7	58	105
8	64	148
9	74	218
10	67	247
11	79	401
12	86	522

Table 2.7 Bivariate Data (Fouls vs Points)

The labeling of the axes in a scatter diagram (i.e., which variable goes on which axis) is usually arbitrary; but if one variable (such as gas mileage in Figure 2.21) can logically be considered as a "dependent" or response variable, it is perhaps easier to relate the two variables if the "dependent" variable is placed on the y or vertical axis and the "independent" variable is placed on the x or horizontal axis. Likewise, scaling of the axes is necessarily somewhat arbitrary, but the reader should keep in mind that the scaling of the axes can depict a relationship as being either stronger or weaker than the true relationship. Lastly, the run charts and control charts in the previous two subsections could be considered as special kinds of scatter diagrams, where the x-axis is usually the variable time. However, while the adjacent points in a run chart or control chart are connected by line segments for the purpose of showing trends over time, the points in a scatter diagram are not connected by line segments.

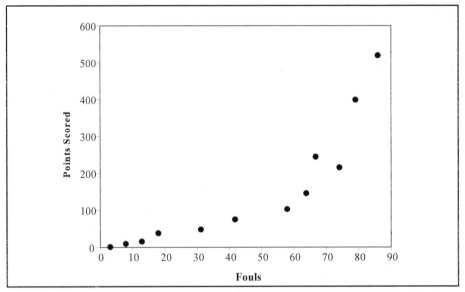

Figure 2.22   Scatter Diagram of Fouls vs Points Scored

2.2.10 Fault Tree Analysis[1]

Fault Tree Analysis (FTA) is a deductive technique used to analyze complex products, systems, or processes. An FTA provides a logical basis for analysis and justification for changes to the product, system, or process being studied. It is a visual tool that logically and graphically presents the various combinations of possible events or failures that could occur in a process or product. While an FTA can be used as a stand alone tool, it is often used in combination with a Failure Mode and Effect Analysis (FMEA) to assist the user in identifying the root cause(s) of possible failure modes. FMEA is discussed in the next section.

The basic structure of the FTA is an inverted tree, with the trunk of the tree at the top of the diagram representing the undesired event or failure as shown in Figure 2.23. The various contributing causes or events are represented by the different branches that emanate from the single failure at the trunk. The nodes at the base of the diagram represent the root causes of the failure. The branches of the tree are connected by the standard "and"/"or" logic symbols to indicate the logical connection between different events. These symbols are not shown in Figure 2.23 but are used in the next example.

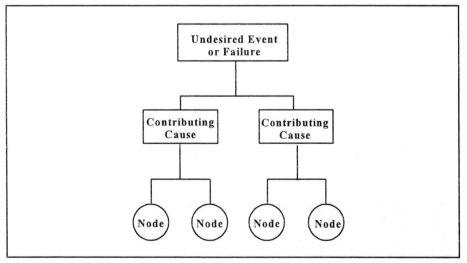

Figure 2.23 FTA Basic Structure

[1]This section was provided by Terry Newton, Air Academy Associates.

Using the simplistic FTA for "Engine Sputters and/or Dies" in Figure 2.24 as an example, the failure ("Engine Sputters and/or Dies") is annotated at the top of the diagram. The diagram tells us that the failure event occurs when any one (or more) of the three events branching out beneath the failure event occurs (i.e., "Out of Gas," "Carburetor Failure," or "Fuel Flow Failure"). The event "Out of Gas" occurs when either there is a "Gage Failure" or the "Son Failed to Fill Tank" events occur. The root cause of the failure event is identified by a node that has no further branches. For our example, let's assume the "Son Failed to Fill Tank" is identified as the event that occurred. Then the root causes for this event would be identified as both "Poor Communication" and "Bad Genes." The "and" connector implies that both of these events must occur concurrently for the event "Son Failed to Fill Tank" to occur.

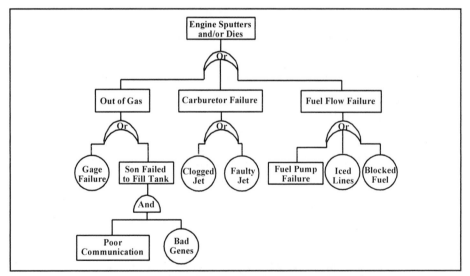

Figure 2.24 FTA for "Engine Sputters and/or Dies"

The FTA is a simple yet very effective technique for determining the root cause of some event. As a visual tool, it provides an efficient and easily understood format for accomplishing a process/product analysis. When combined with an FMEA, an FTA can assist the user in quickly identifying the root cause of a failure and accurately documenting the thought process that was used to produce the given results.

2.3 Failure Mode and Effect Analysis (FMEA)[2]

FMEA is a quality planning tool used to identify and eliminate potential product and process failures or defects. By using an FMEA we can systematically identify, analyze, prioritize, and document potential failure modes, their consequences on a system, project, or process, and prevent future failures.

The applications of FMEAs are almost limitless. Early applications were safety oriented in the aerospace business. Safety FMEAs in some companies today are referred to as Hazard Analysis. FMEA has since spread into manufacturing, design, administrative, accounting and finance processes as an attempt to eliminate errors and reduce failures and customer complaints. Another recent application is in software design. Wherever we apply FMEA, the earlier it is used in the life cycle of a product or process, the greater the potential for positive results.

Prior to beginning an FMEA, we should address each of the following:

1.   We should form an appropriate team (preferably crossfunctional) and select a team leader. All team members must know who the FMEA owner is.

2.   We must thoroughly define the process, product, or defect that will be studied. For a process, this is best accomplished with a process flow diagram (Section 2.2.1). For a product, this can be accomplished with blueprints or a prototype.

3.   We must clearly define the FMEA boundaries to include deadlines, product limitations, budget, usable resources, report format, final report dissemination, or any other limitations.

4.   We must also consider the size of the problem. If the task is too large for one FMEA, it might be advantageous to break the problem into smaller pieces and attend to each piece separately.

---

[2]This section was provided by Terry Newton, Air Academy Associates.

The steps for conducting the FMEA are:

1.   Study the process/product to be analyzed.

2.   Brainstorm the possible failures.

3.   List the potential consequences of each failure mode.

4.   Assign severity (SEV) scores.

5.   Identify the cause(s) of each failure mode.

6.   Assign occurrence (OCC) scores.

7.   Identify current controls to detect the failure modes.

8.   Assign an escaped detection (DET) score for each cause and control.

9.   Calculate the Risk Priority Number (RPN) for each line in the FMEA.

10.   Prioritize the failure modes and causes based on RPN.

11.   Determine the action to be taken.

12.   Recalculate the RPNs based on the action plans.

Ideally, the steps of an FMEA should be accomplished in order and each step should be complete before progressing to the next step. However, this does not preclude the team from backtracking when they deem it necessary due to errors, incomplete steps, or incomplete information. Throughout this section we will use a continuing example of the process of filling your car with self-serve gasoline to demonstrate each step. The complete FMEA form can be seen in Table 2.12. As each step is detailed, the applicable parts of the FMEA will be shown. A process flow diagram for this process is included in Figure 2.25 as a reference for the reader. The continuing example will use *only* the first four steps of the self-serve gas process.

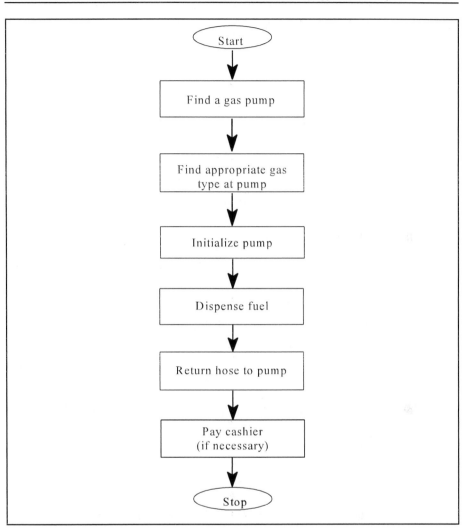

Figure 2.25 FMEA for a Self-Serve Gas Process

Step 1.

Study the process/product to be analyzed. The FMEA process starts with team members studying the process flow chart and/or the product blueprint. For a product, team members should also study the functions of the major product components. For a process FMEA, the team members should actually walk through the process, step by step, so all members of the team are familiar with the details of every step.

Step 2.

Brainstorm the possible failure modes. Depending on the process or product, the generation of possible failure modes may take a lot of time. To facilitate the brainstorming session(s), it is advantageous for the team leader to assign team members the job of developing their own ideas concerning potential failure modes before the team meeting. When the brainstorming meeting begins, team members should already have ideas in mind (hopefully on paper) and be prepared to present their ideas. As individuals present their ideas, other team members should hitchhike on those ideas as they are being presented, leading to a synergistic effect that generates an abundance of ideas.

One possible way to identify potential failure modes in a process FMEA is to study the desired output from each step in the process (from the process flow diagram) and identify how the process could fail to generate the expected outputs. Use the six elements of any process mentioned earlier (people, materials, machine, methods, measurement, and environment) to help facilitate the brainstorming. But the basic question should remain, "What are the potential failures at this step in the process or in this component of the product?" As the brainstorming continues, each potential failure is paired with its associated process step and placed on the FMEA form. When doing a product FMEA, divide the product into major components and list the ways the individual components could fail. At this point, don't worry about the causes of the failure. The current concern is simply how can this component fail?

The end product of this step is an FMEA form with all process steps or product components listed and the potential failure modes paired with their associated process step or product component. For the self-serve gas example, see columns 1 and 2 in Table 2.8 for the listing of the process steps and the potential failure modes. Note that each process step may have more than one failure mode.

1	2	3	4
Product Component or Process Step	**Failure Mode**	**Failure Effects (Consequence of Failure Mode)**	**S E V**
Find gas pump	All pumps busy	Must wait for pump	1
		Find another station	3
	Gas cap on other side of the car	Must move car	2
Find appropriate gas type at pump	Unavailable	Must move car	1
		Buy different octane	2
Initialize pump	Pump won't reset	Can't pump gas	2
	Pump won't read credit card	Must pay cash	4
Dispense fuel	Auto shut-off fails	Spill gas on ground	5
		Pay for gas on ground	2

Table 2.8 Self Serve Gas Process FMEA

Step 3.

List the potential failure effects (subsequently referred to as consequences) of each failure mode. There may be only one consequence for some failure modes while other failure modes may have several consequences. It is helpful to think of this step in terms of "if this failure mode occurs, then what are the consequences to people, the process, the machine, the environment, etc...?" Refer to column 3 of Table 2.8 for a partial listing of the logical consequences of the failure modes in the self-serve gas example.

Step 4.

Assign a severity (SEV) score to each consequence. The severity score is an estimation of how serious the consequence would be, ***assuming*** the given failure mode occurred. If there has been past experience with the failure, then historical data can be used to make the determination. If this is a new product or process, then it may require engineering knowledge and estimation to generate a severity score. However we determine the score, it is imperative that the score be based upon the severity of the ***consequence*** and not the failure mode. This can best be illustrated using an example. If when filling your car with self-serve gasoline the automatic shut-off fails to stop the flow of gas when your tank is full (the failure mode), then there will be several consequences. One consequence will be that you will pay for more gas than you actually put into your tank. While this is not very serious, it is annoying. Another consequence is that the spilled gas poses a potential fire hazard. Both consequences are the result of the same failure mode, but they have different severity scores.

The method used to generate severity scores should be based on Table 2.9. The more severe the consequences are, the higher the severity score. A scale of 1 to 5 is common although a team may use any numeric scale they desire. If the consequence is severe (very bad), then a high score should be given (a 5 using the scale from Table 2.9). If the consequence is negligible, then a low score should be given (a 1 using the scale from Table 2.9). Enter the severity score in the table as shown in column 4 of Table 2.8.

RISK PRIORITY NUMBER (RPN) =
SEVERITY × OCCURRENCE × ESCAPED DETECTION

SCORING:			(Product and Process FMEAs)		
Score \ Category	5 (Very Bad)	4	3	2	1 (Good)
Severity	*Severe* consequence of failure	High	*Moderate* consequence of failure	Minor	*Negligible* consequence of failure
Occurrence	*Very high* probability cause of failure mode will occur	High	*Moderate* probability cause of failure mode will occur	Low	*Very low* probability cause of failure mode will occur
Escaped Detection	*Very high* probability failure will escape to the "customer"	High	*Moderate* probability failure will escape to the "customer"	Low	*Very low* probability failure will escape to the "customer"

Table 2.9 FMEA Scoring System

Step 5.

Identify the cause(s) of each failure mode. Each failure mode was caused by some basic reason and that reason must be discovered. This is a critical step because if the team fails to identify the root cause(s) and instead identifies symptoms, the result will be a solution that treats symptoms and leaves the problem in place to continue causing failures.

One tool that can be used to identify root causes of failures is Fault Tree Analysis (FTA). See Section 2.2.10. FTA focuses on the root causes of a failure by exploring the interdependent relationships involved in complex processes and products [Stam 95]. With the help of FTA, the root cause of each failure mode and consequence can be identified and listed on the FMEA chart. Another tool that can be very helpful in identifying root causes is the cause and effect or fishbone diagram (see Section 2.2.2). The failure mode is the effect or head of the fishbone diagram and the causes are what generated the failure mode. See Table 2.10, column 5, for a continuation of the self-serve gas example. Note that each failure mode can have more than one cause and that filling in the "Causes" column is completely independent of columns 3 and 4 (Failure Effects and SEV). That is, columns 3 and 4 could (and probably should) be covered up when filling in the "Causes" column.

Step 6.

Assign occurrence (OCC) scores. The occurrence score for each cause should be based on actual data. This can be found in failure logs, run charts, or other process/product data. If this is a design FMEA for a product that does not exist, then the team must rely on engineering knowledge and estimations and possibly some preliminary designed experiments. Whatever method is used, the occurrence score should use the same numeric range as used to generate severity scores. See Table 2.9. If the cause of a failure mode is very likely to occur, then a value of 5 should be assigned. If the root cause of the failure is unlikely to occur, then a value of 1 should be assigned. The occurrence score for each root cause represents the likelihood of occurrence of that root cause and its associated failure mode. A special case occurs when we are able to eliminate, through some control mechanism or whatever, the possibility of a cause occurring at all. In this case, the occurrence score for that cause becomes 0. The occurrence scores for the self-serve gas process are entered in column 6 of Table 2.10.

Step 7.

Identify current controls to detect the failure modes. Once the root causes and their occurrence scores have been determined, it is important to consider what controls are *currently* being used to identify or eliminate the cause of the associated failure mode. There are many techniques, including probability, reliability theory, or modeling, for identifying or controlling failures in a product or process. The

1	2	3	4	5	6	7	8	9
Product Component or Process Step	Failure Mode	Failure Effects (Consequence of Failure Mode)	S E V	Causes (of failure mode)	O C C	Controls	D E T	R P N
Find gas pump	All pumps busy	Must wait for pump	1	Fri / pm / rush hour	3	Fill on Thursday	1	3
		Find another station	3	Impatient / hurry	3	More Exercise	2	18
	Gas cap on other side of the car	Must move car	2	Poor planning	2	None	5	20
Find appropriate gas type at pump	Unavailable	Must move car	1	Poor planning	1	None	5	5
		Buy different octane	2	Impatient / hurry	3	Exercise	5	30
Initialize pump	Pump won't reset	Can't pump gas	2	Previous person has not paid	1	Find another station	1	2
			2	Attendant failed to reset	1	Find another station	1	2
	Pump won't read credit card	Must pay cash	4	Credit card dirty	1	Replace card	1	4
Dispense fuel	Auto shut-off fails	Spill gas on ground	5	Pump failure	1	None	5	25
			5	Auto tank design	5	None	5	125
		Pay for gas on ground	2	Auto tank design	5	None	5	50

Table 2.10 Self-Serve Gas Process FMEA

actual tool currently being used may be in the form of checklists, control charts, or published standard operating procedures. Whatever tools are being used must be identified and entered in column 7 of the FMEA chart as shown in Table 2.10.

Step 8.

Assign an escaped detection (DET) score for each cause and control. We will define an escaped detection as the missed detection of some failure mode/cause. When assigning the DET score, the assumption is that failure has occurred, no matter how unlikely its occurrence. The DET score reflects the inability of the current system to detect the failure.

If there is a low likelihood that the problem or failure will escape our control system and get to the customer, then the DET score will be 1. However, if there is a high likelihood of the failure escaping our control system and reaching the customer, then the DET score will be 5. Like the other two assigned scores (SEV and OCC), a higher DET score represents a worse (high risk) condition. If there are no controls in place to detect the failure, then the DET score should be high because there is no capability to identify the failure prior to the product reaching the customer. Enter the DET score for each line in column 8 of the FMEA as shown in Table 2.10.

Step 9.

Calculate the Risk Priority Number (RPN) for each line in the FMEA table. The RPN is simply the product of the Severity (SEV), Occurrence (OCC), and Escaped Detection (DET) scores for each line in the FMEA. After all individual RPNs are calculated and entered in column 9 of the FMEA as has been done in Table 2.10, add the individual RPNs together. The sum of the RPNs in Table 2.10 is 284. RPNs have no specific units, but since the RPN is a measure of risk, the sum of the RPNs is one way to assess total risk. The total RPN will be used in later steps to evaluate the change in total risk.

Step 10.

Prioritize the failure modes and causes based on RPN. The magnitude of the RPN identifies the criticality of the problem. The bigger the RPN, the greater the risk. Since the RPN is a measure of the total risk associated with the product component or process step, the higher the RPN, the higher the priority for fixing the problem. A Pareto diagram (see Section 2.2.4) based on the RPNs can be helpful in prioritizing the failure modes/causes because it will identify the "heavy-hitters."

Step 11.

Determine the action to be taken. The ultimate goal of this step is to eliminate the failure mode completely. This can be accomplished most effectively by attacking the root causes of the failure. Occasionally, this may be as simple as changing the supplier of a specific part. If it is possible to completely eliminate the failure mode, then the revised RPN will be zero since the occurrence score should become zero even though the scale only ranges from 1 to 5. While eliminating a failure mode is the ultimate goal, it seldom occurs. Usually, brainstorming other methods to reduce the RPN will be required.

Starting with the failure mode that has the highest RPN as identified by the Pareto diagram, we should identify ways to reduce the severity, occurrence, and/or escaped detection scores of each failure mode. Any brainstorming tool can facilitate this task. To get started, Table 2.11 [McDe 96] lists some specific actions that can be used to reduce each of the three different ratings.

Usually, the easiest approach is to decrease the escaped detection score. However, this amounts to no more than a band-aid approach to masking the problem from the customer and is usually the least desirable fix. Lowering the escaped detection score (making detection more likely) is usually accomplished through an inspection process to identify the failures before the customer finds them. This is a very common technique used in many industries.

Severity	Occurrence	Escaped Detection
• Personal protective equipment (e.g., hard hats or bump caps, side shields on safety glasses, full force protection, cut-proof gloves, long gloves).	• Increasing the C_{pk} through design of experiments and/or equipment modifications.	• Statistical process control (to monitor the process and identify when the process is going out of control).
• Safety stops/ emergency shut-offs.	• Focus on continuous improvement/problem solving teams.	• Ensure the measuring devices are accurate and regularly calibrated.
• Use different materials such as safety glass that will not cause as severe an injury should it fail.	• Engaging mechanism that must be activated for the product or process to work (e.g., some lawn mowers have handles that must be squeezed in order for them to operate).	• Institute preventive maintenance to detect problems before they occur.
		• Use coding such as colors and shapes to alert the user or worker that something is either right or wrong.

Table 2.11 Specific Actions to Reduce Ratings

The greatest potential for product/process improvement lies in reducing the occurrence score of each failure.   However, identifying a way to reduce the occurrence score can be very difficult.   One example of reducing the occurrence score of a (tragic) failure is the child proof caps that have been placed on pill bottles.   By making the pill bottles inaccessible to children (and to many adults) the occurrence of child poisonings has decreased.

To reduce the severity of a failure, the team could identify protective clothing or devices that will protect people, equipment, and the environment. This is most often applied when safety is a factor. As an example, air bags were placed in automobiles to reduce the severity of automobile accidents. The air bags do not reduce the rate of occurrence of automobile accidents nor do they detect the accident before it happens, but they do reduce the injuries to the individuals involved.

As we generate actions to reduce the three ratings, we must document these actions on the FMEA form and identify a plan of action for implementation. When a score for severity, occurrence, or escaped detection is modified by the planned action, we document the action in the appropriate location on the FMEA form. These new values become the ***proposed*** scores for severity (pSEV), occurrence (pOCC), and escaped detection (pDET) based on the corrective actions proposed by the team. If a score did not change, then use the initial score. Columns 10 - 14 in Table 2.12 would be used for this information.

Step 12.

Recalculate the proposed RPNs (pRPN) based on the revised ratings. That is, pRPN = pSEV ∗ pOCC ∗ pDET. After the proposed RPNs are calculated, place them in column 15 of Table 2.12 and compare them with the initial RPN values for each failure mode/cause. If one or more of the proposed RPNs is still unacceptable, then the team may need to reconsider how to change the RPN for those failure modes. One way to measure the overall effectiveness of the team is to use the sum of the RPN values calculated in step nine and compare it to the sum of the new RPN values. The difference between the two summed RPNs is a measure of how much risk the changes will eliminate ***if*** the action plans are used and are as effective as the team believes they will be.

While the self-serve gas example developed in this section is a process FMEA, the FMEA in Table 2.13 is an example of a product FMEA. Neither example was intended to be a complete analysis of the subject. Rather, they are included to illustrate the steps of an FMEA.

1 Product component or Process Step	2 Failure Mode	3 Failure Effects (Consequence of Failure Mode)	4 S E V	5 Causes (of Failure Mode)	6 O C C	7 Controls	8 D E T	9 R P N	10 Actions	11 Plans	12 pSEV	13 pOCC	14 pDET	15 P R P N
Find gas pump	All pumps busy	Must wait for pump	1	Friday / pm rush hour	3	Fill on Thursday	1	3						
		Find another station	3	Impatient / hurry	3	More exercise	2	18						
	Gas cap on other side of vehicle	Must move car	2	Poor planning	2	None	5	20						
Get correct hose	Unavailable	Must move car	1	Poor planning	1	None	5	5						
		Buy different octane	2	Impatient / hurry	3	Exercise	5	30						
Initialize pump	Pump won't reset	Can't pump gas	2	Previous person has not paid	1	Choose better station	1	2						
			2	Attendant failed to reset	1	Choose better station	1	2						
	Pump won't read credit card	Must pay cash	4	Credit card dirty	1	Replace card	1	4						
Dispense fuel	Auto shut-off fails	Spill gas on ground	5	Pump failure	1	None	5	25						
			5	Auto tank design	5	None	5	125						
		Pay for gas on ground	2	Auto tank design	5	None	5	50						

Table 2.12 Example of a Process FMEA: Self-Serve Gas

1	2	3	4	5	6	7	8	9	10	11	12	13	14	15
Automobile Part/ Component	Failure Mode	Failure Effects (Consequences of Failure Mode)	S E V	Causes (of failure Mode)	O C C	Controls	D E T	R P N	Actions	Plans	pSEV	pOCC	pDET	pRPN
Windshield washer	Does not squirt washer fluid on windows	Safety hazard under certain environmental conditions	2	No fluid in reservoir	2	Check fluid regularly	1	4						
			2	Supply line disconnected	1	Check supply line regularly	3	6						
Battery	Does not retain charge	Car won't start	3	Bad connection	1	Check battery	2	6						
			3	Dead battery	1		3	9						
Brakes	Brakes fail	Can't stop car	5	Lost brake fluid / no pads	1	Check brake pads / fluid	4	20						
Air conditioner	No cool air	Discomfort	1	No freon	2	Check freon	4	8						

Table 2.13   Example of a Product FMEA: Automobile

Summary

One of the most important factors for the successful implementation of an FMEA is timeliness. It is meant to be a "before-the-event" action. However, that does not mean we cannot gain valuable information by applying an FMEA after the fact. To achieve the greatest value, the FMEA should be done before a design or process failure mode has been unknowingly designed into a product. Time spent up front doing a comprehensive FMEA correctly, when product/process changes can be most easily implemented, will alleviate late crises. An FMEA can reduce or eliminate the need to implement a corrective action which could create an even larger concern down the road [Pote 95].

When should we update an FMEA? Properly applied, an FMEA is an interactive process which is never ending. When a change is being considered in a product or process, the FMEA can help identify problems before they are implemented and therefore, the FMEA should be updated at that time. An FMEA should also be updated anytime there is a change in environment, materials, supplier, manufacturing process, or assembly process. In any of these situations, the FMEA can help the user avoid unknowingly designing failure into a product.

Another key question: is the FMEA ever finished? For a design FMEA, the process is over when the item is released for production. However, that does not imply that we want to abandon the FMEA report. Should a problem with the product be discovered downstream, the information from the design FMEA can be very useful in fixing the problem. In fact, the information from the design FMEA can be used as an input for the defect FMEA that should be used to solve the problem. In the case of the process FMEA, it is never complete until the process is permanently removed from use.

A Failure Mode and Effect Analysis is a systematic group of activities designed to recognize and evaluate the potential failure modes of a product or process. Without FMEA, the product designer or process manager may not be able to anticipate the many problems that do not show up until the product is being manufactured, or worse yet, is in the hands of the consumer. When used appropriately, FMEA will help the user avoid designing failures into products and processes, save the user time and money, and document the developmental process for any future changes.   FMEA means mistake-proofing or fool-proofing the system!

2.4 Data Collection

Before an analyst can produce histograms, scatter diagrams, or analyze the data, he or she must have data to work with. Recall that the ultimate purpose of collecting and analyzing data is to aid in the decision making process. Decisions to implement or change a process, to implement a plan of action, or to improve an operation should all in some part be based on data. Additionally, the collection of data can be a time-consuming and expensive proposition. Thus, it behooves the decision maker and analyst to understand the data collection process, for a decision can be no better than the data upon which it is based. This section is provided to remind us of the importance of this initial step in decision making, one that is often overlooked or taken for granted; to give the practitioner some guidelines to consider when collecting data; and to introduce some terminology that will be used throughout the remainder of this text.

The first guideline to be observed is to **understand the purpose of collecting data**. There are several reasons why data is collected:

(1)   To identify and/or verify a problem or problem area.

(2)   To analyze a problem.

(3)   To understand, describe, or monitor a process.

(4)   To test a hypothesis.

(5)   To find a relationship between inputs and outputs of a process.

Identifying the purpose of collecting the data at the outset can help clarify how the data will ultimately be used in the decision making process and can help bring the magnitude of the data collection effort in line with the magnitude of the ultimate decision to be made. It can also help identify the graphical tool(s) to be used in transforming the data into information.

A second guideline is to **properly classify and efficiently collect the data**. How much accuracy is needed? What instrumentation is available? Will new methods and instrumentation have to be developed to collect the data? Basically, there are two kinds of numerical data:

Continuous (or measurement) data: length, height, weight, volume, and time are examples of continuous data.

Discrete (or countable) data: number of defects, number of failures, number of choices, and number of births or deaths are examples of discrete data.

Sometimes the distinction between these two classes becomes blurred. For example, would one categorize a person's "age" as continuous or discrete? Since we usually "count" birthdays, we might tend to classify "age" as discrete, when in fact it is continuous. When Sue records her age on a job application as "23," she is merely telling how many birthdays she has celebrated which is the usually accepted "level of precision" when measuring age in our society. However, at the moment Sue enters her age, she is actually 23 years plus some fraction of another year. Thus, age, despite taking on the appearance of a discrete variable, is really a **continuous** variable. On the other hand, some variables — like shoe size — can take on fractional values, but they are still classified as **discrete**. Even though shoe sizes range from, say, 6, 6½, ..., 11½, 12, there are still only a countable number of sizes. The bottom line is that any "countable units" (e.g., defects, failures) will be discrete.

The third and final guideline mentioned here is **to verify and validate the data** being collected. This may seem to be a strange statement for, after all, isn't the data supposed to tell us about the system and not vice versa? Certainly we expect the data to tell us something about the system or process being investigated, but we also want to be sure we know what that process is. Verifying data is concerned with answering the question, "Was the data actually collected under the desired conditions?" The validation of data is concerned with answering the question, "Is the data representative of the process or system under scrutiny?" The use of statistical methods can aid immensely in guaranteeing that the answer to these questions is in the affirmative. In particular, design of experiments addresses the first question, while random sampling addresses the second question. While entire texts and courses have been dedicated to each of these topics, we devote a chapter later in this text to design of experiments and now briefly address the concept of sampling.

Before one can adequately define a sample, one must be familiar with the term **population**.

A **population** is a set or collection of all possible objects or individuals of interest. It can also be the corresponding set of values which measure a certain characteristic of a set of objects or individuals.

For example, the population of interest might be the set of all males, age 18 or older, in the United States as of January 1, 1997. The number of elements in this population is certainly in the millions, a large population to be sure, but still **finite**. It is finite because the number of males in this category is countable (perhaps via social security number). Typically, however, we will be interested in a particular characteristic (or variable) of this population, like height in inches. Since each person has an associated height, we can also think of the population as being the set of all values (heights in inches) associated with this set of individuals.

Another example of a population is the set of all 12-volt batteries being manufactured by a certain company. When we consider elements of a population to be items (like batteries) coming off of a production line, we assume this population to be **infinite**. This assumption is based on the consideration that a production process, if allowed to continue indefinitely under unchanging conditions, is an infinite process. The characteristic or variable of interest for these batteries could very well be battery lifetime, in which case the population is the set of lifetimes for all of the batteries coming off of the production line.

Recall that the purpose of considering a population in the first place is to draw some conclusion about the characteristic of interest for that population. It might be the average height of males 18 years or older or possibly average battery life. If every element of a population is considered and listed, the result is called a **census**. Unfortunately, censuses are seldom constructed because they are very time-consuming and expensive, if not impossible to construct accurately. Fortunately, probability and statistics provide the framework which allows the judicious use of samples to draw conclusions about a population characteristic.

A **sample** is any subset or subcollection of a population. There are many types of samples, but the sample that is of fundamental importance to us is called a **simple random sample**.

A **simple random sample** (or random sample, for short) of size n is a sample chosen in such a way that every possible sample of size n has an equally likely chance of being chosen.

Consider a batch of 10 television sets coming off of an assembly line to be the population of interest. If a sample of size $n=2$ televisions is to be selected from the original 10 and tested, how can one randomly select the 2 TVs? First, the number of possible samples of size $n=2$ is $(10)(9)/2=45$ (reference Section 3.6, Combinatorics, for the basis of this computation). The definition of a simple random sample implies that each of these 45 possible samples must have an equally likely chance of being chosen. Assuming the 10 TVs are labeled 1, 2, ..., 10, we could make up 45 slips of paper and write one of the following pairs of numbers on each slip: (1,2), (1,3), ..., (1,10), (2,3), (2,4), ..., (2,10), ..., (9,10). Putting these 45 slips in a hat, mixing them thoroughly, and then drawing one slip out would produce a simple random sample. Unfortunately, despite exhibiting the concept of equal likelihood, this process is tedious and becomes very messy quickly as n gets bigger than 2. An easier and much more palatable approach is to use a table of random digits like that in Appendix A. In a table of random digits, every digit from 0 to 9 has an equal chance of appearing in each position. The digits in Appendix A are grouped in columns of 5 for ease of reading. Numbers may be chosen from the table in any fashion as long as a predetermined systematic approach is established prior to actually looking in the table. For example, we will choose two digits (using a "0" to denote TV # 10) from the table starting at Row 21 (chosen randomly) and Column 19 (chosen randomly) and working downward in that column. If successive digits should be the same, we will continue downward until a second distinct digit is obtained, because we are sampling **without** replacement (i.e., we will not test the same TV twice). Now, looking in the table, we see that we have chosen the digits "2" and "6," meaning that TVs #2 and #6 form our sample of size 2.

2.5 Measures of Location

We have already presented some graphical methods for summarizing and describing data sets. Now we address some different kinds of **numerical** summary measures. When speaking of a data set, we admittedly could be referring to either a sample or population. Since our ultimate goal is to use sample data to draw a conclusion about a population, we will concentrate mostly on sample data measures. However, as we progress, the notation and context will make it clear as to which is being used. The most common measures of location are used to describe where the **middle** of a data set lies.

Perhaps the most useful measure of central tendency, and the one most commonly used, is the mean or arithmetic average of the data set. The **sample mean**, x̄ (read "x bar"), of a set of sample data values x_1, x_2, ..., x_n is defined by the following equation:

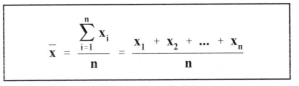

$$\bar{x} = \frac{\sum_{i=1}^{n} x_i}{n} = \frac{x_1 + x_2 + ... + x_n}{n}$$

Equation 2.1 Sample Mean: x̄

There is nothing sacred about the use of the letter (or variable) "x" in this definition. It is only a placeholder for whatever we allow it to represent in our application. We could have as easily used "y" and will, in fact, use them interchangeably. Sometimes equations can be intimidating so it is always a good idea to put equations into words. In words, Equation 2.1 says, "to find the average or mean, just add up all of the values in the data set and then divide by the number of values."

Consider the six data points in Table 2.14. These values represent a sample of six repair times (to the nearest hour) for six different C-130 engines.

Table 2.14 C-130 Engine Repair Times (hrs)

Letting x be the variable associated with engine repair times, the mean of this sample is $\bar{x}$ = (8 + 11 + 6 + 7 + 12 + 7)/6 = 51/6 = 8.5 hours. The physical interpretation of this value is illustrated in Figure 2.26, where it can be seen that the point on the horizontal axis at 8.5 serves as a balance point.

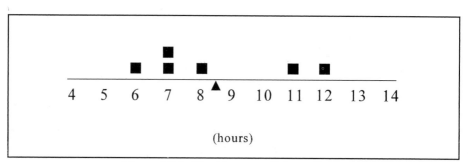

Figure 2.26   The Mean as a Balance Point

If we placed one-pound weights on the horizontal axis at the values in our data set and the horizontal axis was of negligible weight, then the only place we could place a fulcrum to balance the system of weights would be at $\bar{x}$ = 8.5. Note that the sample mean $\bar{x}$ = 8.5 is not any of the original values in the sample. Such is also the case for the gas mileage data in Table 2.3. Letting y be the variable for gas mileage, $\bar{y}$ is found to be 22.88 mpg. Although the sample mean could possibly turn out to be one of the sample data values, it does not necessarily have to, as illustrated by these two examples. While we use $\bar{x}$ or $\bar{y}$ to denote the sample mean, we use the Greek letter μ to denote a **population mean**. If we know all of the population values, then the population mean can be calculated by μ = (sum of all N population values)/N, where N is the number of elements in the finite population.

A second and also very commonly used measure of location is the median. The **sample median** $\tilde{x}$ (read "x tilde") of a set of sample data is defined by the following equation.

Assume that x_1, x_2, ..., x_n is a list of the sample data sorted from smallest to largest. Then

$$\tilde{x} = \begin{cases} \text{middle value, if n is odd} \\ \text{the average of the two middle values, if n is even} \end{cases}$$

Equation 2.2 Sample Median: $\tilde{x}$

Sorting the six values in Table 2.14 gives the following sequence:

$$6 \quad 7 \quad 7 \quad 8 \quad 11 \quad 12$$

Since n = 6 (sample size or number of data values) is even, the two middle values are 7 and 8, and their average is 7.5, so $\tilde{x}$ = 7.5. Sorting the gas mileage data in Table 2.3, one can see that $\tilde{y}$ for the gas mileage data is 21.5 (i.e., the average of 21 and 22, the two middle values in the sorted sample data). We summarize the two measures of location for the two data sets in the following table.

Data Set Name	Variable Name	Sample Size	Sample Mean	Sample Median
C-130 Engine Repair Times	x	6	8.50	7.5
Gas Mileage	y	50	22.88	21.5

Table 2.15 Summary Measures of Location

While the mean and median are both measures of central tendency, the two data sets investigated here serve to illustrate how the mean and median are affected by different aspects of the data. The median is the "middle" value in a sorted data set, and it separates the set of values into two equal parts. As such, the median is typically insensitive to extreme values. For example, even if the "12" in the engine repair time data (Table 2.14) had been a "24," the median still would have been 7.5 hours. Such is not the case for the mean, which is extremely sensitive to large or small values. In the gas mileage data, the large values "41" and "45" had the effect of pulling the mean of the data set to the right of the median, namely $\bar{y} = 22.88 > 21.5 = \tilde{y}$. When the mean of a data set lies to the right of the median (i.e., $\bar{y} > \tilde{y}$), we say that the data is **skewed to the right**. Conversely, if $\bar{y} < \tilde{y}$, we say that the data is **skewed to the left**. If $\bar{y} = \tilde{y}$, then the data is said to be **symmetric**. Figure 2.27 illustrates the concept of skewness by using histograms.

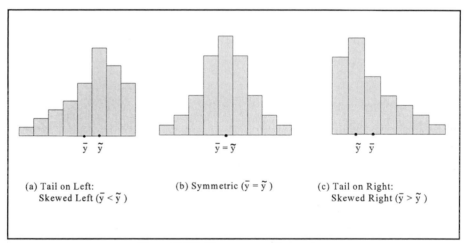

Figure 2.27 Skewness vs Symmetry

A commonly used measure of skewness is Pearson's Coefficient of Skewness which we shall denote as SK. SK is computed as follows.

$$SK = \frac{3(\text{mean} - \text{median})}{\text{standard deviation}}$$

Equation 2.3 Pearson's Coefficient of Skewness: SK

The standard deviation shown in the denominator of Equation 2.3 is a measure of dispersion which is defined in the next section. It is always a non-negative number. Note that if the mean is less than the median, SK is negative, indicating negative skewness (i.e., skewed left); and if the mean is greater than the median, SK is positive, indicating positive skewness (i.e., skewed right). For symmetrical data, SK=0. SK values typically fall between -3 and 3.

Table 2.15 shows that both the engine repair times and the gas mileages are skewed right. Which is the better measure of central tendency? Well, that depends on what the analyst is focusing on because as shown above, the mean and median are affected by and concentrate on different properties of the data. While the sample mean has some very useful properties and will be used more often in making statistical inferences, the median is also a useful measure, particularly when it comes to income or salary data. The Bureau of Labor Statistics always quotes a median national income because of its insensitivity to very large incomes and because it portrays a more accurate picture of the "middle" of American incomes. The reader should find the fallacy (and humor) in a politician's recent statement, "I will not be satisfied until well over half the population exceeds the median income."

The mean and median are not the only measures of location. Instead of using the median to divide a data set into two equal parts, we could also use **quartiles** which divide a data set into four equal parts. To obtain an even finer resolution, we could divide the data set into 100 equal parts called **percentiles**. The second quartile and the 50th percentile are equivalent names for the median. The concept of a percentile is extremely important and will be addressed in greater detail in Section 2.8.

2.6 Measures of Dispersion

As valuable as the numerical measures of location are, they alone do not give a complete summary of a sample data set. Consider the three data sets denoted by the variables x, y, and z in Table 2.16.

x	y	z
30	10	10
30	30	20
30	30	30
30	30	40
30	50	50
$\bar{x} = 30$	$\bar{y} = 30$	$\bar{z} = 30$
$\tilde{x} = 30$	$\tilde{y} = 30$	$\tilde{z} = 30$

Table 2.16 Three Different Data Sets

Certainly the three data sets are different, but one would not be able to detect any differences by looking at only the means and the medians. If we graph the three data sets as shown in Figure 2.28, we see that the data are dispersed or spread differently in each of the data sets. Thus, we see the need for summary measures which quantify the dispersion or variability in the data.

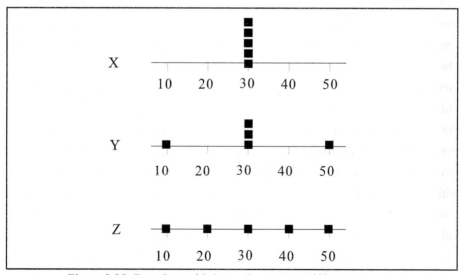

Figure 2.28 Data Sets with Same Center But Different Spreads

One very simple measure of variability that has previously been mentioned in Section 2.2.5 (in the construction of a histogram) is the **range** of the data. This is the difference between the largest and smallest values in a data set. The range of x, denoted R_x, is 30 - 30 = 0. Similarly, R_y = 50 - 10 = 40 and R_z = 50 - 10 = 40. Note that the range is a **value**, not an interval. Range is a numerical measure that is quite intuitive and seems to give us a "feel" for the spread of the data. Unfortunately, the range of a data set depends only on the two most extreme values in the data set; and certainly each of us can comprehend two different data sets that have the same range but are quite different "between" the two extreme values. For example, consider data sets y and z in Table 2.16 above. R_y = R_z = 50 - 10 = 40, but the dispersion characteristics of y and z are different. Fortunately, there is a numerical measure of variability that depends on all the values in a data set, not just two. This measure, no doubt the most commonly used measure of variability, is based on how each value in the data set deviates from the mean of the data set.

The deviation of a data set value x_i from the mean $\bar{x}$ is denoted by $x_i - \bar{x}$. When $x_i - \bar{x}$ is positive, x_i lies to the right of $\bar{x}$ and when $x_i - \bar{x}$ is negative, x_i lies to the left of $\bar{x}$. If a point (or value) is far from the center, then the deviation will be large (either positively or negatively), and if an observation is close to the center, then the deviation will be small (again, either positively or negatively). One might conclude, then, that a reasonable numerical measure of variability would be obtained by averaging all of the deviations (i.e., adding all of the deviations and dividing by n). However, one of those magical properties of $\bar{x}$ (i.e., it is a balance point) is that the sum of all deviations from $\bar{x}$ is 0 (i.e., $\sum(x_i - \bar{x}) = 0$). That is, the negative and positive deviations about $\bar{x}$ will *always* cancel each other out. Perhaps we could use "distances" or "absolute deviations" from $\bar{x}$ instead of just deviations. That would certainly force $\sum|x_i - \bar{x}|$ to be greater than or equal to 0, with equality occurring only when all values are the same, as in data set x above. Unfortunately, there are some mathematical difficulties with using the "absolute value" (Note: those who have had calculus and recall what $f(x) = |x|$ looks like might be able to surmise why "absolute value" could cause problems). We instead use the **squared** deviations $(x_1 - \bar{x})^2$, $(x_2 - \bar{x})^2$,..., $(x_n - \bar{x})^2$. The **sample variance**, s^2, of a set of sample data is defined by the following equation.

$$s^2 = \frac{\sum\limits_{i=1}^{n} (x_i - \bar{x})^2}{n-1}$$

Equation 2.4 Sample Variance: s^2

It is important to note that the denominator used in calculating sample variance is "n - 1," not "n." Thus, sample variance is "not quite" the average of the squared deviations. It is "slightly bigger" than the average squared deviation because "n - 1" is smaller than "n." The theoretical reasons for using "n - 1" instead of "n" are beyond the scope of this text, but perhaps the following motivation for using "n - 1" instead of "n" will suffice for now.

Recall that we are going to be using $\bar{x}$, the sample mean, to estimate the unknown population mean μ. Similarly, we will want to use s^2, the sample variance, to estimate the population variance which is denoted as σ^2 (read "sigma squared"). If the population is finite of size N, and we knew all of the population values, then we could calculate σ^2 as follows:

$$\sigma^2 = \frac{\sum\limits_{i=1}^{N} (x_i - \mu)^2}{N}$$

Equation 2.5 Population Variance: σ^2

This is the average of the squared deviations of population values about the population mean. However, μ is hardly ever known because it is either impossible or too expensive to find. Thus, σ^2 is hardly ever known and will have to be estimated. Now, because the x_i's of the sample tend to cluster more closely about their $\bar{x}$ than they would cluster about the population mean μ, it turns out that using division by "n" in calculating s^2 would produce an s^2 that would on the average underestimate σ^2. In statistical jargon, using "n-1" allows s^2 to be an "unbiased" estimator of σ^2, just like $\bar{x}$ is an "unbiased" estimator of μ.

Using Equation 2.4 to compute the sample variances for the three data sets in Table 2.16, we find that $s_x^2 = 0$, $s_y^2 = 200$, and $s_z^2 = 250$, where the subscripts x, y, and z are used with the s^2 to link a particular variance with its data set. Unlike the numerical measures of central tendency ($\bar{x}$ and $\tilde{x}$) and the range, sample variance is not so intuitive. For example, what does $s_y^2 = 200$ mean when considered by itself? Not much! However, when comparing it to $s_x^2 = 0$ and $s_z^2 = 250$, it becomes a little more meaningful in that data set y has more variability than data set x but not as much variability as data set z, which is also evident in Figure 2.28. For now, it is this comparative nature of the sample variance that should be grasped. Despite its unintuitive and less obvious nature, variability is the heart and soul of the entire quality movement because detecting and correcting the causes of variability are the keys to improved quality.

While using Equation 2.4 was quite tolerable in computing the variances above, not always will the sample data be so humanly bearable when it comes to computing the variance. In general, the following alternative but equivalent equation for computing s^2 is more efficient and accurate. Due to its better round-off error properties, this is the equation that is usually implemented in calculators and statistics packages. For computational purposes, its use is preferred.

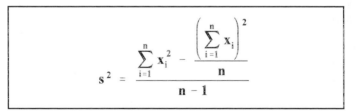

$$s^2 = \frac{\sum_{i=1}^{n} x_i^2 - \dfrac{\left(\sum_{i=1}^{n} x_i\right)^2}{n}}{n - 1}$$

Equation 2.6 Sample Variance: Computational Form

It should be noted that the units involved with variance are "squared units" of the original measurement units associated with the sample data itself. For example, suppose that the unit of measurement for the data in Table 2.16 is feet (ft); then $s_y^2 = 200$ ft^2, which is not usually very interpretable. Hence, it is desirable to give a measure of variability in terms of the data set's original unit of measurement. This leads to the definition of what is called the **sample standard deviation** of a set of sample data values x_1, x_2, ..., x_n.

$$s = \sqrt{s^2} = \sqrt{\frac{\sum_{i=1}^{n} (x_i - \bar{x})^2}{n - 1}}$$

Equation 2.7 Sample Standard Deviation: s

Obviously, the computational form for s^2 could be used in place of the radicand in the rightmost side of Equation 2.7. Like the variance, the standard deviation is a common vehicle for comparing variability between data sets, for it is simply the square root of the variance. The corresponding population standard deviation is given by σ, where $\sigma = \sqrt{\sigma^2}$. From the definitions of variance and standard deviation, it should be clear that neither of these measures can ever be negative.

An important result relating standard deviation to the dispersion of values in *any* data set is called the ***Chebyshev Inequality***. Spellings of this Russian scientist's name vary greatly, so a good way to remember it is to refer to it as the "Chubby Chef" inequality.

No matter what the distribution or pattern of variation for a data set, the proportion of values in the data set that fall more than k standard deviations from the mean is at most $\dfrac{1}{k^2}$

Chebyshev Inequality

Using $k = 3$ as an example, this result says that no more than 1/9 (or about 11%) of the data will be more than 3 standard deviations from the mean. This applies to *any* data set.

In practitioner jargon, the standard deviation is a rough approximation for the average deviation of the data set values about their mean. The standard deviation is a measure of the **absolute** variability in a data set and is the most common measure of dispersion used in statistics. Sometimes, however, the **relative** variability of a data set is more useful. The most common measure of relative variability is the **coefficient of variation**, which is simply the ratio of the standard deviation to the mean[3]. Since the standard deviation and the mean of a data set have the same units, the coefficient of variation (denoted by CV) will be a unitless measure. Therefore, we can use the coefficient of variation to compare the variability of data sets measured in different units. For example, recall our discussion of comparing the sample variances of the three data sets given in Table 2.16. Now suppose that the variables x, y, and z were measured in inches, grams, and seconds, respectively. Which data set contains the largest amount of variation? Well, that is difficult to tell because we are "comparing apples and oranges." Instead, we can calculate the coefficient of variation for each data set as follows:

$$CV = \frac{s}{\bar{x}}$$

Equation 2.8 Coefficient of Variation: CV

Therefore, using Equation 2.8, we can calculate the coefficient of variation for the three data sets in Table 2.16. Using the sample variances we calculated earlier $CV_x = 0$, $CV_y = 0.47$, and $CV_z = 0.53$, indicating that data set Z has the most relative variability as well as absolute variability, not an unexpected result since $\bar{x} = \bar{y} = \bar{z} = 30$ in this example.

2.7 Box Plots

A graphical tool that can be used to effectively provide a visual depiction of a data set is the box plot (sometimes called a box and whisker plot). A box plot summarizes both the location and dispersion of the values in a data set by breaking the data set into quartiles. Recall that a quartile is a "fourth" of the data set. The first quartile of a data set is a value (or number) such that 25% of the values in the data set are less than it and 75% of the values are greater than it. The first quartile

---

[3]The information on the coefficient of variation was provided by John Tomick of the USAF Academy.

is also known as the 25th percentile. The second quartile, or 50th percentile, is the median of the data set. The third quartile is the value such that 75% of values in the data set are less than it and 25% of the values are greater than it. The *interquartile range* is the difference between the third and first quartiles. Figure 2.29 shows the key elements of a box (and whisker) plot. This box plot is displayed horizontally. Box plots can also be displayed vertically.

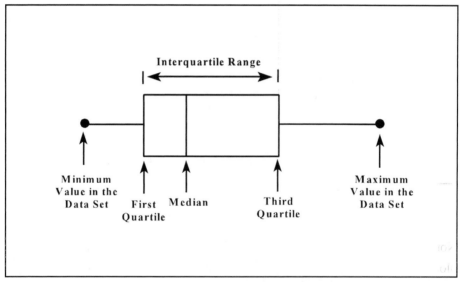

Figure 2.29 Key Elements of a Box Plot

The box plot is especially useful for describing the shape of a distribution for small data sets or when histograms and stem and leaf plots do not convey a meaningful summary. The box plot displays some of the most prominent features of a data set, namely (1) the center, i.e., median; (2) the spread; and (3) the skewness or lack of symmetry. The box plot shown in Figure 2.29 shows some skewness in the data set, in particular to the right, because the longest "whisker" and most of the box lies to the right of the median.

To illustrate the difference between a box plot and a histogram we have superimposed on the same axis the box plot and histogram for the gas mileage data in Table 2.3. Please reference Figure 2.30.

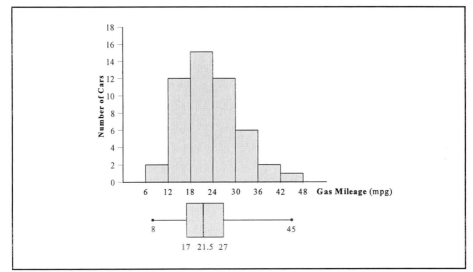

Figure 2.30 Box Plot and Histogram for the Same Data Set

Note that the histogram clearly shows the shape of the distribution, whereas the box plot shows the median explicitly and that the middle half (interquartile range) of the data is between 17 and 27. Both graphics illustrate slight skewness to the right.

2.8 Grouped Data

This section is presented to aid those practitioners who may be faced with analyzing grouped data. Sometimes it is necessary to analyze a set of sample data when it is in grouped form. That is, the individual data values may not be available or they may be too numerous to consider as individual points. In any case, we shall assume in this section that the data is presented to the analyst in the form of a **frequency distribution** and that the individual data values are not known. A frequency distribution is nothing more than a tabular representation of a histogram. Table 2.17 displays a frequency distribution of exam scores in a statistics course consisting of 194 students. For each interval or class of scores, the frequency or number of students who scored in that interval is recorded. For example, from Table 2.17 we can see that 3 students had exam scores somewhere between 30 (including 30) and 40 (not including 40). We don't know if those 3 scores were a 34, 36, and 38, or whatever. All we know is that 3 scores fell in that interval.

Class (i)	Exam Score Interval (%)	Number of Students (f_i)
1	[30, 40)	3
2	[40, 50)	7
3	[50, 60)	24
4	[60, 70)	41
5	[70, 80)	60
6	[80, 90)	44
7	[90, 100]	15
	Total:	194

Table 2.17 Frequency Distribution of Exam Scores

The histogram for the data in Table 2.17 is shown in Figure 2.31 along with what is called a **frequency polygon**. The frequency polygon is a line graph that connects the midpoints of the tops of each bar or rectangle that is associated with that class interval. The frequency polygon shown in Figure 2.31 is said to be "grounded" to the horizontal axis. The two points on the horizontal axis that "ground" the polygon are chosen as the midpoints of imaginary class intervals (of the same width) to the left and right of the leftmost and rightmost bars, respectively. The real purpose of a frequency polygon is to be able to compare it to another polygon, perhaps a polygon from another exam. It is much easier to read two superimposed polygons than it is to interpret two superimposed histograms.

Now that we have a graphical representation of the data, how do we obtain numerical measures for the grouped data? We can approximate the sample mean $\bar{x}$ by using Equation 2.9. Notice the use of the symbol "$\approx$" which means "approximately equal to." How good an approximation Equation 2.9 is for the unknown x̄ depends on how the data is distributed within each class. The assumption built into Equation 2.9 is that we will obtain a good approximation for x̄ as long as M_i, the midpoint of the ith class, is a good estimate of the mean of each class.

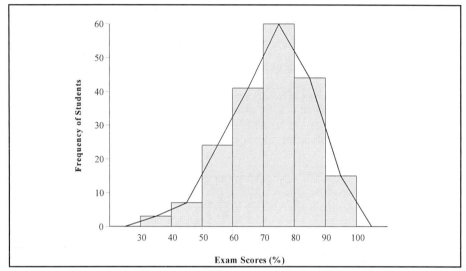

Figure 2.31 Histogram and Frequency Polygon of Exam Scores

$$\overline{x} \approx \frac{\sum\limits_{i=1}^{k} f_i M_i}{n}$$

where:

k	=	number of classes in the frequency distribution,
M_i	=	midpoint of the ith class,
f_i	=	frequency of the ith class, and
$n = \Sigma f_i$	=	total sample size.

Equation 2.9 Approximating x̄ from Grouped Data

Using Equation 2.9 to approximate $\bar{x}$ yields

$$\bar{x} \approx \frac{(3 \cdot 35 + 7 \cdot 45 + 24 \cdot 55 + 41 \cdot 65 + 60 \cdot 75 + 44 \cdot 85 + 15 \cdot 95)}{194}$$

$$= \frac{14070}{194} = 72.53\% \text{ (accurate to 2 places)}.$$

The true value for $\bar{x}$ is 72.76 % (computations not shown) so Equation 2.9 gives a fairly good approximation (to the nearest percent).

The sample variance can be estimated using Equation 2.10.

$$s^2 \approx \frac{\sum\limits_{i=1}^{k} f_i \left(M_i - \bar{x}\right)^2}{n-1}$$

where:

k = number of classes in the frequency distribution,

M_i = midpoint of the ith class,

f_i = frequency of the ith class,

$\bar{x}$ = the approximated mean as given by Equation 2.9, and

$n = \Sigma f_i$ = total sample size.

Equation 2.10 Approximating s^2 from Grouped Data

Using this equation to approximate s^2 yields

$$s^2 \approx \frac{3(35 - 72.53)^2 + 7(45 - 72.53)^2 + ... + 15(95 - 72.53)^2}{193}$$

$$= \frac{34012.37}{193} = 176.23, \text{ which gives } \quad s \approx 13.3\%.$$

The true value of s is 12.4% (computations not shown), indicating that the approximation is once again quite good.

A graphical tool commonly used to describe grouped data is an **ogive** (pronounced o'j$\bar{\text{i}}$v). An ogive is a line graph of the **cumulative frequency distribution**. First, a less-than (this is the most common type of) cumulative frequency distribution is constructed from the frequency distribution in Table 2.17. This is shown in Table 2.18.

Less Than	Cumulative Frequency	Cumulative % Frequency
30	0	0.0
40	3	1.5
50	10	5.2
60	34	17.5
70	75	38.7
80	135	69.6
90	179	92.3
=100	194	100.0

Table 2.18 Cumulative Frequency Distribution of Exam Scores

The construction of this table could also have been based on the histogram in Figure 2.31. The frequencies are accumulated (or summed) in turn as each new class boundary is written down. For example, there are no exam scores less than 30. This will lead to the grounding of the ogive at 30 (see Figure 2.32). Then, since there are 3 students who scored between 30 and 40, there is a total of 3 who scored less than 40. Similarly, since 7 students scored between 40 and 50, there are now 10 (3+7) who scored less than 50, etc. The ogive given in Figure 2.32 shows the graph of this table. Normally, the vertical axis is given in "cumulative %" to facilitate the computation of percentiles, as will be shown shortly.

In large data sets containing an abundance of distinct values the idea of referencing a certain value in the range of the data as a given percentile takes on meaning. We have all been placed "in a percentile" at some point in our lives, whether it was the 10th percentile of weight at birth, or the 65th percentile of Math

SAT scores, etc. If your testing record says that you scored in the 98th percentile of all Math SAT scores, is that good or bad? Furthermore, what does that mean? It means that your score on the Math SAT test was better than approximately 98% of your contemporaries who also took the same exam and that at most only 2% of the total number of exam takers scored higher than you did. If that is you, you did well and you should do well in this course. More formally,

the kth **percentile**, denoted P_k, is a value which divides the data set into two parts such that k % of the observed values are smaller than P_k.

Definition of a Percentile

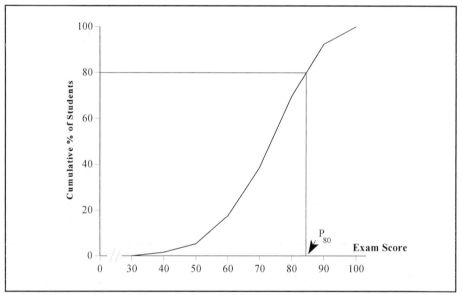

Figure 2.32 Ogive for Exam Scores

It is important to note that a percentile is a value that falls within the range of the data set and has the same unit of measurement (e.g., ft, inches, %) attached to it as all of the observed values do. On the other hand, this percentile or value does not have to be identical to any of the observed data set values. Recall that the median, which is another name for the 50th percentile, of the engine repair times (reference Tables 2.14 and 2.15) was 7.5 hours, when in fact 7.5 was not one of the observed repair times.

Suppose we wish to find the 80th percentile, P_{80}, of the exam scores presented earlier as grouped data. If an ogive is available, like that in Figure 2.32, one can get a very good estimate of what P_{80} is by drawing a horizontal line from 80 on the vertical axis until it intersects the line graph and then drawing a vertical line from that point downward until it intersects the horizontal axis. This point on the horizontal axis, if the graph is constructed fairly accurately, is an estimate of P_{80}. Referencing Figure 2.32, it appears that $P_{80} \approx 84\%$, meaning about 80% of the scores were less than 84%. This is equivalent to saying 4 out of 5 students scored less than 84% on that exam. A more accurate estimate of P_{80} can be obtained from the cumulative frequency distribution in Table 2.18 using linear interpolation as follows:

	Less Than	**Cum % Frequency**
	•	•
	•	•
	80	69.6 (from Table 2.18)
$P_{80} \Rightarrow$	?	80 $\Leftarrow$ k = 80
	90	92.3 (from Table 2.18
	•	•
	•	•

From Table 2.18 we select the two adjacent rows whose cumulative % frequencies encompass the desired percentile level, namely k = 80, as shown in the righthand column above. What we seek is the value between 80 and 90 in the lefthand column that is proportionally the same distance between 80 and 90 as k = 80 is between 69.6 and 92.3. What fraction of the distance from 69.6 to 92.3 does k = 80 represent?

This can be given by the ratio $\dfrac{80 - 69.6}{92.3 - 69.6} = \dfrac{10.4}{22.7} = .46$

Thus, P_{80} must also reside .46 of the distance from 80 to 90, which can be calculated as $P_{80} = 80 + .46\,(90 - 80) = 80 + .46(10) = 84.6$. We have just used "linear interpolation" to find $P_{80} = 84.6\%$, a slightly different answer than we obtained from the ogive. The two methods are equivalent, but the graphical method is easier to use even though it may be slightly less accurate.

One can also find P_{80} from the original frequency distribution in Table 2.17, although it is a bit more tedious. The key to finding P_{80} in this case is to first find the number or position that corresponds to 80% of 194. That number is $(.8)\,(194) = 155.2$. If we were to rank-order the 194 data values in ascending order from 1 to 194 and label their positions from 1 to 194, as shown here, then P_{80} would correspond to the position labeled 155.2.

$$1 \quad 2 \quad 3 \quad . \quad . \quad . \quad . \quad . \quad . \quad . \quad . \quad . \quad . \quad 194$$

$$\Uparrow$$

$$155.2$$

If we knew the values of the 155th and 156th elements, we could move one-fifth (due to the .2) of the way between them (like we moved halfway between the two middle elements when finding the median) and thus obtain P_{80}. Unfortunately, we don't know any individual values because we were given grouped data. However, we make the assumption that the observed values within any interval are equally spaced and proceed based on this. Next we need to identify the class interval in which the position 155.2 falls. Working from the top of Table 2.17 down and accumulating as we go (i.e., $3 + 7 + 24 + 41 + 60 = 135$ and $135 + 44 > 155.2$), we find that the value associated with position 155.2 must reside in the interval [80,90). Assuming the 44 observations in this interval are equally spaced, we again use linear interpolation to compute exactly where position 155.2 lies. We ask the question, how far into the interval [80,90) does position

155.2 reside when there are 135 positions prior to this interval and 44 positions in this interval? This ratio is given by

$$\frac{155.2 - 135}{44} = \frac{20.2}{44} = .46,$$

and by the argument used before,

$$P_{80} = 80 + .46(90 - 80) = 80 + .46(10) = 84.6,$$

which is the same P_{80} we obtained from Table 2.18.

An important interpretation of percentile with which we conclude this section involves the equivalency of area inside the rectangles of a histogram with the percentile level. The histogram in Figure 2.31 is reproduced in Figure 2.33, but the scaling of the vertical axis has been changed to be relative frequency (frequency divided by total number of 194) divided by 10 (because the interval width is 10).

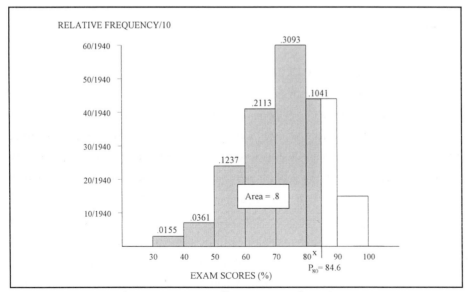

Figure 2.33 Percentile as Area Under a Curve

The reason for the change of scale is to normalize the entire area inside all 7 rectangles to be 1. Hence, finding the 80th percentile, P_{80}, is equivalent to finding the exam score on the horizontal axis that produces a total area of .80 in the rectangles to the left of this score. The decimal values shown are the areas of each of the 6 shaded rectangles that contribute to the total area of .8. For example, the area of the leftmost box is $(3/1940)(10) = (3/194) = .0155$, which is just the relative frequency of those students who scored in the interval [30,40). Adding the areas of the first 5 boxes yields .6959, which is $(.80 - .6959) = .1041$ short of the desired area of .80. Thus, we must take a portion of the sixth box as well in order to reach .8 as a final area. Since the height of the sixth box is 44/1940, we let x be the width of the portion of the sixth box we still need and solve the equation

$$(44/1940) \ x = .1041$$

for x, obtaining $x = 4.6$ (to the nearest tenth of a percent). Hence, the total area in the boxes to the left of 84.6 is .8, indicating that $P_{80} = 84.6\%$. Not unexpectedly, this is the same P_{80} obtained previously. Thus, 80% of the students scored less than 84.6% on the exam. The concept of area under a curve will become paramount in a short time; thus, it is beneficial to understand this example.

2.9 Bivariate Data and Correlation

Bivariate data, or data that describe two different variables at the same time, were discussed briefly in Section 2.2.9 (Scatter Diagrams). Recall that each observation is now an ordered pair (x_i, y_i) of values where x_i is associated with variable x and y_i is associated with variable y. The primary reason for looking at bivariate data is to study the relationship between variables. If there is a strong relationship between two variables, then it may be possible to use the behavior of one variable to predict the behavior of another variable. This section presents one of the most common measures of the strength of a **linear** relationship between two variables: the correlation coefficient. This measure originated from the work of Sir Francis Galton (1822-1911) in England when he tried to quantify the relationship between the heights of fathers and their sons. Actually, the name of the measure we present here will be due to Karl Pearson (1857-1936), one of Galton's disciples.

Before we get to the correlation coefficient, we define what is called the sample covariance. If a sample of n observations is (x_1, y_1), (x_2, y_2), ..., (x_n, y_n), then the **sample covariance**, denoted s_{xy}, is defined by the following equation:

$$s_{xy} = \frac{\sum\limits_{i=1}^{n} (x_i - \bar{x})(y_i - \bar{y})}{n-1}$$

Equation 2.11 Sample Covariance: s_{xy}

This equation should look familiar, for it is the bivariate equivalent of Equation 2.4, which is sample variance for one variable. Note that the factor $(y_i - \bar{y})$ has replaced one of the $(x_i - \bar{x})$ factors under the summation (Σ). However, while the two equations may look very similar, Equation 2.11 is capable of producing a negative number for s_{xy} whereas s^2 could never be negative. A positive covariance will result if the factors $(x_i - \bar{x})$ and $(y_i - \bar{y})$ tend to be either both positive or both negative. This will happen, for example, if there is a tendency for both variables to increase at the same time. That is, when the x_i's are greater than their $\bar{x}$, the y_i's tend to be greater than their $\bar{y}$. See Figure 2.34(a). On the other hand, a negative covariance will result if the two factors $(x_i - \bar{x})$ and $(y_i - \bar{y})$ tend to be of opposite sign. In order for this to happen, the tendency must exist for y to decrease as x increases. See Figure 2.34(b).

The bivariate data for gas mileage (mpg) vs weight (100 lbs) from Table 2.6 is reproduced in Table 2.19, along with the computations needed to find s_{xy}. If one refers back to the scatter diagram in Figure 2.21, the negatively-sloped appearance of the data substantiates the negative covariance found in Table 2.19. While the covariance s_{xy} may seem to be a legitimate measure of the linear relationship between the variables x and y, the most important part of the covariance is the sign, either $+$ or $-$. It signifies whether there is a positive or negative linear relationship between the variables. The magnitude of s_{xy} cannot be used, however, as a direct measure of the strength of the linear relationship because one is able to make this number as large or small as desired by simply changing the units of measurement.

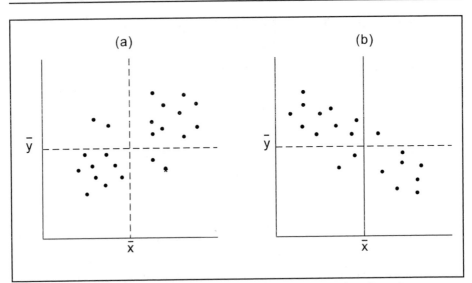

Figure 2.34 (a) Positive Covariance   (b) Negative Covariance

Weight (x_i)	mpg(y_i)	($x_i - \bar{x}$)	($y_i - \bar{y}$)	($x_i - \bar{x}$)($y_i - \bar{y}$)
30	18	1.6	-2.3	-3.68
28	21	-0.4	0.7	-0.28
21	32	-7.4	11.7	-86.58
29	17	0.6	-3.3	-1.98
24	31	-4.4	10.7	-47.08
33	14	4.6	-6.3	-28.98
27	21	-1.4	0.7	-0.98
35	12	6.6	-8.3	-54.78
25	23	-3.4	2.7	-9.18
32	14	3.6	-6.3	-22.68

$\bar{x} = 28.4$ $\bar{y} = 20.3$ -256.20

$s_x = 4.326$ $s_y = 6.865$

$s_{xy} = (-256.20)/9 = -28.47$

Table 2.19 Covariance of Gas Mileage(y) vs Vehicle Weight(x)

The strength of the linear relationship between the variables x and y is more accurately portrayed by the unitless measure called the **sample correlation coefficient**, also known as Pearson's product moment. The sample correlation coefficient, usually denoted r (sometimes also R), is defined by the following equation:

$$r = \frac{s_{xy}}{s_x s_y}$$

Equation 2.12 Sample Correlation Coefficient: r

Note that the numerator is the sample covariance as defined in Equation 2.11 and the denominator is the product of the two sample standard deviations for each of the variables.

For the gas mileage vs vehicle weight data, we can now use Equation 2.12 to compute r as

$$r = \frac{-28.47}{(4.326)\ (6.865)} = \frac{-28.47}{29.698} = -.959.$$

Equation 2.6 (the computational form for sample variance), combined with Equations 2.11 and 2.12 and a little algebra, produce Equation 2.13, the computational form for r. Equation 2.12 gives a good conceptual view of r in that r is really a normalized covariance, i.e., normalized by the product of the two standard deviations. Alternatively, Equation 2.13 gives a good picture of what quantities are needed to compute r, namely $\Sigma x_i y_i$, Σx_i, Σx_i^2, Σy_i, and Σy_i^2. If these 5 quantities are known, together with n, the computation of r is quite straightforward. Of course, it is even more straightforward if one uses a statistical package like SPC KISS, the software that accompanies this text. We suggest this avenue for computing r. We present Equation 2.13 here not to scare the reader, but rather to remind us all how lucky we are to live in the late 1990's.

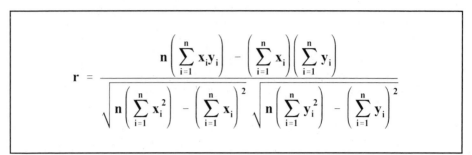

$$r = \frac{n\left(\sum_{i=1}^{n} x_i y_i\right) - \left(\sum_{i=1}^{n} x_i\right)\left(\sum_{i=1}^{n} y_i\right)}{\sqrt{n\left(\sum_{i=1}^{n} x_i^2\right) - \left(\sum_{i=1}^{n} x_i\right)^2}\sqrt{n\left(\sum_{i=1}^{n} y_i^2\right) - \left(\sum_{i=1}^{n} y_i\right)^2}}$$

Equation 2.13 Sample Correlation Coefficient: Computational Form

We conclude this section by stating some important properties of the sample correlation coefficient r.

Properties of r

1. The value of r is restricted to the interval $[-1,1]$. That is, $-1 \le r \le 1$.

2. The only way that r can equal 1 is if *all* pairs (x_i, y_i) lie on a straight line with **positive** slope.

3. The only way that r can equal -1 is if *all* pairs (x_i, y_i) lie on a straight line with **negative** slope.

4. When $r = 1$ or $r = -1$, we say that x and y are perfectly correlated. We sometimes refer to this situation as a functional relationship between the two variables.

5. The sign of r (+ or −) will be the same as the sign of the covariance.

6. The value of r does not depend upon the units in which x and y are measured.

7. The value of r does not depend on which variable is labeled x and which is labeled y.

8. r is a measure of the **linear** relationship between two variables. The closer r is to ± 1, the stronger is the relationship. If $r = 0$, then the variables are **linearly** uncorrelated.

2.10 Problems

2.1   Construct a flow chart detailing the process of determining why your automobile will not start on a cold, winter morning. Include at least 5 diamond shaped decision boxes in your flowchart.

2.2   Your travel time to work tends to vary by several minutes. As a result of this variation problem, you have been late to work on several occasions. Your boss is concerned and has asked you to brainstorm the factors that contribute to travel time variability. Use a fishbone (cause and effect) diagram to organize and document the brainstorming process.

2.3   The Welders' Association of America (WAA) recorded the following data during the last decade. It records the number of injuries, categorized by type, incurred by its members.

	Type of Injury				
Year	**Eye**	**Hand**	**Arm**	**Back**	**Other**
1980	24	13	6	20	3
1981	27	11	7	14	7
1982	31	14	3	8	2
1983	23	15	4	11	6
1984	18	16	6	10	5
1985	25	9	8	12	4
1986	27	11	7	7	5
1987	19	13	9	8	7
1988	23	16	10	13	2
1989	25	12	4	14	6
Total	242	130	64	117	47
(Percent)	(40.3%)	(21.7%)	(10.7%)	(19.5%)	(7.8%)

Construct a Pareto diagram describing the number of injuries by type. What does the Pareto diagram suggest with regard to possible investigations into cause and effect?

2.4   The Federal Aviation Administration records customer complaints from
       passengers on all U.S. airlines.   The machinists, pilots, and flight
       attendants of one major airline were on strike for over a year when the
       most recent numbers were published.   The data below represent total
       complaints for the year before and the year after the strike.

Type of Complaint	Number of Complaints	
	Before Strike	**After Strike**
Late departure	97	103
Canceled flight	13	47
Lost baggage	12	24
Missed connection	2	4
No beverages served	0	2
No food served	1	3
No pillows/blankets on flight	0	4
Total	125	187

       a)   Construct a Pareto diagram patterned after Figure 2.10 depicting
             the types of complaints before the strike.
       b)   Construct a Pareto diagram for the After Strike data.
       c)   Describe the differences in the patterns of complaints.

2.5   Using the gas mileage data in Table 2.3, construct a histogram with k=6
       classes. Compare and contrast Figures 2.14 and 2.15 with your histogram.
       In general, what kind of an effect does the number of classes in a histogram
       have on the data set summary capability? When would it not be possible to
       use equal class widths in the construction of a histogram?

2.6   The yearly salaries of 36 salespersons at a local car dealership are given
       below. Using the rule $k = \sqrt{n}$, construct a histogram for the annual wages
       of the salespersons.

43,800	25,200	22,600	29,900	18,200	13,500
33,000	36,200	32,700	32,500	26,300	34,700
44,600	35,900	34,000	54,100	34,400	26,600
41,700	33,600	36,600	30,300	22,600	28,400
33,700	45,600	26,300	36,900	47,100	46,900
27,700	44,100	27,200	17,100	45,800	26,000

2.7   The National Weather Service records the daily high temperature for most major cities across the country. The high temperatures of one city are as follows:

Temperature	# of Days
below 0	8
0 to under 10	12
10 to under 20	20
20 to under 30	28
30 to under 40	33
40 to under 50	41
50 to under 60	45
60 to under 70	49
70 to under 80	63
80 to under 90	47
90 to under 100	17
above 100	2

Construct a histogram for this set of data. You may assume that all classes have the same width.

2.8   The average spark plug fires over 1000 times per minute. A revolutionary new spark plug which could possibly extend spark plug life was designed and tested. The length of time of continuous operation of the 50 spark plugs tested before they first failed to ignite is given below in hours. Construct a histogram for this data.

244.4	168.4	205.9	293.6	155.7
171.6	159.8	96.1	54.2	47.3
206.8	180.6	113.5	290.3	180.5
147.8	213.1	168.1	182.8	220.4
163.8	156.9	132.9	229.9	107.7
111.6	172.1	108.2	204.6	157.5
157.6	116.6	158.7	137.1	166.4
119.2	168.8	124.1	230.2	193.8
110.1	144.5	140.2	217.6	237.7
123.0	154.9	259.9	234.6	116.4

2.9   The following 25 values represent the wall thickness in millimeters (mm) of 25 metal-casted turbine blades.

3.03	3.16	2.96	3.04	3.21
2.99	3.11	2.91	3.00	3.07
3.01	2.92	3.13	2.98	3.11
2.84	3.08	2.96	2.96	2.98
2.86	3.04	3.00	3.02	3.03

Construct a stem and leaf display for this data. Around which value do the data points tend to cluster?

2.10   The following 35 values are the scores obtained by 35 students in a freshman calculus course.

84	96	71	79	66	91	61
87	79	53	92	96	60	71
70	78	65	66	61	55	66
99	61	45	46	58	76	97
44	60	57	57	61	47	86

Construct a stem and leaf display for this data. Which stem has the most leaves? Where does the middle of the data appear to be (to the nearest tens digit)?

2.11   The closing price of a leading stock on the NYSE was recorded for 15 consecutive days (read left-to-right, then top-to-bottom).

108.50	108.25	107.50	106.25	107.50
106.25	105.50	108.25	110.50	107.25
102.50	108.00	115.00	111.00	108.00

Construct a run chart for the closing price of the stock.

2.12 The number of people in the U.S. who contract a particular life-threatening illness has been recorded for each year since 1980.

1980	1981	1982	1983	1984
114	374	965	1137	2465

1985	1986	1987	1988	1989
7453	14169	27395	41304	73953

Construct a run chart for the number of people who contract this illness.

2.13 The Coca-Cola Bottling Company wants to put 16 ounces of Coke in every can. If the filling machine places less than 15.9 or more than 16.1 ounces in a can, the process is considered out of control. Sixteen cans were sampled after being filled and their contents measured. The number of ounces in each can follows (read left-to-right, then top-to-bottom):

15.99	15.99	15.97	16.03
16.01	16.01	15.98	16.00
16.02	16.00	15.97	15.98
16.01	15.99	16.03	16.03

a) Construct a run chart for the volume in the cans.
b) Construct a control chart for the volume in the cans.
c) What is the difference between the above charts?

2.14 The computer operators at a university computer center recorded the CPU time (y) and number of disk I/O operations (x) for ten computer jobs. The data is as follows:

i	number of I/O operations (x)	CPU Time (y)
1	20	210
2	25	252
3	30	305
4	38	390
5	39	392
6	40	398
7	40	398
8	42	410
9	50	502
10	60	590

Construct a scatter diagram for this data.

2.15   The heights and weights of 12 men are listed below:

Height (inches)	Weight (pounds)
60	110
60	135
60	120
62	120
62	140
62	130
62	135
64	150
64	145
70	170
70	185
70	160

a)   Construct a scatter diagram for this data with height along the x-axis and weight along the y-axis.

b)   Explain why it might be advantageous to place height along the x-axis.

2.16   A sample of five students from a local college class was taken in order to identify characteristics of students who attend this school.

Name	Age	Eye Color	Height
Steve	19	Brown	5'10"
Cory	19	Blue	5'8½"
Rachel	20	Blue	5'9"
James	19	Green	5'7"
Anna	21	Blue	5'6¼"

List the variables in this data set and identify them as discrete or continuous.

2.17   There are 20 students in a classroom and each is assigned a number on the class roster from 1 to 20. Use the table of random digits in Appendix A to select a random sample of 5 students. Explain the procedure you used and identify the 5 students by their numbers.

2.18   A random sample of 12 basketball players was taken and their heights were as recorded below.

5'11"	6'3"	6'8"	6'1"	6'2"	6'11"
6'6"	7'1"	6'9"	6'11"	6'11"	7'1"

a)   Find the mean height for the players in this sample.
b)   Find the median height for these basketball players.
c)   Is this sample data set skewed left or right? If so, which direction?

2.19   Refer to Problem 2.18, the sample of 12 basketball players.

a)   Find the range of the height of the sample of basketball players.
b)   Find the sample variance for the heights of the basketball players.
c)   Find the sample standard deviation of the heights of the basketball players.
d)   What units are used with standard deviation?

2.20   A frequency distribution for the cost of repairing the damage from a broken window on vehicles repaired by ACME Glass Company is given below.

Repair Cost (in dollars)	Number of Repairs
0 to under 100	13
100 to under 200	7
200 to under 300	31
300 to under 400	27
400 to under 500	22

a)   Construct a histogram for this data.
b)   Construct a frequency polygon for this data.
c)   Construct an ogive for this data.

d)   Estimate the mean from the above frequency distribution.

e)   Estimate the median from the above frequency distribution.

f)   Estimate the 50th percentile from the above frequency distribution.

g)   Estimate the 80th percentile from the above frequency distribution.

h)   Estimate the interquartile range (difference between the 75th percentile and 25th percentile) from the above frequency distribution.

i)   Estimate the variance of this repair cost data.

2.21   Refer to problem 2.14, computer operations and CPU time.

a)   Calculate the sample covariance between CPU run time and number of disk I/O operations. What does this number tell us?

b)   Calculate the sample correlation coefficient between CPU run time and number of disk I/O operations.

c)   How would you describe the relationship between CPU time and disk I/O operations?

2.22   Refer to problem 2.15 and the sample of 12 heights and weights.

a)   Calculate the sample covariance between height and weight measurements.

b)   Calculate the variance of the height measurements on the men sampled.

c)   Calculate the variance of the weight measurements on the men sampled.

d)   Calculate the sample correlation coefficient between height and weight for the men sampled.

e)   Is there a relationship between height and weight?

2.23   The following sources of real-life data have been obtained from Wasserman
        and Bernero's *Statistics Sources.*

Almanacs

CBS News Almanac
Information Please Almanac
World Almanac and Book of Facts

Annual Publications

Commodity Yearbook
Facts and Figures on Government
    Finance
Municipal Yearbook
Standard and Poor's Corporation,
    Trade and Securities: Statistics

International Data

Compendium of Social Statistics
Demographic Yearbook
United Nations Statistical Yearbook
World Handbook of Political and
    Social Indicators

US Government Publications

Agricultural Statistics
Digest of Educational Statistics
Handbook of Labor Statistics
Housing and Urban Development
    Yearbook
Social Indicators
Uniform Crime Reports for the
    United States
Vital Statistics of the United States
Business Conditions Digest
Economic Indicators
Monthly Labor Review
Survey of Current Business
Bureau of the Census Catalog

Using one of the sources listed above, a source suggested by your
instructor, or a source of your own choosing, find a quantitative data set
consisting of at least 50 elements. Describe the data set by constructing a
histogram.   Find the mean, median, range, variance, and standard
deviation of the data. Determine the actual number of observations that fall
within 1, 2, and 3 standard deviations of the mean. Is your data skewed
in either direction?

2.24   Construct a cause and effect diagram where quality of life is the output or
        response (performance) measure. Be sure to label each input variable with
        either a C, N, or X, based on how you are controlling these variables
        today. Suggest possible changes, including SOPs, that will reduce the
        variability of the input variables and consequently improve the performance
        measure.

Chapter 3

Probability

3.1 Introduction

We have just completed a basic coverage of some of the graphical and numerical measures used to summarize or describe data sets. For the most part, the data sets we have considered were samples. As stated earlier, our ultimate objective is to be able to use information obtained from a sample to draw some conclusion about a population. We now begin our trek of crossing the bridge that spans descriptive and inferential statistics, and which also links samples to populations.

Many applied statisticians and engineers believe that coverage of probability concepts in an introductory statistical methods text such as this should be minimal. We agree and have tried to follow this guidance to as great an extent as possible. Unfortunately, without probability there would be no statistical inference, and we feel that the motivated student deserves a glimpse at what makes statistical inference "tick." After all, at every level of learning the student should have the opportunity to answer the question "why" as well as "how." Another valid criticism is perhaps the too frequent use of examples dealing with coins, cards, balls in urns, and dice, etc. Here, too, we will try to minimize the use of these examples, but will not avoid them altogether because the learning curve for these examples is minimal. Almost every reader has had direct experience with a coin, deck of cards, or a 6-sided die, whereas the same cannot be said about integrated circuit (IC) board defects or disk search errors. Hence, the occasional use of these examples should not distract the reader. Besides, the origins of probability lie with games of chance, and it may be enjoyable for the student to calculate some of these probabilities.

3.2 Considering the Chances

Consider the following three statements:

(1)   "There is a 10% chance of snow tomorrow."

(2)   "There is a 50% chance that a fair (i.e., properly balanced) coin will turn up heads on the next toss."

(3)   "There is a 40% chance that the next IC board defect detected will be due to soldering."

What are the differences, if any, in these propositions? While statement (1) gives us a feel that we may not have to change over to snow tires just yet, there is a good bit of subjectivity involved on the part of the meteorologist. What she no doubt means is that if atmospheric conditions were identical to what they are now on 100 different days, she would expect snow to occur about 10 times. The problem is that the identical atmospheric conditions are not repeatable, and subjectivity plays a big part in this assessment. Statement (2) is certainly plausible if the assumption of a fair coin is true. We would expect the outcome of a heads or tails to be equally likely. Statement (3) is based on the Pareto Diagram in Figure 2.10, where it can be seen that of the 150 defects, 60 were soldering defects. This statement is based on the **relative frequency** argument, whereby a sample of 150 defective items was taken and of those, 60 were soldering defects. We will concern ourselves throughout this text with statements like (2) or (3). In (2), we assumed something (that the coin was fair) and made an inference based on that assumption. In (3), we assumed nothing, collected data, and then made a statement based on the data.

While the term "probability" has not yet been defined, we have used the term "% chance" in the above statements to connote the meaning of probability. In terms of probability, statement (3) could alternatively be stated as "the probability that the next IC board defect will be a soldering defect is .4." Probability refers to the level of certainty we attach to a particular outcome of an activity or process which has more than one possible outcome.

In subsequent chapters we will be using probability to make conclusions similar to the following (computations for these results are not shown):

(1)   Prior to releasing software for production, it is tested to remove as many discrepancies as possible. Assume we are trying to decide which testing method would identify the most discrepancies for testing all future products. We would select one set of codes as representative of all future software and run method 1 using four different testers and then the same software could be subjected to method 2 using another four testers. Suppose this set of experiments produced the following results:

	Method 1	Method 2
Average # of discrepancies	20	24
Standard Deviation	1.8	1.6

The question is whether the results in favor of method 2 are significantly different from the results for method 1. The use of probability and inferential statistics will allow us to state that we are at least 98% confident that these sample averages are significantly different.

(2)   Given the following data on bath temperature and plating thickness,

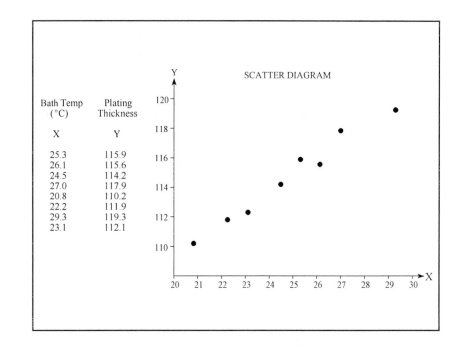

Bath Temp (°C)	Plating Thickness
X	Y
25.3	115.9
26.1	115.6
24.5	114.2
27.0	117.9
20.8	110.2
22.2	111.9
29.3	119.3
23.1	112.1

we would be able to state that we are at least 99.9% confident that there is a significant linear relationship between Y and X and that the model is

$$\hat{Y} = 86.757 + 1.125X$$

(where $\hat{Y}$, pronounced "Y hat", represents a predicted average thickness for some given X value).

Probability theory will also allow us to state that we are at least 99% confident that setting X = 25.1 °C will produce a product with thickness values between 112.75 and 117.24.

(3)   Given that the hospital billing process has monthly errors as stated below:

Month	% Billing Errors
Jan	20.1
Feb	17.6
Mar	21.3
Apr	19.4
May	18.5
Jun	20.4
Jul	18.6
Aug	19.3
Sep	20.2
Oct	16.9
Nov	22.0
Dec	20.5

Probability theory will enable us to state that we are at least 95% confident that the next month's billing error percentage will be between 16.653 and 22.481 unless we change the process.

(4)   Given a history of emergency room data, probability theory can also be used to determine how many staff members are required to be on duty to meet customer needs.

The rest of this chapter is for those seeking more depth into probability theory or for university students who are required to learn more depth. For those who only want to be familiar with probability, we recommend you move on to Chapter 4.

The next section will formalize some of the common terminology and properties of probability.

3.3 Properties of Probability

An experiment can be defined as follows:

An **experiment** is an activity or process that is capable of generating two or more possible outcomes.

Definition 3.1 Experiment

It is the "two or more" aspect of this definition that renders the notion of probability useful, because before the experiment takes place, there are two or more possible outcomes. Consider the activity of throwing a 6-sided, fair die. This experiment has six **basic outcomes**: "1", "2", "3", "4", "5", "6". These are called basic outcomes because they cannot be decomposed into anything simpler than just these six outcomes.

The **sample space** of an experiment, denoted **S**, is the set of all basic outcomes of that experiment.

Definition 3.2 Sample Space

Thus, for the experiment of tossing the die, $S = \{1, 2, 3, 4, 5, 6\}$. We are interested not only in the individual outcomes but also in any combination of outcomes from **S**.

An **event** is any subset of the basic outcomes contained in the sample space.

Definition 3.3 Event

We define the following events:

$$E_1 = \{1, 3, 5\} \quad = \text{"the outcome is odd,"}$$
$$E_2 = \{2, 4, 6\} \quad = \text{"the outcome is even,"}$$

$E_3 = \{1, 2\}$ = "the outcome is in the lower third,"
$E_4 = \{5, 6\}$ = "the outcome is in the upper third,"
$E_5 = \{1\}$ = "the outcome is the smallest possible," and
$E_6 = \{6\}$ = "the outcome is the largest possible."

If the experiment is performed and the basic outcome "i" (i = 1, 2, 3, 4, 5, or 6) results, then any event which has "i" as one of its members (i.e., possible outcomes) is said to have occurred. For example, if "2" were the result of tossing the die, then, from the above list, events E_2 and E_3 both occurred. That is, the outcome was even and in the lower third. New events can be constructed from old events using three very basic constructs from elementary set theory. These concepts are defined in Definition 3.4.

---

1. The **union** of two events A and B, denoted **A ∪ B**, is the event consisting of all basic outcomes that are **either in A or in B or in both**.

2. The **intersection** of two events A and B, denoted **A ∩ B**, is the event consisting of all basic outcomes that are **in both A and B**.

3. The **complement** of an event A, denoted **A*** or **$\bar{A}$**, is the set of all basic outcomes in **S** that are **not in A**.

Definition 3.4 Union, Intersection, and Complement of Events

Using these definitions, it should be clear that $E_1 \cup E_2 = S$, i.e., the union of the odds and evens is the entire sample space. The following statements can also be made.

$E_1 \cup E_3 = \{1, 2, 3, 5\}$ $E_2 \cup E_4 = \{2, 4, 5, 6\}$

$E_1 \cap E_3 = \{1\} = E_5$ $E_2 \cap E_4 = \{6\} = E_6$

$E_3 \cup E_5 = \{1, 2\} = E_3$ $E_4 \cup E_6 = \{5, 6\} = E_4$

$E_3 \cap E_5 = \{1\} = E_5$ $E_4 \cap E_6 = \{6\} = E_6$

$E_3^* = \{3, 4, 5, 6\}$ $E_4^* = \{1, 2, 3, 4\}$

$E_1^* = \{2, 4, 6\} = E_2$ $E_2^* = \{1, 3, 5\} = E_1$

Another important relationship that could possibly exist between events is to have no basic outcomes in common. The following definition applies to this situation.

Two events A and B are said to be **mutually exclusive** or **disjoint** if A and B have no outcomes in common. Notationally, if A ∩ B = Ø (the empty set), then A and B are **mutually exclusive**.

Definition 3.5 Mutually Exclusive Events

As defined above, E_1 and E_2 are mutually exclusive because $E_1 \cap E_2 = \{1, 3, 5\} \cap \{2, 4, 6\} = \emptyset$. That is, events E_1 *and* E_2 could not *both* happen from the same experimental roll of the die because they have no basic outcomes in common.

Venn diagrams are pictorial representations of set operations and can be used to clarify the concepts involved in combining events. For arbitrary events A and B, Figure 3.1 illustrates the concepts of intersection, union, complement, and mutually exclusiveness.

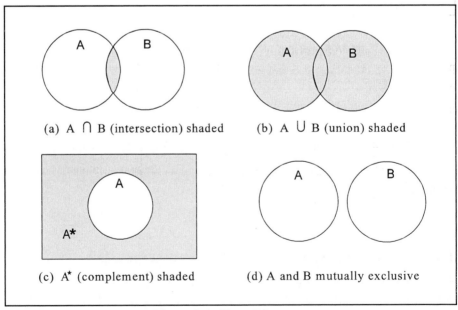

(a) A ∩ B (intersection) shaded   (b) A ∪ B (union) shaded

(c) A* (complement) shaded   (d) A and B mutually exclusive

Figure 3.1 Venn Diagrams

For any event A, the notation P(A) will be used to denote "the probability that event A will occur." In other words, P(A) is a measure of the chance that event A will occur. A probability is always a numerical value between 0 and 1. Namely, for any event A,

Property 1: $0 \le P(A) \le 1$

If the probabilities of the basic outcomes in an experiment are known, the probability that an event A will occur can be obtained by summing the probabilities of each of the basic outcomes contained in A. If $A = \{o_1, o_2, ..., o_k\}$, where the o_i, for $i = 1, 2, ..., k$, represent the k basic outcomes which define event A, then

Property 2: $P(A) = \sum_{o_i \in A} P(o_i)$

where "$o_i \in A$" means "o_i is an element of set A"

Referring to the experiment of tossing a fair, 6-sided die, we do know the probability of each basic outcome. Since the die is assumed to be fair, each basic outcome is equally likely to occur. This scenario is summarized in the following table.

o_i	1	2	3	4	5	6
$P(o_i)$	1/6	1/6	1/6	1/6	1/6	1/6

This table is called a probability distribution because a probability is associated with each possible basic outcome of the experiment. Probability distributions will be considered in detail in Chapter 4. Recalling that $E_1 = \{1, 3, 5\}$, Property 2 tells us that $P(E_1) = P(1) + P(3) + P(5) = 1/6 + 1/6 + 1/6 = 3/6 = 1/2$.

When an experiment takes place, one of the basic outcomes must occur. This fact leads to the following properties.

Property 3:   **P(Ø) = 0**, where Ø = { } = the empty set

Property 4:   **P(S) = 1**, where **S** = the sample space

The probabilities of complementary events are also complementary with respect to "1." For any event A,

Property 5: $\mathbf{P(A^*) = 1 - P(A)}$

The probability of the intersection of two events, P(A ∩ B), is sometimes referred to as a "joint" probability, because it measures the chance of both A **and** B occurring (i.e., their joint occurrence).

Property 6:   If A and B are **mutually exclusive events** then
                P(A ∩ B) = 0

For any events A and B, the probability of their union is

Property 7:   **P(A ∪ B) = P(A) + P(B) − P(A ∩ B)**

Property 7 can be verified (and probably better understood) by considering the Venn diagram in Figure 3.2. The darkest or cross-hatched area is included in both events A and B. Thus, when P(A) and P(B) are each calculated, the darkened area is included in both P(A) and P(B). Consequently, the cross-hatched area,

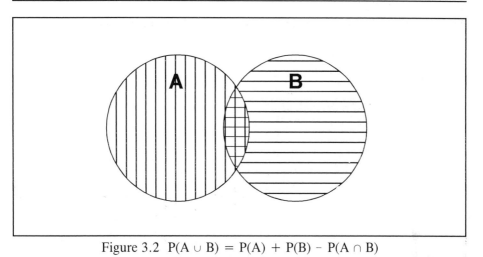

Figure 3.2 $P(A \cup B) = P(A) + P(B) - P(A \cap B)$

$P(A \cap B)$, has been included **twice** in the sum $P(A) + P(B)$. That is precisely why, in Property 7, the $P(A \cap B)$ is subtracted from $P(A) + P(B)$ to get the exact area for $P(A \cup B)$. Combining properties (6) and (7) yields $P(A \cup B) = P(A) + P(B)$ whenever A and B are mutually exclusive.

Example 3.1 Refer to the experiment of tossing a die. Suppose we were to toss a die and that, prior to making the toss, we wanted to calculate the probability of the outcome being odd *or* in the lower third of the possible values. How could we calculate this probability? From our previous event definitions, we have $E_1 = \{1, 3, 5\}$ = "the outcome is odd" and $E_3 = \{1, 2\}$ = "the outcome is in the lower third." What we seek is $P(E_1 \cup E_3)$. From Property 7, we know that:

$$
\begin{aligned}
P(E_1 \cup E_3) &= P(E_1) + P(E_3) - P(E_1 \cap E_3) \\
&= P(\{1,3,5\}) + P(\{1,2\}) - P(\{1\}) \\
&= 3/6 + 2/6 - 1/6 = 4/6 = 2/3.
\end{aligned}
$$

■

    We conclude this section by defining event independence. Two events are independent if the occurrence (or non-occurrence) of one event has absolutely no effect or bearing on the occurrence of the other event. We incorporate this definition into the following property.

Property 8:	Two events A and B are said to be (statistically) **independent** if $P(A \cap B) = P(A) \bullet P(B)$

If two events are not independent, they are said to be dependent.

Example 3.2 Suppose that the probability of a married man voting in a given election is .6 and that the probability of his wife voting is .8. Assume, further, that their actions are **independent**. What is the probability that both a husband and his wife will vote in the given election? If "H" is the event of the husband voting and "W" is the event of the wife voting, then what we seek is $P(H \cap W)$. Since we can assume these events are independent, then by Property 8, $P(H \cap W) = P(H) \cdot P(W)$ = $(.6)(.8) = .48$. What is the probability that at least one of them will vote? In this case, we seek $P(H \cup W)$. By Property 7, $P(H \cup W) = P(H) + P(W) - P(H \cap W)$ = $(.6) + (.8) - (.48) = .92$.

∎

3.4 Conditional Probability and Independence

In the previous section, we defined $P(A)$ to be the probability that event A will occur. The probability assigned to this event is dependent upon what was known about the situation at the time the assignment was made. We can think of this probability as being an "original" probability, or an "unconditional" probability. In this section we address the nature of the probability of an event when considering additional information that is now available which was not known at the time when the "original" probability of an event was assigned. For example, suppose $P(A)$ denotes the probability that an individual selected at random has a certain disease. Let us suppose further that $P(A)$ was assigned the probability .0001, i.e., $P(A) = .0001$, where A is the event "has the disease." This probability was perhaps assigned based on the incidence of the disease in the population at large, that is, the relative frequency approach. In this case, the frequency of the disease is supposedly known to be 1 in every 10,000 people. Consider further the case in which a person is selected at random, tested for the disease, and the result of the test is positive. We now have a situation in which more information is known (i.e., the test came back positive) than was known when $P(A)$ was assigned to be .0001 for any randomly selected individual. We assume that the individual so tested now has a much higher probability (than .0001) of having the disease because of the positive test result. The new probability will be greater than .0001 but will no doubt be less than 1, because seldom are medical tests perfectly trustworthy. If B is the event that an individual tests positive for the disease, then P(A|B) denotes the probability of the individual having the disease *given* that the person already tested positive. The probability P(A|B) is called a **conditional** probability, the definition of which is as follows:

Property 9:   If A and B are any events with P(B) > 0, then the **conditional probability** of A given that B has occurred is

$$P(A|B) \; = \; \frac{P(A \cap B)}{P(B)}$$

The conditional probability P(A|B) is usually read "the probability of A given B." Property 9 establishes the fact that a conditional probability is a ratio of probabilities: namely a "joint" (intersection) probability divided by an "original" probability. In a Venn diagram (reference Figure 3.3), P(A|B) is seen to be the ratio of P(A ∩ B), the cross-hatched area, to P(B), the area of B alone. The reason P(B) is in the denominator is that if B is known to have occurred, then the sample space is reduced to just B.

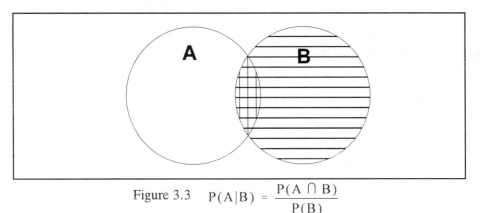

Figure 3.3 $P(A|B) \; = \; \dfrac{P(A \cap B)}{P(B)}$

We illustrate these properties of probability by way of an example involving what is called a bivariate distribution.

Example 3.3 Consider the following (qualitative) bivariate distribution involving 400 students who are classified according to whether or not they were vaccinated and whether or not they came down with the flu. We use the following notation for these 4 events:

$$V \; = \quad \text{"was vaccinated"}$$
$$V^* = \quad \text{"was not vaccinated"}$$
$$F \; = \quad \text{"got the flu"}$$
$$F^* = \quad \text{"did not get the flu"}$$

The bivariate frequency distribution is given in Table 3.1.

	V	**V***	**Marginal Frequency**
F	60	85	145
F*	190	65	255
Marginal Frequency	250	150	Total Freq = 400

Table 3.1 Bivariate Distribution

(a)   What is the probability that a student got the flu *and* was vaccinated? We seek P(F ∩ V) because the word "and" translates into set or event "intersection." This is a **joint** probability where P(F ∩ V) = 60/400 = .15.

(b)   What is the probability that a student got the flu *given* that she was vaccinated? We seek the conditional probability P(F|V). By Property 9,

$$P(F|V) \; = \; \frac{P(F \cap V)}{P(V)} \; = \; \frac{(60/400)}{(250/400)} \; = \; \frac{60}{250} \; = \; .24.$$

We call P(V) a "marginal" probability because it is obtained by using the "marginal" frequency of 250 which is the sum of all of the frequencies in the column labeled "V." Here we see the fact that a conditional probability is the ratio of a joint probability over a marginal probability. We can also see that we could have obtained P(F|V) without using the total frequency of 400 at all. Namely, finding P(F|V) allows us to reduce the entire sample space to just the column labeled "V" because we know the individual was vaccinated. Of the 250 students vaccinated, only 60 got the flu, so P(F|V) = 60/250 = .24.

(c)   What is the probability that a student got the flu *given* that he was not vaccinated? We seek P(F|V*). Restricting ourselves to the column labeled "V*," we see that P(F|V*) = 85/150 = .567. Comparing this result with the result in (b), we see that a student who was not vaccinated was more than twice as likely (.567 vs .24) to get the flu than a student who was vaccinated. Please note that P(F|V) and P(F|V*) are *not* complementary probabilities, whereas P(F|V) and P(F*|V) are complementary probabilities. That is, P(F|V) + P(F|V*) ≠ 1, but P(F|V) + P(F*|V) = 1. In the latter case, the given information was the same (i.e., the student was vaccinated) whereas in the former case, the two conditional probabilities were based upon different known (given) information.

∎

Using the frequency approach to probability, the bivariate frequency distribution in Table 3.1 can be converted into the **bivariate probability distribution** shown in Table 3.2. This table displays 4 **joint** probabilities: P(F ∩ V) = .1500, P(F ∩ V*) = .2125, P(F* ∩ V) = .4750, and P(F* ∩ V*) = .1625. In Table 3.2, the joint probabilities are enclosed within the inner 2 x 2 rectangle. The table also displays 4 **marginal** probabilities: P(V) = .6250, P(V*) = .3750, P(F) = .3625, and P(F*) = .6375. The marginal probabilities are shaded in Table 3.2.

	V	**V***	**Marginal Probability**
F	.1500	.2125	**.3625**
F*	.4750	.1625	**.6375**
Marginal Probability	**.6250**	**.3750**	Total = 1.0000

Table 3.2 Bivariate Probability Distribution

Notice that **conditional** probabilities are *not* direct entries in the bivariate probability table, but that they can be computed from the **joint** and **marginal** probabilities in the table by using Property 9.

Now having a feel for conditional probability, we briefly revisit the concept of independence and expand upon it. Recall from Property 8 that if two events A and B are (statistically) independent, then $P(A \cap B) = P(A) \cdot P(B)$. If we combine this property with Property 9, we see that if A and B are independent, then

$$P(A|B) = \frac{P(A \cap B)}{P(B)} = \frac{P(A) \cdot P(B)}{P(B)} = P(A).$$

We summarize this more intuitive notion of independence in the following property.

Property 10:   If A and B are **independent** events, then
$$P(A|B) = P(A) \text{ and } P(B|A) = P(B).$$

The result $P(A|B) = P(A)$ indicates that the probability of event A occurring is really not affected by whether B does or does not occur. Hence, Property 10, which is equivalent to Property 8, is no doubt a more intuitive representation of the independence of events A and B.

We illustrate what independence means by referring again to Example 3.3 and Table 3.2. What do the probabilities in Table 3.2 tell us about the possible independence of "getting the flu" and "having been vaccinated against the flu?" Hopefully, the data will indicate that the two are *not* independent, for if they were independent, we would have to question the utility of being vaccinated. One way to look at the data to see if vaccination does make a difference is to examine the two conditional probability distributions. That is, compare $P(F|V)$ and $P(F^*|V)$ with $P(F|V^*)$ and $P(F^*|V^*)$, respectively. We summarize these values in Table 3.3.

	Probabilities Conditioned on V: P(|V)	**Probabilities Conditioned on V*: P(|V*)**
F	.240	.567
F*	.760	.433
	1.000	1.000

Table 3.3 Conditional Probability Distributions

We see from Table 3.3 that the two conditional distributions are indeed quite different, indicating that the two variables involved (vaccination and the flu) are *not* independent. Even if these two variables were independent, we would not expect the two distributions to be exactly the same because of natural variability in a sample of 400 students. However, the large difference in Table 3.3 is sufficient to conclude that the variables are not independent.   An equivalent test for the independence of two variables is to use the definition of independence of events (Property 8) directly. That is, every joint probability should equal the product of its corresponding marginal probabilities. A formal test for independence is given in Chapter 6.

We conclude this section with a final property of probability that is equivalent to Property 9 (conditional probability) but whose form is sometimes more convenient to apply. If we solve for $P(A \cap B)$ in the conditional probability property (Property 9), we obtain what are called the **multiplication rules**.

Property 11: For any events A and B,

$$P(A \cap B) = P(A|B) \bullet P(B), \text{ provided } P(B) > 0$$
$$P(A \cap B) = P(B|A) \bullet P(A), \text{ provided } P(A) > 0$$

We will find these rules useful in the next section.

3.5 Bayesian Probability

A simple, but very useful application of conditional probability was introduced by the Reverend Thomas Bayes in a paper that was published in 1763. Bayes' rule is nothing more than a restatement of Property 9, conditional probability. However, its usefulness derives from being able to revise a probability based on additional information obtained after-the-fact. Bayes' rule will allow us to calculate a certain conditional probability when other conditional probabilities (of the reverse form), as well as marginal probabilities, are known. We motivate the use of Bayes' rule through an example and encourage the use of the 4-step tabular solution technique presented in Example 3.4. We also present the formula for Bayes' Rule and demonstrate its application in Example 3.5.

Example 3.4 A certain aircraft manufacturer uses three different subcontractors, call them X, Y, and Z, to manufacture their aircraft tires. Company X produced twice as many tires as Company Y which produced the same number as Company Z (over a certain production period). It is also known that 2% of the tires produced by companies X and Y fail the specifications, while 4% of the tires produced by company Z fail the specifications. If a tire selected at random was found to fail the specification, what is the probability it came from company X? The solution technique for this problem involves several steps. The first step is to establish a notation for the various events and to write down what is known in terms of the notation. This step is no doubt the most time-consuming and difficult, therefore demanding appropriate caution.

Step 1: $\quad$ X = "produced by company X" $\qquad$ $P(X) = .5$
$\qquad\qquad$ Y = "produced by company Y" $\qquad$ $P(Y) = P(Z) = .25$
$\qquad\qquad$ Z = "produced by company Z" $\qquad$ $P(D|X) = .02$
$\qquad\qquad$ D = "defective tire" $\qquad\qquad\qquad$ $P(D|Y) = .02$
$\qquad\qquad$ $D^* =$ "not defective tire" $\qquad\qquad$ $P(D|Z) = .04$

You should understand how these probabilities were determined from the statements in the problem. You may have to read the problem several times to establish everything that is known. The second step is to define what we are asked to find.

Step 2:   Find $P(X|D)$.

Note that what we are asked to find is a conditional probability of the "reverse" form of a conditional probability already known (i.e., $P(D|X) = .02$). $P(X) = .5$, a marginal probability, is also known and it is called a "prior" probability. However, we now have additional information, namely, that the tire randomly selected was defective. Thus, by finding $P(X|D)$, we are revising our original "prior" probability, $P(X) = .5$, to the "posterior" probability $P(X|D)$ based on new information (namely, the tire was defective). The third step is to fill in the "road map," that is, the bivariate probability table.

Step 3:   Set up the following bivariate distribution table:

	D	D*	Marginals
X			
Y			
Z			
Marginals			1.000

Now, insert into the table those probabilities that are known. $P(X) = .5$, and $P(Y) = P(Z) = .25$ are marginal probabilities that can be inserted into the right-hand margin. Besides marginal probabilities, the only other kind of probability that shows up directly in a bivariate probability table is a **joint** probability. Although we do not yet know any of the joint probabilities, we can find some of the joint probabilities by using what we do know: conditional probabilities and marginal probabilities. Recall the definition of conditional probability (Property 9), which says that $P(A|B) = P(A \cap B) / P(B)$. Algebra allows us to find the third probability in this relationship whenever any two of the probabilities are known.

Thus,

$$P(D|X) = \frac{P(D \cap X)}{P(X)} \quad \text{where} \quad P(D|X) = .02 \quad \text{and} \quad P(X) = .5.$$

Therefore,

$$.02 = \frac{P(D \cap X)}{.5} \quad \text{and} \quad P(D \cap X) = (.02)(.5) = .010,$$

which can be put directly into the table. Similarly,

$$P(D \cap Y) = P(D \mid Y) \cdot P(Y) = (.02)(.25) = .005 \text{ and}$$
$$P(D \cap Z) = P(D \mid Z) \cdot P(Z) = (.04)(.25) = .010.$$

Note the use of Property 11 (the multiplication rules) here. These values can also be inserted into the table. A bivariate probability table exhibits the following features:

(a)   the sum of the joint probabilities in any row or column must equal the respective row or column marginal probability.

(b)   the sum of the row marginal probabilities must equal 1, and the sum of the column marginal probabilities must also equal 1.

Knowing these properties of a bivariate probability table, we can now fill in the remainder of the table by just using addition and subtraction. The completed table is shown on the following page. With the table complete, we have all of the information (and more) needed to determine what we were asked to find. The final step uses the table and the definition of conditional probability (again) to obtain the answer. Note that while the original estimate of a randomly selected tire coming from company X was $P(X) = .5$, the revised estimate or probability is $P(X \mid D) = .4$, which is based on the additional knowledge that the tire randomly selected is defective.

	D	D*	Marginals
X	.010	.490	.500
Y	.005	.245	.250
Z	.010	.240	.250
Marginals	.025	.975	1.000

Step 4: $P(X|D) = \dfrac{P(X \cap D)}{P(D)} = \dfrac{.010}{.025} = .4$.

∎

The information included in the 4-step method using the bivariate distribution table can be expressed in another tabular form that systematizes the mechanical process.[1] You may want to consider this method as your computational procedure. By way of explanation, Example 3.4 is repeated using the new method. As a starter the appropriate table is set up identifying conceptually the various probability entries. Additionally, some Bayesian terminology is introduced as column headings.

	Prior Distribution $P(\theta_i)$	New Information $P(D|\theta_i)$	Joint Distribution $P(D\cap\theta_i)$	Revised Distribution $P(\theta_i|D)$
$X = \theta_1$	$P(X)$	$P(D|X)$	$P(D\cap X)$	$P(X|D)$
$Y = \theta_2$	$P(Y)$	$P(D|Y)$	$P(D\cap Y)$	$P(Y|D)$
$Z = \theta_3$	$P(Z)$	$P(D|Z)$	$P(D\cap Z)$	$P(Z|D)$
			$P(D)$	

[1]This tabular approach and accompanying discussion was provided by Dr. Paul Ruud, Visiting Professor at the USAF Academy.

Substituting the numbers from Example 3.4, we have

	Prior	New	Joint	Revised
X	.50	.02	(.50)(.02)=.010	.010/.025 = 0.4
Y	.25	.02	(.25)(.02)=.005	.005/.025 = 0.2
Z	.25	.04	(.25)(.04)=.010	.010/.025 = 0.4
			Total .025	Total 1.0

Recall the original question — a tire selected at random was found to fail specifications; what is the probability that the tire came from company X? The answer is found at the intersection of the X row and Revised column, $P(X|D) = 0.4$. Notice the likelihood of the tire being from company X was revised from 0.5 to 0.4 using the new information that the tire selected failed to meet specifications.

We conclude this section by formally presenting Bayes' rule; by making some comments about how Bayes' rule compares to the tabular solution technique illustrated above; and by giving an example of how Bayes' rule can be used in analyzing the quality of a product.

If A_1, A_2,..., A_n is a collection of mutually exclusive and exhaustive events with $P(A_i) > 0$ for each $i = 1, 2,..., n$, then for any event B, where $P(B) > 0$, and any A_k, $k = 1, 2,..., n$,

$$P(A_k|B) = \frac{P(A_k \cap B)}{P(B)} = \frac{P(A_k \cap B)}{\sum_{i=1}^{n} P(A_i \cap B)} = \frac{P(B|A_k)P(A_k)}{\sum_{i=1}^{n} P(B|A_i)P(A_i)}$$

Equation 3.1 Bayes' Rule

Equation 3.1 could have been used directly, without constructing a bivariate probability table, to solve the problem in Example 3.4. The tabular method, however, is an implementation of Bayes' rule. The first equation in Equation 3.1 is just the definition of conditional probability (Property 9). The denominator of the third expression in Equation 3.1 is another way to write P(B). The tabular presentation illustrates this well because a marginal probability is the sum of joint probabilities. The fourth expression in Equation 3.1 just replaces every joint probability in the third expression with its equivalent product of a conditional probability and marginal probability (Property 11). Recall that in Example 3.4 we found the joint probabilities precisely in this manner. The next example uses the formula (Equation 3.1) for Bayes' rule and illustrates its usefulness in evaluating the quality of a product.

Example 3.5 Consider a firm that manufactures integrated circuits (chips). Based on extensive (and expensive) testing of a sample of 1000 chips, it finds that the defect rate is 10%. That is, it found that 100 out of the 1000 chips were defective. The company concludes that a randomly selected chip from their production process has a probability of .1 of being defective, i.e., $P(D) = .1$ and $P(D^*) = .9$. The firm then examines its inspectors' abilities to differentiate between defective and nondefective chips by subjecting them to the same 1000-chip batch from which exactly 100 chips are known to be defective. It found that the inspectors typically had a 5% error rate on both defective and nondefective chips. That is, if we let "A" be the event of "accepting" the chip as being good, then $P(A|D) = .05$ and $P(A|D^*) = .95$. These two conditional probabilities represent the inspectors' abilities. The company's main objective is to reduce the probability of shipping defective chips, namely $P(D|A)$, because this is the probability that is directly correlated to customer dissatisfaction and potential loss of revenue. Using Bayes' rule, we find that

$$P(D|A) = \frac{P(A|D)\,P(D)}{P(A|D)\,P(D) + P(A|D^*)\,P(D^*)}$$

$$= \frac{(.05)\,(.1)}{(.05)\,(.1) + (.95)\,(.9)} = .0058$$

If one looks carefully at the situation, one can see that the probabilities of .05 and .95 are virtually fixed in the sense that these probabilities are dependent on and limited by human inspection capabilities. Hence, the only way that the company can expect to reduce the .0058 probability of shipping a defective chip is to reduce $P(D) = .1$. That is, it must ultimately reduce the defect rate if it is to have fewer defective parts shipped.

Bayes' rule provides a convenient vehicle for investigating how $P(D|A)$ and $P(D)$ are related. We investigate this relationship by looking at

$$P(D|A) = \frac{(.05)\ \ P(D)}{(.05)\ \ P(D) + (.95)\ \ P(D^*)}$$

and using several values of $P(D)$ to see the corresponding change in $P(D|A)$.

P(D)	P(D|A)
.10	.005814
.05	.002762
.01	.000531

From this table we can see that the firm could cut its shipped defect rate to almost 5 in 10,000 if it could reduce its production defect rate from .1 to .01. In the continuous improvement strategy, once the probability of a defect is driven low enough, costly inspectors will be ineffective and, therefore, can be eliminated. This is the focus of many modern quality improvement initiatives.

■

3.6 Combinatorics

No treatment of probability can be considered adequate if it does not address the notion of combinatorics. Combinatorics is a branch of mathematics that deals with various kinds of counting problems. It involves the enumeration of both ordered and unordered collections of objects from a set of objects. Our treatment of combinatorics will be limited to a few very basic principles, those concepts which will allow us to adequately understand some of the probability distributions and statistical tools to be presented later in this text.

In some experiments, the problem of computing probabilities reduces to counting. If the N possible outcomes of an experiment are equally likely and if we desire to find the probability of an event A which contains exactly n of those outcomes, then $P(A) = n/N$ which is again a relative frequency expression. In the "tossing of a fair die" example, recall that we defined event E_1 to be "the outcome is odd," or $E_1 = \{1, 3, 5\}$. Since the die was assumed to be "fair," each of the 6 basic outcomes was assumed to be equally likely. Thus, since event E_1 contains 3 of the 6 basic outcomes, $P(E_1) = 3/6 = 1/2$. Not all experiments produce a set of possible outcomes that can be so easily listed, because N could be quite large. It is possible to compute probabilities based on counting principles, without having to produce a list of all possible outcomes. The rules we are about to present can also be used in many problems in which the outcomes of an experiment are not equally likely. Some of these rules may be intuitive, but others may not be.

One of the most fundamental principles of combinatorics is given in the following rule.

If task A can be performed in **M** different ways and task B can be performed in **N** different ways, then the sequence of tasks A and B can be performed in **M•N** different ways.

Rule 1. Fundamental Principle

Example 3.6 If the Super Bowl matchup is between an American Football Conference (AFC) team and a National Football Conference (NFC) team, and there are 15 teams in the AFC and 15 teams in the NFC, how many Super Bowl matchups are possible? There are 15 choices for the AFC team and 15 choices for the NFC team; thus, by Rule 1, there are (15)(15) = 225 possible Super Bowl matchups. Note: according to the oddsmakers, these 225 possible matchups are not all equally likely.

∎

Obviously, the fundamental principle in Rule 1 could be extended to more than just two tasks or operations.

The next rule is really a definition of what is meant by a "factorial." It will be used often because of its compact notation.

For any positive integer n, n! is read as "n factorial" and is defined to be **n! = n(n - 1) . . . (2)(1)**. A special case is **0! = 1**.

Rule 2. Definition of n!

A **permutation** is an **ordered** sequence of objects. The next rule gives the number of possible permutations of n objects.

Given n distinct objects, the number of possible permutations (or ordered arrangements) of those n objects is **n! = n(n − 1)(n − 2) . . . (2)(1)**.

Rule 3. Number of Permutations of n Objects

The nature of Rule 3 is quite intuitive. If n objects are to be placed in n slots, one would have n choices to fill the first slot. Once that slot is filled, there are n-1 choices remaining for the second slot, and so on. Thus, the total number of permutations is a direct consequence of Rule 1.

Example 3.7 How many ways can a baseball team manager select a batting order for a starting lineup of 9 players? n! = 9! = (9)(8)(7)(6)(5)(4)(3)(2)(1) = 362,880 ways!

∎

The value of n! grows very rapidly as n gets bigger. Computing n! gets out of hand quickly, and some calculators either may not have an n! key or may not be able to store such large numbers. For the interested reader, a fairly easy-to-use, and good approximation for n! is given by **Stirling's formula**, which states that $n! \approx \sqrt{2\pi} \, n^{(n+1/2)} \, e^{-n}$. This approximation is usually implemented by first finding $\log(n!) \approx \frac{1}{2}\log(2\pi) + (n + \frac{1}{2})\log n - n \log e$ and then exponentiating to get n!.

A special case of Rule 3 is given in the following rule which shows how to count the number of permutations when not all of the n objects are used.

Given n distinct objects, the number of possible permutations of r of those objects is given by

$$_n P_r = n(n-1)(n-2)\ldots(n-r+1) = \frac{n!}{(n-r)!}$$

Rule 4. Number of Permutations of Length r from n Objects

Example 3.8 If a basketball coach has 10 players on her squad, how many ways can she write a starting 5 in the scorebook? In this case, n = 10 and r = 5, and by Rule 4, 10!/5! = (10)(9)(8)(7)(6) = 30,240 ways!

∎

The use of the term "permutation" implies that no object can be used more than once. If this restriction is lifted, we have the following rule.

If repetitions are allowed, the number of arrangements of length r that can be formed from n distinct objects is n^r.

Rule 5. Number of Arrangements if Repetitions are Allowed

Hopefully, you see that this rule is a direct result of Rule 1 because, allowing repetitions, we have n choices for each of r consecutive "tasks."

Example 3.9 Suppose the format for the auto license plates in a certain state is 3 letters followed by 3 digits; and repetitions of both letters and digits are allowed. How many different license plates can be produced? Since repetitions are allowed, we combine Rules 1 and 5 to obtain

$$26^3 \cdot 10^3 = (26)(26)(26)(10)(10)(10) = 17,576,000.$$

∎

Up to now, this section has presented rules that deal with counting "ordered" collections of objects. Despite its usefulness, "order" is not always a necessary or desirable property. For example, in selecting a sample from a population (like we did in Section 2.4 when we selected a sample of 2 TVs from a lot of 10 TVs), the order in which we took the two TVs off of the assembly line was unimportant. We mentioned at that time that there were 45 different ways to choose 2 TVs from 10. We will substantiate that number now by way of a rule that addresses the number of unordered collections of objects. Where the term "permutation" connotes "order makes a difference," we now introduce the term **combination** which deals with an **unordered** collection of objects. For example, from the set of 3 objects {A, B, C}, there are 6 permutations of these 3 objects: ABC, ACB, BAC, BCA, CAB, CBA. However, there is only one combination of 3 elements that can be selected from the original 3 elements. In other words, all 6 permutations constitute the same combination. The following rule is no doubt one of the most important in all of combinatorics and should be mastered.

The number of ways to choose **k** different objects
from a set of **n** distinct objects is given by

$$\binom{n}{k} = \frac{n!}{k! \, (n - k)!}$$

Rule 6. Combinations of **n** Things Taken **k** at a Time

The notation $\binom{n}{k}$ is read "n things taken k at a time" or "n choose k." Another commonly used equivalent notation is $_nC_k$, or a combination of n things taken k at a time.

Example 3.10 How many ways can a sample of 2 TVs be selected from a lot of 10 TVs? From Rule 6,

$$\binom{n}{k} = \binom{10}{2} = \frac{10!}{2!\,8!} = \frac{10 \cdot 9 \cdot 8!}{2!\,8!} = \frac{10 \cdot 9}{2 \cdot 1} = 45,$$

which is what was stated in Section 2.4.

We conclude this section with an example that demonstrates a combinatorial approach to the basketball team problem of Example 3.8. It hopefully will draw the concepts of "permutation" and "combination" together.

Example 3.11 If a basketball coach has 10 players on her squad, how many ways can she choose a starting 5? This is a different problem than stated in Example 3.8. Rule 6 can be used here because "order" is unimportant. There are

$$\binom{10}{5} = \frac{10!}{5!\,5!} = \frac{10 \cdot 9 \cdot 8 \cdot 7 \cdot 6 \cdot 5!}{5!\,5!} = \frac{10 \cdot 9 \cdot 8 \cdot 7 \cdot 6}{5!}$$

$$= \frac{10 \cdot 9 \cdot 8 \cdot 7 \cdot 6}{5 \cdot 4 \cdot 3 \cdot 2 \cdot 1} = \frac{9 \cdot 8 \cdot 7}{2 \cdot 1} = 9(4)(7) = 252$$

ways to choose a starting 5. Now, for each of these 252 possible starting 5's, there are 5! or 120 ways to write the starting 5 in the scorebook, resulting in $(252)(120) = 30,240$ total possible ways to write a starting 5 in the scorebook. This, of course, is what was found in Example 3.8.

3.7 Binomial Coefficients

The previous section introduced the symbol $\binom{n}{k}$ in terms of its combinatorial significance, namely that it represents the number of ways one can choose k distinct objects from n distinct objects. This symbol appears in other branches of mathematics as well. This section shows that $\binom{n}{k}$ also represents the coefficients in the binomial expansion. It is presented here to provide the needed background for later topics in the text.

By binomial expansion, we mean raising the binomial expression (a + b) to various powers. From previous experience we know that

$$(a + b)^0 = 1,$$

$$(a + b)^1 = a + b,$$

$$(a + b)^2 = a^2 + 2ab + b^2, \text{ and}$$

$$(a + b)^3 = a^3 + 3a^2b + 3ab^2 + b^3.$$

The pattern in expanding a binomial is quite clear: if the power is n, then there are n + 1 terms in the expansion. Each term consists of a power of "a" multiplied by a power of "b" multiplied by a constant coefficient. The terms can be written, as we have above, with the powers of "a" descending and the powers of "b" ascending. If we only knew the coefficients, the task of expanding a binomial to any power would be quite simple. Pascal's triangle, shown in Figure 3.4, gives these coefficients in the form of a "Christmas tree" outlined in "1"s. Pascal's triangle is easily constructed by noticing the pattern that every integer in the triangle not equal to 1 is formed from the sum of the two integers immediately above it in the preceding row.

$(a + b)^0$					1					
$(a + b)^1$					1		1			
$(a + b)^2$				1		2		1		
$(a + b)^3$			1		3		3		1	
$(a + b)^4$		1		4		6		4		1
$(a + b)^5$	1		5		10		10		5		1

Figure 3.4 Pascal's Triangle

The entries in Pascal's triangle can also be obtained by using $\binom{n}{k}$. The integers in the nth row (rows are numbered 0, 1, 2,..., where the row number is associated with the exponent) are the values $\binom{n}{k}$, k = 0, 1, . . ., n. Rewriting Pascal's triangle in combinatorial notation results in Figure 3.5. By using Rule 6, one can verify the equivalence of the corresponding entries in Figures 3.4 and 3.5. There are some amazing properties imbedded in this triangle, the detailed analysis of which is beyond the scope of this text.

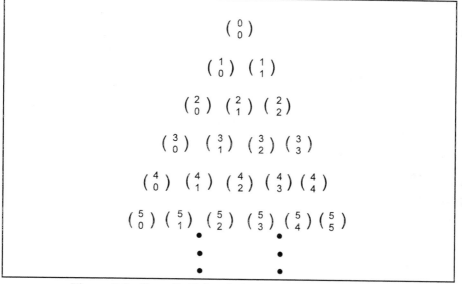

Figure 3.5   Pascal's Triangle in Combinatorial Notation

If one analyzes what is actually done to expand $(a + b)^n$ by longhand methods, it should not be surprising that combinatorials can be used to describe the coefficients. Since $(a + b)^n = (a + b)(a + b) \cdots (a + b)$, a product of n binomial factors, the expansion involves multiplying every term in each binomial factor times every other term in all of the other binomial factors. For example, the final expansion term of the form $a^r b^{n-r}$ consists of the sum of all the products formed by choosing "a" from exactly r of the n binomial factors and "b" from the remaining $(n - r)$ factors.

Since $\binom{n}{r}$ represents the number of ways this can be done, $\binom{n}{r}$ is the coefficient

of the term $a^r b^{n-r}$.

We conclude this section by summarizing some important aspects of binomial coefficients and Pascal's triangle.

(1)   The terms of the binomial, a and b, may be any two expressions and could look much different than a and b.

(2)   The exponent or power of the expansion must always be an integer $n \geq 0$.

(3)   There is symmetry about the vertical centerline in Pascal's triangle. Note that $\binom{n}{0} = \binom{n}{n} = 1$ for every n;

$$\binom{n}{1} = \binom{n}{n-1} = n \text{ and in general, } \binom{n}{r} = \binom{n}{n-r}.$$

The latter equality follows directly from Rule 6.

(4) $(a + b)^n = \displaystyle\sum_{r=0}^{n} \binom{n}{r} a^r b^{n-r}.$

3.8 Problems

3.1   A coin was tossed 100,000 times in an automatic coin tossing machine. Heads were obtained 70,000 times and tails 30,000 times. If the **relative frequency** approach to probability is employed, it is reasonable to conclude in this coin tossing experiment that: (select the best choice)

a)   the probability of a tail is .50.
b)   the probability of a head is .70.
c)   the concept of probability does not apply because of the large number of tosses.
d)   the concept of probability does not apply because the probability of each possible outcome can be determined in advance.
e)   no valid results can be obtained as the coin apparently was not "fair".

3.2   **Sunstroke**
The sample space in a statistical study at Marina Hospital consists of the number of patients {0, 1, 2, . . . } admitted each week for sunstroke. A basic outcome in this study would be: (select the best choice)

a)   no sunstroke patients are admitted.
b)   more than ten sunstroke patients are admitted.
c)   fewer than five sunstroke patients are admitted.
d)   more than one of (a), (b), and (c) are basic outcomes.

3.3   **Tossing a Die**
Consider tossing a fair six-sided die once.

a)   List the Sample Space **S**.
b)   List the outcomes which satisfy the event B = "the outcome is in the lower half."
c)   List the outcomes which satisfy the event C = "the outcome is divisible by 3."
d)   List the outcomes which satisfy the events B or C above.
e)   List the outcomes which satisfy the events B and C above.
f)   List the outcomes which satisfy the event C^*.

3.4   Refer to **Tossing a Die** (Problem 3.3)
        Consider tossing a fair six-sided die once.

   a)   Calculate P(B) = P("the outcome is in the lower half").
   b)   Calculate P(C) = P("the outcome is divisible by 3").
   c)   Calculate P(B ∪ C).
   d)   Calculate P(B ∩ C).
   e)   Calculate P(C*).
   f)   Calculate P(S).

3.5   Refer to **Tossing a Die**, Problems 3.3 and 3.4.

   a)   Show that events B and C are independent.
   b)   Are events B and C* independent? Why or why not?
   c)   Are events B* and C independent? Why or why not?
   d)   Are events B* and C* independent? Why or why not?

3.6   Suppose two events A and B are mutually exclusive and independent.
        Show that either P(A) = 0 or P(B) = 0.

3.7   **City Park**
        The manager of a city park and recreation department has constructed the
        following joint probability distribution for two employee characteristics:

   1)   an employee's class (A, B, C) and
   2)   whether or not the employee is paid more than $15.00 per hour for
        overtime.

Hourly Overtime Pay Rate	Employee Class			Total
	A	B	C	
More than $15.00	.20	.12	.08	.40
$15.00 or less	.20	.18	.22	.60
Total	.40	.30	.30	1.00

   a)   Which, if any, pairs of events comprising joint outcomes in the
        distribution are independent?
   b)   Are the variables "employee class" and "overtime pay rate"
        independent? Why or why not?

3.8   The probability of a snowstorm today is .80.

The probability that your airline flight will take off on time, given that it snows, is .20.

The probability that your flight will take off on time, given that it does not snow, is .90.

Complete the following table, giving all marginal and joint probabilities.

	SNOW	NO SNOW	TOTAL
ON TIME			
LATE			
TOTAL	.80		

3.9   **Election Day**
If the probability that a married man will vote in a given election is 0.60, the probability that a married woman will vote in the election is 0.75, and the probability that a woman will vote in the election given that her husband will vote is 0.95, then

a)   What is the probability that a husband and his wife will both vote in the election?
b)   What is the probability a man will vote in the election given his wife will vote?
c)   What is the probability at least one of the married couple will vote?

3.10   **Car Wash**
The probability of a thunderstorm on an average summer day is .5; the probability Dr. Polinski will wash her car is .1; and the probability it will rain, given Dr. Polinski washes her car, is .8.

a)   What is the probability Dr. Polinski will wash her car and it will rain?
b)   What is the probability Dr. Polinski washed her car, given it has rained?

3.11   **Super Bowl Forecast**
Since the first AFL-NFL Championship Football game in 1967, the direction of the stock market has followed the success of the team from the old National Football League (now the NFC).   Suppose the probability of the NFC team winning the Super Bowl is .60. Further, the probability of the NFC team winning and the stock market going up is .54.   Given the San Francisco 49'ers (an NFC team) win the Super Bowl, what is the probability the stock market will go up?

3.12   We know that, in Suburbia, the type of heating fuel used and the type of air conditioning used are statistically independent random variables in single family homes. These probability distributions are given below:

Fuel System	Probability
Heating oil	.30
Natural gas	.70
	1.00

Type of Air Conditioning	Probability
Central air	.70
Window units	.20
No A/C	.10
	1.00

a)   Find the probability a randomly selected house has natural gas *and* central air.

b)   If a specific home has natural gas, what is the probability this home has central air?

c)   What is the probability a randomly selected home has central air *or* natural gas?

3.13   The probability that your instructor will be absent from class is .05. The probability no students will be absent is .05. The probability all students and the instructor will be in class is .50. Do the students and instructors act independently regarding their attendance in class? Why or why not?

3.14   **Cancer**

In a small community near a plant which produces agricultural chemical products, the occurrence of breast cancer in women age 30 to 49 has been observed to be slightly higher than the general public. It is known that 14% of the women living in the same county as the plant develop breast cancer, 7% of the women in the neighboring counties develop breast cancer, and 5% of the women in the state who do not live within one county of the plant develop breast cancer. We know 500 women between ages 30 and 49 live in the same county as the plant; 7,500 women live in the neighboring counties; and 42,000 women live within this state but not near the plant.

Let
C = "woman has cancer"
C^* = "woman does not have cancer"
P = "woman lives in the same county as the plant"
N = "woman lives in a nearby county"
S = "woman lives in the same state but not in a neighboring county"

a)   Complete the following table.

	C	C*	Total
P			.01
N			.15
S			.84
Total			1.00

b)   If the first patient seen by a doctor today is properly diagnosed as having breast cancer, what is the probability she lives in the same county as the chemical plant?

3.15   On a Saturday evening at a sobriety checkpoint, 100 persons were tested with a new breathalyzer test to determine if they were legally intoxicated. We know the breathalyzer is 90% accurate when testing a person who is intoxicated and 98% accurate when testing a person who is not intoxicated. If 16 of the tested were in fact intoxicated, what is the probability a particular person is not intoxicated given that the breathalyzer said he or she was intoxicated?

3.16   **Car Alarm**
Your neighbor's automobile has a theft alarm system. If someone breaks into this car, the probability the alarm will sound is .95. Four times in the last two years this car's alarm has sounded when no one was around. From state police reports, the probability an automobile is broken into on any given night is .002. If your neighbor's car alarm sounds tonight, what is the probability someone is burglarizing the car? Assume there are 365 days (or nights) per year.

3.17   **License Plates**
Suppose the format for the auto license plates in a certain state is 3 letters followed by 3 digits. If no repetitions of letters or numbers are allowed, how many different license plates can be produced?

3.18   **Baseball Lineup**
If a baseball coach has 25 players, how many different batting orders of 9 players can he choose from?

3.19   Refer to Problem 3.17, **License Plates**
If repetitions of letters and numbers are allowed but none of the five vowels are allowed, how many different license plates can be produced?

3.20   If there were 600 students enrolled in a statistics course, how many different ways could we choose a class of 20 students?

3.21   If a baseball coach has 50 players try out for his team and he can choose 24 for his team, how many different ways could he choose his team?

3.22   a)   If an eighteen-wheel truck has 18 identical wheels, how many different ways can the wheels be put on the truck?
        b)   If the two front wheels are uniquely different from the other sixteen wheels, how many different ways can the wheels be placed on the truck?

3.23   **Cards**
In a deck of 52 playing cards, how many different hands of 5 cards are there?

3.24   **Indy**
        There are 33 cars at the start of the Indianapolis 500. Sixteen have
        American engines and 17 have British engines. How many ways can
        they finish if we are only interested in the engine origin?

3.25   Show the following:

$$\binom{n+1}{k} = \binom{n}{k} + \binom{n}{k-1}$$

3.26   The incidence of a certain disease in the population is estimated to be
        0.5%. That is, the probability of a randomly selected person having the
        disease is .005. A test for this disease is 90% accurate (i.e., it will
        positively identify a person who has the disease with a probability of .9).
        On the other hand, the test produces false positives with a probability of
        .02. If a person tests positive for the disease, what is the probability that
        the individual actually has the disease?

3.27   The Colorado lottery chooses 6 numbers at random from a total of 42
        numbers. How many ways can 6 numbers be chosen from the 42? What
        is the probability of winning the Colorado lottery on any given ticket?

Chapter 4

Probability Distributions

4.1 Introduction

We continue toward the goal of crossing the bridge that will allow us to make inferences about a population based on information obtained from a sample. A major pillar of that bridge is the notion of a probability distribution. In order to introduce the concept of a probability distribution and apply that concept to a variety of naturally occurring phenomena in our world, we need to briefly review the idea of an experiment and introduce a few new terms. The terminology is introduced to more compactly represent a concept or idea, thus allowing us to communicate more effectively in less time or space.

An experiment sometimes has outcomes that are numerical values (e.g., measuring the amount of salt in a bar of soap), and sometimes it has outcomes that are not numerical (e.g., inspecting a product and concluding it is either defective or not defective). Furthermore, sometimes the basic outcomes of an experiment are not directly of interest, but instead some function or combination of the basic outcomes is of interest. Whatever the case, we need a mechanism that will allow us to describe what it is that we are interested in when considering a certain experiment. Consider the experiment of "tossing a pair of dice." Perhaps Dimitri is interested in "the **sum** of the two faces landing up." Maybe Mike is concerned with "the **difference** between the larger and smaller numbers." Suppose, too, that Susan is looking to see "**how many sixes** land face up." Each of these three individuals is interested in a different aspect or function of basic outcomes of the same experiment. The vehicle we use to define exactly what is of importance regarding the results of an experiment is called a **random variable**.

A **random variable (RV)** simply converts any experimental outcome variable to a numerical variable that can be analyzed with statistical techniques.

Definition 4.1 Random Variable

We shall typically use upper case letters from the end of the alphabet to denote random variables. Consider the following random variables (from Dimitri, Mike, and Susan above) defined on the experiment of "tossing a pair of dice."

X = the sum of the two values.
Y = the difference between the larger and smaller values.
Z = the number of sixes that result.

Other examples might include the thickness of a plated material, the number of software discrepancies detected, or the percentage of billing errors per month.

The set of values that a RV can take on is called the **range** of the RV.

Definition 4.2 Range of a Random Variable

For the RV X above, we write R_x = X = {2, 3, 4, . . ., 12} to denote the range or set of values that X can possibly take on. We will typically not use the subscripted notation because context usually makes it clear that the range or set of values is implied. Similarly, Y = {0, 1, 2, 3, 4, 5} and Z = {0, 1, 2}. Other examples include plating thickness values between 110 and 118 micro inches, 0 to 20 software discrepancies detected, or 0 to 25% billing errors per month. The random variables X, Y, Z, and number of software errors detected are discrete random variables, whereas the thickness of a plated material and percent of billing errors per month are continuous random variables.

One can now see why this mechanism is called a random variable. The outcome of the experiment is clearly uncertain or variable and, without any further assumptions being made, it is difficult to associate any measure of certainty (i.e., probability) with any of the possible outcomes. In other words, each observed outcome numerical value is subject to chance.

To get a visual feel of distributions, let us consider the plating process where we are measuring plating thickness (a continuous random variable). It is highly unlikely that any two products are exactly the same when measured out to several decimal places. Let us further assume that our measure is limited to a few decimal places, where now some of the product will look alike and some of it will obviously be different. Figure 4.1a shows how multiple products will vary in plating thickness. The graphic on the left of Figure 4.1b depicts numerous products stacked on top of each other forming a histogram as described in Chapter 2. The graphic on the right of Figure 4.1b has a smooth curve overlayed on the histogram. This smooth curve represents a distribution for a continuous random variable. Figure 4.1c depicts three different ways that distributions can differ from one another.

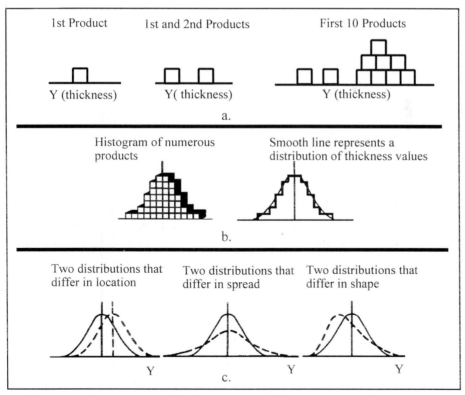

Figure 4.1 Developing a Distribution and Different Types of Distributions

Probability theory is used to determine the proportion of products that would occur within a region of the distribution. For example, see the distribution in Figure 4.2. Using probability theory we can determine the proportion of products whose critical measure Y falls within the specification limits, shown as the shaded region in Figure 4.2. The proportion of products *outside* the spec limits can be multiplied by 10^6 to determine the defects per million (dpm) being produced.

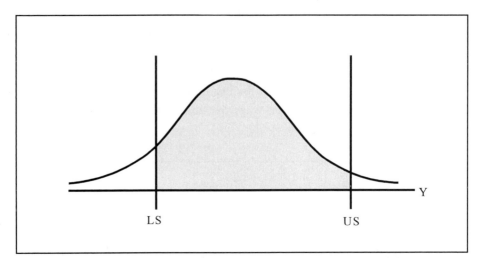

Figure 4.2 Use of Probability To Determine Process Yield

We now tackle the problem of assigning probabilities to each of the possible outcomes of a discrete RV.

A **probability distribution** of a discrete random variable is an assignment of probabilities to each of the possible values that the random variable can take on. Each probability must be a value between 0 and 1, and the sum of all of the probabilities must equal 1.

Definition 4.3 Probability Distribution of a Discrete RV

Typically, the assignment of probabilities to each possible value the RV can take on is based on some assumption(s). If assumptions are not made, then the relative frequency approach to probability would have to be taken and this involves repeated experimentation under identical circumstances. For the experiment of "tossing a pair of dice," we make the assumption that the dice are fair and that each die is equally likely to come up a "1", "2", . . ., "6". To assign probabilities, we must go back to the basic outcomes of the experiment which can be represented by a 6 x 6 grid. We then use the definition of the RV to associate one of the possible outcomes of the RV with each basic outcome of the experiment. This representation for RV X is shown in Table 4.1.

		Basic Outcome for 2nd Die					
		1	2	3	4	5	6
Basic	**1**	2	3	4	5	6	7
Outcome	**2**	3	4	5	6	7	8
for 1st Die	**3**	4	5	6	7	8	9
	4	5	6	7	8	9	10
	5	6	7	8	9	10	11
	6	7	8	9	10	11	12

Table 4.1 Associating a Value of X (sum) with each Basic Outcome

The contents of the table verify that $X = \{2, 3, 4, . . ., 12\}$. Based on our assumption, each of the 36 basic outcomes of the experiment is equally likely with a probability of 1/36. Constructing the probability distribution for X is now straightforward because we need only count the number of occurrences of each possible value for X in the table and multiply by 1/36. Table 4.2 shows the probability distribution for the RV X.

x	2	3	4	5	6	7	8	9	10	11	12
P(X = x)	$\frac{1}{36}$	$\frac{2}{36}$	$\frac{3}{36}$	$\frac{4}{36}$	$\frac{5}{36}$	$\frac{6}{36}$	$\frac{5}{36}$	$\frac{4}{36}$	$\frac{3}{36}$	$\frac{2}{36}$	$\frac{1}{36}$

Table 4.2 Probability Distribution for X (sum of both dice)

While upper case letters denote RVs, we will use lower case letters to denote the possible values that an RV could take on. The notation $P(X = 6) = 5/36$ is read "the probability that X (the sum) equals 6 is 5/36." The interested reader may want to verify that the probability distribution for the RVs Y and Z as defined earlier are as shown in Tables 4.3 and 4.4, respectively. Perhaps constructing a table such as Table 4.1 may help in verifying this.

y	0	1	2	3	4	5
$P(Y = y)$	$\dfrac{6}{36}$	$\dfrac{10}{36}$	$\dfrac{8}{36}$	$\dfrac{6}{36}$	$\dfrac{4}{36}$	$\dfrac{2}{36}$

Table 4.3 Probability Distribution for Y (difference)

z	0	1	2
$P(Z = z)$	$\dfrac{25}{36}$	$\dfrac{10}{36}$	$\dfrac{1}{36}$

Table 4.4 Probability Distribution for Z (# of sixes)

Probability distributions are not always presented in tabular form as we have done for the RVs X, Y, and Z above. Sometimes they are presented in graphical form or in an equation mode. We present the probability distribution for the RV Y in graphical format in Figure 4.3. It is equivalent to the probability distribution in Table 4.3. Notice that Figure 4.3 is simply a histogram for outcome values y_i where the vertical axis is scaled such that the sum of the vertical line values is 1.0.

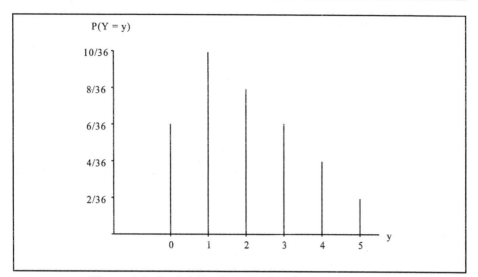

Figure 4.3 Probability Distribution for Y (difference)

Thinking of a probability distribution for a discrete random variable in this way will also facilitate the connection with the continuous distribution displayed in Figure 4.1.

Similarly, we use the format in Equation 4.1 to present the probability distribution of the RV Z. This equation is equivalent to Table 4.4. Computations for determining Equation 4.1 are not shown. The interested reader should fit a quadratic through the points (0, 25), (1, 10), and (2, 1) to verify that Equation 4.1 is correct.

$$P(Z = z) = (1/36) (3z^2 - 18z + 25) \text{ for } z = 0, 1, 2$$

Equation 4.1 Probability Distribution for Z (# of sixes)

A probability distribution for a given discrete random variable is valid if each probability is between 0 and 1, and the sum of all of the probabilities equals 1. This assumes, of course, that the RV is defined such that the values that the RV can take on are mutually exclusive and exhaustive. That is, the RV spans all possible outcomes of the experiment and each basic outcome of the experiment can be associated with only one value of the RV. We conclude this section with some examples which demonstrate how to compute various probabilities from probability distributions.

Example 4.1 Use Table 4.2 to compute the following probabilities:

(a)   What is the probability that the sum of two dice will be greater than 10?   We first translate the written question into mathematical notation, namely, $P(X > 10)$. From the table, we see that

$$P(X > 10) = P(X = 11) + P(X = 12)$$

$$= \frac{2}{36} + \frac{1}{36} = \frac{3}{36} = \frac{1}{12}.$$

(b)   What is the probability that the sum will be no larger than 4? We seek $P(X \le 4) = P(X = 2) + P(X = 3) + P(X = 4)$

$$= \frac{1}{36} + \frac{2}{36} + \frac{3}{36} = \frac{6}{36} = \frac{1}{6}.$$

Note that one must be careful to construct the proper probability statement from the wording of the problem.

∎

Example 4.2 Use Tables 4.3 and 4.4 to compute the following probabilities:

(a)   What is the probability that the difference between the largest and smallest values on two dice will be odd? We seek

$P(\{Y = 1\} \cup \{Y = 3\} \cup \{Y = 5\})$. Since the events $\{Y = 1\}$, $\{Y = 3\}$, and $\{Y = 5\}$ are mutually exclusive, we have

$$P(\{Y = 1\} \cup \{Y = 3\} \cup \{Y = 5\}) = P(Y = 1) + P(Y = 3) + P(Y = 5)$$

$$= \frac{10}{36} + \frac{6}{36} + \frac{2}{36} = \frac{18}{36} = \frac{1}{2}.$$

(b)   What is the probability that the tossing of two dice will result in at least one "6"? We seek $P(Z \geq 1)$.

$$P(Z \geq 1) = P(Z = 1) + P(Z = 2) = \frac{10}{36} + \frac{1}{36} = \frac{11}{36}.$$

(c)   Given that no sixes were obtained in the tossing of two dice, what is the probability that the difference (between larger and smaller) will be greater than 4? We seek the conditional probability $P(Y > 4|Z = 0)$. This may seem to be a more difficult problem, but it is not very difficult if one recognizes that the events $\{Y > 4\}$ and $\{Z = 0\}$ are mutually exclusive. By the definition of conditional probability,

$$P(Y > 4|Z = 0) = \frac{P(\{Y > 4\} \cap \{Z = 0\})}{P(Z = 0)},$$

but the numerator of this expression is 0 because $Y > 4$ and $Z = 0$ cannot occur simultaneously. That is, if no sixes were obtained, it is impossible to get a difference of 5, since the only way to obtain a difference of 5 is with a "6" and "1". Hence,

$$P(Y > 4 \mid Z = 0) = 0.$$

∎

Example 4.3 What is the probability that the difference between the larger and smaller values will exceed the smaller value? To answer this question, we revisit the sample space of the experiment and count the number of basic outcomes for which the difference exceeds the smaller value. The entries in Table 4.5 contain two pieces of information for each basic outcome: the difference is to the left of the slash and the smaller of the two values is to the right of the slash. We count those basic outcomes for which the difference exceeds the smaller value, indicating those basic outcomes by shading them. The answer is $12/36 = 1/3$.

∎

2nd Die						
	1	**2**	**3**	**4**	**5**	**6**
1st die **1**	0/1	1/1	2/1	3/1	4/1	5/1
2	1/1	0/2	1/2	2/2	3/2	4/2
3	2/1	1/2	0/3	1/3	2/3	3/3
4	3/1	2/2	1/3	0/4	1/4	2/4
5	4/1	3/2	2/3	1/4	0/5	1/5
6	5/1	4/2	3/3	2/4	1/5	0/6

Table 4.5   Entries are D/S: D = difference, S = smaller of the values

4.2 Expected Value and Variance

In Chapter 2 we found that two very important descriptive measures of a sample data set were the mean ($\bar{x}$) and the variance (s^2). The mean or average value is a measure of where the center of the data lies, while the variance is a measure of how the data values are dispersed about the mean. These two measures apply equally well to any probability distribution, and we shall be as interested in the mean and variance of a probability distribution as we were in the mean and variance of a data set. However, the perspective of the mean and variance of a probability distribution is from a slightly different angle. While we calculated the mean and variance of a sample of data values from **actual** values, we will have to find the mean and variance of a probability distribution from **possible** values because of the nature of the discrete random variable associated with the probability distribution. Recall from the last section that a probability distribution is an assignment of probabilities to each of the possible values that a discrete random variable can take on. Thus, we will speak of the mean and variance of a random variable as being the same as the mean and variance of the probability distribution associated with that random variable.

Consider the discrete random variable X (the sum of two dice) and its probability distribution, as defined in Table 4.2 of the previous section. If we were going to toss a pair of dice, what would we expect to see as the sum of the two dice? By looking at Table 4.2, one might conclude that we would expect to get "7" most of the time because a "7" is the most likely outcome. One might also base that conclusion partly on the observation that the probabilities of the other possible outcomes are symmetrically distributed about the center value of "7." The method used to conclude that "7" is an expected value is analogous to concluding that the mean of the data set {21, 22, 23, 24, 25} is 23. The mean value of "23" can be obtained without computation because the data set happens to be "nice" (i.e., symmetric about 23). We will show more formally below why the expected value for the discrete random variable X is indeed "7." Not all probability distributions are so "pleasant."

Consider the discrete random variable Y whose probability distribution is given both in Table 4.3 and Figure 4.3 of the previous section. Although "1" is the most likely "difference" to occur when tossing two dice, we hesitate to say that "1" is the expected result, partly due to the lack of symmetry shown. Obviously we need a more formal, precise way of finding the mean or expected value of a random variable or probability distribution. For random variables (RVs) of the kind we have considered thus far, namely RVs that can take on only a countable number of values, we can think of the probability associated with each possible value of the RV as a weighting of that value. Thus, the mean or expected value of an RV is really a weighted average of the possible values the RV can take on. Therefore, the expected value of the RV Y, denoted E(Y), can be computed by multiplying each value by its weight (or probability) and summing. We define the expected value of any discrete RV Y as follows.

$$E(Y) \; = \; \mu_Y \; = \; \sum_i y_i P(y_i)$$

Definition 4.4 Expected Value of a Discrete RV

The term "discrete" is used here to describe RVs having only a "countable" number of possible values. We will address the other (i.e., continuous) kind of RV later. $P(y_i)$ is a shorter notation for its equivalent, $P(Y = y_i)$. The summation is assumed to be over all possible values y that the RV Y can take on. The notation μ_Y (or just μ, if the RV is contextually clear) is equivalent to E(Y) but has the additional connotation that it is a "population" mean. Thus, the mean of an RV or probability distribution is considered to be the mean of a population. Close examination of Definition 4.4 should indicate that we are simply taking each discrete value y_i and multiplying it times the proportion of y_i values that can occur. This approach is identical to finding any population mean.

Using Definition 4.4, the expected value of Y (the difference between two dice) can be computed via Table 4.3 or Figure 4.3 as

$$E(Y) = (0)(6/36) + (1)(10/36) + (2)(8/36) + (3)(6/36) + (4)(4/36) + (5)(2/36)$$

$$= \frac{0 + 10 + 16 + 18 + 16 + 10}{36} = \frac{70}{36} = \frac{35}{18} = 1.9\overline{444}.$$

This RV demonstrates that the expected value of an RV, like the mean of a sample data set, need not necessarily be one of the values that the RV can possibly assume. The physical interpretation of E(Y) is that of a "balance point" or "center of gravity." $E(Y) = 1.9\overline{444}$ is the "balance point" on the y-axis in Figure 4.3 Again, $\mu = 1.9\overline{444}$ is considered to be the mean of a conceptually infinite population consisting of the values 0, 1, . . ., 5 whose frequencies of occurrence in the population are given by the ratios (or probabilities or weights) shown in Table 4.3. Using Tables 4.2 and 4.4 along with Definition 4.4, the reader should now verify that $E(X) = 7$ and $E(Z) = 12/36 = 1/3$.

Just as we used s^2, the sample variance, to measure the variability in a sample, we will measure the variability of an RV called X by what we call the variance of X, which is defined as follows.

$$V(X) = \sigma_x^2 = \sum_i (x_i - \mu_x)^2 P(x_i)$$

Definition 4.5 Variance of a Discrete RV

V(X) is the sum of the "weighted" squared deviations of each possible value of the RV about its expected value. As in the definition of E(X), $P(x_i)$ is the "weight" or probability associated with the value x_i, and the summation is assumed to be over all possible values that X can take on. μ_x and σ_x^2 denote population values. σ_x is used to denote the standard deviation of the RV X. Close examination of Definition 4.5 should reveal that we are taking each possible discrete outcome x_i, computing its squared deviation from the population mean, multiplying its squared deviation times the proportion of time each x_i will occur, and then summing.

Again, consider the RV X, defined as the "sum" of two dice, whose probability distribution is given in Table 4.2. We saw previously that $E(X) = \mu_x = 7$. We now use Definition 4.5 to find V(X).

$$
\begin{aligned}
V(X) \ &= (2-7)^2 \, (1/36) + (3-7)^2 \, (2/36) + (4-7)^2 \, (3/36) + (5-7)^2 \, (4/36) \\
&+ (6-7)^2 \, (5/36) + (7-7)^2 \, (6/36) + (8-7)^2 \, (5/36) + (9-7)^2 \, (4/36) \\
&+ (10-7)^2 \, (3/36) + (11-7)^2 \, (2/36) + (12-7)^2 \, (1/36) \\
&= (1/36)[(25)(1) + (16)(2) + (9)(3) + (4)(4) + (1)(5)][2] \\
&= (1/18) \, [25 + 32 + 27 + 16 + 5] = (1/18) \, [105] = 105/18 \\
&= 5.8333.
\end{aligned}
$$

Thus, $\sigma_x^2 = 5.8333$ and $\sigma_x = 2.415$.

Recall that there is a computational form for s^2 that is more amenable to computing sample variance. Similarly, the following equivalent form of Definition 4.5 is usually more convenient when computing V(X).

$$
V(X) = \sigma_x^2 = \left[\sum_i x_i^2 \, P(x_i) \right] - \mu_x^2 = E(X^2) - [E(X)]^2
$$

Definition 4.6 Variance of X (computational form)

The values μ_x and σ_x are the mean and standard deviation of the RV X, respectively. The form of $V(X) = E(X^2) - [E(X)]^2$ is especially useful if $\mu_x = E(X)$ has already been computed. This form is also applicable to both discrete and continuous RVs.

If we use Definition 4.6 to find V(X), we have

$$V(X) = [(2)^2(1/36) + (3)^2(2/36) + \ldots + (12)^2(1/36)] - (7)^2$$

$$= (1/36)[(4)(1) + (9)(2) + \ldots + (144)(1)] - 49$$

$$= (1/36) [1974] - 49$$

$$= 54.8333 - 49$$

$$= 5.8333,$$

which is the same result obtained previously using Definition 4.5.   Table 4.6 summarizes the expected values, variances, and standard deviations of the three RVs X, Y, and Z which were defined on the experiment of "tossing the dice."

	μ	σ^2	σ
X (sum)	7.000	5.833	2.415
Y (difference)	1.944	2.052	1.433
Z (# of Sixes)	.333	.278	.527

Table 4.6 Summary of Expected Values and Variances for X, Y, Z

The reader should be able to reproduce these values using the definitions of this section and the probability distributions of Section 4.1.

Even though the concept of variance of an RV may not be quite as intuitive as the expected value of an RV, its importance is certainly the equal to that of expected value. Variance, too, can be interpreted in the same physical scenario as expected value. If we have a set of five masses, m_1, m_2, . . . , m_5 positioned along a weightless axis as shown in Figure 4.4, we can think of these masses as a probability distribution with, as pointed out earlier, μ representing the center of gravity or balance point. If we rotate this axis about the point at μ, then the quantity $\sum r_i^2 m_i$, where r_i is the distance that mass m_i is from μ along the axis, is what physicists call **moment of inertia**. Variance, however, is defined precisely in the same manner. Hence, moment of inertia is one physical interpretation of variance.

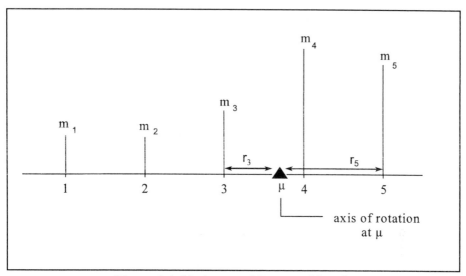

Figure 4.4 Variance as Moment of Inertia

4.3 Discrete Uniform Distribution

We now present several of the more important probability distributions for discrete random variables. These statistical distributions are deemed "important" because they occur so frequently in the real world that it is impossible to do anything statistically without encountering or requiring one of these distributions. The first of these distributions is the discrete uniform distribution.

Perhaps the simplest of all discrete distributions is the distribution associated with a random variable whose outcomes are all equally likely to occur. Clearly, the tossing of a fair six-sided die is an example of a process whose possible outcomes, $\{1,2,3,4,5,6\}$, are each equally likely, namely $1/6$. The roulette wheel is a disk containing 38 slots, each of which has an equally likely chance of capturing the bouncing ball. The chemist who is about to extract a vial from a high speed 20-slot centrifuge that has just come to a stop is equally likely to extract a vial from any of the 20 slots. Each of these is an example of a process which generates a discrete uniform distribution. They are discrete because there are a countable number of outcomes in each scenario: 6, 38, and 20, respectively. They are uniform because each outcome is equally likely.

The discrete uniform distribution has two parameters, a and n. Parameter "a" denotes the smallest outcome (e.g., $a = 1$ in the die example) and "n" denotes the total number of distinct outcomes ($n = 6$ in the die example). A summary of the discrete uniform random variable is given in Table 4.7.

DISCRETE UNIFORM RANDOM VARIABLE X

X = a specific integer from a finite list of consecutive integers

$X = \{a, a+1, a+2, ..., a+(n-1)\}$

Parameters: a = integer representing the smallest number in the list

$\qquad\qquad\quad$ n = integer representing the total number of items on the list

Probability Mass Function (pmf): $P(X = x) = \dfrac{1}{n}$

$E(X) = a + \dfrac{n-1}{2}$

$V(X) = \dfrac{n^2 - 1}{12}$

Shape: symmetric

Table 4.7 Summary of Discrete Uniform RV

The discrete uniform RV is particularly useful in sampling, as the next example illustrates.

Example 4.4 Suppose that we have a roster of 80 names from which we would like to select a random sample of 8 names. Our intent is to send surveys to each of these eight randomly selected individuals. We will use Appendix A of this text to select a random sample of eight individuals. Appendix A is a table of random digits. That is, each position in the table is equally likely to be a 0, 1, 2, ..., 9. Each position represents a discrete uniform RV, where $a = 0$ and $n = 10$.

However, since the roster or frame, as it is sometimes called, contains 80 numbers (1, 2, ..., 80) and 80 is a 2-digit number, we will need to look at 2-digits simultaneously when making our selections from the table. Each 2-digit quantity is thus equally likely to be a number from 00, 01, ..., 99. First, we arbitrarily select a position, or a particular row and column, in the table. Suppose we have blindly pointed to row 25 and column 11. Since we need 2-digits at a time, we elect to use adjacent columns 11 and 12, starting in row 25. We see a "02" in those positions. Starting there and reading downward (this is also arbitrary because every position is equally likely to be a 0, 1, ..., 9), we encounter the sequence 02, 07, 84, 97, 09, 63, 36, 94, 41, 34, 95, 00, 47, 68, ... From this sequence, we select the first 8 numbers that fall between 01 and 80. These would be 02, 07, 09, 63, 36, 41, 34, and 47. The names on the roster associated with these 8 numbers represent our random sample of 8.

∎

The applications of the discrete uniform distribution are boundless. A common problem involving this distribution is the estimation of the parameter n. One of the first recorded attempts at estimating the parameter n comes from the American and British armies who during WWII wanted to quantify the German army war fighting capability in terms of the total number of tanks in the German inventory. Whenever the Allies would capture/destroy a German tank, they would record the serial number of that tank. Equipped with a list of serial numbers which were assumed to have come from a uniform distribution, the Allies were able to estimate the total number of tanks in the Germany inventory. Rumor has it that the Soviets did a similar analysis on U.S. U-2 spy plane tail numbers to estimate the size of the U.S. U-2 inventory. These applications just cited for estimating n are analogous to watching the Lotto results on television and using the 6 numbers drawn to estimate the total number of balls in the barrel from which the 6 numbers were drawn. Those readers interested in developing estimates (i.e., the formulas) for estimating the parameter n are referred to problem 4.52 at the end of this chapter.

4.4 Binomial Distribution

Suppose we must address the problem of determining the probability that a batch of 1000 integrated circuits (chips) contains no more than one defective chip. Suppose, too, that we know from previous purchases from this manufacturer that one out of every 100 chips has turned out to be defective. This is typical of many questions that must be addressed in industrial applications, and one for which the binomial distribution is of considerable use in solving. There are many experiments which satisfy either exactly or approximately the following conditions:

1. The experiment consists of **a sequence of n Bernoulli trials**. A Bernoulli trial is itself an experiment in which there are only two possible outcomes, which we call success or failure.

2. The n **trials are identical**, i.e., the same trial is performed under identical conditions.

3. The **trials are independent**. That is, the outcome of any one trial neither influences nor is influenced by the outcome of any other trial.

4. The **probability of success on every trial is the same**; we call that probability p.

If a process satisfies these four conditions, we call this process a **binomial experiment**, and the binomial distribution is applicable in this situation. The use of the labels "success" and "failure" for the two possible outcomes of a Bernoulli trial is for classification purposes only and many times they may in fact connote the opposite meaning. For example, the "success/failure" tandem could be associated with "defective/nondefective," where it is obvious that "defective" is not being "successful."

Example 4.5 Consider the "process" of a student showing up for an exam completely unprepared and having absolutely no idea what the material is about. The exam, however, consists of 10 multiple choice questions each of which has four choices. Supposing the student guesses at each question, this process can be considered a binomial experiment. In this case, n = 10 and p = .25, because each question can be considered a Bernoulli trial with the same probability of success (i.e., answering the question correctly). What is the probability of the student passing the exam, if he needs to answer 6 questions correctly to pass? We can use the binomial distribution to answer this question and will do so below.

∎

Example 4.6 A more realistic scenario in which a binomial experiment could be assumed is the batch of 1000 chips being produced, as mentioned at the beginning of this section. Considering the batch of 1000 chips as being a sequence of n = 1000 Bernoulli trials in which each chip has a probability p = .01 of being defective, we will be able to use the binomial distribution to answer the question about the probability of no more than one chip out of the 1000 being defective.

∎

The two previous examples illustrate binomial experiments, but since the numbers involved are not really ideal, we will use the following example to demonstrate the details of the binomial distribution. We will then return to answer the questions posed in these two examples.

Example 4.7 Suppose that boys and girls are equally likely to be born. What is the probability that in a randomly selected family of 5 children, 3 will be boys? Assuming that a binomial experiment applies in this case, we see that n = 5 and p = .5. That is, each child is considered a Bernoulli trial with p = .5 based on the equal likelihood of a boy or girl. The first thing that must be done to solve this problem is to define the random variable. Since the probability we are asked to find involves a number of boys, we define the RV X as

X = the number of boys out of a family of 5 children.

Obviously X = {0, 1, 2, 3, 4, 5}, i.e., X can take on only one of these six values. We next establish the values of what are called the parameters of the binomial distribution, namely n and p:

$$n = 5 \qquad p = .5$$

We now write in mathematical notation the probability to be found, which is

$$P(X = 3).$$

If we had the probability distribution for X, we could solve this problem. The probability distribution for the binomial RV X having parameters n and p can be obtained from what is called the binomial probability mass function (pmf):

$$\mathbf{P(X=x)} = \binom{\mathbf{n}}{\mathbf{x}} \mathbf{p^{x}(1-p)^{n-x}} \quad \textbf{for x} = \mathbf{0, 1, ..., n}$$

Definition 4.7 Binomial Probability Mass Function (pmf)

The binomial pmf gives a rule for finding the probability associated with each possible value of the binomial RV. To find $P(X = 3) = P(3)$, we use Definition 4.7 with $x=3$, $n=5$, and $p=.5$ to get

$$P(X = 3) = P(3) = \binom{5}{3} (.5)^3 (.5)^2 = \frac{5!}{3!\,2!}(.5)^5 = 10(.5)^5 = .3125.$$

Although we have now answered the question posed at the beginning of the example, we can use Definition 4.7 to completely construct the probability distribution for X, i.e., assign probabilities to the remaining values that X can assume. Using the binomial pmf, we compute

$$P(X=0) = P(0) = \binom{5}{0} (.5)^0(.5)^5 = (1)(.5)^5 = .03125$$

$$P(X=1) = P(1) = \binom{5}{1} (.5)^1(.5)^4 = (5)(.5)^5 = .15625$$

$$P(X=2) = P(2) = \binom{5}{2} (.5)^2(.5)^3 = (10)(.5)^5 = .3125$$

$$P(X=4) = P(4) = \binom{5}{4} \ (.5)^4(.5)^1 = (5)(.5)^5 = .15625$$

$$P(X=5) = P(5) = \binom{5}{5} \ (.5)^5(.5)^0 = (1)(.5)^5 = .03125$$

The probability distribution for X now looks like

x	0	1	2	3	4	5
P(x)	.03125	.15625	.3125	.3125	.15625	.03125

This is a valid probability distribution since each probability is between 0 and 1, and the sum of all of the probabilities is 1.

∎

We now have the tools needed to answer the questions posed in Examples 4.5 and 4.6. From Example 4.5 we have the question, "What is the probability of the student passing the exam if he needs to answer at least 6 of the 10 multiple choice questions correctly?" Letting

Y = the number of questions answered correctly out of 10,

we seek $P(Y \geq 6)$.

$$P(Y \geq 6) = P(6) + P(7) + P(8) + P(9) + P(10),$$

i.e., the student will pass if he gets either 6, 7, 8, 9, or 10 questions correct. We now use the binomial pmf in Definition 4.7 to compute

$$P(Y \geq 6) = \sum_{k=6}^{10} \binom{10}{k} (.25)^k (.75)^{10-k}$$

$$= \binom{10}{6} (.25)^6 (.75)^4 + \binom{10}{7} (.25)^7 (.75)^3$$

$$+ \binom{10}{8} (.25)^8 (.75)^2 + \binom{10}{9} (.25)^9 (.75)^1$$

$$+ \binom{10}{10} (.25)^{10} (.75)^0$$

$$= .0162 + .0031 + .0004 + .0000 + .0000$$

$$= .0197,$$

indicating the chances of the guessing student passing the exam are quite slim. At this point, you are probably wondering if you must get through all of these computational hurdles, even involving combinatorics, to solve a problem like this.

Fully aware that life is not always easy in the problem-solving business, we do acknowledge that computers are better suited for these computations than humans. Thus, we have included computer-generated tables in Appendix B that give the binomial probability distributions for a wide variety of values for n and p. Fortunately, the binomial distribution for $n = 10$ and $p = .25$ is given in Appendix B, and the reader should verify at this time that the five probabilities shown above did come from these tables. Note that the tables give probabilities rounded to 4 decimal places. Thus, when one sees .0000, as we did twice above, the probability is admittedly small but not necessarily 0. For example, if you did happen to calculate $P(Y = 10) = (.25)^{10} = .00000095367$, you would have obtained a more precise answer than the table shows. For most practical purposes, 4 decimal place accuracy is more than sufficient. Unfortunately, as the next problem illustrates, not always are the desired values for n and p in the tables. Thus, the student should be proficient at using the binomial pmf.

From Example 4.6, if we let Z = the number of defective chips out of 1000, we seek $P(Z \leq 1)$. The parameters are n = 1000 and p = .01.

$$P(Z \leq 1) \;=\; P(0) + P(1)$$

$$= \binom{1000}{0} (.01)^0(.99)^{1000} + \binom{1000}{1} (.01)^1(.99)^{999}$$

$$= .00004317125 + .0004360732$$

$$= .000479$$

$$\approx .0005,$$

which is a very small probability. We conclude from this result that, based on the company's past performance (from which we have p = .01), it is extremely unlikely that there would be at most one defective chip in the batch of 1000. The complementary probability, $P(Z > 1) \approx .9995$, is, on the other hand, extremely high.

We have shown how one can use the binomial pmf to construct the binomial distribution for a given n and p. We are also interested in the expected value and variance of the binomial distribution. If we have the complete probability distribution, we could find μ and σ^2 for the binomial RV by using Definitions 4.4 and 4.5 (or 4.6), respectively. However, when n is large, just constructing the probability distribution is tedious, let alone computing μ and σ^2 for that distribution. Therefore, it would be very helpful if we could find μ and σ^2 for a binomial RV without having to build the entire probability distribution. Thanks to mathematics, this is possible. Using some algebra and the definitions, mathematical manipulation can show that the expected value and variance of X are given by

---

$$E(X) = \mu = np$$

$$V(X) = \sigma^2 = npq, \text{ where } q = (1 - p)$$

Definition 4.8 Expected Value and Variance of a Binomial RV

The expected value of np is quite intuitive. For example, if n = 1000 chips, and p = .01 is the defect rate, we would expect np = (1000)(.01) = 10 chips to be defective. The use of q = 1-p is used throughout the remainder of the text to denote the complementary probability of p.

The binomial distribution has extremely wide applicability and has been used extensively in quality improvement work, especially for attribute data (defective vs nondefective). It is the underlying distribution in statistical process control when constructing p and np control charts. We conclude this section with a summary of some of the more important properties of the binomial random variable that have been discussed in this section.

BINOMIAL RANDOM VARIABLE X

X = number of successes out of n trials

X = {0, 1, ..., n}

Parameters: n = number of trials

p = probability of success on any trial (q = 1 - p)

These two parameters uniquely determine the particular binomial distribution out of an entire family of binomial distributions.

Probability Mass Function (pmf):

$$P(X = x) = \binom{n}{x} p^x q^{n-x} \quad \text{for} \quad x = 0, 1, \ldots, n$$

$$E(X) = \mu = np$$

$$V(X) = \sigma^2 = npq$$

Shape: if p < .5, then skewed right

if p > .5, then skewed left

if p = .5, then symmetric

Table 4.8 Summary of Binomial RV

4.5 Hypergeometric Distribution

A distribution closely related to the binomial distribution that has wide applicability in statistical acceptance sampling is the hypergeometric distribution. To illustrate the applicability of the hypergeometric distribution to statistical sampling, we revisit Example 4.6 (the manufacturing of 1000 chips) and consider the problem from a slightly different perspective.

Example 4.8 Suppose a manufacturer receives a shipment of 1000 chips from a supplier. The first decision facing the manufacturer is whether to accept the shipment in the sense that the chips meet her specifications. She could, of course, inspect every chip, but this would be virtually impossible due to the cost and time involved. Instead, she decides to take a random sample from the shipment, inspect each chip in the sample, and then accept the shipment only if the number of defective chips discovered is sufficiently small. The manufacturer creates (perhaps arbitrarily) her game plan as follows: select a random sample of size $n=10$ chips, test each of these chips, and accept the shipment only if she finds no defects. What is the probability that she will accept the shipment if the supplier is known to have a 10% defect rate? If we let

X = the number of defective chips found in the sample of size 10,

then $P(X=0)$ represents the probability that she will accept the shipment. To find $P(X=0)$ we first establish the notation for what we know.

N = 1000 (N represents the size of the population sampled),

D = 100 (D is the number of defects we can expect in the population since the supplier is assumed to have a 10% defect rate), and

n = 10 (n represents the size of the sample selected).

Clearly, $\binom{N}{n} = \binom{1000}{10}$ represents the total number of ways that a sample of

size n = 10 can be chosen from 1000 items. Now, assuming that there are 900 nondefective chips and 100 defective chips in the population, $\binom{D}{0} = \binom{100}{0}$ represents the number of ways we can choose 0 defectives from

a total of 100 defectives, and $\binom{N-D}{n-0} = \binom{900}{n} = \binom{900}{10}$ represents the

number of ways we can choose 10 good chips from the 900 good chips. Thus, from Rule 1 in Section 3.6, $\binom{100}{0}\binom{900}{10}$ represents the total number of

possible samples of size 10 that have no defects. It follows, then, that

$$P(X=0) = \frac{\binom{100}{0}\binom{900}{10}}{\binom{1000}{10}}$$

since $\binom{1000}{10}$ represents the total number of possible samples of size 10 (of any

makeup). Therefore,

$$P(X=0) = \frac{(1)(9.1384 \times 10^{22})}{2.6341 \times 10^{23}} = .347.$$

The computations that produced the large numbers in the numerator and denominator above were performed on a computer and are not shown here. This probability indicates that the manufacturer has a relatively low chance (about 1 in 3) of accepting the shipment under her current approach. If she altered her criteria for acceptance to "no more than 1 defect," then $P(X \le 1)$ represents the probability of acceptance. $P(X \leq 1) = P(X = 0) + P(X = 1)$. We already know $P(X = 0) = .347$ and we can find $P(X = 1)$ by

$$P(X = 1) = \frac{\binom{100}{1}\binom{900}{9}}{\binom{1000}{10}}.$$

Again using a computer (computations not shown here), we find

$$P(X = 1) = \frac{(100)(1.0256 \times 10^{21})}{2.6341 \times 10^{23}} = .389, \text{ and thus}$$

$$P(X \leq 1) = .347 + .389 = .736.$$

The manufacturer, under the new criteria, has almost a 3 out of 4 chance of accepting the shipment. The hypergeometric distribution can also be used to generate what is called an **operating characteristic curve** as shown in Figure 4.5. This figure shows the acceptance probability as a function of the assumed defective rate.

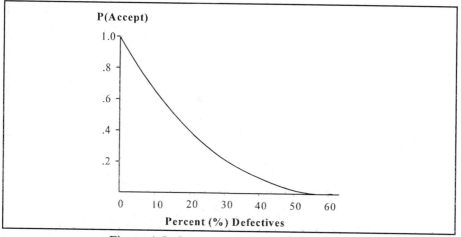

Figure 4.5 Operating Characteristic Curve

The acceptance criteria used was "no more than 1 defect" in the sample of 10. The curve shows the sensitivity of the sampling plan to changes in product quality. This graph shows that even at a 40% defect rate (which most consider intolerable), the manufacturer still would accept the shipment nearly 10% of the time. Ultimately, we would like to make this curve steeper. The ways to do this will be investigated later.

∎

The hypergeometric distribution has three parameters which we shall call N (the size of the population sampled), n (the size of the sample), and D (the size of the subpopulation of interest). The use of "D" for the last parameter is chosen because "number of defects" is the most commonly occurring scenario. The probability mass function for this distribution, where the RV X denotes the number of elements falling into the "subpopulation of interest," is given by

$$P(X=x) = \frac{\binom{D}{x}\binom{N-D}{n-x}}{\binom{N}{n}}$$

$$\text{for } x = 0, 1, ..., \min(n, D); \quad n \le N - D$$

Definition 4.9 PMF for Hypergeometric Random Variable

The relationship between the binomial and hypergeometric distributions can be described as one in which the binomial distribution results from "sampling *with* replacement" whereas the hypergeometric distribution is associated with "sampling *without* replacement." Another way to describe this relationship is that the hypergeometric distribution does *not* have the **independence** between Bernoulli trials that is required in a binomial experiment. We illustrate these points in the following example.

Example 4.9 Consider a regular deck of cards consisting of 52 cards (i.e., 4 suits: hearts, diamonds, clubs, spades; and 13 cards in each suit: A, K, Q, J, 10, 9, . . . , 2). Suppose we consider a sequence of Bernoulli trials, where each trial consists of drawing one card from the deck. We are concerned with "getting a heart" (success) or "not getting a heart" (failure). On the first trial, P(Heart) = 13/52 = 1/4. On the second trial, what is the probability of getting a heart? Well, that **depends**. "Depends" is the key word here because if we put the card drawn on the first trial back into the deck, P(Heart) = 1/4 again on the second draw. However, if we did not put the first card drawn back into the deck prior to the second draw, we really don't know what P(Heart) is for the second draw. P(Heart) = 12/51 or P(Heart) = 13/51, **depending** on whether or not a heart was selected on the first draw, respectively. Replacing the card after each draw and reshuffling results in a sequence of **independent** Bernoulli trials for which P(Heart) = 1/4 on each trial or draw. This satisfies a binomial experiment and the binomial RV applies. If we don't replace the card after each draw, P(Heart) for subsequent draws is **dependent** upon what happened previously, and the hypergeometric distribution is applicable for modeling this scenario.

∎

We conclude this section by noting that the hypergeometric distribution has a very wide range of applicability. For example, estimating the size of a wildlife population using the capture/recapture method is based on the hypergeometric distribution [Lars 85]. Another area where this distribution is often used is in discrimination cases [Degr 86]. The following simple example shows how the hypergeometric distribution can be used to calculate probabilities that may indicate the presence or lack of discrimination.

Example 4.10 A certain college faculty consists of 20 members, 4 of whom are women. Two tenure positions are open and will be filled from the 20-member faculty. If all 20 faculty members are equally qualified and the selection is done randomly, what is the probability that both positions will be filled by men? We can use the hypergeometric distribution to find the probability. Let the RV X = the number of men selected for tenure. Obviously X = {0, 1, 2}. We seek $P(X=2)$. Using the pmf for the hypergeometric distribution (Definition 4.9) with the following parameters:

   N = 20 (population size),

   n = 2 (sample size–the number of positions open), and

   D = 16 (size of subpopulation of interest),

we get $P(X=2) = \dfrac{\dbinom{16}{2}\dbinom{4}{0}}{\dbinom{20}{2}} = \dfrac{(120)(1)}{190} = .632.$

Note that the subpopulation of interest in this case is the set of all men because the probability question being asked involved a number of men.

■

We summarize the hypergeometric random variable in Table 4.9. Note that E(X) in Table 4.9 is the same as the mean for the binomial RV. However, the variance of the hypergeometric distribution differs from the variance of the binomial RV (npq) by the factor (N - n)/(N - 1). Since this factor is less than 1, the variance of the hypergeometric RV is smaller than the variance of the binomial RV.

HYPERGEOMETRIC RANDOM VARIABLE X

X = number of elements selected from some subpopulation of interest

X = {0, 1, ..., min(n, D)}

Parameters: N = size of the population
n = size of sample selected
D = size of subpopulation of interest

PMF: $P(X=x) = \dfrac{\dbinom{D}{x}\dbinom{N-D}{n-x}}{\dbinom{N}{n}}$

for x = 0, 1, ..., min(n, D) and where n ≤ N - D

E(X) = np, where p = D / N

$V(X) = \dfrac{(N-n)}{(N-1)}\,npq,$ where q = 1 - p

Shape: if p < .5, then skewed right

if p > .5, then skewed left

if p = .5, then symmetric

Table 4.9 Summary of Hypergeometric RV

4.6 Poisson Distribution

Another discrete probability distribution that can be applied to many randomly occurring phenomena over time or space is the Poisson distribution or Poisson RV. Unlike the binomial or hypergeometric distributions, which relate to a special kind of experimental scenario, the Poisson RV is not based on a particular experimental setting. However, its relationship to the binomial will be pointed out.

Consider the following scenarios:

(1)   The number of speeding tickets issued in a certain county per week.

(2)   The number of disk drive failures per month for a particular kind of disk drive.

(3)   The number of calls arriving at an emergency dispatch station per hour.

(4)   The number of flaws per square yard in a certain type of fabric.

(5)   The number of typos per page in a technical book.

Each of these situations can typically be modeled by a Poisson RV. The Poisson RV is characterized by the form "the number of occurrences per unit interval," where an occurrence could be a flaw or defect, a mechanical or electrical failure, an arrival, or a departure, etc. The unit interval could be a unit of time, like seconds, hours, weeks, etc., or a unit of space or area, like m^2, yd^2, etc. The Poisson RV or distribution has one parameter, λ, which is a positive constant denoting "the **average** number of occurrences per unit interval." The pmf for the Poisson RV is given in the following definition.

$$P(X=x) \; = \; \frac{\lambda^x e^{-\lambda}}{x!} \; \text{ for } x = 0, 1, 2, ...; \quad \lambda > 0$$

Definition 4.10 PMF for Poisson Random Variable

The letter "e" represents the base of the natural logarithm and it is approximately equal to 2.71828. Note that this probability mass function is defined for **all** nonnegative integer values of x, which is not the case for the binomial or hypergeometric RVs.

Example 4.11 Suppose X = the number of suicides per month in a certain city, and it is known (from historical data) that the **average** number of suicides per month is 5. We can use the Poisson RV with λ = 5 to model this scenario. What is the probability that there will be less than 3 suicides in the next month? Seeking P(X < 3) = P(X $\leq$ 2), we use the Poisson pmf (Definition 4.10) to find P(X = 0), P(X = 1), and P(X = 2) and then add these probabilities as follows:

$$P(X \leq 2) = P(0) + P(1) + P(2) = \frac{(5)^0 \, e^{-5}}{0!} + \frac{(5)^1 \, e^{-5}}{1!} + \frac{(5)^2 \, e^{-5}}{2!}$$

$$= e^{-5} [1 + 5 + 12.5] = e^{-5} [18.5] = .125,$$

or about only 1 in 8 chances of having fewer than 3 suicides next month. ∎

What is it about this example and the five scenarios mentioned above that allows us to use the Poisson probability model? To illustrate the commonality, we refer to scenario (3), the number of emergency calls arriving at a dispatch station per hour. Let us divide a typical hour into n equal, but very small time intervals (like n = 3600 seconds). We want n to be so large that it is not likely that more than one call will arrive during any mini-interval. If λ is the average number of calls per hour, then it should seem reasonable to assign a probability of λ/n to the event that there is one call per mini-interval. Since each mini-interval is so small, we assume that the probability of two or more calls in any mini-interval is 0. Thus, there are only two possible outcomes for each mini-interval: either one call (with probability λ/n) or no calls (with probability $1 - \lambda/n$). If we can also assume that the outcomes (either 1 call or no calls) in each mini-interval are independent with probabilities λ/n and $1 - \lambda/n$, we again have a binomial experiment. Thus, the probability that x calls occur in n trials (i.e., 1 hour) can be given by the binomial pmf as

$$P(X = x) = \binom{n}{x} \left(\frac{\lambda}{n}\right)^{x} \left(1 - \frac{\lambda}{n}\right)^{n-x}.$$

But as n gets bigger and bigger (i.e., $n \to \infty$), it can be shown that the right side of the equation is actually equal to $\lambda^{x}e^{-\lambda}/x!$, which is the Poisson pmf. What we have just shown is that the Poisson distribution is the limiting form of a binomial distribution. In fact, the Poisson distribution can be used to approximate the binomial under the following conditions: n is big ($n \geq 100$), p is small ($p \leq .01$), and $np = \lambda \leq 20$.

The reason that the Poisson RV can be applied in so many instances is that many real world phenomena lend themselves nicely to the process described in the previous paragraph. That is, the interval can be divided into many subintervals in which either 1 or no occurrences are the only alternatives, the probabilities are the same for each subinterval as well as proportional to the size of the subinterval, and the occurrences in each subinterval are independent of one another. The suicide example in Example 4.11 satisfies these requirements. Certainly the month can be broken into many small equal-sized subintervals in which either 1 or 0 suicides occur, with equal probabilities for each subinterval and independence between the subintervals. The Poisson distribution is sometimes referred to as the "disaster distribution" because so many of the naturally occurring disasters (e.g., floods, tornadoes, earthquakes) can be modelled by the Poisson RV.

A very important aspect in using the Poisson RV to solve probability problems is to make sure that the time interval given in the definition of the RV matches exactly the time interval for λ. Sometimes it is necessary to adjust λ prior to solving. We illustrate this in the next example.

Example 4.12 Continuing with the scenario of Example 4.11, if there is an average of 5 suicides per month, what is the probability that on any randomly selected day there will be at least 1 suicide? As always, to solve this problem we first must define the RV. It is recommended that one define the RV in terms of the interval specified in the question being asked. In this case, the question is about the number of suicides per **day**. Thus, we define the RV Y as

$$Y = \text{the number of suicides per } \textbf{day}.$$

Now we must convert the average of 5 suicides per **month** to an equivalent number per **day**. Assuming 30 days per month, we can expect on the average 5/30 or 1/6 suicides per day. Thus, we have $\lambda = 1/6$ suicides per **day**. Be aware that the assumption of proportionality allowed this conversion to take place. Now that the interval (**day**) for λ matches the interval (**day**) in the definition of the RV, we can use the Poisson pmf to find $P(Y \geq 1)$. At first glance, it might appear as if we must find $P(Y \geq 1) = P(1) + P(2) + P(3) + \ldots$, meaning a good number of applications of the Poisson pmf. However, noting (from complementary events) that $P(Y \geq 1) = 1 - P(Y < 1) = 1 - P(Y = 0)$, we have a simple problem in that we need only find $P(Y = 0)$ and subtract this result from 1 to get the desired result.

$$P(Y \geq 1) = 1 - P(Y = 0) = 1 - \frac{\left(\frac{1}{6}\right)^0 e^{-\frac{1}{6}}}{0!}$$

$$= 1 - e^{-\frac{1}{6}} = 1 - .846 = .154.$$

We conclude this section with an example detailing a historical application of the Poisson RV from which the Poisson RV received its name. A summary of the Poisson distribution follows the example.

Example 4.13 During the 19th century, the Prussian army enlisted the help of Simeon Poisson (1781-1840), a noted mathematician and physicist, to study the number of deaths due to horsekicks. A 20-year study ensued in which the number of deaths in 10 different cavalry regiments was recorded. Twenty years for each of 10 different units yielded 200 unit-years. The frequency distribution for this data is shown in Table 4.10 [Bort 98].

x = number of deaths	number of unit-years in which x deaths occurred
0	109
1	65
2	22
3	3
4	<u>1</u>
	200

Table 4.10 Frequency Distribution of Deaths due to Horsekicks

First, we let the RV X = number of deaths per unit-year. Next, we must estimate λ, the average number of deaths per unit-year. We use the frequency distribution to get

$$\lambda = \frac{1}{200} [0 \cdot 109 + 1 \cdot 65 + 2 \cdot 22 + 3 \cdot 3 + 4 \cdot 1] = \frac{122}{200} = .61.$$

Note that this is just a "weighted average" (like finding the mean of a probability distribution). Using $\lambda = .61$ and the Poisson pmf, we have

$$P(X = x) = \frac{(.61)^x e^{-.61}}{x!} \quad \text{for } x = 0, 1, \dots ,$$

which yields

$$
\begin{aligned}
P(X = 0) &= .5434 \\
P(X = 1) &= .3314 \\
P(X = 2) &= .1011 \\
P(X = 3) &= .0206 \\
P(X = 4) &= .0031 \\
P(X \geq 5) &= .0004
\end{aligned}
$$

By multiplying each of these probabilities by 200, we are able to obtain the **expected** number of unit-years in which there were "x" deaths. The following table lists this **expected** number of unit-years, as well as the **observed** or **actual** number of unit-years in which there were "x" deaths.

x: number of deaths	Expected # of unit-years	Observed # of unit-years
0	108.7	109
1	66.3	65
2	20.2	22
3	4.1	3
4	.6	1
≥ 5	.1	0

Note that the values in the "expected" column are extremely close to the values in the "observed" or "actual" column. Since the Poisson pmf was used to generate the values in the "expected" column, we conclude that the Poisson distribution is a very good model for the phenomenon of Prussian soldiers being kicked to death.

■

POISSON RANDOM VARIABLE X

X = number of occurrences per unit interval (time or space)

X = {0, 1, 2, ...}

Parameter: λ = **average** number of occurrences per unit interval
(Note: this unit interval must match the unit interval given in the definition of the RV)

PMF: $P(X = x) = \dfrac{\lambda^x e^{-\lambda}}{x!}$ for x = 0, 1, 2, ... and $\lambda > 0$

$E(X) = \lambda$

$V(X) = \lambda$

Shape: skewed right

Table 4.11 Summary of Poisson RV

4.7 Normal Distribution

We have thus far examined four very important probability distributions: the discrete uniform, binomial, hypergeometric, and Poisson distributions. Each of these is discrete in the sense that the random variable could take on only a countable number of values, most notably integer values. For instance, there could possibly be 3 suicides per month, but never could we consider the possibility of 3.451 suicides per month. We now take up the other kind of distribution, a **continuous** distribution. The normal, exponential, and continuous uniform distributions that we cover in this and the next two sections, respectively, are continuous distributions. No doubt the most important distribution in all of probability and statistics, the normal distribution is the gateway to statistical inference. Before we delve headlong into the normal distribution, however, we make a brief transition to continuous distributions.

Although we have not yet specifically stated it as such, probabilities are essentially equivalent to areas under a curve. Looking at the binomial RV of Example 4.7, where X = the number of boys out of a family of 5 children, the discrete probability distribution for X can be shown graphically as in Figure 4.6.

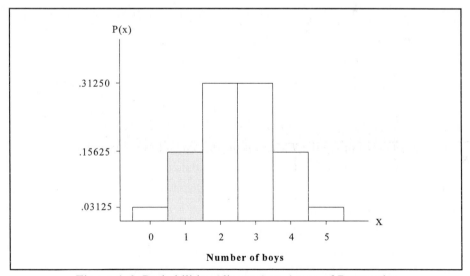

Figure 4.6 Probabilities (discrete) as Areas of Rectangles

For example, $P(X = 1) = .15625$ is the area of the shaded rectangle, where the width of each rectangle is 1. Similarly, $P(X > 3) = P(4) + P(5) = .15625 + .03125 = .1875$, which is the area of the two rightmost rectangles. Thus, areas of rectangles represent the probabilities of discrete random variables.

Areas of rectangles can also be used to **approximate** probabilities associated with a continuous RV. Consider the RV Y = gas mileage of a randomly selected vehicle. Y is a continuous RV because there exists a continuum of possible gas mileages measured in mpg. For example, 21.164 mpg is a possible value for Y, although we would seldom measure mileage to that degree of accuracy simply because our measuring devices (pumps at service stations and odometers) do not justify that kind of precision. Recall the sample of 50 vehicle mileages that was described in the histogram of Figure 2.14. If we consider the areas of all rectangles in that histogram to sum to 1 (like we do in the discrete case), we have a basis for estimating probabilities. Figure 4.7 is a reproduction of Figure 2.14, but the frequencies (vertical axis values) have been normalized by 300 = (50 vehicles) x (6 mpg width for each interval) to insure that the entire area is 1. Based on the empirical or sample data that produced Figure 4.7, we can now estimate probabilities like $P(6 \le Y \le 18)$. This probability, the probability that a randomly selected vehicle has a gas mileage between 6 and 18 mpg, is estimated by the sum of the areas of the two leftmost rectangles. That is, $P(6 \le Y \le 18) \approx (2/300)6 + (12/300)6 = 84/300 = .28$.

If we had taken a much larger sample than 50 vehicles, say a sample of size 1000, then the histogram of gas mileages might appear as in Figure 4.8. With a larger sample size and more empirical data, we would be able to get better estimates of the true probabilities. Many times empirical data allow us to make conclusions or assumptions about the distribution of the population from which the data was obtained. In such cases, we are able to assume that a continuous curve such as the one superimposed on the histogram in Figure 4.8 is an adequate representation of the population distribution. A continuous curve for a continuous RV is sometimes called a probability **density** function (pdf), just as the term probability **mass** function (pmf) is used for a discrete RV. The entire area under a pdf equals 1.

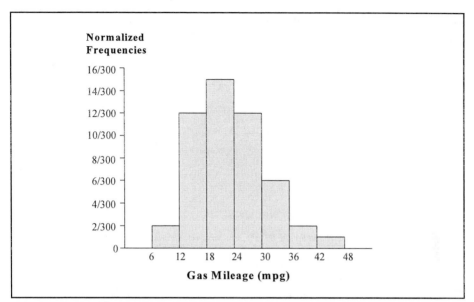

Figure 4.7 Modified Histogram of Gas Mileage Data

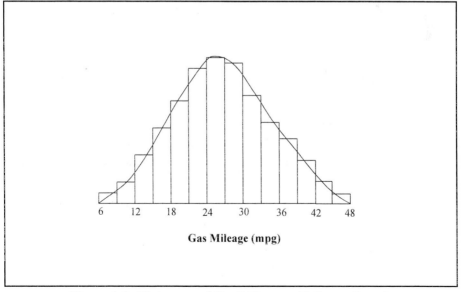

Figure 4.8 Histogram of Gas Mileage with Large Sample

We shall no longer use areas of rectangles to obtain probabilities for continuous RVs. Instead, we shall use areas under the continuous pdf to obtain probabilities for a continuous RV. Figure 4.9 illustrates a pdf for some continuous RV X. The shaded area represents P(a ≤ X ≤ b), the probability that the RV X takes on a value on the interval between a and b. The key word here is "interval." Unlike probabilities for discrete RVs, the probability that a continuous RV takes on any particular value is 0. For example, P(Y = 21.164) = 0, where Y is the gas mileage RV. Probabilities for continuous RVs are measured over **intervals**.

Note: the reader who has had calculus should be able to relate area under a curve as the limiting process of sums of areas of rectangles (Riemann sums), as well as recall that the area under a curve at a point is zero, e.g., P(Y = 21.164) = 0.

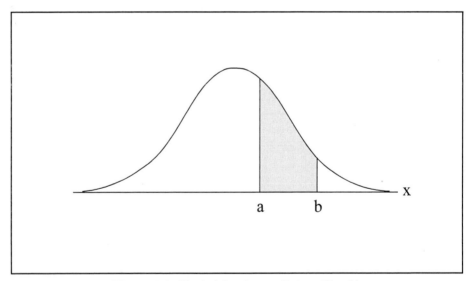

Figure 4.9 Shaded Region = P (a ≤ X ≤ b)

A pdf for some continuous RV X, denoted f(x), must satisfy the following two criteria if it is to be a valid pdf.

(1)   f(x) ≥ 0 for all x (graphically, this means that the continuous curve must lie on or above the x axis everywhere).

(2)   the total area between the curve and the x axis must be 1.

These criteria are analogous to those for a discrete RV, namely (1) each probability must be a value between 0 and 1, and (2) the sum of all probabilities must equal 1.

Example 4.14 Assume the RV X = waiting time in minutes for a bank customer to obtain a teller, and X has the pdf

$$f(x) = \begin{cases} 1 - 0.5x & \text{for } 0 \le x \le 2 \\ 0 & \text{elsewhere.} \end{cases}$$

The graph of this pdf is shown here.

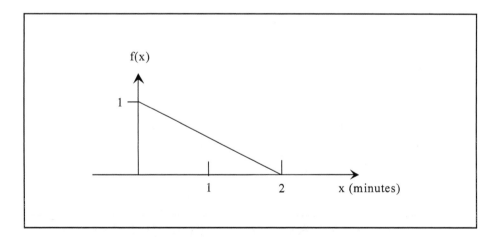

Is this a valid pdf? By inspection, the graph of f(x) is clearly on or above the x axis everywhere. Secondly, the total area between the curve and the x axis is the triangular area A = (1/2)(b)(h) = (1/2)(2)(1) = 1. Hence, f(x) is a valid pdf. What is the probability that a randomly selected customer will have to wait more than 1 minute for a bank teller? We seek $P(X > 1)$. This probability is represented by the area under $f(x)$ to the right of $x = 1$, as shown here:

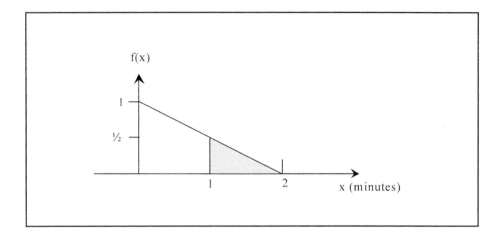

Again, using $A = (1/2)(b)(h)$,

$$P(X > 1) = (1/2)(2 - 1)(f(1)) = (1/2)(1)(1/2) = 1/4.$$

■

Typically, calculus is required to find the mean $\mu = E(X)$ and variance $\sigma^2 = V(X)$ for a continuous RV X. However, μ and σ^2 can be approximated by using narrower and narrower rectangles to approximate the area under the pdf. Using a set of rectangles (or histogram), μ and σ^2 can be approximated by using the formulas previously given for discrete RVs (Definitions 4.4 - 4.6).

The normal distribution, also called the Gaussian distribution or bell-shaped curve, dates back to the 18th century German mathematician Gauss, who found that repeated measurements of the same astronomical quantity produced a pattern similar to the continuous curve in Figure 4.9. This pattern has since been found to occur almost everywhere in life. Heights, weights, IQs, shoe sizes, chest sizes, various standardized test scores, economic indicators, and a host of measurements in service and manufacturing are all examples of where the normal distribution applies. Even for distributions that are known not to be normal, it turns out that under suitable conditions their sums and averages tend to follow a normal distribution. The mathematical equation for this famous bell-shaped curve is given by

$$f(x) = \frac{1}{\sqrt{2\pi}\,\sigma}\; e^{-\frac{1}{2}\left(\frac{x-\mu}{\sigma}\right)^2} \quad \text{for} \; -\infty < x < \infty.$$

The number e $\approx$ 2.71828 is the same irrational number that occurs in the Poisson pmf. We present this equation not to frighten the reader, but rather to point out the two parameters that are present in the equation: μ and σ. Mu (μ) is the mean (center or balance point) of the distribution. The highest point on the symmetric bell-shaped curve occurs at the point μ on the horizontal axis. Sigma (σ) is the standard deviation of the distribution and, graphically, it is the distance between the center and a point of inflection on the curve. A point of inflection is where the concavity of the curve changes. Sigma (σ) will determine how narrow or how flattened out the bell-shaped curve will be. Each μ, σ pair determines a unique normal distribution. Figure 4.10 illustrates how the values of μ and σ impact the shape and position of the curve. It is important to note that the normal curve is symmetric about its center point μ and that it extends indefinitely in both directions, approaching the x axis from the top in both directions. Although not verified here, it can be shown that the entire area under the normal curve, for any values of μ and σ, is equal to 1.

In our work with the normal distribution, it will be necessary to find various areas under the normal curve (i.e., calculate probabilities). To do this we will use a table of values which will allow us to compute areas under the curve. Since it is both impossible and unwise to produce a table of values for every possible normal distribution, we use a table of values for one specific normal distribution and convert between the original distribution and the tabulated distribution. We use a table of normal curve areas for the distribution having $\mu = 0$ and $\sigma = 1$.

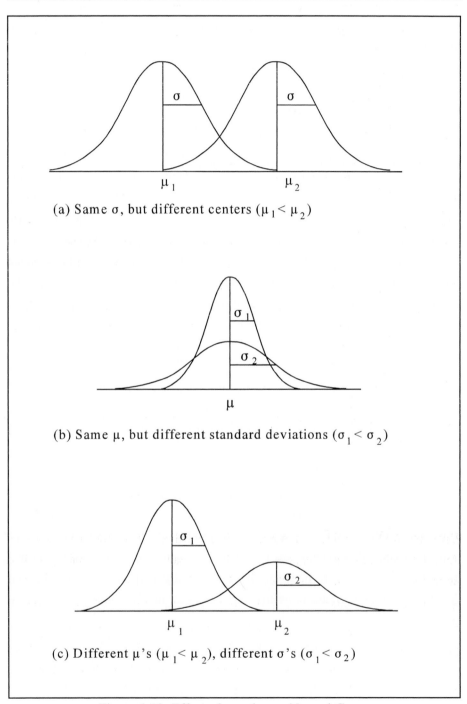

(a) Same σ, but different centers ($\mu_1 < \mu_2$)

(b) Same μ, but different standard deviations ($\sigma_1 < \sigma_2$)

(c) Different μ's ($\mu_1 < \mu_2$), different σ's ($\sigma_1 < \sigma_2$)

Figure 4.10 Effect of μ and σ on Normal Curve

This distribution is called the standardized normal distribution, or the Z distribution. The conversion formula which allows movement between the original X distribution and the Z distribution is given by

$$Z = \frac{X - \mu}{\sigma}$$

Definition 4.11 Standardized Value Transformation (X to Z)

where μ and σ are the mean and standard deviation, respectively, of the X distribution. It is important to be able to cross the bridge between the X and Z distributions both ways. That is, assuming μ and σ are known, if the X value is given one must be able to find the corresponding Z value, and if the Z value is given one should be able to find the corresponding X value. An equivalent form of Definition 4.11 is given in Definition 4.12.

$$X = \mu + Z\sigma$$

Definition 4.12 Standardized Value Transformation (Z to X)

We illustrate the conversion process in the following example.

Example 4.15 Consider the normally distributed RV X = height in inches of American adult males. Suppose that X has a mean of 70 inches and a standard deviation of 2 inches. A shorthand notation for displaying this information is $X \sim N(70, 4)$. Note that the second element within the parentheses is the **variance**, not standard deviation. Based on this information, how would we calculate the probability of a randomly selected male being less than 72 inches tall? We seek $P(X < 72)$. This probability is the shaded area in Figure 4.11. Since no table of normal curve areas exists for the μ, σ combination of $\mu = 72$ and $\sigma = 2$, we must convert to standardized values. To do this we proceed as follows:

$$P(X < 72) = P\left(\frac{X-u}{\sigma} < \frac{72-70}{2}\right) = P(Z < 1.0).$$

.8413

64   66   68   70   72   74   76   X

-3   -2   -1   0   1   2   3   Z

Figure 4.11   P(X < 72) = P(Z < 1) is the Shaded Area

In Figure 4.11, we have also drawn the Z axis and have noted that P(Z < 1) is the same area as P(X < 72). P(Z < 1) can be found in Appendix D by looking up the z value of 1.00 and finding the area completely to the left of z = 1 as .8413, which is the desired probability. It is important to know that the table in Appendix D gives cumulative probabilities for z values greater or equal to 0. That is, P(Z ≤ z) is given for z values ≥ 0. P(Z ≤ z) is the area under the standardized normal curve completely to the left of z on the Z axis. One should also note that a standardized (or z) value is nothing more than a measure of "the **number** of standard deviations." For example, a person who is 72 inches tall is 1 standard deviation above the mean. Alternatively, a person who is 67 inches tall has a standardized value of -1.50, i.e., he is 1.5 standard deviations below the mean. For any continuous distribution X, it is also true that P(X ≤ 1) = P(X < 1). Knowing exactly what the table values in Appendix D mean, knowing that a normal distribution is symmetric about μ, and knowing that the total area under the curve is 1 are the keys to solving many problems.

Example 4.16 Continuing with the same RV X, what is the probability that a randomly selected male will be between 73 and 75 inches tall? We seek $P(73 \le X \le 75)$. Figure 4.12 shows the desired area. We rewrite $P(73 \le X \le 75)$ as $P(X \le 75) - P(X \le 73)$. Now, converting to standardized values, we have

$$P(X \le 75) - P(X \le 73) = P(Z \le 2.5) - P(Z \le 1.5)$$

$$= .9938 - .9332 = .0606.$$

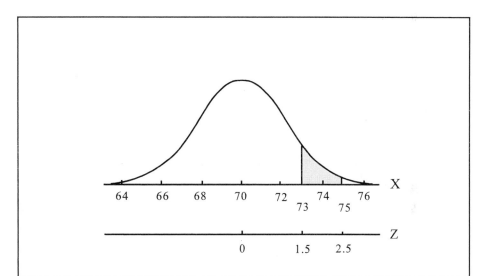

Figure 4.12 $P(73 \le X \le 75) = P(1.5 \le Z \le 2.5)$ is the Shaded Area

Example 4.17 Assuming the same RV X, namely X ~ N(70, 4), what is the probability a male selected at random will be shorter than 66 inches? We seek $P(X < 66)$, which is the shaded region on the left in Figure 4.13. Converting to standardized values, $P(X < 66) = P(Z < -2.0)$. However, negative z values such as -2.0 are not in Appendix D. We compensate for this by using the symmetry of the normal curve and noting that $P(Z < -2.0) = P(Z > 2.0)$, which is the equivalent shaded region in the right tail. Thus, we have

$P(X < 66)$ $=$ $P(Z < -2.0)$ $=$ $P(Z > 2.0)$ $=$ $1 - P(Z \leq 2.0)$

$=$ $1 - .9772$ $=$ $.0228.$

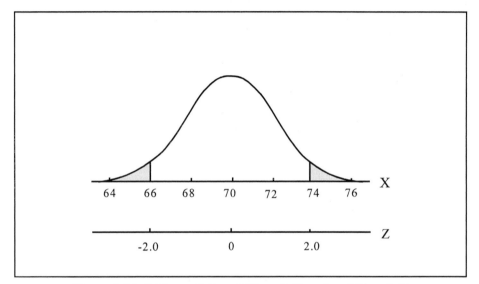

Figure 4.13 $P(Z < -2.0) = P(Z > 2.0) = 1 - P(Z \leq 2.0)$

■

Figure 4.14 shows the area under the normal curve within 1, 2, and 3 standard deviations of the mean, respectively. From the three areas shown in Figure 4.14 and using two decimal places only, the numbers 68, 95, and 99 will reappear throughout this text, but here is where they are mentioned first and the reader is encouraged to retreat to here if so needed later. The examples above illustrated how one can use the tabular values to find an area when given a particular z value. The next example shows how to find a given z value (and consequently, an x value) when given a cumulative area under the curve.

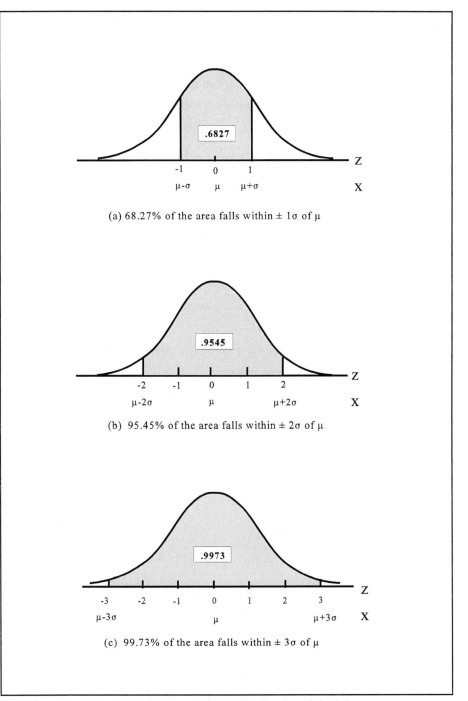

(a) 68.27% of the area falls within ± 1σ of μ

(b) 95.45% of the area falls within ± 2σ of μ

(c) 99.73% of the area falls within ± 3σ of μ

Figure 4.14 Areas within 1σ, 2σ, 3σ of the Mean

Example 4.18 Suppose we wish to find the 95th percentile of the RV X, where X ~ N(70, 4). That is, we desire to find the height in inches such that 95% of American adult males fall below it and 5% are above it. Thus, we want to find the value on the X axis such that exactly 95% of the area under the normal curve lies to the left of this value. The scenario is depicted in Figure 4.15. We want to find X(.95), but before we can find that, we must find Z(.95). Now, looking for ".9500" in the body (i.e., not margins) of Appendix D, we see that there is no ".9500" shown in the table. However, there is a ".9495" which corresponds to a z value of 1.64, and there is a ".9505" which corresponds to a z value of 1.65. Since ".9500" is halfway between ".9495" and ".9505", then Z(.9500) is halfway between Z(.9495) = 1.64 and Z(.9505) = 1.65. Therefore, Z(.95) = Z(.9500) = 1.645.

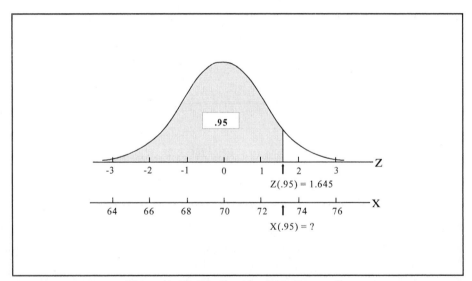

Figure 4.15 Finding the 95th Percentile

This means that the 95th percentile of X, X(.95), is 1.645 standard deviations to the right of μ = 70, where one standard deviation is worth 2 inches. Using Definition 4.12 to cross the bridge from Z to X, we have X(.95) = μ + Z(.95)σ = 70 + 1.645(2) = 73.29 inches. **Note:** Some of the more commonly used percentiles of the Z distribution are compiled at the end of Appendix D for quick reference. Since Z(.95) is one of these, the interpolation process demonstrated above would not have been necessary.

∎

The examples in this section emphasize the use of sketches of a normal curve to solve probability problems. We cannot overemphasize this approach. Besides allowing us to clarify exactly what it is we must find (an area or a value on the x axis), the sketch allows us to obtain a "reasonable" magnitude for our answer, even before we perform any calculations. One should get into the habit of always making a sketch when solving normal distribution probability problems. We conclude this section on the normal distribution by presenting an example of its typical use in industry.

Example 4.19 The diameter of a fiber optics cable is targeted to be exactly .5 inches. The specifications stipulate that the cable diameter must be within the range of .35 and .65 inches. If empirical data collected indicates that the cable diameters, X, are normally distributed with X ~ N(.45, .01), what percentage of the cables produced do not meet the specifications? We seek $P(X > .65)+P(X < .35)$. The shaded area in Figure 4.16 represents the desired probability, where $\mu = .45$ and $\sigma = .10$.

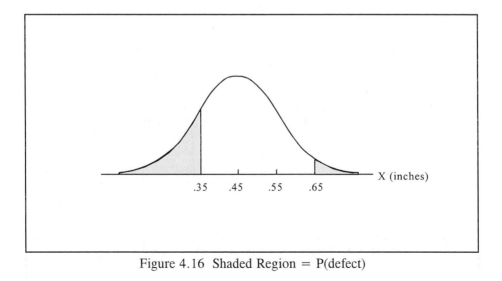

Figure 4.16 Shaded Region = P(defect)

$$P(X > .65) = 1 - P(X \le .65) = 1 - P\left(Z \le \frac{.65 - .45}{.10}\right)$$

$$= 1 - P(Z \le 2) \quad = 1 - .9772 = .0228, \quad \text{and}$$

$$P(X < .35) = P\left(Z < \frac{.35 - .45}{.10}\right) = P(Z < -1) = P(Z > 1)$$

$$= 1 - P(Z \le 1) \quad = 1 - .8413 = .1587.$$

Thus, $P(X > .65) + P(X < .35) = .0228 + .1587 = .1815$, which means that about 18% of the cables will not meet specifications. Clearly an 18% defect rate is unacceptable. What could be done to decrease the defect rate and increase the proportion of cables that meet specifications? One obvious quirk is that the target value of .5 is not the center of the distribution. If we were somehow able to shift the distribution center to .5 inches, would that help? We can check by finding $P(X > .65) + P(X < .35) = 2\ P(X > .65)$, since we now would have a symmetric distribution about $\mu = .5$, as shown in Figure 4.17.

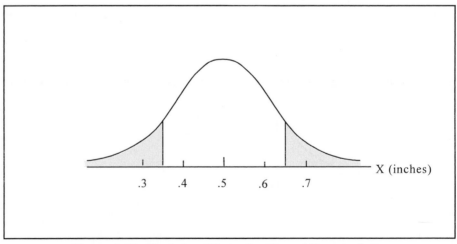

Figure 4.17 Shaded = P(reject) with μ at Target of .5

$$2P(X > .65) = 2[1 - P(X \le .65)] = 2[1 - P(Z \le 1.5)] = 2[1 - .9332] = 2[.0668] = .1336,$$

which still leaves more than 13% of the cables not meeting spec. Centering the distribution at the target value of .5 inches helped but it was not the panacea one may have expected. While one might argue that the specification limits should be reduced from [.35, .65] to, say, [.4, .6], this is only going to exacerbate the problem because more, not fewer, cables will be rejected. The only way that we can reduce the number of rejected cables is to **change the process** so that σ is decreased. The culprit here is $\sigma = .10$ and any shifting of μ will not reduce the reject rate lower than 13.36%. If we could reduce σ by 25% to $\sigma = .075$, while retaining the same specification range of [.35, .65], we would have

$$P(X < .35) + P(X > .65) = 2P(X > .65) = 2[1 - P(X \le .65)]$$

$$= 2[1 - P(Z \le 2)] = 2[1 - .9772]$$

$$= 2[.0228] = .0456,$$

indicating that a 25% reduction in σ resulted in about a 67% drop in the defect rate! Design of Experiments (DOE) is a tool that can be positively applied to reduce σ, as well as putting μ on target. This valuable tool will be addressed in Chapter 8.

∎

The following table summarizes the key aspects of the normal distribution.

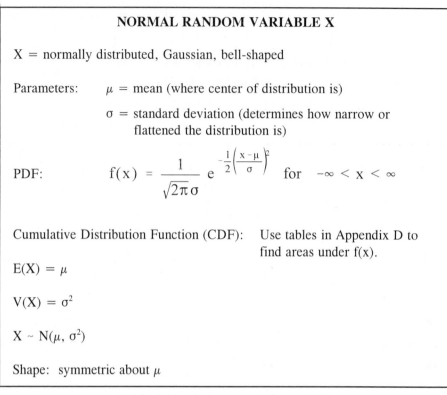

NORMAL RANDOM VARIABLE X

X = normally distributed, Gaussian, bell-shaped

Parameters: μ = mean (where center of distribution is)

σ = standard deviation (determines how narrow or flattened the distribution is)

PDF: $f(x) = \dfrac{1}{\sqrt{2\pi}\,\sigma}\, e^{-\frac{1}{2}\left(\frac{x-\mu}{\sigma}\right)^{2}}$ for $-\infty < x < \infty$

Cumulative Distribution Function (CDF):   Use tables in Appendix D to find areas under f(x).

$E(X) = \mu$

$V(X) = \sigma^{2}$

$X \sim N(\mu, \sigma^{2})$

Shape: symmetric about μ

Table 4.12   Summary of Normal RV

4.8 Exponential Distribution

A second continuous distribution that has wide applicability is the exponential distribution. The exponential RV is typically used to represent "time or space between occurrences." More specifically, the exponential RV is used often in the study of reliability of parts and systems and it represents "time until failure." The exponential distribution is very useful for modeling processes that tend to last forever except for some external force interrupting that process. Many electronic parts or components fall into this category because surges are the primary reason for component failure. Some examples of an exponential RV are:

(1)   time until a light bulb burns out;

(2)   service time at a bank teller;

(3)   time between calls to an emergency dispatcher;

(4)   time between electric generator failures;

(5)   survival time for patients diagnosed with a certain disease; and

(6)   distance between potholes.

The exponential RV has a pdf of the form

$$\mathbf{f(x) = \lambda e^{-\lambda x} \text{ for } x \geq 0 \text{ and } \lambda > 0}$$

Definition 4.13 Exponential pdf

and has one parameter, λ. This is the same λ that appears in the Poisson pmf. The "e" is the same irrational number that we have seen previously in both the Poisson pmf and normal pdf. The parameter λ determines the shape of the graph of the exponential pdf. Figure 4.18 shows three different exponential distributions, those for $\lambda = 2$, $\lambda = 1$, and $\lambda = .5$. It can be shown (calculus needed) that the area under any exponential pdf is exactly 1.

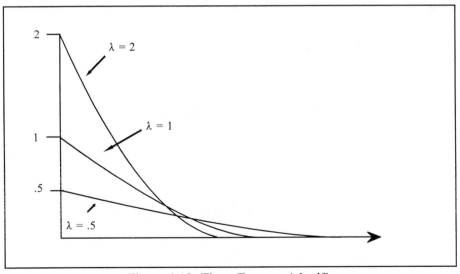

Figure 4.18 Three Exponential pdf's

Finding areas under any portion of an exponential curve is easy because, unlike the normal distribution for which we must use a pre-determined table to find areas, there is a simple formula or equation available. That equation is called a **cumulative distribution function (CDF)** because the functional value for a given x value represents the cumulative (or entire) area under the exponential curve that lies to the left of the x value. The following equation specifies this function and Figure 4.19 graphically portrays the corresponding area under the curve.

$$\mathbf{F(x) \; = \; P(X \leq x) \; = \; 1 - e^{-\lambda x} \; for \; x \; \geq \; 0}$$

Definition 4.14 CDF for Exponential RV X

Note the use of "F" for the CDF, while "f" is used for the pdf. Since the area under the entire curve is 1, the area under the curve to the "right" of x must be $e^{-\lambda x}$. This concept, illustrated in Figure 4.19, will simplify probability (area) calculations for the exponential distribution.

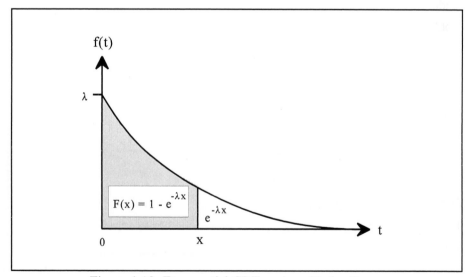

Figure 4.19 Exponential CDF as Area Under Curve

Two other important properties of the exponential distribution are the mean and variance. If X is an exponential RV, then

$$E(X) = \mu = \frac{1}{\lambda} \quad \text{and} \quad V(X) = \sigma^2 = \frac{1}{\lambda^2}$$

which makes $\sigma = \frac{1}{\lambda}$.

Knowing that $E(X) = 1/\lambda$ is extremely valuable because when one knows what $E(X)$ is, then one also knows what λ is. The next example illustrates this and other useful properties of the exponential distribution.

Example 4.20 A certain manufacturer of small electric motors obtains empirical data from sampling that shows the service life of her electric motors to be exponentially distributed with an average (service) life of 100 hours. That is, the average time until motor failure is 100 hours. What is the probability that a randomly selected electric motor will fail within the first 50 hours? To solve this problem, we define the random variable as

X = time (in hours) until motor failure,

and we seek $P(X \leq 50)$. Since $E(X) = 100$, $\lambda = 1/100$ and we have

$$P(X \leq 50) = 1 - e^{-(1/100)(50)} = 1 - e^{-.50} = .393.$$

Figure 4.20 graphically displays the area of interest. This area indicates that nearly 40% of the electric motors fail within the first 50 hours of service and about 60% last longer than 50 hours.

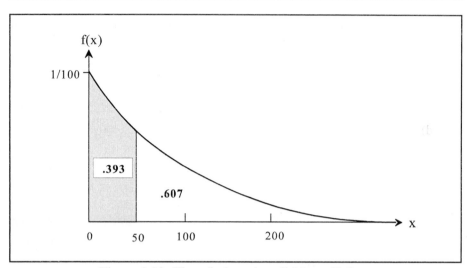

Figure 4.20 Time (in hours) until Motor Failure

Let us assume further that the manufacturer warrants these electric motors for the first 50 hours of service life. In this case, the manufacturer is performing warranty work on 40% of her products, a clearly unacceptable position to be in. What can the manufacturer do to alleviate this problem? Rather than altering the manufacturing process to produce a more reliable motor, which is the ultimate solution but which could take some time, she decides to begin "burning the motors in" for, say, 25 hours prior to shipping them. Finding

$$P(X \le 25) = F(25) = 1 - e^{-(1/100)(25)} = 1 - e^{-.25} = .221,$$

she correctly concludes that about 22% of the motors will fail within the first 25 hours of operation. She believes that, if she performs preventive maintenance on these and puts them back into circulation after the maintenance, she will cut the "warranty problem" down considerably because more than half of the failures within the first 50 hours will occur within the first 25 hours. The manufacturer's line of reasoning is correct, provided the preventive maintenance program corrects a systemic problem and provides a more reliable product than the product coming directly off the assembly line. A more reliable product means one that has a lower failure rate (i.e., λ is smaller). If the preventive maintenance program does not result in a more reliable product, the manufacturer has lulled herself into a false sense of security because of the following property of the exponential distribution. If an electric motor has already functioned properly for 25 hours (the burn-in period), what is the probability that it will function another 50 hours (the warranty period)? This probability is a conditional probability and can be written as

$$P(X > 75 \,|\, X > 25) = \frac{P(X > 75 \cap X > 25)}{P(X > 25)} = \frac{P(X > 75)}{P(X > 25)}$$

$$= \frac{e^{-\left(\frac{1}{100}\right)(75)}}{e^{-\left(\frac{1}{100}\right)(25)}} = \frac{e^{-.75}}{e^{-.25}} = e^{-.50} = .607$$

$$= e^{-\left(\frac{1}{100}\right)(50)} = P(X > 50) \, ,$$

which is exactly the probability $(1 - .393)$ obtained previously for "passing" the warranty period (see Figure 4.20). Hence, the only true way to reduce the number of motors requiring warranty work is to build a better manufacturing process.

■

The property illustrated in the latter part of this example is sometimes called the **memoryless** property of the exponential distribution. Mathematically, it states that

$$P(X > a+b \,|\, X > a) = P(X > b).$$

Graphically, it says that the ratios of the "shaded" area to the "bracketed" area in each of the two graphs in Figure 4.21 are equal. Intuitively, it means that if a component has lasted so long, it is at that point in time as good as new!

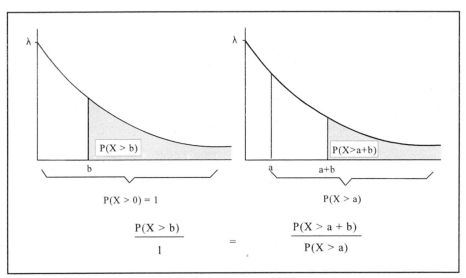

Figure 4.21 Graphical Interpretation of Memoryless Property

Since the λ used in both the Poisson and exponential distributions is the same, there must be some relationship between these two distributions. We state that relationship in the following definition.

If X is a Poisson RV denoting the **number of occurrences per time (space) interval** and having on the average λ occurrences per time (space) interval, then Y, the RV denoting the **time (space) between occurrences**, is exponential with parameter λ and has an average time (space) between occurrences of $1/\lambda$.

Definition 4.15 Relationship between Poisson and Exponential RVs

We demonstrate this relationship in the following example.

Example 4.21   A Colorado ski resort has a certain chairlift that operates continuously for 8 hours per day and on the average there are 12 lift stoppages per day. If the ride on this chairlift is a 10-minute ride, what is the probability that your next ride on this lift will not be affected by a lift stoppage? We demonstrate two solutions to this problem, using the Poisson and exponential distributions, respectively.

Solution (1)   We let the Poisson RV X be

X = the number of lift stoppages per **10-minute period**.

To get the appropriate λ, we convert 12 stops/day as follows:

12 stops/day = 12 stops/8 hours

= 12 stops/480 minutes

= .25 stops/**10 min**.

Thus, λ = .25 stops/10-minute period. Since we are concerned with no stops on the next ride, we must find P(X = 0). Using the Poisson pmf with λ = .25, we have

$$P(X=0) = \frac{\lambda^x e^{-\lambda}}{x!} = \frac{(.25)^0 e^{-.25}}{0!} = e^{-.25} = .7788.$$

Thus, we have better than a 3 out of 4 chance to make it to the top of the run in 10 minutes without a stop.

Solution (2)   We let the exponential RV Y be

Y = the time (in minutes) until the next lift stoppage (i.e., time until failure).

Since the unit of time used in the definition of Y is a minute, we will use λ = .025 stops/minute. We must find P(Y $\geq$ 10) because the lift must operate for at least 10 minutes continuously if we are to have no stoppages on our next ride. Using the exponential CDF with $\lambda = .025$, we have

$$P(Y \geq 10) = e^{-\lambda y} = e^{-.025(10)} = e^{-.25} = .7788,$$

the same result obtained using the Poisson distribution. Note that $P(Y \geq 10)$ represents the area to the right of $y = 10$, and that is why we used $e^{-\lambda y}$, the complement of the exponential CDF $F(y) = 1 - e^{-\lambda y}$.

EXPONENTIAL RANDOM VARIABLE X

$X = $ time or space between occurrences (could also be time until failure)

Parameter: $\lambda = $ **average** number of occurrences per unit interval (the same parameter as in the Poisson distribution) Note: the unit interval used here must match the time or spatial unit used in the definition of the RV X.

PDF: $f(x) = \lambda e^{-\lambda x}$ for $x \geq 0$ and $\lambda > 0$

CDF: $F(x) = 1 - e^{-\lambda x}$ for $x \geq 0$

$E(X) = \mu = 1/\lambda$

$V(X) = \sigma^2 = 1/\lambda^2$

Shape: skewed right

Table 4.13 Summary of Exponential RV

4.9 Continuous Uniform Distribution

The final distribution we present is one of the simplest continuous distributions, the uniform distribution. It is the distribution that applies to any situation where all possible values that a random variable can take on lie in an interval [a, b] on the real number line and all are equally likely. The uniform distribution has two parameters, a and b. Parameter "a" is the smallest possible value and parameter "b" is the largest possible value for the uniform RV. Graphically the continuous uniform distribution looks like a rectangle and is sometimes even called the rectangular distribution. The rectangular or uniform pdf is shown in Figure 4.22.

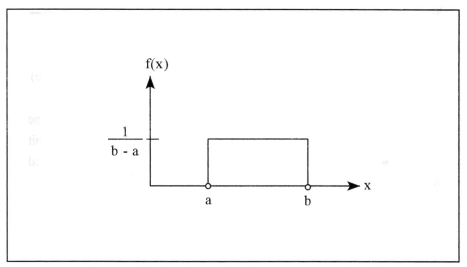

Figure 4.22 Continuous Uniform or Rectangular pdf

This pdf is a constant function on [a, b] and is a valid pdf because $f(x) = \dfrac{1}{b-a} > 0$

and the area under the curve is Area $=$ Length x Width $= (b - a)\left(\dfrac{1}{b-a}\right) = 1$.

The following table summarizes the key attributes of the continuous uniform distribution.

CONTINUOUS UNIFORM RANDOM VARIABLE X

X = any number on a given interval [a, b], all values presumed equally likely

Parameters: a = smallest possible value of X.

b = largest possible value of X.

PDF: $f(x) = \dfrac{1}{b - a}$ for $a \le x \le b$

CDF: $F(x) = \dfrac{x - a}{b - a}$ for $a \le x \le b$

$E(X) = \dfrac{a + b}{2}$

$V(X) = \dfrac{(b - a)^2}{12}$

$X \sim U(a, b)$

Shape: symmetric

Table 4.14 Summary of Continuous Uniform or Rectangular RV

Example 4.22 Sam is leaving his apartment to catch a bus into the city. He knows that buses run every half hour, but he has lost his bus schedule and can't recall on which minutes after the hour the bus stops at his location. What is the probability that Sam will have to wait longer than 20 minutes? To solve this problem we define the random variable Y as

Y = # of minutes Sam will have to wait.

The parameters are a = 0 and b = 30. We will find P(Y > 20). The graphical distribution with the desired shaded area is shown in Figure 4.23.

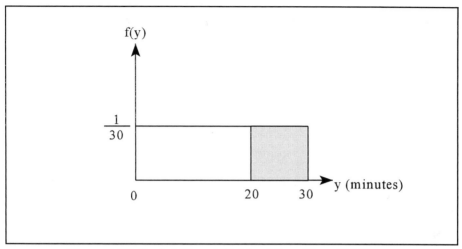

Figure 4.23 Probability of Waiting More Than 20 Minutes

From the graph, $P(Y > 20) = \dfrac{30 - 20}{30 - 0} = \dfrac{1}{3} = .333.$

Since the pdf is constant, we need only deal with values on the horizontal axis. We could have also used the CDF from Table 4.14 to get

$P(Y > 20) = 1 - P(Y \le 20) = 1 - F(20) = 1 - \dfrac{20 - 0}{30 - 0} = 1 - \dfrac{2}{3} = \dfrac{1}{3} = .333.$

■

The application of this distribution is useful whenever we deal with a quantity that randomly varies over a given continuous interval. Perhaps one of the most useful applications of this distribution is in generating random variables from other distributions. Since the CDF for any valid pdf is uniform on the interval (0, 1), we can use this fact to generate (via inverse transformations) random variables from other distributions. We provide three algorithms that use uniform random variables to generate normal, exponential, and Poisson random variables, respectively. Simulation modeling often requires the use of these three types of random variables so the following algorithms may come in handy.

Normal

The algorithm below actually produces normal random variables in pairs. However, they are independent so they give the user two random variables in each pass.

Step 1. Generate $U_1 \sim U(0,1)$ and $U_2 \sim U(0,1)$.

Step 2. Set $V_1 = 2U_1-1$ and $V_2 = 2U_2 - 1$.

Step 3. Set $W = V_1^2 + V_2^2$.

Step 4. If $W \geq 1$ then reject and go to step 1.

Step 5. Otherwise calculate $Y = \sqrt{-2(\ln(W))/W}$.

Step 6. Return $Z_1 = V_1Y$ and $Z_2 = V_2Y$.

This algorithm yields two standardized normal (Z) RVs. To get a normally distributed random variable with mean μ and variance σ^2, use the following formula for each Z value:

$$X = \mu + Z\sigma.$$

Exponential

To generate an exponential random variable X with parameter λ:

Step 1. Generate $U \sim U(0,1)$.

Step 2. Set $X = (-\ln U) / \lambda$.

Poisson

To generate random variables for a Poisson distribution with a parameter λ, we can use the following very simple algorithm:

Step 1. Set $N = 0$ and $k = 1$.

Step 2. Generate $U \sim U(0,1)$.

Step 3. Set $k = kU$.

Step 4. If $k > e^{-\lambda}$ then set $N = N + 1$ and return to step 2.

Step 5. Otherwise return N.

4.10 Problems

4.1   The number of patients visiting an emergency room during the graveyard shift (11:00 pm - 7:00 am), denoted by the random variable X, has the following probability distribution:

x	0	1	2	3	4	5	6
P(X=x)	.05	.20	.25	.20	.10	.15	.05

a)   Find the probability there will be three or more emergency room patients during a randomly selected 8-hour graveyard shift.

b)   Find the probability there will be an even number of patients.

c)   Find the probability there will be at least two but not more than five patients.

d)   Find the probability there will be less than two patients.

e)   How do you suppose the probability distribution above was obtained?

4.2   The number of insurance policies a salesman will sell on any one day is denoted by the random variable Z and has the following probability distribution:

z	0	1	2	3
P(Z=z)	.4	.2	.3	.1

a)   Find the probability that on a randomly selected day he will sell at least one policy.

b)   Find the expected value of Z, E(Z).

c)   The salesman makes $800.00 on each policy sold. Find his expected earnings.

d)   Find the probability the salesman will make more than $1500 on any given day.

4.3   The number of injections using a needle contaminated with the AIDS virus that is required to detect an exposure to the virus is denoted here as the random variable X, and it follows the following probability distribution:

x	1	2	3	4	5
P(X=x)	.45	.35	.10	.05	.05

a)   Find the expected number of injections required for detectable exposure of the virus, E(X).

b)   Find the variance of the random variable X.

c)   Let X and Y denote the number of injections using AIDS contaminated needles for each of two people, respectively, where X and Y both follow the distribution above. Further, assume that X and Y are statistically independent. What is the probability that both exhibit a detectable exposure after only one injection?

4.4   A car salesperson notes that she sells either 1, 2, 3, 4, 5, or 6 cars per week with equal frequency.

a)   Find the probability she sells 3 or more cars in a given week.

b)   Find the average number of cars sold in a given week.

c)   Find the standard deviation for the number of cars sold in a given week.

4.5   The number of goals scored in a hockey match between the home team (H) and the visiting team (V), respectively, follows each of the probability distributions given below:

h	0	1	2	3	4	5
P(h)	.25	.25	.20	.15	.10	.05

v	0	1	2	3	4	5
P(v)	.30	.25	.20	.10	.10	.05

Suppose that the random variables H and V are statistically independent. Represent the total number of goals scored by $T = H + V$ and represent the difference in goals scored by $D = H - V$.

a)   Find the probability distribution for T.
b)   Find E(T) and the variance of T.
c)   Find E(D) and the variance of D.
d)   Find the probability that the home team will emerge victorious.

4.6   A manufacturer of semiconductors has a 20% defect rate in the semiconductors produced. Every hour, ten semiconductors are selected at random and inspected for defects. If more than three defectives in ten are observed, the process is stopped. If the probability a semiconductor is defective is not affected by whether the other semiconductors are defective, find the probability the manufacturing process must be stopped within the next hour. What is the probability that the process will be stopped within the next two hours?

4.7   Ten rockets used for placing satellites into orbit around the earth have been manufactured and shipped to NASA. The rocket manufacturer claims that the probability of successfully placing a satellite into orbit on any one rocket launch is .90. Assuming the satellite launches are independent, what is the probability of no failures in ten satellite launches?

4.8   On a day when 600 stocks listed on the New York Stock Exchange rise in price and 400 stocks fall in price, you decide to look up the prices in the newspaper of the ten stocks you own. What is the probability that more than half of your stocks went up in price?

4.9   On the average, only 40% of all new medicines are profitable. A leading drug company developed 15 new medicines last year.

a)   How many of the 15 new medicines would you expect to eventually become profitable?
b)   Calculate the variance for the number of profitable medicines developed last year.
c)   What is the probability that more than half of the medicines developed last year will be profitable?

4.10   In determining if aerial spraying is necessary to eradicate the Mediterranean fruit fly, health inspectors sample and inspect all types of fruit. We know that a piece of fruit has a .90 probability of being free of the fruit fly. If 10 pieces of fruit are inspected from a lot of 1000 pieces, what is the probability we discover at most one piece of fruit infested by the Mediterranean fruit fly? Suppose that, from the sample of size 10, we found 5 pieces of fruit infested by the fruit fly. Based on the .90 probability given above, the probability of finding exactly 5 infested pieces of fruit is very small (like .0013, computations not shown). What does this information tell us?

4.11   As contracting officer for a company that manufactures lifters, you must decide whether to accept a shipment of machine parts from a supplier for use in the construction of lifters. Each shipment contains 100 new parts and you sample 10 and inspect them for defects. Your supervisor instructs you to accept a shipment only if you find no more than 1 defective part.

   a)   What is the probability you will accept a shipment which has 10% defective parts?
   b)   On the average, how many defectives would you expect in a sample of size 10 if the defect rate is 10%?
   c)   What would be the variance of the number of defectives for a sample of size 10 (again, assume a 10% defect rate)?

4.12   Consider the draft lottery of the Vietnam era. There are 366 possible selections (birthdays). *Note: this lottery was held in 1970 for males turning 18 that year, meaning their birth year was 1952, a leap year; hence, the 366 birthdays.* If four males from your high school have different birthdays, what is the probability that exactly two of them will be selected in the first 20 birthdays chosen? Unlike the actual lottery that took place, assume that each birthday is equally likely to be chosen.

4.13   On the average, a 4-foot wide sheet of plywood has a knothole every 5 feet in length.

   a)   Find the probability that a 4' x 8' sheet of plywood has no knotholes.
   b)   Find the probability that the 4' x 8' sheet of plywood has exactly one knothole.
   c)   Find the probability that the 4' x 8' sheet of plywood has more than one knothole.

4.14   A genetic abnormality associated with an irregular heartbeat occurs once in every 1,000,000 births.

    a)   If 12,000,000 children are born in the United States this year, what probability mass function could you use to find the exact probability that at least 5 will have this genetic abnormality?
    b)   Find this probability approximately by using the Poisson distribution.

4.15   We know that the average number of red blood cells visible under the Reichert-Jung Standard Microscope is 5 per square millimeter.

    a)   Find the probability that less than 5 red blood cells per square millimeter are visible under this microscope.
    b)   Find the probability that more than 5 red blood cells per square millimeter are visible.
    c)   Why do the probabilities in (a) and (b) above sum to less than 1?

4.16   The average number of aircraft arriving at Travis Air Force Base is 24 per hour during normal daytime traffic. The Travis AFB runways are capable of handling 20 aircraft per half hour. Find the probability of exceeding the 20 aircraft in a given half-hour period and thus overloading the air traffic threshold.

4.17   The valet at a prestigious restaurant is told by the manager to expect an average of 15 cars per hour to arrive at the restaurant between 5 pm and 9 pm.

    a)   What is the probability that exactly 15 cars will arrive in a given hour?
    b)   What is the probability that no more than 5 cars arrive in a given hour?

4.18   The number of seconds of continuous spray yielded by cans of a brand of outdoor fogger is normally distributed with a mean of 260 seconds and a standard deviation of 15 seconds.

    a)   Find the probability a can will yield between 250 and 275 seconds of continuous spray.
    b)   Find the length of time in seconds for which 90% of the cans will exceed in terms of continuous spray time.
    c)   What is the value found in (b) called?

4.19   A primary contractor for the Delta Rocket purchases bolts from a machine shop. The contractor requires the bolts to have a diameter between 2.6 and 2.7 cm. The machine shop produces bolts whose diameters are normally distributed with a mean diameter of 2.65 cm and a standard deviation of .02 cm. Find the proportion of bolts outside the specifications.

4.20   The Dual Probe Precision Digital Thermometer is used to measure the temperature of any liquid solution. The difference between the true temperature and the measured temperature is a normally distributed random variable with a mean of $0°$ Centigrade(C) and a standard deviation of $0.5°$ C. Find the probability that a measured temperature is inaccurate by more than $1°$ C.

4.21   The Decade Resistance Box is used to set resistance in a laboratory circuit. If we set the box for 110 ohms, the distribution of ohm settings is normal with a mean of 110 ohms and a variance of .01 $ohms^2$. Find the probability that the box actually set the resistance to more than 110.2 ohms.

4.22   The GRE combined scores for students applying to an engineering school's Graduate Program are normally distributed with a mean of 1000 and a standard deviation of 100.

   a)   An applicant needs a minimum score of 1300 to gain admission. What percentage of applicants meet or exceed the minimum qualification?
   b)   The Director of Admissions wants to change the minimum standard GRE score so that 25% of applicants are eligible. What should the new minimum score be?
   c)   Find the proportion of applicants within one standard deviation of the mean score, within two standard deviations of the mean, and within three standard deviations of the mean.

4.23   Discretionary household income in a certain state is exponentially distributed with a mean of $4,800 per year (i.e., $\lambda = 1/4800$). Approximately what proportion of households have discretionary income greater than the mean of $4,800 per year?

4.24   The number of in-flight emergencies (IFEs) for an Air Force T-37 Squadron is Poisson distributed with a mean of 3 IFEs per month. Thus, the length of time between IFEs is exponentially distributed.

a) Find the probability of having no IFEs in the next month. Use the **Poisson** distribution to solve this problem.

b) Find the probability of one month passing before the next IFE. Use the **exponential** distribution to solve this problem.

4.25 The laser used in a Compact Disc player has a time to failure which follows an exponential distribution with a mean of 1250 hours of playing time.

a) Find the probability that the laser will fail within one year if it is used one hour every day.

b) Given that the laser has already lasted 2 years, find the probability that it will last a total of 3 years. Assume the Compact Disc player is used one hour every day.

4.26 Two different kinds of treatment (call them A and B) for a particular kind of cancer at a given stage have both been shown to provide an average survival time after treatment of 5 years. Treatment A has been shown to provide a survival time for its patients that is normally distributed with a standard deviation of 2 years. Treatment B, however, has been shown to be exponentially distributed. If a person diagnosed with this cancer has as a primary goal to survive for 7 more years, which treatment would offer the greatest chance to achieve that goal?

4.27 A major insurance company has determined that the probability of a person being involved in an auto accident is .01 during any year. Assuming independence between years, what is the probability of having 3 or more accident-free years during a 5-year period?

4.28 A certain company makes actuators for the F-15 aircraft. Each actuator has a probability of .9 of meeting all military specifications. From a group of 20 randomly selected actuators, what is the probability that at least 19 of these will meet all specifications?

4.29 An electronics company produces resistors whose service lives are normally distributed. 93.7% of the resistors have service lives less than 3150 hours, and 1.7% of the resistors have service lives greater than 3622 hours. What are the mean and standard deviation of the resistors' service lives?

4.30   An automobile manufacturer warrants the power train of its vehicles for one year. Empirical data indicate that the time until failure of the power train is exponentially distributed with an average time until failure of 4 years. Suppose the manufacturer's profit on a new car is $1500 and that warranty work on power train failures is $500 for each failure.

    a)   What percentage of cars will experience a power train failure during the warranty period?
    b)   Assuming that the manufacturer provides warranty work on any given vehicle at most once, what is the manufacturer's expected profit per vehicle?

4.31   Your company has a 30% share of the worldwide market for a particular product having a very large demand. If you randomly select the names of 15 purchasers of this product, what is the probability that between 4 and 6 of these customers, inclusive, will be your customers?

4.32   A company produces a cylindrically shaped rod that is to be assembled into a cylindrically shaped casing which is also produced by this company. Suppose the diameter of the casings is a normally distributed RV X with a target value of 1 inch. That is, $X \sim N(1.0, .0001)$. Y, the diameter of the rods, is also a normally distributed RV, independent of X, with $Y \sim N(.98, .0004)$. Manufactured rods and casings are paired up, and the fit is acceptable only if the rod fits into the casing and there is no more than .03 inches in clearance between the casing and the rod. Find the proportion of rod/casing pairs that are acceptable.

    Hint: Define a new RV W as W = X - Y. W then represents the difference between the diameters of the casing and the rod or the clearance between the casing and the rod. Now use the fact (proof not shown) that if X and Y are normally distributed and independent, then $W \sim N(\mu_x - \mu_y , \sigma^2_x + \sigma^2_y)$.

4.33   The IRS uses statistical criteria to determine which returns will be audited. For example, one criterion might be that a person's income tax return will be audited if that return shows an itemized deduction amount that exceeds 2.5 standard deviations above the mean of the population from which it was taken. Assume that the itemized deductions for a certain tax bracket (population) are normally distributed with a mean of $20,000 and a standard deviation of $2,000. That is, let $X \sim N(20,000, 2,000^2)$. Using the criterion given above,

a) What percentage of the returns will be audited?

b) What is the minimum amount of itemized deductions that would trigger an audit?

4.34 Let X have a Poisson distribution with parameter $\lambda = 5$. Use Appendix C to find the following probabilities:

a) $P(X = 5) =$
b) $P(X < 5) =$
c) $P(5 \le X < 12) =$
d) $P(X > 12) =$
e) $P(X \ge 8 | X \ge 4) =$

4.35 Consider writing onto a computer disk and then sending it through a certifier that counts the number of missing pulses. Suppose this number has a Poisson distribution with parameter $\lambda = 0.2$.

a) What is the probability that a disk has exactly one missing pulse?

b) What is the probability that a disk has at least two missing pulses?

c) If two disks are independently selected, what is the probability that neither contains a missing pulse?

4.36 Check all of the following that could be a Poisson trial.

_____ The thickness of a coat of paint on a metal sheet.
_____ The number of trout in a lake.
_____ The number of blue pairs of socks among the next five pairs sold in a store.
_____ The number of women among six scholarship winners.
_____ The length of a crack in a timber.
_____ The pollen count in a cubic centimeter of air.

4.37 The number of irregularities per kilometer of optical fiber is a constant rate of 0.9.

a) What is the probability of finding no irregularities in a 1 kilometer strand of optical fiber?

b) What is the probability of finding at most 5 irregularities in a 10 kilometer strand of optical fiber?

c) What is the expected number of irregularities in a 5 kilometer strand of optical fiber?

4.38   The number of typographical errors per page is a Poisson random variable
        with an average rate of 0.8 errors per page.

    a)   What is the expected number of errors on a page and the
        coefficient of variation of the probability distribution?

    b)   If the number of errors on different pages are independent, what
        is the expected total number of errors on 400 pages?

    c)   What is the probability of finding more than 2 errors on any given
        page?

    d)   What is the probability of finding more than 2 errors on at least 2
        out of 10 pages? (Hint: use Binomial Distribution.)

4.39   Given $X \sim N(10,25)$ find

    a) $P(X \leq 10) =$

    b) $P(X > 5) =$

    c) $P(3 \leq X \leq 18) =$

    d) $P(X \geq 18 | X \geq 12) =$

4.40   The weekly amount spent for maintenance and repairs in a certain company
        was observed over a long period of time to be approximately normally
        distributed with a mean of $400 and a standard deviation of $20. If $450
        is budgeted for next week, what is the probability that the actual costs will
        exceed the budgeted amount?

4.41   Wires manufactured for use in a certain computer system are specified to
        have resistances between 0.12 and 0.14 ohms.   The actual measured
        resistances of the wires produced by Company A have a normal probability
        distribution with a mean of 0.13 ohm and a standard deviation of 0.005
        ohm.

    a)   What is the probability that a randomly selected wire from
        Company A's production will meet the specifications?

    b)   If four such wires are used in the system and all are selected from
        Company A, what is the probability that all four will meet the
        specifications?

4.42   Weights of canned hams are normally distributed with a mean of 4.15 kilo-
        grams and standard deviation of 0.12 kilogram. The label weight is stated
        as 4.00 kilograms. Which one of the following two modifications in the
        canning process would lead to a greater reduction in the proportion of hams

below the label weight? (1) Increase the process mean weight to 4.20 kilograms while keeping the standard deviation unchanged. (2) Decrease the process standard deviation to 0.10 kilogram while keeping the mean unchanged.

4.43   The width of bolts of fabric is normally distributed with a mean of 950 millimeters and a standard deviation of 10 millimeters. What is the probability that a randomly chosen bolt has width between 947 and 958 millimeters?

4.44   The automatic opening device of a military cargo parachute has been de-signed to open when the parachute is 200m above the ground. Suppose opening altitude is normally distributed with a mean value of 200m and a standard deviation of 30m. Equipment damage will occur if the parachute opens at an altitude of less than 100m. What is the probability that there is damage to the payload of at least one of five independently dropped parachutes?

4.45   A machine fills 100-pound bags with white sand. The actual weight of the sand when the machine operates at its standard speed of 100 bags per hour follows a normal distribution with a standard deviation of 1.5 pounds. The mean of the distribution depends on the setting of the machine. At what mean weight should the machine be set so that only 5 percent of the bags are underweight?

4.46   A standardized test produces scores having a mean of 70 and standard deviation of 10.

a)   Find the 90th percentile score.
b)   If 2% of the examinees are to receive a failing grade, what is the minimum passing score?

4.47   The service lives of electron tubes produced by Tensor Corporation are normally distributed; 92.51% of the tubes have lives greater than 2160 hours, and 3.92% have lives greater than 17,040 hours. What are the mean and standard deviation of the service lives?

4.48   Given X is exponentially distributed with $\lambda = 2.5$, find

  a) $P(X > 0.5) =$
  b) $P(X \leq 2.0) =$
  c) $P(X \geq 2.0 | X \geq 1.5) =$
  d)   What property tells you that the answers to (a) and (c) above should be equal?
  e)   What is the coefficient of variation for this probability distribution? Can you generalize your result?

4.49   The time between customer arrivals to McDonald's drive-through window at lunchtime is exponentially distributed with a mean of 1.5 minutes.

  a)   If you just arrived at the menu board, what's the probability that the next customer will arrive in less than 1 minute?
  b)   What is the 90th percentile of the time between arrivals at lunchtime?
  c)   What is the probability that more than 3 customers will arrive during a five-minute interval at lunchtime? (Hint: use Poisson Distribution.)

4.50   On the average, a salesman sells one car per week.

  a)   What is the probability that the salesman makes exactly 3 sales in a two-week period?
  b)   What is the probability that the salesman has no sales in a two-week period? (Solve this problem using the Poisson Distribution first and then using the Exponential Distribution second.)

4.51   A bank teller can service an average of 20 customers per hour.

  a)   What is the probability that the person in front of you will take longer than 4 minutes to complete her banking transactions?
  b)   What is the expected time for this bank teller to service a customer?
  c)   You are the next customer to be serviced and you have waited for 5 minutes already. What's the probability you will have to wait another 5 minutes before being served?
  d)   What is the probability the bank teller serves at least 25 customers during the next hour?

4.52   With reference to the German tank problem given in Section 4.3, suppose that the serial numbers of 6 captured German tanks are 240, 1447, 1514, 1692, 3068, and 3378. Develop two different estimators or algorithms for estimating n, the parameter in the discrete uniform distribution representing population size, which in this case represents total inventory or production of German tanks. Apply each estimator to the data above and compare the estimates. Explain why you developed those two estimators and what is the difference between them.

4.53   The measuring instruments used to time the 100 meter dash during world-class track events are precise to within $\pm.02$ seconds. That is, if an individual was timed at 10.04 seconds, the true reading could equally likely have been anything between 10.02 and 10.06 seconds. The current world record is 9.73 seconds. Donavon Bailey just ran the 100 meters in a recorded time of 9.71 seconds. What is the probability that this recorded time of 9.71 seconds is truly *not* a world record?

Chapter 5

Sampling Distributions and Estimation

5.1 Sampling Distribution of the Mean

We now begin our trek into the area of inferential statistics. Recall that our goal has been to use information obtained from a sample to conclude something about a population. In addition, we would like to be able to make that conclusion with some degree of accuracy or precision. In this chapter, we will concentrate primarily on using a sample mean, x̄, to estimate the population mean, μ, which is unknown. The value x̄ obtained from a sample is known as a **statistic**. Statistics are values obtained from sample data. The median (x̃), as well as the variance (s^2) and standard deviation (s), of a sample are also statistics we have worked with, but we shall concentrate primarily on x̄. A **parameter**, on the other hand, is a population value which more often than not is unknown. In fact, if the parameter μ is known, then there is really no need to take a sample for the purpose of estimating μ. Notationally, we continue the convention of using Greek letters (like μ, σ) to designate parameters, and Roman letters (like x̄, s) to refer to sample statistics. Clearly, there is nothing that prevents us from taking a sample, calculating x̄, and using it to estimate μ. The important question, however, is how good an estimate of μ is x̄? The idea of a sampling distribution is the heart and soul of being able to associate precision with an estimate. The conscientious student would do well to master the concept of a sampling distribution, for it is the key to understanding inferential statistics.

As an illustration of a sampling distribution, consider the distribution described in Figure 5.1 for some continuous random variable X. We will sometimes refer to this X distribution as the "parent" population because it is the distribution from which we will sample. If we were to take 4 different samples, each of size n = 4, as displayed in Figure 5.2, we see that the sample averages differ and therefore have their own distribution. We will refer to this distribution as the $\overline{X}$ (or child) distribution, because it is derived from the parent (X) distribution. Obviously, the spread in sample averages is less than the spread in individual X values. The question is how much do the $\bar{x}$'s vary? As we study the sampling distribution of $\overline{X}$, we will show that $\sigma_{\bar{x}} = \dfrac{\sigma}{\sqrt{n}}$, where n = sample size and $\sigma_{\bar{x}}$ denotes the standard deviation of the $\overline{X}$ (or child) distribution.

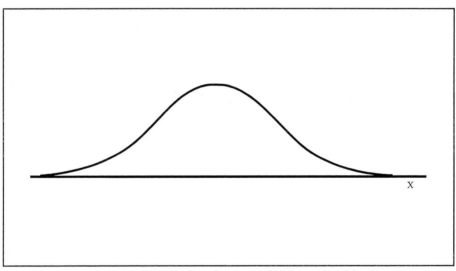

Figure 5.1 Distribution for Some Random Variable X

Sometimes we will take only one sample from which we use the $\bar{x}$ to estimate the population mean, μ. However, knowledge of the underlying distribution of all $\bar{x}$'s will allow us to build a confidence interval within which the unknown mean, μ, should fall. That is why in Figure 5.2 we show several samples with an $\bar{x}$ from each sample. Using the following examples, we introduce the concept of a confidence interval.

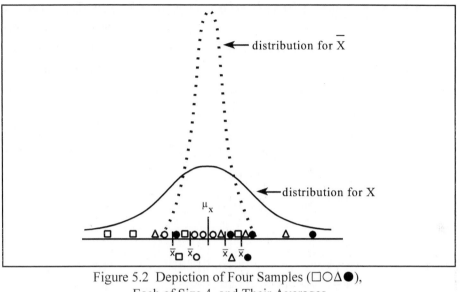

Figure 5.2 Depiction of Four Samples ($\square\bigcirc\triangle\bullet$),
Each of Size 4, and Their Averages

Example 5.1 Assume a sample of 30 bills were processed in an average of 7.8 days with a standard deviation of 1.4. If our sample is representative of the types of bills typically processed we can conclude with approximately 95% confidence that the population mean, μ, for our bill processing time is between 7.289 and 8.311 days.

This range was computed as follows:

$$\overline{x} \pm 2\left(\frac{s}{\sqrt{n}}\right) = 7.8 \pm 2\left(\frac{1.4}{\sqrt{30}}\right) = (7.289, \ 8.311).$$

To build an approximate 99% confidence interval we would compute

$$\overline{x} \pm 3\left(\frac{s}{\sqrt{n}}\right) = 7.8 \pm 3\left(\frac{1.4}{\sqrt{30}}\right) = (7.033, \ 8.567).$$

Notice that higher levels of confidence will produce wider confidence intervals.

Example 5.2 A sample of 50 similar software products had 8 delayed releases due to misunderstood user requirements. If our sample is representative of the population of all future similar products and if the process remains unchanged, we can be approximately 95% confident that the true proportion of delayed products will be within

$$p \pm 2\sqrt{\frac{p(1-p)}{n}} = \frac{8}{50} \pm 2\sqrt{\frac{\left(\frac{8}{50}\right)\left(\frac{42}{50}\right)}{50}} = (.056, .264).$$

An approximate 99% confidence interval would be

$$p \pm 3\sqrt{\frac{p(1-p)}{n}} = \frac{8}{50} \pm 3\sqrt{\frac{\left(\frac{8}{50}\right)\left(\frac{42}{50}\right)}{50}} = (.004, .316).$$

■

The multipliers 2 and 3 used in examples 5.1 and 5.2 are considered to be approximations of the exact values that will be determined later in this chapter. If we are needing a quick Rule of Thumb (ROT) for confidence, we recommend these numbers. It is, however, advised that $n \geq 30$ and, if using proportions, that $.1 \leq p \leq .9$.

We now turn to an example which further illustrates the sampling distribution for a sample mean.

Example 5.3 Suppose we have a population (called X) of size $N = 5$ whose elements are 1, 2, 3, 4, and 5. Clearly,

$$\mu = \frac{1+2+3+4+5}{5} = 3, \quad \text{and}$$

$$\sigma^2 = \frac{(1-3)^2 + (2-3)^2 + (3-3)^2 + (4-3)^2 + (5-3)^2}{5} = 2.$$

Suppose we take a random sample of size $n = 2$ from this population. How many different samples of size $n = 2$ are possible? Combinatorics tells us that there are

$$\binom{5}{2} = 10 \quad \text{possible samples.}$$

These 10 samples, together with the mean of each sample, are listed here:

Sample of Size n=2	$\bar{x}$
(1, 2)	1.5
(1, 3)	2.0
(1, 4)	2.5
(1, 5)	3.0
(2, 3)	2.5
(2, 4)	3.0
(2, 5)	3.5
(3, 4)	3.5
(3, 5)	4.0
(4, 5)	4.5

If we tally each of the possible $\bar{x}$'s that can result from a sample of size n = 2, together with its relative frequency of occurrence, we have the following table.

$\bar{x}$	$P(\bar{x})$
1.5	1/10
2	1/10
2.5	2/10
3	2/10
3.5	2/10
4	1/10
4.5	1/10

Since $\bar{x}$ in this example can only take on 7 discrete values, it is a discrete random variable (RV). The probability distribution given in this table is the **sampling distribution of the mean for samples of size n = 2**. In other words, the sampling distribution of the mean is the distribution of all possible means that can arise from all possible samples of a given size, in this case, n = 2. Since the sample mean is dependent on the random sample chosen, prior to sampling we will refer to the sample mean as the random variable $\bar{X}$. The distribution for the discrete RV $\bar{X}$ is given in the table above. Since $\bar{X}$ is an RV, it has a mean and variance.

$E(\overline{X})$ $= 1.5(1/10) + 2(1/10) + 2.5(2/10) + 3(2/10) + 3.5(2/10) + 4(1/10)$

$+ 4.5(1/10) = 3.$

$V(\overline{X})$ $= \sigma^2(\overline{X}) = E(\overline{X}^2) - [E(\overline{X})]^2$

$= (1.5)^2(1/10) + (2)^2(1/10) + (2.5)^2(2/10) + (3)^2(2/10) + (3.5)^2(2/10)$

$+ (4)^2(1/10) + (4.5)^2(1/10) - [3]^2$

$= 9.75 - 9 = .75,$

and thus $\sigma^2(\overline{X}) = .75.$

The important points to notice about this example are (1) the mean of the sampling distribution of $\overline{X}$, namely $E(\overline{X})$, turned out to be **the same as** μ, i.e., $E(\overline{X}) = \mu$ $= 3$; and (2) the variance of the RV $\overline{X}$ was much **smaller than the variance of the population**, i.e., $\sigma^2(\overline{X}) = .75 < 2 = \sigma^2$. That is, the distribution of $\overline{X}$ will be centered at the same point as the center of the population, but the variance of $\overline{X}$ will be smaller than the variance of X.

∎

While the previous example illustrates sampling from a very small population for the sake of enumerating all possible samples, the next example demonstrates how a sampling distribution can be obtained from an infinite population.

Example 5.4 Suppose we wish to sample from an infinite population consisting of only 3 different kinds of elements, namely 1, 2, or 3, where each of these elements is equally likely (*Note:* another way of looking at this is that we have a finite population consisting of only the three elements 1, 2, and 3, and we sample *with replacement*). We now investigate the sampling distribution of the mean for several different sample sizes: n = 1, 2, and 3.

(a)   **Sample size is n = 1.** In this case, there are only three possible samples: (1), (2), and (3), the means of which are obviously 1, 2, and 3, respectively. Since each of the three samples is equally likely, the probability distribution for the sample mean looks like the following:

	x̄	P(x̄)
for **n = 1**:	1	1/3
	2	1/3
	3	1/3

A graph of this distribution is shown in Figure 5.3.

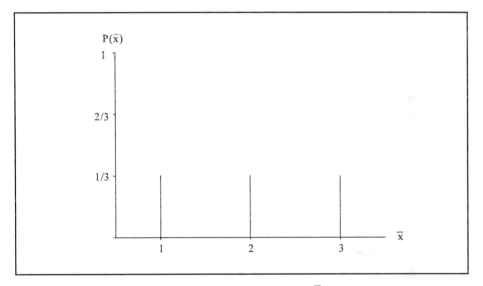

Figure 5.3 Sampling Distribution of $\overline{X}$ for n = 1

This case of n = 1 is a trivial case, but it does point out that the sampling distribution of the mean for n = 1 is identical to the population distribution. Since the mean and variance are measures of interest, we calculate them as follows:

$$E(\overline{X}) = 1(1/3) + 2(1/3) + 3(1/3) = 2 = \mu$$

$$V(\overline{X}) = \sigma^2(\overline{X}) = E(\overline{X}^2) - [E(\overline{X})]^2$$

$$= 1^2(1/3) + 2^2(1/3) + 3^2(1/3) - [2]^2$$

$$= 1/3 + 4/3 + 9/3 - 4$$

$$= 14/3 - 4 = 2/3 = \sigma^2$$

(b)   **Sample size is n = 2**. In this case, there are 9 possible samples, each of
      which is equally likely. These, along with the mean of each sample, are
      as follows:

Sample of Size n=2	Sample Mean (x̄)
(1, 1)	1
(1, 2)	3/2
(1, 3)	2
(2, 1)	3/2
(2, 2)	2
(2, 3)	5/2
(3, 1)	2
(3, 2)	5/2
(3, 3)	3

The samples are given as ordered pairs, with the first element of each pair
representing the result of the first draw, and the second element representing the
second draw. From the nine samples we see five different means, some occurring
more frequently than others. Since each of the nine samples is equally likely to be
selected, we construct the sampling distribution of $\overline{X}$ by observing the frequencies
of each of the five possible means.

	x̄	P(x̄)
	1	1/9
	3/2	2/9
for **n = 2**:	2	3/9
	5/2	2/9
	3	1/9

The graph of this distribution is shown in Figure 5.4.

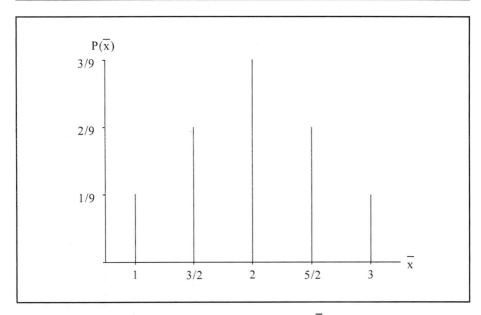

Figure 5.4 Sampling Distribution of $\overline{X}$ for n $=$ 2

Calculating the mean and variance of $\overline{X}$, we have

$$E(\overline{X}) \quad = 1(1/9) + (3/2)(2/9) + 2(3/9) + (5/2)(2/9) + 3(1/9) = 2$$

and

$$\sigma^2(\overline{X}) \quad = 1^2(1/9) + (3/2)^2(2/9) + 2^2(3/9) + (5/2)^2(2/9) + 3^2(1/9) - [2]^2$$

$$= 1/9 + \frac{1}{2} + 4/3 + 25/18 + 1 - 4 = 60/18 + 1 - 4$$

$$= 10/3 - 3 = 1/3.$$

Note that while $E(\overline{X}) = 2$ is the same as μ, $\sigma^2(\overline{X}) = 1/3$ is smaller than $\sigma^2 = 2/3$, the population variance.

(c)   **Sample size is n $=$ 3.** There are 27 equally likely possible samples when n $=$ 3. These, along with their respective means, are as follows:

Sample	Sample Mean ($\bar{x}$)
(1, 1, 1)	1
(1, 1, 2)	4/3
(1, 1, 3)	5/3
(1, 2, 1)	4/3
(1, 2, 2)	5/3
(1, 2, 3)	2
(1, 3, 1)	5/3
(1, 3, 2)	2
(1, 3, 3)	7/3
(2, 1, 1)	4/3
(2, 1, 2)	5/3
(2, 1, 3)	2
(2, 2, 1)	5/3
(2, 2, 2)	2
(2, 2, 3)	7/3
(2, 3, 1)	2
(2, 3, 2)	7/3
(2, 3, 3)	8/3
(3, 1, 1)	5/3
(3, 1, 2)	2
(3, 1, 3)	7/3
(3, 2, 1)	2
(3, 2, 2)	7/3
(3, 2, 3)	8/3
(3, 3, 1)	7/3
(3, 3, 2)	8/3
(3, 3, 3)	3

Compiling the frequencies of each of the 7 possible means, we obtain the following sampling distribution of the mean when n = 3.

	$\bar{x}$	$P(\bar{x})$
	1	1/27
	4/3	3/27
	5/3	6/27
for **n = 3**:	2	7/27
	7/3	6/27
	8/3	3/27
	3	1/27

The graph of this distribution is shown in Figure 5.5.

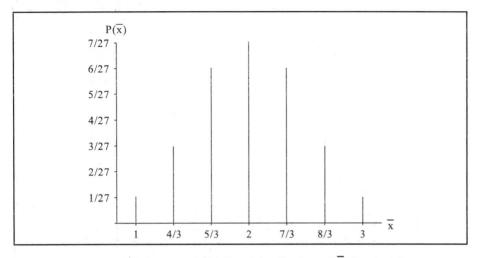

Figure 5.5 Sampling Distribution of $\overline{X}$ for n = 3

The mean and variance for the RV $\overline{X}$ when n = 3 are (computations not shown, see Problem 5.1) $E(\overline{X}) = 2$ and $\sigma^2(\overline{X}) = 2/9$.

We summarize the results (means and variances) of these 3 cases in the following table.

	n = 1	n = 2	n = 3	n = 4
$E(\overline{X})$	2 (μ)	2	2	?
$\sigma^2(\overline{X})$	2/3 (σ^2)	2/6	2/9	?

Table 5.1 Summary Measures for the Sampling Distributions
of $\overline{X}$ for Three Different Sample Sizes (n)

The values in the shaded column are the same as the population parameters, i.e., $\mu = 2$, and $\sigma^2(X) = \sigma^2 = 2/3$. The student is encouraged to establish a pattern, if possible, and determine what values should be placed in the column labeled n = 4 (feedback will be provided in the next section). Also, see Problem 5.2.

We summarize the results of these two examples in the following definition and comments.

The **sampling distribution of the mean** ($\overline{X}$) for samples of a given size n is the distribution of all possible means that result from all possible samples of size n.

Definition 5.1 Sampling Distribution of the Mean

In the two previous examples we have actually listed *all* of the possible samples of a given size (like n = 2 or n = 3) that could be taken from the population. We did that in order to illustrate the concept of a sampling distribution. In reality, however, usually only **one** sample is drawn and inferences are based on that one sample. The important point about looking at the distribution of all possible means is that it will allow us to make some predictions (i.e., probability estimates) about where we would expect the mean of the **one** sample to be. For example, if we revisit Figures 5.4 (sampling distribution of $\overline{X}$ for n = 2) and 5.5 (sampling distribution of $\overline{X}$ for n = 3), we see that a greater proportion of the means clusters about $E(\overline{X}) = \mu = 2$ when n = 3 than when n = 2. More formally, if we arbitrarily use an interval of 1/3 on either side of μ, we see that for n = 3,

$$P(5/3 \leq \overline{X} \leq 7/3) = 6/27 + 7/27 + 6/27 = 19/27 \quad \text{(Fig 5.5, n = 3)}$$

whereas, for n = 2,

$$P(5/3 \leq \overline{X} \leq 7/3) = 3/9 = 9/27 \qquad \text{(Fig 5.4, n = 2)}.$$

In other words, if we took a sample of size n = 3, we would expect to get a sample mean closer to $\mu = 2$ than if we took a sample of size n = 2. That is not to say that a specific sample of size n = 2 could not result in a mean that is closer to $\mu = 2$ than a specific sample of size n = 3. Certainly, the sample (2, 2) has a mean of exactly 2, whereas the sample (3, 3, 3) has a mean of 3. All we are saying at this point, however, is that a sample of size n = 3 is *more likely* to produce a mean closer to $\mu = 2$ than will a sample of size n = 2. This notion of closeness to the population mean μ, which will become more formalized in the next few sections, is not counterintuitive. Certainly our intuition tells us that an average of, say, 10 values from a population will typically give us a better estimate of the population mean than will an average of, say, only 2 values.

We conclude this section with a brief clarification of terminology that is often confusing to students. When we use the phrase "sample of size n = 10," we mean **one** sample consisting of 10 randomly selected items (or values) from a population. Students often mistakenly restate this phrase by saying "take 10 samples from a population," which means nothing because it says nothing about the size of each sample. Furthermore, it obliterates the fact that seldom do resources (like time or money) allow more than **one** sample to be taken. The student is strongly encouraged to verbalize these concepts because correct verbalization enhances retention. We reiterate that the only reason we looked at **all** possible samples of a given size was to strengthen the student's understanding of a sampling distribution. Again, knowledge about a sampling distribution will allow us to make statements about the precision of our estimates. The next sections will elaborate on this point.

5.2 Standard Error of the Mean

The examples in the preceding section illustrated two very important results:

(1)   the mean of the sampling distribution of $\overline{X}$ **is the same as** the mean of the population sampled, i.e., $E(\overline{X}) = \mu$.

(2)   the variance of the sampling distribution of $\overline{X}$ **is smaller than** the variance of the population sampled, i.e., $\sigma^2(\overline{X}) < \sigma^2$.

The result given in (1) was quite apparent in Example 5.4, and the student should have had no difficulty in filling in the last column of row 1 in Table 5.1. That is, $E(\overline{X}) = \mu = 2$ for n = 4. In particular, it would not have mattered what the size of the sample was; $E(\overline{X}) = \mu$ for any sample size n. We restate this result as an important property of the sampling distribution of the mean.

> The expected value of the random variable $\overline{X}$ is
> $$E(\overline{X}) = \mu$$
> regardless of the sample size.

Property 5.1 Mean of the Sampling Distribution of $\overline{X}$

The result given in (2) above, namely $\sigma^2(\overline{X}) < \sigma^2$, was also apparent from Table 5.1, but the exact relationship between $\sigma^2(\overline{X})$ and σ^2 was more difficult to pinpoint. Perhaps the student who has worked Problem 5.2 and filled in the last column of Table 5.1 has already deduced the relationship. Although not difficult, the derivation of this relationship is beyond the scope of this text. We present it here as another important property of the sampling distribution of the mean.

> The variance of the random variable $\overline{X}$ is dependent
> on the size of the sample taken and is given by
> $$\sigma^2(\overline{X}) = \frac{\sigma^2}{n}$$

Property 5.2 Variance of the Sampling Distribution of $\overline{X}$

You can now verify your entry in Table 5.1. For n = 4,

$$\sigma^2(\overline{X}) = \sigma^2/n = (2/3)/4 = 2/12 = 1/6.$$

Since the standard deviation of a random variable is the value used to standardize a population value, it is a very important quantity in inferential statistics. In fact, the standard deviation of the random variable $\overline{X}$ is used so often that it merits a special name: the **standard error of the mean**. We present it as a property of the $\overline{X}$ distribution.

> The standard deviation of $\overline{X}$, or the **standard error of the mean** is
> $$\sigma(\overline{X}) = \frac{\sigma}{\sqrt{n}}$$

Property 5.3 Standard Error of the Mean

Note that the standard error of the mean is dependent on the standard deviation of the population (σ) *and* the sample size (n).

The formulas presented in Properties 5.2 and 5.3 are for sampling distributions taken from either infinite populations or from finite populations in which n/N is very small (i.e., .05 or less). If n/N > .05, then a finite population correction factor of $(N-n)/(N-1)$ is usually applied to σ^2/n to get $\sigma^2(\bar{X})$. Normally, this correction factor need not be applied, so we have formally stated only $\sigma^2(\bar{X})$ and $\sigma(\bar{X})$ for the cases in which the correction factor is not needed, as shown in Properties 5.2 and 5.3. In order to duplicate the $\sigma^2(\bar{X})$ obtained in Example 5.3 of the preceding section, one would have to apply the correction factor of

$$(N-n)/(N-1) = (5-2)/(5-1) = 3/4 \text{ to } \sigma^2/n = 2/2 = 1$$

to obtain the result of .75 shown in that example.

Thus far in this chapter we have emphasized that the sampling distribution of the mean, or the random variable $\bar{X}$, has its own distribution apart from the original population or random variable X. As we will see shortly, sometimes the differences between the X and $\bar{X}$ distributions will be astounding. What we have seen thus far is the fact that the $\bar{X}$ distribution has a decidedly smaller variance (and standard deviation) than the original distribution, with this difference being a function of the size of the sample being taken. We conclude this section with an example that illustrates this fact.

Example 5.5 Consider a population of 200 leading mutual funds for which we are interested in their price-earnings ratio (X). Suppose that X has a mean of $\mu = 20$ and a standard deviation of $\sigma = 9$. Describe the distribution of all possible means from all possible samples of size n = 4. We call this new distribution $\bar{X}$ and know at the outset that $\bar{X}$ is a much bigger population than X. Why? Recall from combinatorics that there are $\binom{200}{4}$ = 64,684,950 (computations not shown) different possible samples of size n = 4 that can be taken from the original population of 200. If we had the time to calculate the means of each of these almost 65 million samples we would find that the average of all of these 65 million means would be 20. That is, Property 5.1 guarantees us that $E(\bar{X}) = \mu = 20$, and we do not have to perform any calculations! We don't even have to know that there are almost 65 million different samples to conclude that $E(\overline{X}) = 20$. Furthermore, Property 5.3 tells us that $\sigma/\sqrt{4} = 9/2 = 4.5$. This means that the standard deviation of the new population $\overline{X}$ is half (4.5 vs 9) of the standard deviation of the old population X. This result says that, in order to cut the standard deviation in **half**, we will have to **quadruple** the sample size (i.e., from n = 1 to n = 4). This fact, which is a direct result of the "ubiquitous" $\sqrt{n}$ in the denominator of Property 5.3, has serious ramifications for future variance reduction strategies.

∎

5.3 Central Limit Theorem

The key to unraveling the mysteries of inferential statistics is contained in the central limit theorem. No doubt the most important result in all of theoretical statistics, the central limit theorem is the major link between the normal distribution and sampling distributions. Recall that we stated in Chapter 4 that the normal distribution was certainly the most important distribution of all. Hopefully, the central limit theorem will make that statement come alive for the reader. Since the proof of this theorem is beyond the scope of our discussion, we state the central limit theorem without proof.

Central Limit Theorem (CLT): For almost all populations, the sampling distribution of the mean can be approximated closely by a normal distribution, provided the sample size is sufficiently large.

Central Limit Theorem

The CLT states that, no matter what kind of distribution (or population) we sample from, if the sample size is big enough, the sampling distribution of $\overline{X}$ is approximately normal. This is an amazing result which allows us to know the form of the distribution from which a sample mean x̄ is extracted. We now address each of the two "cautions" imbedded in the CLT, the first being "for almost all populations." The CLT applies to both infinite and finite populations, as long as the population has a finite mean μ and variance σ^2 [Ostl 75]. Pathological examples of density functions for which μ or σ^2 do not exist, such as the Cauchy distribution [Rice 88], can be constructed. However, for all practical purposes, the CLT applies to all sampled populations.

The second caveat, "provided the sample size is sufficiently large," is a bit nebulous because the term "sufficiently large" is not definitive. The size of the sample needed to produce a sampling distribution of the mean that is approximately normal really depends on the functional form of the population being sampled. A result of computer simulation, Figure 5.6 shows how the size of the sample affects the "normalization" of the sampling distribution of the mean for each of four different kinds of parent populations (i.e., X distributions). Figure 5.6(a) shows that, if the parent population is normal, all of the sampling distributions of the mean are also (exactly) normal, regardless of the sample size. However, as the sample size n increases, the sampling distribution of the mean becomes taller and narrower. That should not be surprising since, by Property 5.3, $\sigma(\overline{X}) = \sigma/\sqrt{n}$. Figure 5.6(b) shows the trend toward normality when samples of various sizes are taken from a uniform or rectangular parent population. Approximate normality is already apparent when n = 5. The exponential parent population in Figure 5.6(c) generates sampling distributions which retain some skewness (to the right) as n gets bigger, but approximate normality is clearly evident at n = 30. Even the extreme parabolic parent population in Figure 5.6(d) exhibits approximate normality by the time n = 30. Although it is difficult to say exactly how large n must be to guarantee approximate normality for the sampling distribution of $\overline{X}$, n = 30 is usually more than adequate. Hence, we shall use n ≥ 30 as being "sufficiently large." Alternatively, when n < 30 we shall refer to the sample as being a "small" sample. A restatement of the Central Limit Theorem in everyday language is given here to help the reader remember this important result.

No matter what the parent looks like, the child will be normal, especially by the age of 30.

The Central Limit Theorem Restated

The CLT should be a comforting thought for parents!

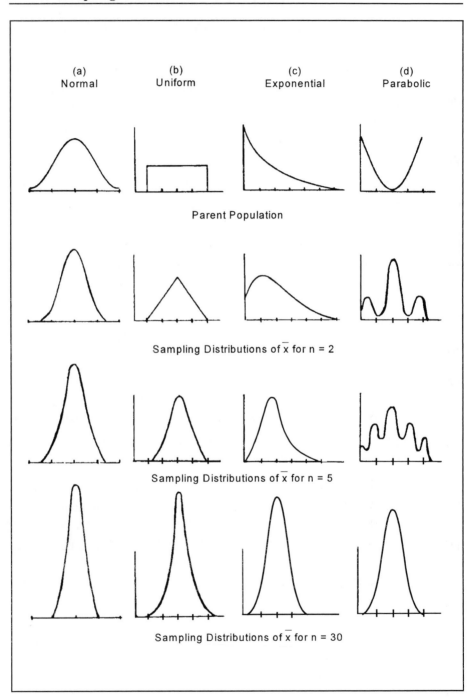

Figure 5.6 Sampling Distributions of $\overline{X}$ for Various Sample Sizes

Recall from Definition 4.11 that, if X is a normal random variable, we can standardize that RV X by

$$Z = \frac{X - \mu}{\sigma}.$$

Now, by the CLT, we know that the RV $\bar{X}$ is approximately normal whenever n is sufficiently large. Using this fact, together with the standard error of the mean (Property 5.3), we are able to standardize the RV $\bar{X}$ as shown in the following property.

$$\mathbf{Z} \;=\; \frac{\mathbf{\bar{X}} - \mathbf{\mu}}{\dfrac{\sigma}{\sqrt{\mathbf{n}}}}$$

Property 5.4 Standardized $\bar{X}$ when $\bar{X}$ is Normal

Property 5.4 says that the quantity $\dfrac{\bar{X} - \mu}{\sigma/\sqrt{n}}$ is distributed as the standardized normal (Z) RV whenever $\bar{X}$ is normally distributed. If $\bar{X}$ is approximately normal, then $\dfrac{\bar{X} - \mu}{\sigma/\sqrt{n}}$ is approximately Z-distributed. We can use this property to calculate probabilities involving $\bar{X}$, as shown in the following examples.

Example 5.6 Suppose the weight (X) of adult males is normally distributed with a mean of 170 pounds and a standard deviation of 20 pounds. That is, X ~ N(170, 20²). If we were to take a sample of 16 adult males, what is the probability that the sample mean will fall between 160 and 180 lbs? We seek

$$P(160 \le \bar{X} \le 180) \;=\; P(\bar{X} \le 180) - P(\bar{X} \le 160)$$

$$= \; P\!\left(Z \le \frac{180 - 170}{20/\sqrt{16}}\right) - P\!\left(Z \le \frac{160 - 170}{20/\sqrt{16}}\right)$$

$$= \; P(Z \le 2) - P(Z \le -2)$$

$$= \quad .9772 - [1 - P(Z \leq 2)]$$

$$= \quad .9772 - [1 - .9772]$$

$$= \quad .9772 - .0228$$

$$= \quad .9544.$$

If we calculate the probability that a randomly selected adult male will weigh between 160 and 180 lbs, we have

$$P(160 \leq X \leq 180) \quad = \quad P(X \leq 180) - P(X \leq 160)$$

$$= \quad P\left(Z \leq \frac{180 - 170}{20} \right) - P\left(Z \leq \frac{160 - 170}{20} \right)$$

$$= \quad P(Z \leq .5) - P(Z \leq - .5)$$

$$= \quad .6915 - [1 - .6915]$$

$$= \quad .6915 - .3085$$

$$= \quad .3830.$$

These two probabilities (or areas) can be seen in the following diagram. The shaded area is .3830, the area under X between 160 and 180. The area under the taller curve ($\overline{X}$) between 160 and 180 is .9544, as computed earlier. Comparing these two values, one can see that the probability of a sample mean (from a sample of size 16) falling between 160 and 180 is two and a half times the likelihood that an individual weight will fall between 160 and 180. The effect of the standard error of the mean ($\sigma/\sqrt{n}$) in narrowing the sampling distribution of $\overline{X}$ is clearly evident.

■

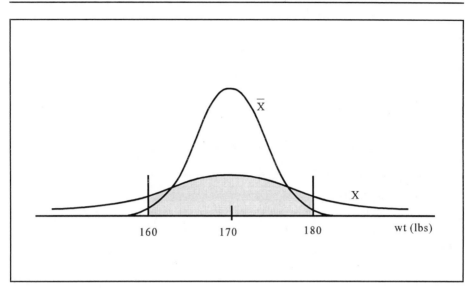

Because n = 16 < 30, the critical reader may have questioned the use of Property 5.4 to standardize $\overline{X}$ in computing P(160 ≤ $\overline{X}$ ≤ 180) = .9544 in the previous example. Recall that if the parent population is normal (which was the case in Example 5.6), then the sampling distribution of $\overline{X}$ is also normal for *any* sample size n. Reference Figure 5.6(a). If the form of the parent population (i.e., RV X) had not been known, then the use of Property 5.4 when n < 30 certainly should be questioned.

Example 5.7 A brick production process has historically produced bricks having weights that are normally distributed with μ = 4 lbs and σ = .05 lbs. That is, X (brick weight) ~ N(4, .05²). Based on this process information, if we consider samples of size 10, what is the symmetric interval around μ within which $\overline{X}$ will fall 95% of the time? The next diagram shows the shaded area of .95 that symmetrically encompasses E($\overline{X}$) = μ = 4. What we seek are the two values, X_1, and X_2, on the $\overline{X}$-axis. Note that X_1 and X_2 are two values from the $\overline{X}$ distribution and that they are the same distance from 4. We have also drawn in the Z-axis, because the Z distribution will be our link to the $\overline{X}$ distribution via Property 5.4. We already know that 0 on the Z-axis corresponds to 4 on the $\overline{X}$-axis (use the fact that μ = 4 and property 5.4 to verify this).

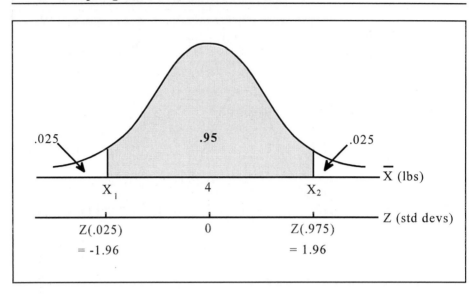

We also know that the point on the Z-axis that corresponds to X_2 on the $\overline{X}$-axis is the 97.5[th] percentile of the Z distribution because the area under the curve *completely* to the left of this point is .975. Notationally, we use Z(.975) to denote the 97.5[th] percentile of the Z distribution. Using the selected percentiles portion of Appendix D, we see that Z(.975) = 1.96. This means that 97.5% of the area under a normal curve lies to the left of the 1.96 standard deviation mark. Recall that a value in the Z distribution represents a "number of standard deviations." Well, how many pounds does one standard deviation in the $\overline{X}$ distribution represent? This, of course, is given by $\sigma(\overline{X}) = \sigma/\sqrt{n}$ which is what we call the standard error of the mean, which also happens to be the denominator in Property 5.4. Hence, if we seek the point X_2 which is 1.96 standard deviations to the right of center in the $\overline{X}$ distribution and we know that each standard deviation is worth $\sigma/\sqrt{n} = .05/\sqrt{10} = .0158$ lbs, then

$$X_2 = 4 + 1.96(.0158) = 4 + .031 = 4.031 \text{ lbs}.$$

Likewise, X_1 will be .031 lbs to the left of 4, so

$$X_1 = 4 - .031 = 3.969 \text{ lbs}.$$

We have thus answered the question posed in this example. That is, 95% of the time we would expect a sample of 10 bricks to produce a mean weight between 3.969 and 4.031 lbs. Note that Property 5.4 serves as the bridge that allows us to move between the Z distribution (for which we have areas and percentiles in Appendix D) and the $\overline{X}$ distribution (for which we have no tables). If we solve for $\overline{X}$ using Property 5.4, we obtain

$$\overline{X} \;=\; \mu + Z(.975)[\sigma/\sqrt{n}]$$

which is precisely what we used to obtain X_2 and X_1 above. For example,

$$X_2 \;=\; \mu + Z(.975)(\sigma/\sqrt{n})$$

$$=\; 4 + 1.96(.05/\sqrt{10})$$

$$=\; 4 + .031$$

$$=\; 4.031.$$

Similarly,

$$X_1 \;=\; \mu + Z(.025)(\sigma/\sqrt{n})$$

$$=\; 4 - 1.96(.05/\sqrt{10})$$

$$=\; 4 - .031$$

$$=\; 3.969.$$

∎

The principles illustrated in the previous example are extremely important, both in statistical process control and in estimation, the latter of which will be addressed in the next section. Suppose that a process for which you are responsible is designed to produce a product with a target or mean weight of 4 pounds. Suppose that you also know this process has some variability in it and that when the process is in control, you expect 95% of the samples (of size 10) to produce a mean weight between 3.969 and 4.031 lbs (the result in Example 5.7). If two consecutive samples produced mean weights of 3.91 and 3.85 lbs, respectively, what would your reaction be?

5.4 Estimating the Population Mean

In the previous sections of this chapter, we have dealt with examples in which the population mean μ was known. Unfortunately, the population mean μ is often not known, and one of the most common problems in statistical inference is to estimate μ from sample data. The fact that this is an important, as well as common, problem is evident in the following scenarios which indicate that finding μ would be either impractical (costwise or timewise) or impossible.

(a)   The regional manager of a supermarket chain would like to know the average checkout time of customers after having installed new scanning devices in her stores.

(b)   The Air Force would like to be able to estimate the average lifetime of the engines on the F-15 aircraft.

(c)   A company concerned about its employees and their stress levels would like to be able to estimate average blood pressure.

(d)   A traffic planning committee would like to estimate the average daily traffic (in number of cars) at a given intersection.

(e)   A fleet manager for a rental car company would like to estimate the average gas mileage for the vehicles in his inventory.

(f)   A hospital would like to estimate the average proportion of bills which are delayed due to errors made by the hospital staff and administrators.

(g)   A software company would like to estimate the average number of discrepancies made per 1000 lines of modified code.

We illustrate how sample data can be used to estimate a population mean by considering the supermarket scenario given in (a) above. Taking a random sample of 36 customers and recording the time (in minutes) that it took for each to check out at the supermarket cashier, we have the data as shown in Table 5.2.

15	9	21	14	13	20	14	21	12
13	7	19	9	16	10	14	12	21
11	13	15	6	11	18	17	16	15
14	6	15	17	20	17	8	14	11

Table 5.2 Supermarket Checkout Times (min), n = 36

From this sample data we would like to estimate the population mean μ, i.e., the true but unknown average checkout time for all customers. The sample mean of this data is $\bar{x} = 14$ minutes, and it certainly can be used to estimate μ. Such an estimate for μ is called a **point estimate** because it is a single number or point on the number line. Although it is not developed here, it can be shown that the sample mean is an unbiased, consistent estimator of the population mean, and that it is a more efficient estimator for μ than, say, the sample median [Devo 87].

While the sample mean, $\bar{x}$, is a convenient value to use as an estimate for μ, there are some obvious shortcomings in using a point estimate like $\bar{x}$ to estimate μ. First, a point estimate gives no indication of how much information was used to obtain μ. Second, and even more important, a point estimate leaves one void of any feeling about how close $\bar{x}$ is to μ. With the sampling distribution of $\bar{X}$ and the CLT as tools, we now have the ability to quantify the error that is associated with using $\bar{x}$ to estimate μ. We do this by developing the notion of a confidence interval for μ, i.e., an **interval estimate** for μ.

The CLT tells us that the distribution from which we obtain a single mean, like $\bar{x} = 14$ in the supermarket scenario, is normal. Furthermore, we know from Property 5.1 that $E(\bar{X}) = \mu$ so the $\bar{X}$ distribution is centered at μ as shown in Figure 5.7, although we do not know what μ is. Our intuition tells us that if we select a random element (i.e., the mean of the random sample selected) from the $\bar{X}$ distribution, it is more likely to fall near the center of the distribution (i.e., near μ) than it is in either of the two tails of the distribution. We formalize this intuitive feeling by partitioning the area under the normal curve into 3 areas, as shown in Figure 5.7.

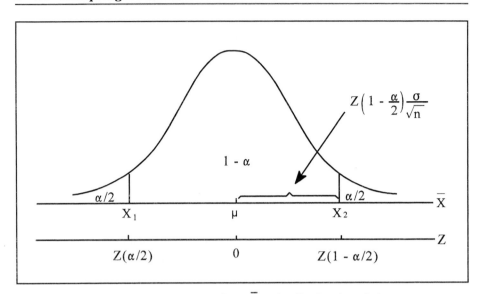

Figure 5.7 Partitioning of the $\overline{X}$ Distribution into 3 Areas

The area labeled $1 - \alpha$ is an area that is symmetric about μ. Each of the tails has an area of $\alpha/2$, which is a direct result of (1) the normal curve is symmetric about μ, and (2) the total area under the curve must be 1. We partition the area under the curve in this manner, using the variable α, so that we can increase or decrease the areas (probabilities) at our discretion simply by altering α. If $\alpha = .05$, we have precisely the situation shown in Example 5.7 of the previous section. In general, Figure 5.7 tells us that

$$P(X_1 \leq \overline{X} \leq X_2) = P(Z(\alpha/2) \leq Z \leq Z(1- \alpha/2))$$

$$= P(Z \leq Z(1 - \alpha/2)) - P(Z \leq Z(\alpha/2))$$

$$= (1 - \alpha/2) - (\alpha/2)$$

$$= 1 - \alpha.$$

A question of concern is, how far is X_2 from μ in Figure 5.7? Since X_2 is $Z(1 - \alpha/2)$ standard deviations to the right of μ and one standard deviation in the $\overline{X}$ distribution is worth $\sigma/\sqrt{n}$ (this is the standard error of the mean), it follows that the distance from μ to X_2 in Figure 5.7 is

$$Z(1 - \frac{\alpha}{2}) \, \frac{\sigma}{\sqrt{n}} .$$

This expression represents the maximum error in estimating μ with $\bar{x}$ that could result if $\overline{X}$ falls anywhere in the interval $[X_1, X_2]$. However, we have just seen that

$$P(X_1 \le \overline{X} \le X_2) = 1 - \alpha.$$

Combining these two statements results in a probability of $1 - \alpha$ that the maximum error is no more than $Z(1 - \frac{\alpha}{2}) \, \frac{\sigma}{\sqrt{n}}$. Suppose that, in our supermarket scenario, we choose $\alpha = .05$ and assume that $\sigma = 5$ is known from past experience. Thus,

$$Z\left(1 - \frac{\alpha}{2}\right) \frac{\sigma}{\sqrt{n}} = Z(.975) \, \frac{5}{\sqrt{36}} = 1.96 \left(\frac{5}{6}\right) = 1.63,$$

and we can be 95% certain that our error (i.e., the difference between $\bar{x}$ and μ) will be no more than 1.63 minutes.

Interestingly, the statement just made is independent of the sample mean obtained. In obtaining the error estimate of no more than 1.63 minutes, we did not use $\bar{x} = 14$ anywhere. The error estimate depends on only these three values:

(1)   The choice of α. As α gets smaller, $1 - \alpha$ gets bigger and $Z(1 - \alpha/2)$ also gets bigger.

(2) σ, where σ is a measure of the variability in the population. The more variability in the population, the bigger the error estimate.

(3)   n, where n is the sample size. As n gets bigger, the error estimate will decrease, provided the other components (σ and α) are not varied.

We have just seen that, given an arbitrary choice for α, we are able to make a probability statement (involving α) about the magnitude of error in estimating μ with x̄.

As shown above, the error estimate could be obtained by choosing α, knowing σ, and selecting n. Actual sampling may not have been accomplished at this point in time. Suppose we now collect the random sample of 36 values that are shown in Table 5.2 and obtain x̄ = 14. What kind of statement can now be made? Since the error estimate of

$$ Z\left(1 - \frac{\alpha}{2}\right) \frac{\sigma}{\sqrt{n}} = 1.63 $$

could be applied to either side of x̄ = 14, we are inclined to say that μ lies between 14 ± 1.63 or

$$ 12.37 \leq \mu \leq 15.63 $$

with probability .95. Knowing that probability statements must involve random variables, and since μ is an unknown **constant**, not a random variable, we elect to use the term "confidence" rather than "probability." The proper terminology is thus, "we are 95% **confident** that the unknown population mean μ falls between 12.37 and 15.63 minutes."

The arbitrarily selected quantity **1 – α** is called the **confidence coefficient**, and α is referred to as the **level of significance**. The confidence coefficient designates the probability that the interval estimate methodology will generate a correct interval estimate for μ. A correct interval estimate is one which captures μ. The interval **L** ≤ μ ≤ **U** is called a **confidence interval** for μ. The values **L** and **U** are called the **lower** and **upper confidence limits**, respectively. These limits are computed by

$$ \begin{matrix} U \\ L \end{matrix} = \bar{x} \pm Z\left(1 - \frac{\alpha}{2}\right) \frac{\sigma}{\sqrt{n}} $$

Equation 5.1 Upper and Lower Confidence Limits for μ

The most commonly used confidence coefficients are .90, .95 and .99. Sometimes a confidence interval is referred to as a $100(1 - \alpha)$ **percent** confidence interval. For example, a 99% confidence interval is a confidence interval with a confidence coefficient of .99.

The precise meaning of a confidence interval is typically not easy to understand for those being exposed to it for the first time. We conclude this section by looking at the supermarket example from a slightly different perspective. As before, we assume that $\sigma = 5$ is known from historical data, that we use a sample size of $n = 36$, and that $1 - \alpha = .95$. Thus, the maximum possible error or confidence interval half-width is 1.63 minutes, as computed previously. Recall that this value was computed independent of the sample chosen (because σ was known). The sampling distribution of $\overline{X}$ with $n = 36$ is reproduced in Figure 5.8 with the half-interval width of 1.63 annotated.

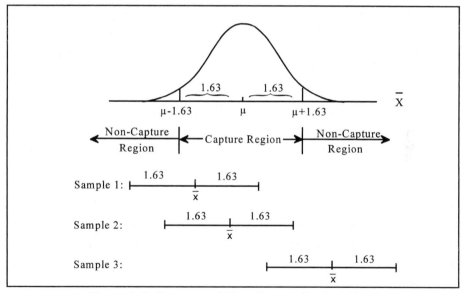

Figure 5.8 Capture and Non-Capture Regions for $\bar{x}$

Figure 5.8 shows the capture and non-capture regions for $\overline{X}$. If the sample selected results in an $\bar{x}$ which falls in the capture region, then the resulting confidence interval with limits $\bar{x} \pm 1.63$ will overlap or "capture" μ when it is superimposed on the $\overline{X}$-axis. The first two samples shown in Figure 5.8 will result in confidence intervals that "capture" μ. The third sample, however, produces a confidence interval that does not "capture" μ because the x̄ of this sample falls outside of the capture region. In this scenario (with σ known), the width of *every* confidence interval is the same as the width of the capture region, namely (2)(1.63) = 3.26. The true meaning of a 95% confidence interval is that, if sampling were repeated over and over again (which in reality it is not) with a confidence interval constructed for every sample, we would expect 95% of the confidence intervals so constructed to capture μ. We emphasize that, in actuality, only one sample is taken, and the confidence interval, $L \le \mu \le U$, is constructed from the data in this sample. We then say that "we are 95% confident that μ lies between L and U." The word "confident" is reserved for post-sampling statements, i.e., after x̄, L, and U are computed. Prior to sampling, $\overline{X}$ is a random variable and probability statements involving $\overline{X}$, like $P(X_1 \le \overline{X} \le X_2) = 1 - \alpha$, are legitimate.

The development of the upper and lower confidence limits, U and L, in this section focused on first finding the error, $Z(1 - \alpha/2)[\sigma/\sqrt{n}]$, and then adding and subtracting it from x̄ to get the U and L shown in Equation 5.1. It should be noted that the equations for U and L in Equation 5.1 can be obtained directly from Property 5.4 with a little algebra. The interested reader is referred to Problem 5.3 at the end of this chapter.

5.5 Confidence Intervals for the Population Mean

In the previous section, we developed the notion of a confidence interval for the purpose of estimating the population mean μ. Equation 5.1, which gives the computational forms for U and L in terms of x̄, $Z(1 - \alpha/2)$, σ, and n, is a direct result of Property 5.4 which says that the expression

$$Z = \frac{\overline{X} - \mu}{\dfrac{\sigma}{\sqrt{n}}}$$

is distributed as the standardized normal (Z) distribution whenever $\overline{X}$ is normally distributed (and the CLT tells us when $\overline{X}$ is at least approximately normal). Unfortunately, the above expression and Equation 5.1 both involve the population standard deviation, σ. Why is this unfortunate? It is unfortunate because σ is seldom, if ever, known. This section addresses this situation, introduces a new distribution, and expands upon how we compute confidence intervals under varying circumstances.

If the population standard deviation, σ, is not known, we will do what hopefully by now comes naturally. That is, we will use the sample standard deviation, s, as an approximation for σ. If we replace σ by s in Property 5.4, we obtain the expression

$$\frac{\overline{X} - \mu}{s/\sqrt{n}},$$

which looks innocent enough. However, W.S. Gosset, an Irish brewery employee who published under the name "Student," showed in the early 1900s that this expression does not follow a Z distribution. In fact, when n is small, the distribution of $\frac{\overline{X} - \mu}{s/\sqrt{n}}$ is sufficiently different from the Z distribution that its use is warranted in certain cases. This distribution came to be known as Student's t distribution or what we shall call the t distribution. The quantity $\frac{\overline{X} - \mu}{s/\sqrt{n}}$ is known as the t statistic and follows the t distribution, provided $\overline{X}$ is the mean of a random sample from a normal parent population with mean μ. We state this in the following property, which is analogous to the Z statistic of Property 5.4.

$$t = \frac{\overline{X} - \mu}{\dfrac{s}{\sqrt{n}}}$$

which should be used only when X is normal with mean μ and σ is unknown.

Property 5.5 t Statistic Follows a t Distribution with (n - 1) df

Before showing how we use the t distribution in computing confidence limits, we digress for a moment to expand upon the properties of the t distribution. The t distribution is really a family of distributions because there is a separate t distribution for each sample size n. Figure 5.9 shows the t distributions for n = 2 and n = 8, as well as the Z distribution.

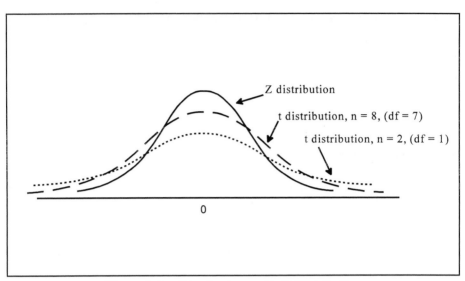

Figure 5.9 t Distributions with Z Distribution

The major properties of the t distribution are as follows:

1)   t distribution is centered at 0 and is symmetric about its mean (like the Z distribution).

2)   t distribution has a variance greater than 1, but as n increases, the variance approaches 1.

3)   t distribution is characterized by less area in the middle and more area in the tails than the Z distribution.

4)   t distribution approaches the Z distribution as n increases.

Note that in Figure 5.9, each particular t distribution in the family of t distributions is indexed by what is called its degrees of freedom (df). The "degrees of freedom" for a particular t distribution is given by $df = n - 1$, where n is the sample size. The term "degrees of freedom" is one of the most difficult in all of statistics to define. For the purpose of the t distribution, one can think of df as an "index" to a particular member of the entire family of t distributions. The t distribution has one parameter, and it is $df = n - 1$. It will be necessary to know what df is for the particular t distribution under consideration because the values in the t table (Appendix E) are "indexed" on df.   The following very cursory approach to understanding what "degrees of freedom" is and why df $= n - 1$ is provided. Suppose we were going to choose a sample of size n $= 5$ elements and we already knew that $\bar{x} = 10$. How many "free" choices would we have in choosing the 5 values that constitute our sample? We could choose the first 4 values in any manner we desire. For example, we freely choose $x_1 = 4$, $x_2 = 16$, $x_3 = -8$, and $x_4 = 20$. Our fifth choice, however, is constrained by the fact that $\bar{x} = 10$. Hence,

$$\bar{x} = \frac{4 + 16 - 8 + 20 + x_5}{5} = 10$$

implies that

$$x_5 = 50 - 4 - 16 + 8 - 20 = 18.$$

Thus, we were free to choose $n - 1 = 5 - 1 = 4$ of the elements, while the fifth element was fixed due to the $\bar{x} = 10$ constraint. In this case, then, df $= n - 1 = 5 - 1 = 4$. Before we return to computing confidence limits, we present the following example which shows how to find a certain percentile of a given t distribution.

Example 5.8   If the sample size is n $= 10$, find the 95th percentile of the corresponding t distribution. As shown in the following diagram, we seek a value on the horizontal axis of the t distribution with df $= 10 - 1 = 9$. The value, denoted t(.95,9), is the point on the t-axis such that 95% of the area under the curve lies to its left. In this case, all that is needed is a simple table look-up: row 9 under the column headed by .95 in Appendix E. We see that t(.95, 9) $= 1.833$. What is the 10th percentile of this same t distribution? We seek t(.10, 9) which, by symmetry, is $-t(.90, 9)$. Thus, t(.10, 9) $= -t(.90, 9) = -1.383$.

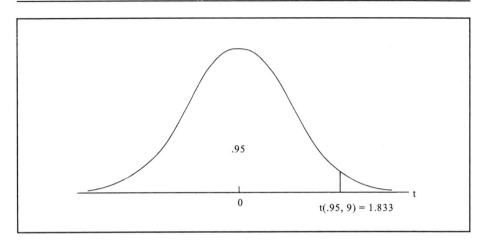

To compute the confidence limits for a $100(1-\alpha)\%$ confidence interval, we have the following template which is an extended version of Equation 5.1.

$$\begin{array}{c} U \\ L \end{array} = \bar{x} \pm \begin{bmatrix} Z\left(1 - \dfrac{\alpha}{2}\right) \\[2ex] \text{or} \\[2ex] t\left(1 - \dfrac{\alpha}{2}, \, n - 1\right) \end{bmatrix} \begin{bmatrix} \dfrac{\sigma}{\sqrt{n}} \\[2ex] \text{or} \\[2ex] \dfrac{s}{\sqrt{n}} \end{bmatrix}$$

Equation 5.2 Computational Template for Confidence Limits

The form of this template should be recognized as being a

point estimate ± error,

where the error is simply a Z or t table look-up times the standard error or estimated standard error of $\bar{X}$. There are basically only two decisions to make. One decision is whether to use σ or s. This is usually very easy, since σ is seldom known. If σ

is not known, we must use s, the sample standard deviation. However, if by chance σ is known, then one should use σ. A second choice rests with using either the Z or t distribution. If σ is known, we shall always use the Z distribution. If only s is known, then we can use the t distribution as long as the underlying parent population is normal. For large n such as n ≥ 30, the t and Z distributions are almost identical so we will typically use the Z distribution whenever n ≥ 30.

As noted previously, σ is seldom known, so s must be used. In this case, the t distribution becomes the workhorse for *small* samples, as long as the underlying normality assumption holds. Our underlying assumption through all of this and the previous section has been that the sampling distribution of $\overline{X}$ is normal (or close to normal). Also recall that it was the CLT that guaranteed $\overline{X}$ to be normal (or close to normal) for any parent population, provided the sample size was large enough (say n ≥ 30). Thus, for small samples (say n < 30), we need to know that the parent population (i.e., the population being sampled) is either normal or approximately normal in order to apply Equation 5.2. Actually, the $\overline{X}$ distribution will be very close to normal even for small sample sizes and some quite gross deviations from normality on the part of the parent (X) distribution. Witness columns (b) and (c) in Figure 5.6, where the $\overline{X}$ distribution is quite normal for n = 5. Thus, the number n = 30 is very conservative when defining a "small" sample. We illustrate the computation of confidence limits in a series of examples.

Example 5.9 A random sample of 64 customers at a local supermarket showed that their average shopping time was 33 minutes with a sample standard deviation of 16 minutes. Find a 90% confidence interval for the true average shopping time.

Since n = 64 > 30, we will use the Z distribution. We have $\bar{x}$ = 33, s = 16, and 1 - α = .90 so α = .10 and α/2 = .05. Using Equation 5.2,

$$\begin{array}{c} U \\ L \end{array} = \bar{x} + Z(.95)\left(\frac{s}{\sqrt{n}}\right)$$

$$= 33 \pm 1.645\left(\frac{16}{\sqrt{(64)}}\right)$$

$$= 33 \pm 1.645(2)$$

$$= 33 \pm 3.29.$$

Thus, we are 90% confident that $29.71 \le \mu \le 36.29$ minutes.

∎

Example 5.10 A test on a random sample of 9 cigarettes yielded an average nicotine content of 15.6 milligrams and a standard deviation of 2.1 milligrams. Construct a 99% confidence interval for the true but unknown average nicotine content of this particular brand of cigarette. Assume that nicotine content is normally distributed. Since $n = 9 < 30$, we will use the t distribution. We have

$$\bar{x} = 15.6, \quad s = 2.1, \quad \text{and} \quad 1 - \alpha = .99.$$

By Equation 5.2,

$$\begin{matrix} U \\ L \end{matrix} = \bar{x} \pm t(.995, 8)\left(\frac{s}{\sqrt{n}}\right)$$

$$= 15.6 \pm 3.355 \left(\frac{2.1}{\sqrt{9}}\right)$$

$$= 15.6 \pm 3.355 \, (.7)$$

$$= 15.6 \pm 2.35,$$

so we are 99% confident that $13.25 \le \mu \le 17.95$ milligrams. Was the assumption that nicotine content is normally distributed necessary in this situation? Why or why not?

∎

The following example demonstrates how we can use the error term in Equation 5.2 to determine the sample size needed to produce an estimate having a specified maximum error.

Example 5.11 A national retail association wants to estimate the average amount of dollars lost each month due to theft in its member stores. If it wants to be 95% confident that the error in its estimate is no more than $100, how many stores would need to be included in the sample to produce an estimate of the desired accuracy.

Previous years' data indicate that $\sigma \approx \$325$. This problem specifies an error ($100) and asks us what n must be to produce a maximum error of $100.

We know from Equation 5.2 that

$$\text{error} = Z\left(1 - \frac{\alpha}{2}\right) \frac{\sigma}{\sqrt{n}}.$$

Therefore,

$$100 = \frac{Z(.975)\ (325)}{\sqrt{n}}, \text{ and}$$

solving this equation for n results in

$$n = \left[\frac{Z(.975)(325)}{100}\right]^2$$

$$= [(1.96)(3.25)]^2 = 40.58.$$

Thus, the retail association must survey 41 stores. Note that we must round our result (40.58) *up* to a whole number to assure our desired confidence of 95%.

∎

Until now we have concentrated entirely on estimates for which we considered an error on both sides of the estimate. That is, we have dealt with only two-sided confidence intervals. Sometimes we are concerned with an error on only one side of the point estimate. We demonstrate the use of one-sided confidence intervals via the following examples.

Example 5.12 Suppose that a bank manager wants to estimate the average amount of time (in minutes) that a teller spends on each customer. She decides to observe 64 randomly selected customers and she finds $\bar{x} = 3.2$ minutes and $s = 1.2$ minutes. Since she is primarily concerned about an upper bound on service time, she decides to find a 95% **upper** confidence limit for the true average service time μ. What we seek is

$$\mu \leq U,$$

where U is computed as shown in Equation 5.2, with one **major** difference. Since all of the error (α) is concentrated on one side of the point estimate ($\bar{x}$), we use Z(1 - α) instead of Z(1 - α/2). Using Equation 5.2 with Z(1 - α) instead of Z(1 - α/2), we have

$$U = \bar{x} + Z(1 - \alpha)\left(\frac{s}{\sqrt{n}}\right)$$

$$= 3.2 + Z(.95)\left(\frac{1.2}{\sqrt{64}}\right)$$

$$= 3.2 + (1.645)\left(\frac{1.2}{8}\right)$$

$$= 3.2 + .247$$

$$= 3.447.$$

Thus, the bank manager can be 95% confident that $\mu \leq 3.447$ minutes.

∎

Example 5.13 A company that manufactures a certain kind of aircraft engine wants to find a 90% **lower** confidence limit for the true average engine lifetime. Knowing that engine lifetimes are normally distributed, the company wants to be 90% confident that the average engine lifetime is **at least** a certain number of hours. The company takes a sample of n = 8 engines, tests them, and finds $\bar{x}$ = 8,275 hours and s = 565 hours.

Using Equation 5.2 and replacing t(1 - α/2, 7) with t(1 - α, 7), we find

$$L = \bar{x} - t(1 - \alpha, 7)\left(\frac{s}{\sqrt{n}}\right)$$

$$= 8275 - t(.90, 7)\left(\frac{565}{\sqrt{8}}\right)$$

$$= \quad 8275 - 1.415 \ (199.76)$$

$$= \quad 8275 - 282.66$$

$$= \quad 7992.$$

Thus, the company can be 90% confident that the average lifetime for these engines is at least 7992 hours, i.e., $7992 \le \mu$.

∎

The following template summarizes the computation of **one-sided** confidence limits.

$$\begin{matrix} U \\ L \end{matrix} = \bar{x} \pm \begin{bmatrix} Z(1-\alpha) \\ or \\ t(1-\alpha, n-1) \end{bmatrix} \begin{bmatrix} \dfrac{\sigma}{\sqrt{n}} \\ or \\ \dfrac{s}{\sqrt{n}} \end{bmatrix}$$

Equation 5.3 Computational Template for One-Sided Confidence Limits

5.6 Confidence Intervals for a Population Proportion[1]

In the last few sections we have been interested in estimating the population mean μ. Another common population parameter of interest is the population proportion π. For example, a politician is probably interested in the proportion of registered voters that will vote for her. A plant manager may be interested in the proportion of defective parts that the plant produces. If p is the sample proportion from a sample of size n and n is "sufficiently large," then we can compute the $(1 - \alpha)100\%$ confidence limits for π as follows [Mend 88]:

---

[1]This section was provided by Jim Rutledge of the USAF Academy.

$$\begin{matrix} U \\ L \end{matrix} = p \pm Z\left(1 - \frac{\alpha}{2}\right)\sqrt{\frac{p(1-p)}{n}} \, ,$$

where

p = sample proportion

n = sample size

Rule of Thumb (ROT): n is "sufficiently large" if the interval

$p \pm 2\sqrt{\dfrac{p(1-p)}{n}}$ does not contain 0 or 1. This

ROT is roughly equivalent to both $np \geq 4$ and $n(1-p) \geq 4$. A more conservative ROT commonly practiced is that both $np \geq 5$ and $n(1-p) \geq 5$.

Equation 5.4 Upper and Lower Confidence Limits for Proportions

The quantity $\sqrt{\dfrac{p(1-p)}{n}}$ should look familiar if you recall the binomial random variable. It is one standard deviation in the $Y = X/n$ distribution where X is binomial with parameters n and p.

Example 5.14 An engineering student is testing a motion detector that he built. He desires to get an idea of how well the motion detector works. He decides to drop an object through the motion detection field and record whether or not his machine detects the motion. He does this 100 times and 90 times his machine correctly detects the motion. Construct a 95% confidence interval for the true but unknown proportion of times that the machine will detect motion.

By the Rule of Thumb, we have $.9 \pm 2\sqrt{\dfrac{(.9)(.1)}{100}} = (.84, .96)$ which does not contain 0 or 1; therefore, n is "sufficiently large."

$$\begin{matrix} U \\ L \end{matrix} = .90 \pm (1.96) \sqrt{\frac{(.9)(.1)}{100}}$$

$$= .90 \pm .0588$$

$$= (.8412, .9588)$$

The engineering student can be 95% confident that the machine will detect motion between 84.12% and 95.88% of the time.

∎

5.7 Determining Sample Size

An important question closely related to the concept of a confidence interval is, "How large a sample do I need?" The answer to this question has not only statistical ramifications, but economic implications as well because it is not prudent to sample 500 elements from some population when a sample of size 100 would suffice.

How large the sample size should be depends on the requirements of the customer. Recall that the confidence limits for a population mean (continuous data) are computed as

$$\begin{matrix} U \\ L \end{matrix} = \bar{x} \pm Z \frac{\sigma}{\sqrt{n}},$$

where we use the multiplier $Z = 2$ or 3 if we desire 95% or 99% confidence, respectively (recall the use of these multipliers as a general rule of thumb). Since the width of the confidence interval depends on (1) desired confidence level, (2) standard deviation, and (3) sample size n, we can solve for n in the following equation:

$$h = \frac{Z\sigma}{\sqrt{n}},$$

where h is the required half-interval width (or maximum error). Thus, if the customer specifies h and Z and we have an estimate of the population standard deviation σ, call it $\hat{\sigma}$, then we can solve for n, obtaining

$$n = \left\lceil \left(\frac{Z\hat{\sigma}}{h} \right)^2 \right\rceil$$

where

Z = 2 or 3 for 95% or 99% confidence, respectively (user specifies),

h = half-interval width (user specifies),

$\hat{\sigma}$ = estimated population standard deviation, and

n = sample size.

Note: $\lceil x \rceil$ means to round x up to the next integer value.

Equation 5.5 Sample Size for Continuous Data

Example 5.15   Recall that in Example 5.10 we found a 99% confidence interval for the true but unknown average nicotine content of a particular brand of cigarette. The upper and lower confidence limits were determined to be

$$\begin{matrix} U \\ L \end{matrix} = 15.6 \pm 2.35,$$

where the half-interval width (or error) is obviously 2.35 milligrams. Suppose, instead, we wanted a 99% confidence interval that produced an interval half-width of only 1 milligram. Using $\hat{\sigma}$ = 2.1 milligrams (from the sample of size n = 9 in Example 5.10), we can use Equation 5.5 to obtain

$$n = \left\lceil \left(\frac{(3)\,(2.1)}{1} \right)^2 \right\rceil = \lceil (6.3)^2 \rceil = \lceil 39.69 \rceil = 40.$$

Thus, 40 cigarettes would be needed to produce the narrower confidence interval of half-width = h = 1 milligram.

Note: $\hat{\sigma}$ is usually obtained from historical data or from a pre-sample as shown in this example.

■

If we are dealing instead with binomial or proportion data, the equation for determining sample size is similar to Equation 5.5, except that $\hat{\sigma}^2$ becomes pq as follows:

$$n = \left[\left(\frac{Z^2 pq}{h^2} \right) \right]$$

where

Z = 2 or 3 for 95% or 99% confidence, respectively (user specifies),

h = half-interval width (user specifies),

p = proportion of sample which is defective (or some other category of interest),

q = 1 - p, and

n = sample size.

Equation 5.6 Sample Size for Binomial Data

Example 5.16 Suppose the engineering student in Example 5.14 wanted a half-interval width of only .03 (i.e., 3%), instead of the .0588 (5.88%) he obtained. What should the sample size be? Using Equation 5.6 with Z = 2 (95% confidence required), p = .9 and q = .1 (from the pre-sample in Example 5.14), he finds

$$n = \left[\frac{(2)^2 \ (.9) \ (.1)}{(.03)^2} \right] = 400.$$

Thus, a sample of size 400 (instead of 100) should allow him to cut the error almost in half (5.88% to 3%). Where have we seen these ratios before (i.e., quadrupling the sample size in order to cut the error in half)?

■

If the practitioner has absolutely no idea or estimate of what p or q are, a conservative approach to sample size determination is to allow p and q to be the values that make the product pq as large as possible. That is, p = q = .5. Thus pq = (.5)(.5) = .25. This combination of p and q has special appeal, especially when Z = 2. Note that Equation 5.6 simplifies in this case to

$$ n \; = \; \left[\; \frac{1}{h^2} \; \right] $$

because $Z^2pq = (2)^2(.5)(.5) = 1$. If this approach had been used in Example 5.16,

$$ n \; = \; \left[\; \frac{(2)^2 \; (.5) \; (.5)}{(.03)^2} \; \right] \; = \; 1112 \; , $$

almost 3 times the sample size he found using p = .9 and q = .1 .

The foregoing discussion on sample size determination assumed that the population from which we choose our sample is either infinite or, at the very least, extremely large compared to the sample chosen. If we are instead sampling from a finite population of size N, then a finite population correction factor could be used in order to more accurately estimate the required sample size. This correction factor is built into the following equations which are the finite population counterparts for Equations 5.5 and 5.6, respectively.

$$ \mathbf{n} \; = \; \left[\; \left(\frac{\mathbf{NZ^2\hat{\sigma}^2}}{\mathbf{Z^2\hat{\sigma}^2 \; + \; Nh^2}} \right) \; \right] $$

where

$\mathbf{Z}$ = 2 or 3 for 95% or 99% confidence, respectively (user specifies),

$\mathbf{N}$ = population size,

$\hat{\sigma}$ = estimated population standard deviation,

$\mathbf{h}$ = half-interval width (user specifies), and

$\mathbf{n}$ = sample size.

Equation 5.7 Sample Size for Continuous Data (finite population)

$$n = \left[\frac{NZ^2pq}{Z^2pq + Nh^2} \right]$$

where

Z = 2 or 3 for 95% or 99% confidence, respectively (user specifies),

N = population size,

h = half-interval width (user specifies),

p = proportion of sample which is defective (or some other category of interest),

q = 1 - p, and

n = sample size.

Equation 5.8 Sample Size for Binomial Data (finite population)

Example 5.17 A human resources manager is preparing a survey to send out to a randomly selected sample of employees from her organization. There are 320 employees in her organization from which she can select the sample. The questions on the survey form require responses on a 6-point Likert scale (i.e., 1 = "strongly agree", 2 = "moderately agree", ..., 6 = "strongly disagree", etc.). The specialist would like to determine the sample size needed to obtain a 95% confidence interval for the unknown population mean (μ) for each of the survey questions. She would like the margin of error to be no more than .25. Previous survey data indicate $\hat{\sigma}$ to be at most 1. Using this data (Z = 2, N = 320, $\hat{\sigma}$ = 1, and h = .25), we can use Equation 5.7 to estimate n as

$$n = \left[\frac{320\,(2)^2\,(1)^2}{(2)^2\,(1)^2 + 320\,(.25)^2} \right] = \left[\frac{1280}{4 + 20} \right] = \lceil 53.33 \rceil = 54.$$

Thus, 54 sample surveys must be received in order to generate the desired confidence intervals. If past data indicate only a 75% return rate on surveys, it will be necessary for the manager to send out 72 surveys because she expects only 75% of these, namely 54, to be returned.

∎

5.8 Problems

5.1   Use the sampling distribution of $\overline{X}$ for n = 3 which is given in Figure 5.5 (or its equivalent tabular form) to verify the values in the column labeled "n = 3" in Table 5.1.

5.2   Using the population given in Example 5.4 and the technique demonstrated in that example, construct the sampling distribution of $\overline{X}$ for n = 4 (either in tabular or graphical form) and use this distribution to compute $E(\overline{X})$ and $\sigma^2(\overline{X})$ for n = 4. These values should be inserted into the column labeled "n = 4" in Table 5.1.

5.3   Consider Property 5.4, which says $Z = \dfrac{\overline{X} - \mu}{\sigma/\sqrt{n}}$.

If we replace $\overline{X}$ with x̄ and Z with Z(α/2) and Z(1 - α/2), respectively, in Property 5.4, verify that solving for μ results in the upper and lower confidence limits, U and L, that are given in Equation 5.1. That is,

$$\begin{matrix} U \\ L \end{matrix} = \overline{x} \pm Z\left(1 - \frac{\alpha}{2}\right) \frac{\sigma}{\sqrt{n}} .$$

5.4   If X ~ N(25, 121) and a random sample of 64 is taken from X, what is the probability that the sample mean will be less than 24.5?

5.5   The following data represent the amount of red meat consumed (in pounds) in a household during five randomly selected weeks:

     4.3   5.2   3.4   2.7   6.2

Assuming that the amount of red meat consumed in this household per week is normally distributed, find a *lower* 95% confidence limit for the true mean consumption, μ.

5.6   A sample of 100 batteries has been selected from a production line that produces batteries with a mean lifetime of 40 months and a standard deviation of 10 months. What is the standard error of $\bar{X}$? If, instead, a sample of size 400 is taken from the same production line, what would the standard error of $\bar{X}$ be? What can you deduce by considering both of these results simultaneously?

5.7   A steel company produces steel bars whose lengths (X) are normally distributed with μ = 31.5 feet and σ = 1.40 feet.

    a)   What is the minimum sample size for which the sampling distribution of $\bar{X}$ becomes normal?

    b)   What is the probability that a sample of size n = 9 will produce a sample mean that is within 1 foot of the population mean μ?

5.8   A rancher would like to estimate the average weight gain of the cattle in his herd between the time of purchase and the time he takes them to market. If he selects a sample of size n = 16 animals and finds $\bar{x}$ = 240 lbs and s = 32 lbs, construct a 95% confidence interval for the true average weight gain per animal. Assume weight gain is normally distributed.

5.9   Suppose the parent population Y from which we are about to sample is exponential with E(Y) = 2, or λ = .5. Consider the sampling distribution of $\bar{Y}$ when the sample size is n = 36.

    a)   Find $E(\bar{Y})$.

    b)   Find $\sigma(\bar{Y})$.

    c)   What do we know about the form of the sampling distribution of $\bar{Y}$?

    d)   What distribution does the quantity $\dfrac{\bar{Y} - \mu}{\sigma/\sqrt{n}}$ follow?

5.10   As in Problem 5.9, suppose the parent population Y is exponential with $E(Y) = 2$, or $\lambda = .5$. Consider the sampling distribution of $\bar{Y}$ when the sample size is n = 5.

    a)   Find $E(\bar{Y})$.

    b)   Find $\sigma(\bar{Y})$.

    c)   What do we know about the form of the sampling distribution of $\bar{Y}$?

    d)   What distribution does the quantity $\dfrac{\bar{Y} - \mu}{\sigma/\sqrt{n}}$ follow?

    e)   Compare and contrast your answers for 5.10 (a) - (d) with your answers for 5.9 (a) - (d). Explain any differences.

5.11   An insurance company surveys 25 households at random and finds that the average amount of whole life insurance owned by these 25 households is $60,000 with a standard deviation of $20,000. Assume that the amount of whole life insurance is normally distributed. Find a 95% confidence interval for the true average amount of whole life insurance per household. Interpret your result. Was the assumption of normality necessary? Why or why not?

5.12   An advertising company wishes to estimate the true average amount of TV viewing time per household. Suppose it wants to estimate the true average viewing time within $\pm$ 10 minutes and be 99% confident that this is a correct estimate. How many households will have to be part of the sample they select? You may assume that a previous year's sample had a standard deviation of s = 45 minutes.

5.13   The manufacturer of a small twin-engine jet airplane is concerned about the difference between the thrusts produced by the port (left) and starboard (right) engines. Sixteen aircraft were tested, and the difference between the port and starboard thrusts (in pounds) was recorded for each case as shown here.

-60	30	120	80
50	-90	40	-70
-110	20	130	-50
-90	100	-80	20

Assuming that thrust differences are approximately normally distributed, find a 90% confidence interval for the true average difference between engine thrusts.

5.14   A construction company subcontracts the installation of hot tubs when hot tubs are installed in the homes they build. The company would like to estimate the (true) average time it takes to install a hot tub to within $\pm$ 10 minutes with 95% confidence. Since the company has no idea about the variability of this process, it takes a sample of n = 12 jobs and finds s = 30 minutes. How many more jobs must it survey (or witness) to obtain the confidence interval it desires? *Note*: this is an example of two-stage sampling, where the first stage was used to determine a planning value, s, for the variability in the process.

5.15   A coffee vending machine fills 8 oz cups. The vending machine distributor is hearing complaints about her machines not filling cups sufficiently full. Thus, she takes a random sample of 10 fillings, finding that $\bar{x}$ = 7.8 oz and s = .5 oz. Find a 90% *lower* confidence limit for the true average amount of coffee dispensed. You may assume that the amount of coffee dispensed follows a normal distribution.

5.16   As a purchase agent, you have just received a shipment of computer memory chips. You take a sample of 100 chips and find 8 to be defective. Find a 95% confidence interval for the true proportion of defective chips contained in the shipment.

5.17   Given the same scenario as in 5.16, find a 99% *upper* confidence limit for the true proportion of defectives in the shipment. *Note*: Just use $Z(1 - \alpha)$ instead of the $Z(1 - \alpha/2)$ that you should have used in 5.16.

5.18   A company produces ball bearings whose diameter X (in mm) is normally distributed with μ = .5 mm and σ = .1 mm. That is, $X \sim N(.5, .1^2)$.

   a)   What percentage of ball bearings produced would you expect to have a diameter between .4 and .6 mm?
   b)   Suppose you take a sample of size n = 5. What is the probability that the sample mean diameter will fall between .4 and .6 mm?
   c)   What is the probability that the mean diameter from a sample of size n = 5 will be less than or equal to .3 mm?
   d)   If, in fact, you did take a sample of size n = 5 and found that the sample mean diameter, $\bar{x}$, was .3, what might you suspect? Why?

5.19   What factors affect the width of a confidence interval?

5.20   A commuter who regularly rides the city bus to and from work is becoming more and more distraught over the habitual tardiness of the bus. She collects data for the next month (twice a day for 20 working days) and obtains a sample of 40 data points (in minutes, late). She finds $\bar{x} = 4.1$ minutes with s = 5.1 minutes. Help the commuter find a 95% confidence interval for the true average tardiness (in minutes) of the bus.

5.21   Memorial Hospital would like to estimate the mean age of all its patients.

   a)   How big must the sample be to estimate this age within $\pm$ 2 years with 95% confidence? A planning value or estimate of σ is needed to determine n. If Memorial assumes that the *range* of the patients' ages represents 6σ, and the oldest patient they have had in recent months was 102 years old, then $\sigma = 102/6 = 17$ years. Use 17 as an estimate of σ.
   b)   If a sample of that size is then obtained, and the sample mean $\bar{x}$ is 44 years old, then what is the desired confidence interval? Interpret the meaning of that interval.
   c)   Suppose you also obtain s = 14.6 years from that sample. Using s = 14.6 and n from (a), as well as the maximum error of $\pm$ 2 years, can you be more than 95% confidence that $42 \leq \mu \leq 46$, which should have been obtained in (b)? How? Why?

5.22   Assume $\bar{X}$ is normally distributed.

   a)   How is the quantity $\dfrac{\bar{X} - \mu}{\sigma/\sqrt{n}}$ distributed?

   b)   How is the quantity $\dfrac{\bar{X} - \mu}{s/\sqrt{n}}$ distributed?

   c)   If X is the parent population, what must happen if we are to conclude that $\bar{X}$ is normal (or close to normal)?

5.23   An engineer desires to estimate the hardness of metal castings produced by his company. If he takes a sample of 50 castings and finds that this sample produces an average hardness of $\bar{x}$ = 226 and a standard deviation of s = 14 (these values are Brinell hardness numbers), find a 95% *lower* confidence limit for the average hardness of all metal castings produced.

5.24   Tire World received a shipment of tires from a certain manufacturer. Within 3 months of receiving this shipment, Tire World had sold 100 tires of which 7 were returned as defective. Find a 95% *upper* confidence limit for the proportion of defective tires in the shipment.

5.25   In 1988 just prior to the Bush-Dukakis presidential election, a major pollster took a random sample of 300 voters and found that 164 of the voters were planning to vote for Bush.

   a)   Find a 99% confidence interval for the proportion of voters who were planning to vote for Bush.
   b)   If the pollster wanted an error margin (i.e., half-interval width) of ± 4%, what would the percent confidence coefficient be reduced to?

5.26   Light bulbs have lifetimes that are known to be approximately normally distributed. Suppose a random sample of 35 light bulbs was tested, and $\bar{x}$ = 943 hours and s = 33 hours.

   a)   Find a 90% confidence interval for the true mean life of a light bulb.
   b)   Find a 95% *lower* confidence limit for the true mean life of a light bulb.
   c)   Are the results obtained in (a) and (b) the same or different? Explain why.

5.27   The compressive strength of lunar concrete is being analyzed by a civil engineer. The engineer tests a sample of 8 specimens and obtains the following compressive strengths (in psi).

3850	3640	3720	3400
3350	3575	3500	3650

Assuming the compressive strength of lunar concrete is normally distributed,

a)   Find a 95% confidence interval for the true mean compressive strength of this lunar concrete.

b)   Find a 99% *lower* confidence interval for the true mean compressive strength of this lunar concrete.

5.28   The strength of a steel beam is measured in the size of deflection (in micrometers, μm) that results when subjecting the beam to a force of 10,000 pounds. The strength of steel beams is believed to be normally distributed. Suppose a materials engineer selects a random sample of 5 beams, tests them, and finds their deflections as follows (in μm):

72   78   68   73   75

a)   Find a 95% confidence interval for the true mean strength of beams of this kind (as measured in μm of deflection).

b)   How many more beams would need to be tested in order to construct a 95% confidence interval which would estimate the true mean strength with a maximum error of $\pm$ 2 μm in deflection?

5.29   A leading tire manufacturer performs accelerated testing on 36 of its steel belted radial tires. It finds that this sample of tires produced a mean useful life of $\bar{x}$ = 40,200 miles and a standard deviation of s = 1800 miles.

a)   Find a 95% *lower* confidence interval for the mean useful life of this kind of tire.

b)   With what percent level of confidence could we state that $\mu \geq 40,000$ miles?

5.30   Air Force pilots have a higher than normal hearing loss due to the noise levels they are exposed to in cockpits. In particular the T-37 aircraft is notorious for high decibel readings. A sample of 30 T-37s produced the following 95% *lower* confidence interval for the true mean noise level (in decibels) of all T-37s: $\mu \geq 85$ decibels. A young pilot interprets this confidence interval as meaning 95% of all T-37 aircraft have a cockpit decibel reading of 85 or greater. Is this interpretation correct? Why or why not?

Chapter 6

Hypothesis Testing

6.1 Introduction

A statistical hypothesis is a statement or claim about some unrealized true state of nature. Some examples might be:

1. Average gas mileage differs depending on type A or B gas.
2. Type of aspirin determines the amount of pain relief.
3. Probability of death in auto accidents differs depending on whether seat belts are worn.
4. Filtration of toxic chemicals is enhanced by method 1 or method 2.
5. Variability of machined thickness of a part depends on the type of tool.
6. Students with urban backgrounds perform better in college than those with a rural background.
7. Etch rate for an integrated circuit manufacturing process is affected by changes in pressure settings.
8. Compressive strength for a type of concrete is within specification.
9. Product quality is independent of raw material supplier.
10. A quality improvement effort produced significant results with regard to mean and/or standard deviation performance.

The actual hypotheses to be tested consist of two complementary statements about the true state of nature. For example, given an oxide coating process used in the development of integrated circuits, the hypotheses might appear as follows:

H_0: Population mean oxide thickness is equal to 200 angstroms.

H_1: Population mean oxide thickness is not equal to 200 angstroms.

These two complementary statements are defined as the null hypothesis (H_0) and the alternative hypothesis (H_1). Since the true state of nature is seldom (if ever) known with 100% certainty, the two statements serve as strawmen to be tested. Probability and statistics are combined with the sample data results to make inference about the entire population (the true state of nature) with a measurable amount of uncertainty.

An analogy of hypothesis testing can be drawn from our legal system where an accused on trial is presupposed to be innocent unless the prosecution presents overwhelming evidence to convict him. In this example, the hypotheses to be tested are stated as:

H_0: Defendant is Innocent

H_1: Defendant is Guilty

Regardless of the jury's conclusion, they are never really sure of the true state of nature. Concluding "H_0: Defendant is Innocent" does not mean that the defendant is in fact innocent. An H_0 conclusion simply means that the evidence was not overwhelming enough to justify a conviction. On the other hand, concluding H_1 does not prove guilt; rather, it implies that the evidence is so overwhelming that the jury can have a high level of confidence in a guilty verdict.

Since verdicts are concluded with less than 100% certainty, either conclusion has some probability of error. Consider Table 6.1. The probability of committing a Type I error is defined as α ($0 < \alpha < 1$) and the probability of committing a Type II error is β ($0 < \beta < 1$).

In the courtroom example, α, the probability of convicting an innocent person, is of critical concern. To minimize the risk of such an erroneous conclusion, our court system requires overwhelming evidence to conclude H_1. Although minimizing α has its advantages, it should be obvious that requiring overwhelming evidence to conclude H_1 will in turn increase β, the probability of a Type II error. To resolve this dilemma, statistical hypothesis tests are designed such that

(1)   The most critical decision error is a Type I error.

(2) α is set at a minimum level, usually .05, .01, or .001, depending on the degree of criticality of such an error (i.e., in academia and social sciences α is usually .05; whereas, in hospital tests or other critical areas of testing, α is either .01 or .001).

(3)   Based on (1) and (2) above, the hypothesis statement to be tested for at least $(1 - \alpha)100\%$ confidence is placed in H_1.

(4)   The nature of most statistical hypothesis tests requires that the equality condition be placed in H_0.

(5)   To minimize β while holding α constant requires increased sample sizes.

		True State of Nature	
		H_0	H_1
Conclusion Drawn	H_0	Conclusion is Correct	Conclusion results in a Type II error
	H_1	Conclusion results in a Type I error	Conclusion is Correct

Table 6.1 Type I or II Error Occurs if Conclusion Not Correct

6.2 One-Sample Hypothesis Tests of the Mean (μ)

6.2.1 Two-Sided Tests (n ≥ 30)

To test whether a population response mean, μ, differs from a specified standard, μ_0, we conduct a one-sample, two-sided test. Consider the following example.

Example 6.1 In a silicon wafer oxide coating process we want to test if the population average thickness differs from the required 200 angstroms. The hypotheses for this test are:

$$H_0: \mu = 200$$

$$H_1: \mu \neq 200$$

Obviously, all of the oxide thickness measurements in the population are not available because every day we process more wafers. Therefore, we must take a sample which is representative of the overall population. Using the sample data, a sample mean, $\bar{y}$, and a sample variance, s^2, are calculated. Based on the sample data, the difference between $\bar{y}$ and μ_0 is investigated to determine if it is large enough to provide overwhelming evidence for rejecting H_0 with at least $(1 - \alpha)100\%$ confidence. If $|\bar{y} - \mu_0|$ is not sufficiently large, the conclusion is to fail to reject H_0, i.e., the evidence is not overwhelming.

Assume the sample average oxide thickness, $\bar{y}$, is found to be 196 angstroms with n = 100 and s^2 = 400. If H_0 is true, the sampling distribution of $\overline{Y}$ appears as shown in Figure 6.1 with $\bar{y}$ and the standardized value for $\bar{y}$, referred to as Z_0, annotated in the figure.

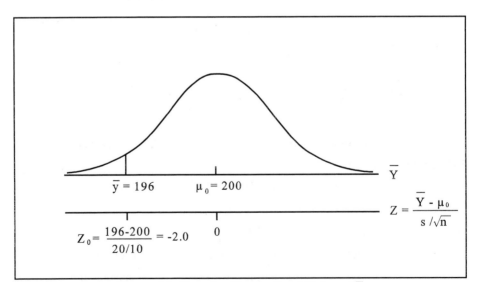

Figure 6.1 Z_0 as the Standardized Value of $\bar{y}$

Given H_0 is true, the probability, P, of obtaining a sample mean at least as extreme in either direction as the one observed $\bar{y}$ is shown as the shaded area in Figure 6.2. These values were determined by looking up the percentiles associated with $Z_0 = 2$ and $Z_0 = -2$ in the standardized normal tables. See Appendix D. Therefore, if we were to conclude H_1 based on our sample data, the risk (probability) of committing a Type I error would be approximately 0.0456. Another correct interpretation is that, based on the sample data, we can be approximately $(1 - P)100\% = 95.44\%$ confident in concluding H_1.

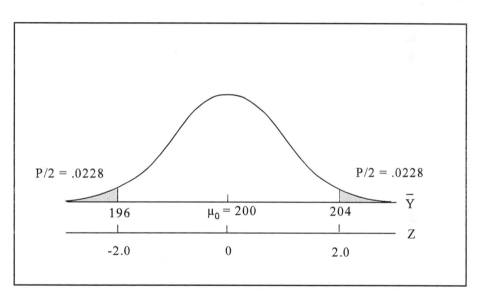

Figure 6.2 Probability of Obtaining a $\bar{y}$ at Least as Extreme as $\bar{y} = 196$

If we compare the value of P with a previously determined value of α, the following decision rule can be used:

i)   if $P \geq \alpha$, then fail to reject H_0

ii)   if $P < \alpha$, then reject H_0 (i.e., conclude H_1) with $(1 - P)100\%$ confidence.

A summary of a one-sample, two-sided hypothesis test of the mean (μ) for $n \geq 30$ is shown in Table 6.2.

Step 1: State the hypotheses to be tested, e.g.,

$$H_0: \mu = \mu_0$$

$$H_1: \mu \neq \mu_0.$$

Step 2: Determine a planning value for α (e.g., .05, .01, or .001).

Step 3: Take a representative sample of size $n \geq 30$ and compute $\bar{y}$ and s^2.

Compute the standardized test statistic $Z_0 = \dfrac{\bar{y} - \mu_0}{\dfrac{s}{\sqrt{n}}}$.

Note: Since s is used in place of σ, the standardized test statistic follows a t distribution. However, the Z distribution is a good approximation when $n \geq 30$.

Step 4: Using a standardized normal distribution, compute the area in the tail beyond $|Z_0|$.

Step 5: Compute P as 2 times the area found in Step 4.

Step 6: If P is sufficiently small (i.e., $P < \alpha$), conclude H_1 with $(1 - P)100\%$ confidence; else fail to reject H_0 based on the sample evidence. Note that for values of P close to α, one should use experience, common sense, and (if possible) additional data to reach a sound conclusion. (See note below for Rule of Thumb Test.)

Note: For a quick Rule of Thumb (ROT), if $|Z_0| \geq 2$, we can be approximately 95% confident in H_1. Similarly, if $|Z_0| \geq 3$, we can be approximately 99% confident in H_1.

Table 6.2 Summary of One-Sample, Two-Sided Tests of the Mean (μ)

for $n \geq 30$

The α and P values in a hypothesis test sometimes tend to confuse the student. The α value is the maximum level of risk that an experimenter is willing to take in making a "reject H_0" or "conclude H_1" conclusion (i.e., it is the maximum risk in making a Type I error). The α value is established *prior* to actually collecting data and is in many cases quite arbitrary. It serves as a benchmark for actual test results. The P-value, on the other hand, comes from the data itself and represents the exact probability of making a Type I error. Thus, if P turns out to be less than α, the risk of making a Type I error is really less than what we were willing to accept at the outset when we established the α benchmark. Since P is derived from the data, it represents the "voice of the process" and is of considerably more value than is α. Almost all statistical software packages will print out P-values, so it is important to know what this value represents. If P is less than α and H_1 is concluded, the student should be able to ascertain that the statements "concluding H_1 with $(1 - \alpha)100\%$ confidence," "concluding H_1 with at least $(1 - \alpha)100\%$ confidence," and "concluding H_1 with $(1 - P)100\%$ confidence" are all correct, with the latter statement being the most precise. In fact, one can always conclude H_1 with $(1 - P)100\%$ confidence, regardless of the value of P. One cannot associate a level of confidence in an H_0 conclusion because we are controlling for the Type I error (α risk) only.

Example 6.2 In an effort to reduce electricity consumption during peak periods, a public utilities company has increased its residential rates between 8 am and 5 pm. Historically, the average residential consumption has been 2000 kw hours per month. A hypothesis test is used to determine if the new pricing policy has changed consumption. A sample of 200 households produced a mean of 1994.35 and a variance of 1565.5.

Step 1: $H_0: \mu = 2000$

$H_1: \mu \neq 2000$

Step 2: $\alpha = .05$

Step 3: $\bar{y} = 1994.35$, $s^2 = 1565.5$

$$Z_0 = \frac{1994.35 - 2000}{\dfrac{39.566}{\sqrt{200}}} = -2.02$$

Step 4:

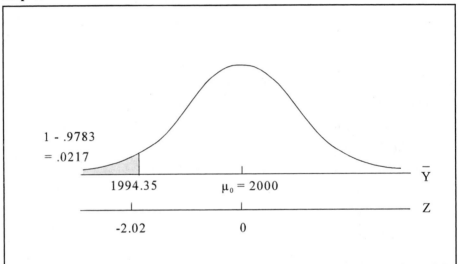

Step 5: $P = 2(.0217) = .0434$

Step 6:   Since $[P = .0434] < [\alpha = .05]$ we can reject H_0 (i.e., conclude H_1) with $(1 - .0434)100\% = 95.66\%$ confidence. Furthermore, since $\bar{y} < [\mu_0 = 2000]$ our conclusion is that we are 95.66% confident that the new pricing policy has reduced μ below 2000 kw hours. (Notice that the ROT stated in Table 6.2 would have made a similar decision. For example, since $|Z_0| = 2.02 > 2$ we are at least 95% confident in H_1.

6.2.2 One-Sided Tests (n ≥ 30)

Example 6.2 produced a P value fairly close to α and in some instances this would imply that maybe more data should be gathered before a conclusion is made. This particular example, however, lends itself to another type of hypothesis testing strategy, one-sided tests. Since the new pricing policy had little if any possibility of producing a $\mu > 2000$, a more appropriate set of hypotheses would be

$$H_0: \mu \geq 2000$$

$$H_1: \mu < 2000.$$

This hypothesis is labeled "one-sided" because of the specific direction of the H_1 statement. When $H_1: \mu < \mu_0$, the test is referred to as a lower-tail test; and when $H_1: \mu > \mu_0$, then it is called an upper-tail test. Revisiting Example 6.2, we are interested in testing whether or not the new policy produces a population mean *less than* 2000. This one-sided test results in all the risk of making a Type I error being located in only one tail of the normal distribution. Therefore, P is equal to the shaded area found in Step 4 of the previous example. Reworking Example 6.2 as a one-sided problem results in the following steps.

Step 1: $H_0: \mu \geq 2000$

$H_1: \mu < 2000$

Step 2: $\alpha = .05$

Step 3: $\bar{y} = 1994.35, \quad s^2 = 1565.5, \quad Z_0 = -2.02$

Step 4:

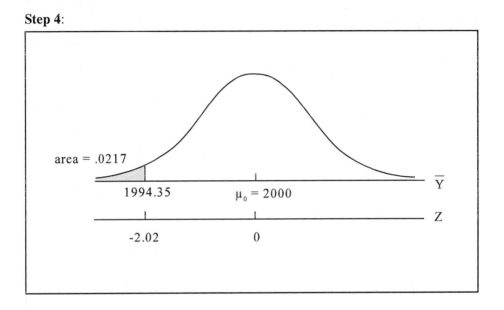

Step 5:   P = .0217

Step 6:   Since [P = .0217] < [α = .05] reject H_0 (i.e., conclude H_1)
with (1 − .0217)100% = 97.83% confidence. That is we can be
97.83% confident that the new pricing policy has reduced
electricity consumption below 2000 kw hours. See ROT in
Table 6.3 for a quick and fairly accurate analysis.

One-tail tests consolidate all of the risk of a Type I error in one tail;
therefore, the confidence in concluding H_1 will be larger than similar two-sided
tests. It is important to be sure that a one-sided test is appropriate before pursuing
something which might artificially inflate the confidence level. In general, use a
one-sided test if μ values in one direction appear highly unlikely or infeasible. The
following table summarizes one-sample, one-sided tests of the mean (μ) for n ≥ 30.

Step 1: Determine the hypotheses to be tested.

Lower-Tail   or   Upper-Tail

$H_0: \mu \geq \mu_0$ $H_0: \mu \leq \mu_0$

$H_1: \mu < \mu_0$ $H_1: \mu > \mu_0$

Step 2: Determine a planning value for α.

Step 3: From the sample data compute $\bar{y}$, s^2, and the standardized test statistic

$$Z_0 = \frac{\bar{y} - \mu_0}{\dfrac{s}{\sqrt{n}}}$$

Note: Since s is used in place of σ, the standardized test statistic follows a t distribution. However, the Z distribution is a good approximation when $n \geq 30$.

Step 4: Using the standardized normal distribution, compute the area to the left of Z_0 (if lower-tail test) or to the right of Z_0 (if upper-tail test).

Step 5: The area in Step 4 is equal to P.

Step 6: For $P \geq \alpha$, fail to reject H_0; and for $P < \alpha$, reject H_0 with $(1 - P)100\%$ confidence in H_1. (See note below for ROT)

Note: *The ROT for one-sided tests would be $|Z_0| \geq 2$ implies approximately 97.5% confidence in H_1 whereas $|Z_0| \geq 3$ implies approximately 99.5% confidence in H_1.*

Table 6.3 Summary of One-Sample, One-Sided Tests of the Mean (μ)
for $n \geq 30$

6.2.3 Two-Sided and One-Sided Tests for Small Samples (n < 30)

When the sample size is less than 30, the Z distribution is no longer a good approximation of the t distribution, and the $\bar{Y}$ distribution may not necessarily be normal. Therefore, tests for small samples when the parent population is approximately normal should follow the steps below:

Step 1: Determine the hypotheses to be tested.

Two-Tail	Upper-Tail	Lower-Tail
$H_0: \mu = \mu_0$	$H_0: \mu \le \mu_0$	$H_0: \mu \ge \mu_0$
$H_1: \mu \ne \mu_0$	$H_1: \mu > \mu_0$	$H_1: \mu < \mu_0$

Step 2: Determine a planning value for α.

Step 3: If the parent population is approximately normal, use sample data to determine the sample mean $\bar{y}$ and the sample variance s^2, and then compute the standardized test statistic

$$t_0 = \frac{\bar{y} - \mu_0}{\dfrac{s}{\sqrt{n}}}.$$

Step 4: Using a t distribution with $v = n - 1$ degrees of freedom, estimate the area in the tail beyond $|t_0|$ (for 2-tail tests); to the left of t_0 (for lower-tail tests); or to the right of t_0 (for upper-tail tests).

Step 5: For two-tail tests, P is 2 times the area estimated in Step 4.
For one-tail tests, P is equal to the area in Step 4.

Step 6: For $P \ge \alpha$, fail to reject H_0; and for $P < \alpha$, reject H_0 with $(1 - P)100\%$ confidence in H_1. (See note below for ROT test.)

Note: *It is not a good practice to use the Rule of Thumb from Table 6.2 or Table 6.3 when we have small samples.*

Table 6.4 Two-Sided and One-Sided Tests for Small Samples (n < 30)

Example 6.3 Historically, time to repair a downed radar system has been normally distributed with an average of 79 minutes. A new checklist has been instituted and it is desired to test whether the average time has changed, using an α risk of 0.05. The following sample data has been collected.

Time to repair data (in minutes)									
59	78	75	77	81	60	76	80	79	65

Step 1: $H_0: \mu = 79$

$H_1: \mu \neq 79$

Step 2: $\alpha = .05$

Step 3: $\bar{y} = 73, \quad s^2 = 70.222, \quad t_0 = \dfrac{73 - 79}{\dfrac{8.38}{\sqrt{10}}} = -2.264$

Step 4:

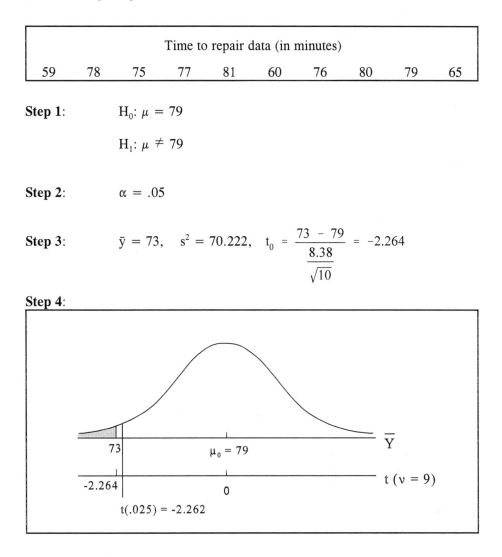

The area in the tail is estimated to be approximately 0.024.

Step 5: $P = 2(.024) = .048$

Step 6:   Since P < α, reject H_0, (i.e., conclude H_1) with approximately 95.2% confidence. That is, the average time to repair a radar system has changed.

■

Example 6.4 In an attempt to improve quality throughout the total organization, company XYZ is currently interested in reducing the time to process customer orders. The historical database indicates a normal distribution with an average processing time of 4.6 hours. Management has reduced process complexity and introduced new forms which should reduce the overall processing time. To test the hypothesis of reduced time, the following hypothesis test is conducted.

Step 1: $H_0: \mu \geq 4.6$

           $H_1: \mu < 4.6$

Step 2: $\alpha = .05$

Step 3:   A sample of n = 16 is taken with the following results:

$$\bar{y} = 3.8, \quad s^2 = 1.44 \text{ and } \quad t_0 = \frac{3.8 - 4.6}{\dfrac{1.2}{\sqrt{16}}} = -2.667.$$

Step 4:

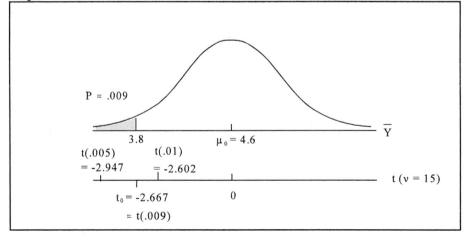

Step 5:   Based on the graphic in Step 4, P is approximately equal to 0.009.

Step 6:   Since $[P \approx .009] < [\alpha = .05]$, the conclusion of H_1 can be made with approximately 99.1% confidence. Thus, there exists a high degree of confidence that process changes have significantly reduced processing time.

■

6.3 Two-Sample Hypothesis Tests of the Mean

Assume we are working on a new composite material and we desire to test the effect of two different processing temperature settings on the strength of the material. Based on engineering knowledge, the autoclave temperature settings to be tested are 200°F and 300°F. Samples of 9 composite materials are taken at each setting. The sample results are shown in Table 6.5.

Strength Measurements										
Temp	y_1	y_2	y_3	y_4	y_5	y_6	y_7	y_8	y_9	$\bar{y}$	s^2
200°	2.8	3.6	6.1	4.2	5.2	4.0	6.3	5.5	4.5	4.6889	1.3761
300°	7.0	4.1	5.7	6.4	7.3	4.7	6.6	5.9	5.1	5.8667	1.1575

Table 6.5 Composite Material Data (Low = 200°, High = 300°)

The hypotheses to be tested are:

$$H_0: \ \mu_H = \mu_L$$

$$H_1: \ \mu_H \neq \mu_L$$

What we are testing is if the two $\bar{y}$'s (4.6889 vs 5.8667) are different enough for us to conclude H_1. Concluding H_1 means the "low temperature (L)" process and the "high temperature (H)" process have different centers or means. These hypotheses are portrayed graphically in Figure 6.3. Note that this graphic illustrates the assumption of equal variances in either case and that both populations are normal.

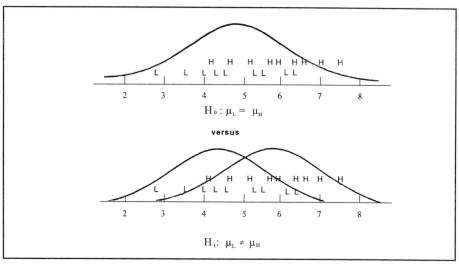

Figure 6.3 Illustration of a Two-Sample Test

Figure 6.3 seems to indicate that H_1 is very likely; however, without conducting the statistical test, it is difficult to determine if the evidence is overwhelming. We conduct the actual test as follows:

Step 1: $H_0: \mu_H = \mu_L$ or $H_0: \mu_H - \mu_L = 0$

$H_1: \mu_H \neq \mu_L$ $H_1: \mu_H - \mu_L \neq 0$

Step 2: $\alpha = .01$

Step 3: $\bar{y}_L = 4.6889$ $\bar{y}_H = 5.8667$

$s^2_L = 1.3761$ $s^2_H = 1.1575$

$n_L = 9$ $n_H = 9$

$$t_0 = \frac{\bar{y}_H - \bar{y}_L}{s_p \sqrt{\dfrac{1}{n_H} + \dfrac{1}{n_L}}} \quad \text{where} \quad s_p = \sqrt{\frac{\left(n_H - 1\right)s^2_H + \left(n_L - 1\right)s^2_L}{\left(n_H - 1\right)+\left(n_L - 1\right)}}$$

S_p represents the **pooled** standard deviation of the two samples. It is just the square root of a weighted average of the two variances.

Therefore,

$$t_0 = \frac{5.8667 - 4.6889}{1.126\sqrt{\dfrac{1}{9} + \dfrac{1}{9}}}, \quad \text{where } s_p = \sqrt{\frac{8(1.1575) + 8(1.3761)}{8 + 8}} = 1.126$$

$$= 2.22$$

The standardized test criteria will be from a t distribution with

$$v = (n_H - 1) + (n_L - 1) = 8 + 8 = 16 \text{ degrees of freedom.}$$

Step 4:

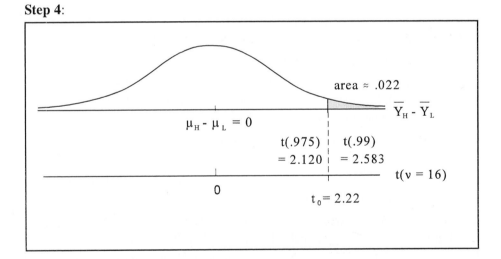

Step 5:   Based on the graphic in Step 4, $P \approx 2(.022) = .044$.

Step 6:   Since $[P \approx .044] > [\alpha = .01]$, fail to reject H_0 at the 0.01 level. Thus, the sample evidence does not allow us to be 99% confident in an H_1 conclusion. However, even though we cannot be 99% confident in H_1, we can be $(1 - P)100\% = 95.6\%$ confident in H_1.

The following table summarizes a two-sample test.

Step 1:	State the hypotheses to be tested.		
	Two-Tail	Upper-Tail	Lower-Tail
	$H_0: \mu_1 = \mu_2$	$H_0: \mu_1 \le \mu_2$	$H_0: \mu_1 \ge \mu_2$
	$H_1: \mu_1 \ne \mu_2$	$H_1: \mu_1 > \mu_2$	$H_1: \mu_1 < \mu_2$

Step 2: Determine a planning value for α.

Step 3: Obtain information from two samples. For example, n_1, $\bar{y}_1$, s_1^2, n_2, $\bar{y}_2$, s_2^2, and compute the test statistic

$$t_0 = \frac{\bar{y}_1 - \bar{y}_2}{s_p \sqrt{\dfrac{1}{n_1} + \dfrac{1}{n_2}}}, \quad \text{where } s_p = \sqrt{\frac{\left(n_1 - 1\right)s_1^2 + \left(n_2 - 1\right)s_2^2}{n_1 + n_2 - 2}}$$

Step 4: Use percentiles of the t distribution with $\nu = n_1 + n_2 - 2$ degrees of freedom to estimate the area in the tail beyond $|t_0|$ (for 2-tail tests); to the left of t_0 (for lower-tail tests); or to the right of t_0 (for upper-tail tests).

Note: If $(n_1 + n_2) \ge 30$, use the Z distribution.

Step 5: For a two-tail test, P = 2 times the area in Step 4. For a one-tail test, P = area in Step 4.

Step 6: If P < α, conclude H_1 with (1 - P)100% confidence. If P $\ge$ α, fail to reject H_0. Note that if $n_1 + n_2 \ge 30$, we can use the ROT from Table 6.2, i.e., if $|Z_0| > 2$ we are at least 95% confident in H_1 and if $|Z_0| > 3$ we are at least 99% confident in H_1.

Note: *This 2-sample test assumes that the two population variances are not significantly different and that, if the samples are small, the parent populations are approximately normal. See the ROT in Table 6.9 for a quick test to check for equal variances.*

Table 6.6 Summary of a Two-Sample Test of the Mean (μ)

A quick test referred to as the "End Count Method" or "Tukey Quick Test" [Schm 94] can also be performed on this type of problem. To conduct the End Count Method to determine if two sample means are significantly different, use the following steps:

1.   Plot the values from each of the two samples on a number line, indicating each sample value with a symbol for the respective sample. Using the data from Table 6.5 we plot the values using an L for the 200° response values and an H for the 300° response values.

```
                     H   H   H  H H    HH H H
             L    L  L L L     L L    L L
        _____
         2.0     3.0    4.0    5.0    6.0    7.0    8.0
```

2.   Count from the left tail the number of identical symbols of one kind until the first opposite symbol is encountered.

```
              L     L  L
        _____
         2.0     3.0    4.0    5.0    6.0    7.0    8.0
```

In our example the count from the left is 3, i.e., there are three "L"s before encountering an "H."

3.   Count in from the right the number of identical symbols of the second kind until the first opposite symbol is encountered.

```
                                     HH H H
        _____
         2.0     3.0    4.0    5.0    6.0    7.0    8.0
```

In our example the right end count is 4, i.e., there are four "H"s before encountering an "L."

(NOTE: If the left and right tails start with the same symbol, the total end count will automatically be zero.)

4.   Sum the left and right end count to obtain the total end count. In this example the total end count is 7.

5.   Compare the total count with the tabled values below for significance.

Total End Count	Confidence in H_1
≥ 6	$\geq 90\%$
≥ 7	$\geq 95\%$
≥ 10	$\geq 99\%$
≥ 13	$\geq 99.9\%$

Since our example had a total end count of 7 we are at least 95% confident in concluding that the two sample means are significantly different and thus $H_1: \mu_L \neq \mu_H$. Notice how these results are very close to those determined previously through actual P value determination. The "End Count Method" is statistically based and very accurate. It also has the advantage of not requiring an underlying normal distribution. See [Schm 94] for more details.

6.4 Hypothesis Tests for Proportions

When data comes from a binomial experiment (see Chapter 4), e.g., each individual outcome is either a 0 for non-defective or a 1 for defective, then tests of hypothesis are based on proportions. We shall use the Greek letter π to denote the true but unknown proportion of a population. Using the Central Limit Theorem, tests of proportions can be conducted with a standardized test statistic, Z_0, which is normally distributed. This is based on the fact that the sampling distribution of the statistic p (i.e., the sample proportion) is approximately normal for sufficiently large sample sizes. A common Rule of Thumb is that both $np \geq 5$ and $n(1 - p) \geq 5$ must hold in order to have a sufficiently large sample size. This section addresses both the one-sample and two-sample tests for proportions.

6.4.1 One-Sample Test (n ≥ 30)[1]

We outline a six-step procedure for conducting a one-sample test of a proportion and then illustrate this procedure with an example.

Step 1:   State the hypotheses to be tested.

Two-Tail	Upper-Tail	Lower-Tail
$H_0: \pi = \pi_0$	$H_0: \pi \leq \pi_0$	$H_0: \pi \geq \pi_0$
$H_1: \pi \neq \pi_0$	$H_1: \pi > \pi_0$	$H_1: \pi < \pi_0$

Step 2:   Determine a planning value for α.

Step 3:   Compute the sample proportion $p = X/n$, where $X = $ number of items in the category of interest in a sample of size n. Then compute

$$Z_0 = \frac{p - \pi_0}{\sqrt{\dfrac{\pi_0 \left(1 - \pi_0\right)}{n}}}$$

Step 4:   Using a standardized normal (Z) distribution, compute the area in the tail beyond $|Z_0|$ (for 2-tail tests); to the left of Z_0 (for lower-tail tests); or to the right of Z_0 (for upper-tail tests).

Step 5:   Compute P as follows:

For two-tail tests, P is two times the area in Step 4.

For one-tail tests, P is the area found in Step 4.

---

[1]For n < 30, see [Walp 85] for the appropriate test statistics.

Step 6:   For $P \geq \alpha$, fail to reject H_0.

For $P < \alpha$, reject H_0 with $(1 - P)100\%$ confidence in H_1.

(See Note 2 below for a ROT test)

Note 1:   *The "lower case" p is used to designate a sample proportion. The "upper case" P, or P-value we have been using in Steps 5 and 6 of hypothesis testing, is an area under the curve which represents the exact probability of making a Type I error. In other words, the P-value is the smallest α value for which we would reject H_0 at the α level of significance.*

Note 2:   *Using a quick Rule of Thumb (ROT): i) for two-tail tests, $|Z_0| \geq 2$ implies approximately 95% confidence in H_1, whereas $|Z_0| \geq 3$ implies approximately 99% confidence in H_1; ii) for one-tail tests where Z_0 is in the appropriate tail, $|Z_0| \geq 2$ implies approximately 97.5% confidence in H_1, whereas $|Z_0| \geq 3$ implies approximately 99.5% confidence in H_1.*

Example 6.5   Historically, a leakage test of the product for which we are responsible has detected a defect rate of 0.12. We have just implemented changes in the process and we desire to test if a difference in defect rate has occurred. We take a sample of 300 products from the new process and find that 45 of these are defective. An appropriate hypothesis test for a proportion is as follows:

Step 1: $H_0: \pi = .12$

$H_1: \pi \neq .12$

Step 2: $\alpha = .01$

Step 3: $n = 300$, $X = 45$, $p = 45/300 = .15$, and thus

$$Z_0 = \frac{.15 - .12}{\sqrt{\dfrac{(.12)(.88)}{300}}} = 1.60$$

Step 4:

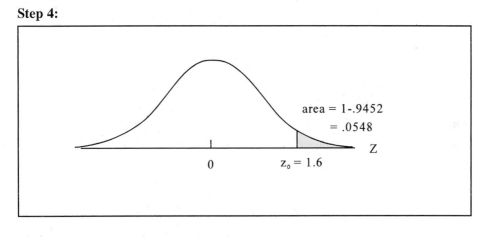

Step 5: $P = 2(.0548) = .1096$

Step 6:   Since $[P = .1096] > [\alpha = .01]$, fail to reject H_0. Notice how these results are supported by the ROT discussed previously in Note 2 on the preceding page.

∎

6.4.2 Two-Sample Test ($n_1 + n_2 \geq 30$)

Similarly, we present a six-step procedure for and give an example of the two-sample test for proportions.

Step 1:   State the hypotheses to be tested.

Two-Tail	Upper-Tail	Lower-Tail
$H_0: \pi_1 = \pi_2$	$H_0: \pi_1 \leq \pi_2$	$H_0: \pi_1 \geq \pi_2$
$H_1: \pi_1 \neq \pi_2$	$H_1: \pi_1 > \pi_2$	$H_1: \pi_1 < \pi_2$

Step 2: Determine a planning value for α.

Step 3: Compute the sample proportions $p_1 = X_1/n_1$ and $p_2 = X_2/n_2$.

Compute the standardized test statistic

$$Z_0 = \frac{p_1 - p_2}{\sqrt{\left(\dfrac{X_1 + X_2}{n_1 + n_2}\right)\left(1 - \dfrac{X_1 + X_2}{n_1 + n_2}\right)\left(\dfrac{1}{n_1} + \dfrac{1}{n_2}\right)}}$$

Step 4: Using a standardized normal (Z) distribution, compute the area in the tail beyond $|Z_0|$ (for 2-tail tests); to the left of Z_0 (for lower-tail tests); or to the right of Z_0 (for upper-tail tests).

Step 5: Compute P as follows.

For two-tail tests, P is two times the area in Step 4.

For one-tail tests, P is the area in Step 4.

Step 6: If $P \geq \alpha$, fail to reject H_0. If $P < \alpha$, reject H_0 with $(1 - P)100\%$ confidence in H_1.

Note: *For a quick ROT, use $|Z_0| \geq 2$ and $|Z_0| \geq 3$ for 95% and 99% confidence, respectively, in $H_1 : \pi_1 \neq \pi_2$.*

Example 6.6 A random sample of 60 Republicans and 60 Democrats showed 68.3% and 40.0%, respectively, in favor of the death penalty for major crimes. Conduct a hypothesis test to determine if these sample proportions provide overwhelming evidence against bipartisan agreement.

Step 1: $H_0: \pi_1 = \pi_2$

$H_1: \pi_1 \neq \pi_2$

Step 2: $\alpha = .05$

Step 3: $p_1 = .683$ $p_2 = .400$

                   $n_1 = 60$ $n_2 = 60$

Therefore, since $p_i = X_i / n_i$, we have

$$X_1 = (.683)(60) = 41 \text{ and } X_2 = (.400)(60) = 24.$$

$$Z_0 = \frac{.683 - .400}{\sqrt{\left(\frac{41 + 24}{60 + 60}\right)\left(1 - \frac{41 + 24}{60 + 60}\right)\left(\frac{1}{60} + \frac{1}{60}\right)}} = 3.11$$

Step 4:

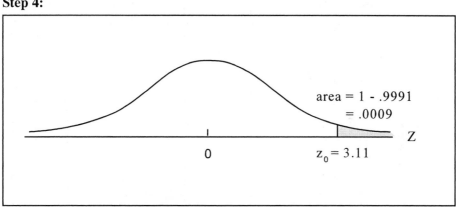

area = 1 - .9991
= .0009

0

$z_0 = 3.11$

Z

Step 5: $P = 2(.0009) = .0018.$

Step 6:   Since $[P = .0018] < [\alpha = .05]$, reject H_0 with 99.82% confidence in H_1. That is, there is a difference between the Republican and Democratic view of the death penalty. Notice how these results are very similar to those obtained using the ROT discussed in the previous note.

6.5 Hypothesis Tests of Variation

Excessive variation of critical product dimensions is a major contributor to poor quality. Today's customers demand uniform, defect-free products which necessitates testing processes in order to identify parameter settings which minimize variability. In this section we consider both one-sample and two-sample tests of population variance. The reader should realize that ideally, sample sizes should be greater than or equal to 17 to get good estimates of variance [Schm 94].

6.5.1 One-Sample Test ($\sigma^2 = \sigma_0^2$)

In designing solid rocket motors, achieving the desired average burnrate is only part of the problem. Another important objective is to minimize the burnrate variability around this average. In Figure 6.4, the A curve indicates the historical distribution of burnrates for many test burns. The B curve indicates the desired burnrate distribution. As the chief engineer you desire to test if a new additive will significantly reduce the burnrate variability below a current value of 16.

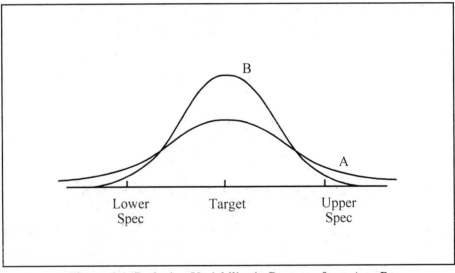

Figure 6.4 Reducing Variability in Burnrate from A to B

To conduct this test you take sample product burnrates which are assumed to be approximately normally distributed. You then calculate the sample variance, s^2, and either construct a $(1 - \alpha)100\%$ confidence interval for the unknown σ^2 and/or perform a hypothesis test. The confidence interval approach is shown in Table 6.7.

$$\frac{(n - 1)s^2}{\chi_U^2} \leq \sigma^2 \leq \frac{(n - 1)s^2}{\chi_L^2}$$

where χ_U^2 and χ_L^2 are values from the chi-square distribution in Appendix G. Specifically, the larger chi-square value is $\chi_U^2 = \chi^2(1 - \alpha/2, n - 1)$ and the smaller chi-square value is $\chi_L^2 = \chi^2(\alpha/2, n - 1)$. The χ^2 distribution will be discussed in greater detail in the next section.

Table 6.7 $(1 - \alpha)100\%$ Confidence Interval for σ^2

In our solid rocket motor example, the sample data provided the following evidence: $n = 15$, $\bar{x} = 50$ (burnrate units * 100) and $s^2 = 4.612$. Using the sample data, a 95% confidence interval for σ^2 is

$$\frac{14(4.612)}{26.12} \leq \sigma^2 \leq \frac{14(4.612)}{5.63}$$

which simplifies to $2.472 \leq \sigma^2 \leq 11.469$. Thus, you can be 95% confident that the unknown σ^2 lies in the range [2.472, 11.469]. Since the old σ^2 was 16, we can conclude with at least 95% confidence that the new additive has significantly reduced burnrate variance.

The hypothesis test approach for the one-sample test of variation is shown in Table 6.8.

Step 1:	Two-Tail	Upper-Tail	Lower-Tail
	$H_0: \sigma^2 = \sigma_0^2$	$H_0: \sigma^2 \leq \sigma_0^2$	$H_0: \sigma^2 \geq \sigma_0^2$
	$H_1: \sigma^2 \neq \sigma_0^2$	$H_1: \sigma^2 > \sigma_0^2$	$H_1: \sigma^2 < \sigma_0^2$

Step 2: Determine a planning value for α.

Step 3: Test Statistic: $\chi_0^2 = \dfrac{(n-1)s^2}{\sigma_0^2}$

Step 4: Using the χ^2 table (Appendix G), estimate the area in the appropriate tail outside of χ_0^2.

Step 5: For two-tail tests, P is equal to 2 times the area in Step 4. For one-tail tests, P is equal to the area in Step 4.

Step 6: If $P \geq \alpha$, fail to reject H_0; and if $P < \alpha$, reject H_0 with $(1 - P)100\%$ confidence in H_1.

Table 6.8 Hypothesis Approach to Testing Variance (One-Sample)

Using our solid rocket motor example, the steps for a hypothesis test are as follows:

Step 1: $H_0: \sigma^2 = 16$
        $H_1: \sigma^2 \neq 16$

Step 2: $\alpha = .05$

Step 3: $\chi_0^2 = \dfrac{14(4.612)}{16} = 4.035$

Step 4:

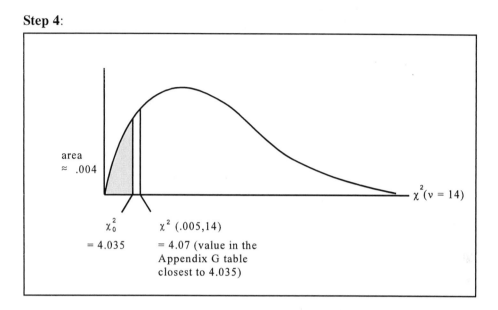

area
$\approx$ .004

χ_0^2
= 4.035

χ^2 (.005,14)
= 4.07 (value in the
Appendix G table
closest to 4.035)

$\chi^2(v = 14)$

Step 5: $P = 2(.004) = .008.$

Step 6:   Since $[P = .008] < [\alpha = .05]$, we can be approximately 99.2% confident that $\sigma^2 \neq 16$. In fact, since $s^2 < \sigma_0^2$ we can state with approximately 99.2% confidence that the new burnrate variance is significantly less than 16.

6.5.2 Two-Sample Test ($\sigma_1^2 = \sigma_2^2$)

Variability of thickness in a nickel plating process has shown to be a major contributor to defective products. The process engineer has decided to test whether the processing temperature has an effect on thickness variability. The engineer has decided to test 25 sample products at each temperature. The test results are as follows:

Sample 1 (Low)	**Sample 2 (High)**
$n_1 = 25$	$n_2 = 25$
$s_1^2 = 6.51$	$s_2^2 = 1.70$

The sample variances are obviously different; however, the question is whether they are significantly different. The two-sample test for variance is shown in Table 6.9.

Step 1:   Two-Tail   Upper-Tail   Lower-Tail

$H_0: \sigma_1^2 = \sigma_2^2$ $H_0: \sigma_1^2 \leq \sigma_2^2$ $H_0: \sigma_1^2 \geq \sigma_2^2$

$H_1: \sigma_1^2 \neq \sigma_2^2$ $H_1: \sigma_1^2 > \sigma_2^2$ $H_1: \sigma_1^2 < \sigma_2^2$

Step 2:   Determine a planning value for α.

Step 3:   Test Statistic: $F_0 = \dfrac{s_1^2}{s_2^2}$

Step 4:   Using an F distribution with $v_1 = n_1 - 1$ and $v_2 = n_2 - 1$, estimate the area in the tail. See Appendix F for details on the F distribution.

Step 5:   For two-tail tests, P is equal to two times the area in Step 4.

   For one-tail tests, P is equal to the area in Step 4.

Step 6:   If $P \geq \alpha$, fail to reject H_0, and

   if $P < \alpha$, reject H_0 with $(1-P)100\%$ confidence in H_1.

Note: *For a quick ROT, if* $\dfrac{S_{max}^2}{S_{min}^2} \sqrt{\dfrac{n_1 + n_2}{2}} > 10,$ *we will*

typically have approximately 95% confidence in H_1. This is a good ROT when the sample sizes n_1 and n_2 are approximately equal (the smaller being no less than 70% of the larger) and

$\dfrac{n_1 + n_2}{2}$ *is less than 60. This ROT will produce a quick*

analysis without having to be proficient in F table lookup.

Table 6.9   Hypothesis Tests for Variance (Two-Sample)

Using the nickel plating example, the hypothesis test is as follows:

Step 1: H_0: $\sigma_1^2 = \sigma_2^2$

H_1: $\sigma_1^2 \neq \sigma_2^2$

Step 2:   Set $\alpha = .01$.

Step 3: $F_0 = 6.51/1.70 = 3.83$.

Step 4:

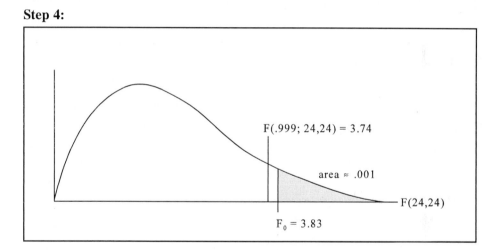

$F(.999; 24,24) = 3.74$

area $\approx .001$

$F(24,24)$

$F_0 = 3.83$

Step 5: $P = 2(.001) = .002$.

Step 6:   Since $[P = .002] < [\alpha = .01]$ reject H_0 with approximately 99.8% confidence in H_1. Therefore, we can conclude that $\sigma_2^2 < \sigma_1^2$ with approximately 99.8% confidence, indicating that the higher temperature setting will significantly reduce the variability in plating thickness.

Note: *Using our ROT from Table 6.9,*

$$\left(\frac{6.51}{1.70}\right) * \sqrt{\frac{25 + 25}{2}} = (3.829 * 5) = 19.145 > 10.$$

Therefore, we can quickly conclude $H_1 : \sigma_1^2 \neq \sigma_2^2$ with at least 95% confidence.
This conclusion can be made without doing an F table lookup and it is fairly
accurate.

6.6 Analysis of Variance (ANOVA)

The previous sections in this chapter have dealt with hypothesis tests for
one-sample and two-sample sets of data. For example, we outlined both a one-
sample t test and a two-sample t test for population means. What happens if we
have more than two independent samples? Unfortunately it is impossible to set up,
say, a 4-sample t test. When faced with more than two independent samples, we
must resort to an entirely different analysis procedure known as the Analysis of
Variance (ANOVA). Table 6.10 shows the ANOVA approach to testing multiple
independent samples for the equality of population means.

Step 1: $H_0: \mu_1 = \mu_2 = \mu_3 = \ldots = \mu_L$

H_1: Not all μ_i are equal

Step 2:   Determine a planning value for α

Step 3:   Calculate an F_0 value using the following two estimates
of variance:

i)   Mean Square Between (MSB)* =

$$\frac{\sum\limits_{1}^{L} n_j\, (\bar{y}_j - \bar{y})^2}{L-1}$$

Table 6.10   (continued on next page)

ii)   Mean Square Error (MSE)** =

$$\frac{\displaystyle\sum_{1}^{L} (n_j - 1)\, S_j^2}{\displaystyle\sum_{1}^{L} (n_j - 1)}, \qquad \text{where}$$

L   = number of different samples to be compared

n_j = sample size for the jth sample

S_j^2 = variance for the jth sample

$\bar{y}_j$ = average of the jth sample

$\bar{y}$ = grand average or weighted average of all samples

The F test value is calculated as $F_0 = \dfrac{MSB}{MSE}$.

Step 4:   Use percentiles of the F distribution (see Appendix F) to estimate the area in the tail of the distribution beyond F_0. The degrees of freedom for the $F(v_1, v_2)$ distribution are

$$v_1 = L - 1 \qquad \text{and} \qquad v_2 = \sum_{1}^{L} (n_j - 1)$$

Step 5:   Set P equal to the area found in Step 4.

Step 6:   If $P \geq \alpha$ fail to reject H_0; and for $P < \alpha$ reject H_0 with

(1 - P) 100% confidence.

*   MSB is an estimate of background noise or experimental error which is calculated by way of the variance of sample averages.

**   MSE is an estimate of background noise or experimental error which is calculated by way of the average sample variances.

NOTE: If H_0 is true, MSB and MSE are estimating the same thing and should therefore be approximately equal. If H_1 is true, MSE will not likely change, but MSB should be significantly larger than MSE. This is the basis of the F test described earlier.

Table 6.10 ANOVA Approach to Testing Multiple Independent Samples

When comparing only two samples, the ANOVA approach is equivalent to the two-sample t test. We illustrate this by applying the ANOVA approach to the two samples shown in Table 6.5.

Step 1: $H_0: \mu_L = \mu_H$

$H_1: \mu_L \neq \mu_H$

Step 2: $\alpha = .01$

Step 3: $\text{MSB} = \dfrac{\sum\limits_{1}^{2} n_j(\bar{y}_j - \bar{y})^2}{2 - 1} = \dfrac{9(4.6889 - 5.2778)^2 + 9(5.8667 - 5.2778)^2}{2 - 1}$

$= 6.2425$

$\text{MSE} = \dfrac{\sum\limits_{1}^{2} (n_j - 1)S_j^{\,2}}{\sum\limits_{1}^{2} (n_j - 1)} = \dfrac{8(1.3761) + 8(1.1575)}{8 + 8} = 1.2668$

$F_0 = \dfrac{\text{MSB}}{\text{MSE}} = \dfrac{6.2425}{1.2668} = 4.93$

Step 4:

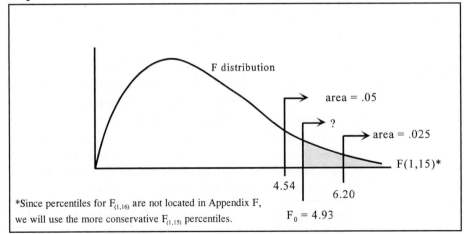

*Since percentiles for $F_{(1,16)}$ are not located in Appendix F, we will use the more conservative $F_{(1,15)}$ percentiles.

The approximate area to the right of 4.93 is 0.04.

Step 5: Set P = .04

Step 6: Since (P = .04) > (α = .01) fail to reject H_0 at the .01 level. This is the same result obtained from the t test in Section 6.3.

Applying the ANOVA approach to a variable A which has been tested at 5 levels (L = 5), we obtain the following results:

	Factor A				
	A_1	A_2	A_3	A_4	A_5
$\bar{y}_j$	18	16	17	22	20
s_j^2	4.888	7.333	12.222	9.333	10.667
n_j	10	10	10	10	10

Step 1: H_0: $\mu_1 = \mu_2 = \mu_3 = \mu_4 = \mu_5$

H_1: not all μ_i are equal

Step 2: α = .01

Step 3: $\bar{y} = \dfrac{\sum n_j \bar{y}_j}{\sum n_j} = \dfrac{10\,(18 + 16 + 17 + 22 + 20)}{50} = 18.6$

$MSB = \dfrac{\sum_1^5 n_j(\bar{y}_j - \bar{y})^2}{5 - 1} = \dfrac{10(-.6)^2 + 10(-2.6)^2 + 10(-1.6)^2 + 10(3.4)^2 + 10(1.4)^2}{5 - 1} = 58$

$$MSE = \frac{\sum_{1}^{5}(n_j - 1)S_j^2}{\sum_{1}^{5}(n_j - 1)} = \frac{9(4.88) + 9(7.333) + 9(12.222) + 9(9.333) + 9(10.667)}{9 + 9 + 9 + 9 + 9} = 8.887$$

$$F_0 = \frac{MSB}{MSE} = \frac{58.000}{8.887} = 6.53$$

Step 4:

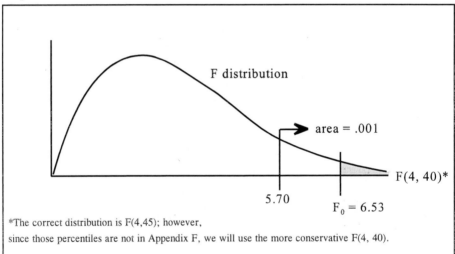

*The correct distribution is F(4,45); however, since those percentiles are not in Appendix F, we will use the more conservative F(4, 40).

The area to the right of 6.53 is much smaller than 0.001.

Step 5:   Set P = .001 as a rough estimate.

Step 6:   Since (P = .001) < (α = .01) conclude H_1 with at least 99.9% confidence.

In this example the conclusion is H_1: μ_i are not all equal. The next question is, "Which levels are different and which are not?" Using a regular t test or F test on each possible pair of treatments, i.e., H_0: $\mu_i = \mu_j$ for all i < j, would result in an increase in the overall α known as the experiment-wise α, α_{EW}. The α_{EW} is also referred to as the probability of committing at least one Type I error when conducting all of the individual pairwise tests. A procedure which allows for testing all pairwise levels at a specified α_{EW} is the Tukey Post Hoc comparison test. Use of this test is valid only when the overall F_0 is significant, as it is in the previous example. The test is fairly simple and requires the use of the studentized ranges provided in Table 6.11. The Tukey test will be illustrated using the previous example.

v	P								
	2	**3**	**4**	**5**	**6**	**7**	**8**	**9**	**10**
1	17.97	17.97	17.97	17.97	17.97	17.97	17.97	17.97	17.97
2	6.085	6.085	6.085	6.085	6.085	6.085	6.085	6.085	6.085
3	4.501	4.516	4.516	4.516	4.516	4.516	4.516	4.516	4.516
4	3.927	4.013	4.033	4.033	4.033	4.033	4.033	4.033	4.033
5	3.635	3.749	3.797	3.814	3.814	3.814	3.814	3.814	3.814
6	3.461	3.587	3.649	3.680	3.694	3.697	3.697	3.697	3.697
7	3.344	3.477	3.548	3.588	3.611	3.622	3.626	3.626	3.626
8	3.261	3.399	3.475	3.521	3.549	3.566	3.575	3.579	3.579
9	3.199	3.339	3.420	3.470	3.502	3.523	3.536	3.544	3.547
10	3.151	3.293	3.376	3.430	3.465	3.489	3.505	3.516	3.522
11	3.113	3.256	3.342	3.397	3.435	3.462	3.480	3.493	3.501
12	3.082	3.225	3.313	3.370	3.410	3.439	3.459	3.474	3.484
13	3.055	3.200	3.289	3.348	3.389	3.419	3.442	3.458	3.470
14	3.033	3.178	3.268	3.329	3.372	3.403	3.426	3.444	3.457
15	3.014	3.160	3.250	3.312	3.356	3.389	3.413	3.432	3.446
16	2.998	3.144	3.235	3.298	3.343	3.376	3.402	3.422	3.437
17	2.984	3.130	3.222	3.285	3.331	3.366	3.392	3.412	3.429
18	2.971	3.118	3.210	3.274	3.321	3.356	3.383	3.405	3.421
19	2.960	3.107	3.199	3.264	3.311	3.347	3.375	3.397	3.415
20	2.950	3.097	3.190	3.255	3.303	3.339	3.368	3.391	3.409
24	2.919	3.066	3.160	3.226	3.276	3.315	3.345	3.370	3.390
30	2.888	3.035	3.131	3.199	3.250	3.290	3.322	3.349	3.371
40	2.858	3.006	3.102	3.171	3.224	3.266	3.300	3.328	3.352
60	2.829	2.976	3.073	3.143	3.198	3.241	3.277	3.307	3.333
120	2.800	2.947	3.045	3.116	3.172	3.217	3.254	3.287	3.314
∞	2.772	2.918	3.017	3.089	3.146	3.193	3.232	3.265	3.294

Table 6.11 Studentized Range Values, q_T, for $\alpha = .05$

Reprinted with permission from *Probability and Statistics for Engineers and Scientists*, Walpole and Myers

Since we concluded H_1, we can perform the Tukey test and determine which levels differ and which do not. The procedure is as follows:

(a)   Compute the critical Tukey distance $d_T = q_T \sqrt{\dfrac{MSE}{n_j}}$ where q_T is found in Table 6.11 and all n_j are assumed equal.

(b)   In Table 6.11, p is the number of variable levels or different samples to be tested. Thus, for our example, p = 5. Since $df_{MSE} = 40$ and $\alpha = .05$, $q_T = 3.171$. Therefore,

$$d_T = 3.171 \sqrt{\frac{8.887}{10}} = 2.989.$$

(c)   Compute all variable level (sample) mean pairwise differences and conclude those differences greater than d_T as significant at the chosen α level (see the table shown next).

	$\bar{y}_1$	$\bar{y}_2$	$\bar{y}_3$	$\bar{y}_4$	$\bar{y}_5$
$\bar{y}_1$	0				
$\bar{y}_2$	2	0			
$\bar{y}_3$	1	1	0		
$\bar{y}_4$	4*	6*	5*	0	
$\bar{y}_5$	2	4*	3*	2	0
* indicates this pairwise difference exceeds $d_T = 2.989$					

Thus we should conclude the following:

i) μ_4 is significantly different than μ_1, μ_2 and μ_3.

ii) μ_5 is significantly different than μ_2 and μ_3.

iii)   fail to reject H_0 for all other pairwise comparisons.

6.7 Chi-Square (χ^2) Goodness-of-Fit Test

In previous sections we have seen extensive use of the normal, t and F distributions as sampling distributions. That is, they are distributions which arise in the investigation of certain test statistics. Another common sampling distribution is the Chi-Square distribution. Its density function is skewed to the right and has a shape as shown in Figure 6.5. The χ^2 distribution has one parameter, ν, which is referred to as the distribution's "degrees of freedom." The mean and variance of the χ^2 distribution are ν and 2ν, respectively. Selected percentiles for the χ^2 distribution are given in Appendix G.

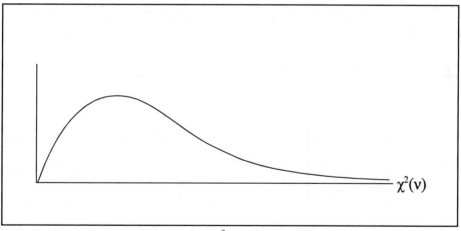

Figure 6.5 $\chi^2(\nu)$ Distribution

While we have seen briefly in the previous section the use of the χ^2 distribution in the one-sample test of σ^2, the χ^2 distribution also arises in situations where data are categorized into cells. Each cell contains a count of the number of random events that fall within the defined limits of the cell. For each of k cells, we denote the observed frequency in each cell by f_i, for i = 1, . . ., k. We then compare the observed frequency, f_i, to an expected frequency, F_i, and base the χ^2 test statistic on the relative squared difference between the two.

Example 6.7 The first 100 Class B accidents at five Air Force bases are observed and classified by base. The following table lists the number observed at each base.

Base	Number of Accidents
1	20
2	23
3	12
4	15
5	30
Total	100

Assuming that accidents are evenly or uniformly distributed among the bases, we would expect an equal number of accidents to occur at each base. Therefore, for the 100 accidents, we would have expected 20 to occur at each base. This (20) is the value of each F_i. The χ^2 test statistic is given by:

$$\chi_0^2 = \sum_{i=1}^{5} \frac{(f_i - F_i)^2}{F_1} = \frac{(20 - 20)^2}{20} + \frac{(23 - 20)^2}{20} + \ldots + \frac{(30 - 20)^2}{20} = 9.9.$$

If the accidents truly are equally distributed among the bases, then the values of f_i would be "close" to those of F_i and the value of the test statistic would be small. However, if the observed frequencies are much different than the expected frequencies, the test statistic will be large and we would reject the hypothesis that the accidents are evenly distributed among the bases.

Assuming the accidents are uniformly distributed among the bases, the test statistic χ_0^2 comes from a χ^2 distribution with four degrees of freedom. If we compare the value for our test statistic with the values of the χ^2 percentiles from Appendix G, we see that

$$\chi^2(.95, 4) = 9.49 \text{ and } \chi^2(.975, 4) = 11.14.$$

Thus, $\chi_0^2 = 9.9$ is bounded by 9.49 and 11.14 and we would conclude that .025 < P < .05, with P more likely to be closer to 0.05 than it is to 0.025.

The degrees of freedom are found by the simple formula:

$$\text{degrees of freedom} = \text{number of cells} - \begin{bmatrix} \text{number of constraints} \\ \text{satisfied by the data} \\ \text{in calculating the} \\ \text{expected frequencies.} \end{bmatrix}$$

In this case there are five cells and one constraint. The constraint is that the sum of the cell frequencies must equal the total sample size. That is, $f_1 + f_2 + f_3 + f_4 + f_5 = n$. Therefore, the number of degrees of freedom for this test is $5 - 1 = 4$.

∎

Two assumptions are required in the development of a χ^2 test:

1.   The sample is a simple random sample from an infinite population.

2.   The sample size is large. "Large" is defined in terms of the expected frequencies. The general rule is that the sample size is large enough if

   (a)   All expected cell frequencies (F_i) are greater than or equal to 2, and

   (b)   At least half of the expected cell frequencies (F_i) are greater than or equal to 5.

If either of these conditions is not satisfied, it will be necessary to pool classes or cells. This technique is demonstrated in Example 6.10.

A χ^2 goodness-of-fit test is an upper-tail hypothesis test that is used to test whether an underlying distribution (or population) from which sample data has been taken is of a specified form. The sample data will be organized in a frequency distribution with each class becoming a cell as discussed previously. We will use the following six-step procedure for conducting the goodness-of-fit test.

Step 1: The null and alternative hypotheses will be given by

H_0: The population distribution is _____

H_1: The population distribution is not _____

The null hypothesis will specify the distribution used in calculating the **expected** frequencies. For example, if the hypotheses are

H_0: The population distribution is Poisson with $\lambda = 2.0$

H_1: The population distribution is not Poisson with $\lambda = 2.0$,

then the Poisson distribution with $\lambda = 2.0$ is used to find the expected frequencies. However, if the hypotheses are

H_0: The population distribution is Poisson

H_1: The population distribution is not Poisson,

then we must estimate λ with $\overline{X} \approx E(X) = \lambda$ and calculate the expected frequencies from a Poisson distribution based on $\lambda \approx \overline{X}$.

Step 2: Determine a planning value for α. As before, the choice for α will be based on the experimenter's risk of a Type I error.

Step 3:   Calculate the test statistic given by

$$\chi_0^2 = \sum_{i=1}^{k} \frac{(f_i - F_i)^2}{F_i}$$

where

k	=	the number of classes (after pooling),
f$_i$	=	observed frequency for cell i, and
F$_i$	=	expected frequency for cell i, calculated under the assumption that the null hypothesis is true.

Step 4:   Using the $\chi^2(v)$ distribution, compute the area in the tail beyond χ_0^2. In order to use the tables in Appendix G, we must first determine the appropriate degrees of freedom. The degrees of freedom are given by the formula

$$v = k - m - 1,$$

where   k = the number of classes (cells) after pooling, and

m = the number of parameters estimated from the data.

The number of parameters estimated from the data is determined by the distribution being tested and also by the hypothesis. For example, if the hypotheses are

H_0: The distribution is normal

H_1: The distribution is not normal,

then the two parameters of the normal distribution, μ and σ, must be estimated by $\bar{x}$ and s, respectively, in order to calculate the expected frequencies. In this case, m = 2. However, if the hypotheses specify the parameters, then the parameters μ and σ are not estimated from the data when calculating the expected frequencies. For example, if

H_0: The distribution is normal with $\mu = 5$, $\sigma = 1$

H_1: The distribution is not normal with $\mu = 5$, $\sigma = 1$,

then the values for μ and σ given in the hypotheses are used to calculate the expected frequencies. In this case, m = 0.

Step 5:   The area in Step 4 is equal to P.

Step 6:   If $P \geq \alpha$ then conclude H_0.

If $P < \alpha$ then conclude H_1.

It is important to remember that the α risk is the probability of rejecting H_0 given that H_0 is actually true; and it is this risk that is controlled. Therefore, when we reject H_0, we have at least $(1 - \alpha)$ 100% confidence in our conclusion. That is, we can be fairly sure (with a significance level of α) that our data does not come from the specified distribution. However, when we fail to reject H_0, we cannot attach a level of significance to that conclusion. The best we can say is that we do not have overwhelming evidence to support the hypothesis (H_1) that the data does not come from the specified distribution. In most cases, we may be willing to assume that the distribution specified in the null hypothesis is adequate to model the data. A χ^2 test is often used to build and/or validate simulation models and to verify population assumptions for further use of statistical inference procedures.

Example 6.8 Suppose we wish to test the following data to see if it can be adequately fit by a uniform distribution. A uniform distribution is one in which all values are equally likely, i.e., its probability density function is a horizontal line. We supplement the frequency distribution with columns p_i, F_i, and

$\dfrac{(f_i - F_i)^2}{F_i}$, where p_i is the probability that a randomly selected value falls within the interval. The p_i's are calculated based on the assumption that the null hypothesis is true (i.e., the distribution is uniform). F_i and f_i are the expected and observed frequencies, respectively, where $F_i = np_i = (105)(1/5) = 21$ for each class (i).

CLASS	f_i	p_i	F_i	$\dfrac{(f_i - F_i)^2}{F_i}$
0 to < 2	17	1/5	21	.762
2 to < 4	20	1/5	21	.048
4 to < 6	30	1/5	21	3.857
6 to < 8	22	1/5	21	.048
8 to < 10	16	1/5	21	1.191
Total	105	1.00	105	$\chi_0^2 = 5.906$

Step 1: H_0: The distribution is uniform on (0, 10)

H_1: The distribution is not uniform on (0, 10)

Step 2:   We will use $\alpha = .05$

Step 3: $\chi_0^2 = \displaystyle\sum_{i=1}^{5} \dfrac{(f_i - F_i)^2}{F_i} = 5.906$

Step 4:   The number of classes is $k = 5$ and the number of parameters estimated from the data is $m = 0$. Therefore, the degrees of freedom are $v = k - m - 1 = 4$ and we would like to find the area in the tail of the $\chi^2(4)$ distribution beyond 5.906. Since the area to the right of 7.78 is 0.10, the area to the right of $\chi_0^2 = 5.906$ is greater than 0.10.

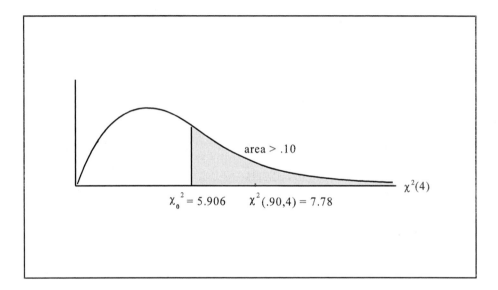

Step 5:   From the previous step, $P > .10$.

Step 6:   Since $P > \alpha$, we fail to reject H_0, concluding that there is not enough evidence to refute the assumption that the uniform distribution is an adequate fit for the data.

∎

Example 6.9 Suppose, however, that we wished to test the same set of data to see if it could be modeled by a normal distribution. The following table would result. In this table we have extended the range of the first and last classes because the range of the normal RV is $(-\infty, \infty)$. This is now a valid distribution to which we may apply the Chi-Square test.

CLASS	f_i	p_i	F_i	$\dfrac{(f_i - F_i)^2}{F_i}$
$-\infty$ to < 2	17	.1093	11.477	2.658
2 to < 4	20	.2135	22.418	.261
4 to < 6	30	.2989	31.385	.061
6 to < 8	22	.2382	25.011	.362
8 to $< \infty$	16	.1401	14.711	.113
Total	105	1.0000	105.002	$\chi_0^2 = 3.455$

The sample statistics calculated from the raw data are $\bar{x} = 5.2$ and $s = 2.6$ (computations not shown). In this case, the p_i values are calculated from the normal distribution, where we use 5.2 as an approximation for μ and 2.6 as an approximation for σ. For example,

$$p_1 = P(X < 2) = P\left(Z < \frac{2 - 5.2}{2.6} \right)$$

$$= P(Z < -1.23)$$

$$= .1093$$

Step 1: H_0: The distribution is normal

H_1: The distribution is not normal

Step 2:   Again, we will use $\alpha = .05$

Step 3: $\chi_0^2 = 3.455$

Step 4: The number of classes is k = 5 and the number of parameters estimated from the data is m = 2 (recall we used 5.2 to estimate μ and 2.6 to estimate σ). Therefore, the degrees of freedom are $v = k - m - 1 = 2$ and we seek the area under the $\chi^2(2)$ distribution to the right of $\chi_0^2 = 3.455$. Since the area under the curve to the right of 4.61 is 0.10, the area beyond $\chi_0^2 = 3.455$ is much more than 0.10.

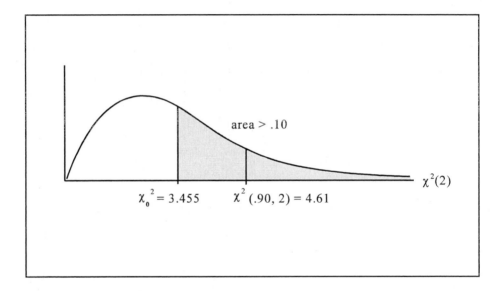

$\chi_0^2 = 3.455$ $\chi^2(.90, 2) = 4.61$

Step 5: From the previous step, P > .10.

Step 6: Since P > α, we fail to reject H_0, concluding there is not enough evidence to refute the normal distribution as being an adequate fit for the data.

∎

In the next example we demonstrate the pooling of classes because the criteria for minimum expected frequencies are not met. In this example, we will test to see if the data can be fit by an exponential distribution.

Example 6.10

Step 1: H_0: The distribution is exponential

H_1: The distribution is not exponential

CLASS	f_i	p_i	F_i
0 to < 1	45	.4866	46.227
1 to < 2	25	.2498	23.731
2 to < 3	15	.1283	12.189
3 to < 4	7	.0658	6.251
4 to < 5	2	.0338	3.211
5 to < 6	0	.0174	1.653
6 to < 7	0	.0089	0.846
7 to < 8	1	.0046	0.437
8 to ∞	0	.0048	0.456
Total	95	1.0000	95.001

From the frequency distribution, we find $\bar{x} \approx 1.5$ and $s^2 \approx 1.6$. Here, the p_i values are calculated from the exponential distribution because we assume that H_0 is true. Since we are not given a value for λ in the null hypothesis, we will estimate λ by $1/\bar{x} = 1/1.5 = .6667$ (recall that $\lambda = 1/E(X)$ when X is exponentially distributed). An example of a calculation for p_i is as follows:

$$p_2 = P(1 \le X < 2) = P(X < 2) - P(X < 1)$$

$$= (1 - e^{-2\lambda}) - (1 - e^{-1\lambda})$$

$$= .7364 - .4866$$

$$= .2498$$

Thus, $F_2 = np_2 = (95)(.2498) = 23.731$ which is the entry shown in the table for the class $[1, 2)$. This means that, if the parent population were exponential (H_0 is true), we would expect 23.731 of the 95 in the sample to fall in the class $[1, 2)$. At this time we notice that several expected frequencies (F_i) are less than 2 and more than half are less than 5. Either of these criteria would invalidate the χ^2 test. By pooling the last 4 classes we can obtain the following frequency distribution, which is a valid distribution for the Chi-Square test.

CLASS	f_i	p_i	F_i	$\dfrac{(f_i - F_i)^2}{F_i}$
0 to < 1	45	.4866	46.227	.0326
1 to < 2	25	.2498	23.731	.0679
2 to < 3	15	.1283	12.189	.6483
3 to < 4	7	.0658	6.251	.0897
4 to < 5	2	.0338	3.211	.4567
5 to ∞	1	.0357	3.392	1.6868
Total	95	1.0000	95.001	$\chi_0^2 = 2.9820$

Step 2:   Again, we will use $\alpha = .05$

Step 3: $\chi_0^2 = 2.982$

Step 4:   In this example, we use the number of classes after pooling so we have $k = 6$ and $m = 1$ (recall that we approximated λ with $1/\bar{x}$). Thus, the correct degrees of freedom are 4. We use the following sketch to estimate the area under the $\chi^2(4)$ distribution beyond $\chi_0^2 = 2.982$.

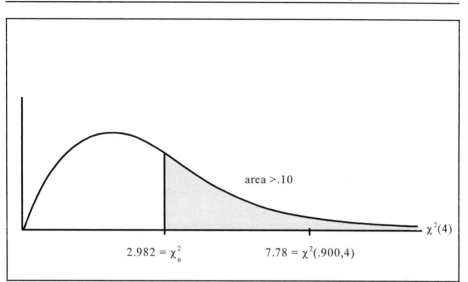

Step 5:   From the previous step, $P > .10$.

Step 6:   Since $P > \alpha$, we fail to reject H_0, concluding that the data can be adequately fit by an exponential distribution (i.e., there is not enough evidence to the contrary).

■

Note: *When estimating the parameters for a goodness-of-fit test, there is often more than one choice for an estimator. The following list gives the maximum likelihood estimators as well as unbiased estimators. Both have the necessary properties to be used in a χ^2 goodness-of-fit test. We provide this list as a table of reference only. Not all of the elements in this table have been fully developed in the text.*

Distribution	Parameter	Max. Likelihood Estimator	Unbiased Estimator
Uniform	a	$X_{(1)}$	$X_{(1)} - \left(\dfrac{X_{(n)} - X_{(1)}}{n - 1} \right)$
	b	$X_{(n)}$	$X_{(n)} + \left(\dfrac{X_{(n)} - X_{(1)}}{n - 1} \right)$
Normal	μ	$\bar{X}$	$\bar{X}$
	σ^2	$\dfrac{\sum (X_i - \bar{X})^2}{n}$	$\dfrac{\sum (X_i - \bar{X})^2}{n - 1}$
Exponential	λ	$\dfrac{1}{\bar{X}}$	$\dfrac{n - 1}{\sum X_i}$
Poisson	λ	$\bar{X}$	$\bar{X}$
Binomial (n fixed)	p	$\dfrac{y}{n}$	$\dfrac{y}{n}$

Note: $\mathbf{X_{(n)}}$ = max $(X_1, X_2, \ldots X_n)$

$\mathbf{X_{(1)}}$ = min $(X_1, X_2, \ldots X_n)$

$\mathbf{y}$ = total number of "successes"

$\mathbf{n}$ = sample size

6.8 Test for Independence

The χ^2 distribution can also be used to analyze data that are grouped in tabular form. The resulting table is called a contingency table. A primary application of the contingency table is the test for independence of two random variables.

Suppose we have a population that is cross-classified by two variables. Examples include machine parts classified by number of defects and service life; chicken eggs classified by shell thickness and weight; engine parts classified by amount of carbon build-up and type of engine oil used; and cities classified by population and percentage of carbon monoxide in the air. The χ^2 test for independence is a procedure used to determine if the two classification variables are related. Strictly speaking, we are testing for the statistical independence of the two random variables. We will consider a random sample from a population that is cross-classified by two variables. The data will be organized into a table as follows:

		Y					
		y_1	y_2	y_3	$\cdots$	y_c	
	x_1	f_{11}	f_{12}	f_{13}	$\cdots$	f_{1c}	$f_{1.}$
	x_2	f_{21}	f_{22}	f_{23}	$\cdots$	f_{2c}	$f_{2.}$
X	x_3	f_{31}	f_{32}	f_{33}	$\cdots$	f_{3c}	$f_{3.}$
	.	.	.	.		.	.
	.	.	.	.		.	.
	.	.	.	.		.	.
	x_r	f_{r1}	f_{r2}	f_{r3}	$\cdots$	f_{rc}	$f_{r.}$
		$f_{.1}$	$f_{.2}$	$f_{.3}$	$\cdots$	$f_{.c}$	N

where f_{ij} is the observed frequency in row i and column j. The marginal frequencies are given by

$$f_{i.} = f_{i1} + f_{i2} + f_{i3} + \ldots + f_{ic}, \text{ where c = number of columns,}$$

$$f_{.j} = f_{1j} + f_{2j} + f_{3j} + \ldots + f_{rj}, \text{ where r = number of rows, and}$$

$$N = \sum_{i=1}^{r} f_{i.} = \sum_{j=1}^{c} f_{.j} = \sum_{i=1}^{r}\sum_{j=1}^{c} f_{ij}$$

Now imagine an underlying bivariate probability distribution (as described in Chapter 3) for the two random variables. The corresponding distribution could be constructed as follows:

Y

		y_1	y_2	y_3	$\cdot\ \cdot\ \cdot$	y_c	
	x_1	p_{11}	p_{12}	p_{13}	$\cdot\ \cdot\ \cdot$	p_{1c}	$p_{1.}$
	x_2	p_{21}	p_{22}	p_{23}	$\cdot\ \cdot\ \cdot$	p_{2c}	$p_{2.}$
X	x_3	p_{31}	p_{32}	p_{33}	$\cdot\ \cdot\ \cdot$	p_{3c}	$p_{3.}$
	$\cdot$	$\cdot$	$\cdot$	$\cdot$		$\cdot$	$\cdot$
	$\cdot$	$\cdot$	$\cdot$	$\cdot$		$\cdot$	$\cdot$
	$\cdot$	$\cdot$	$\cdot$	$\cdot$		$\cdot$	$\cdot$
	x_r	p_{r1}	p_{r2}	p_{r3}	$\cdot\ \cdot\ \cdot$	p_{rc}	$p_{r.}$
		$p_{.1}$	$p_{.2}$	$p_{.3}$	$\cdot\ \cdot\ \cdot$	$p_{.c}$	1.00

where $p_{ij} = P(X = x_i \cap Y = y_j)$ is a joint probability. The marginal probabilities are given by

$$p_{i.} = P(X = x_i) \quad \text{and} \quad p_{.j} = P(Y = y_j).$$

The following formulas provide estimates for the marginal probabilities:

$$p_{i.} \approx \frac{f_{i.}}{N} \quad \text{and} \quad p_{.j} \approx \frac{f_{.j}}{N}.$$

Therefore, if the two variables X and Y are independent, we have

$$p_{ij} = (p_{i.})(p_{.j}) \quad \text{for all possible i, j combinations.}$$

An estimate for the expected frequency for cell i, j is given by

$$F_{ij} = N(p_{ij}) = N(p_{i.})(p_{.j}) \approx N \cdot \frac{f_{i.}}{N} \cdot \frac{f_{.j}}{N}, \quad \text{so}$$

$$F_{ij} = \frac{f_{i.} \cdot f_{.j}}{N}$$

Equation 6.1 Computational Formula for Expected Frequencies

In words, Equation 6.1 says that to get the expected cell frequencies, just multiply the respective marginal frequencies and divide by the total sample size.

We will use the following six-step procedure for conducting the test for independence.

Step 1:   The null and alternative hypotheses will be given by

H_0: The classification variables are independent

H_1: The classification variables are not independent

Step 2:   Determine a planning value for α.

Step 3:   Calculate the test statistic as follows:

$$\chi_0^2 = \sum_{i=1}^{r} \sum_{j=1}^{c} \frac{\left(f_{ij} - F_{ij}\right)^2}{F_{ij}} \quad \text{where}$$

f_{ij} = observed frequency for cell i, j

F_{ij} = expected frequency for cell i, j (from Equation 6.1)

Step 4:   Using the χ^2 distribution tables in Appendix G, estimate the area to the right of χ_0^2 in a χ^2 distribution having the correct number of degrees of freedom.

For a test of independence, the degrees of freedom are given by the formula

$$(r - 1)(c - 1)$$

where   r = the number of rows and

c = the number of columns.

Step 5:   The P value is the area determined in Step 4.

Step 6:   If P $\geq$ α, then conclude H_0.

If P < α, then reject H_0 and conclude H_1 with (1 - P)100% confidence.

Note that we can be (1 - P)100% confident that the variables are *not* independent when we conclude H_1, but we cannot assign any level of confidence to the conclusion that the data are independent (H_0).

Example 6.11   A placement agency sent surveys to 2500 university students selected at random from a list of all students enrolled in four-year universities throughout the United States. They were interested in knowing whether the college in which students enroll (within the university) is independent of gender. The following table summarizes the results of the survey involving the 1820 students who replied to the survey.

		COLLEGE			
		Engineering	**Business**	**Education**	
G E N D E R	**Male**	$f_{11} = 512$	$f_{12} = 357$	$f_{13} = 127$	$f_{1.} = 996$
	Female	$f_{21} = 215$	$f_{22} = 220$	$f_{23} = 389$	$f_{2.} = 824$
		$f_{.1} = 727$	$f_{.2} = 577$	$f_{.3} = 516$	N = 1820

Step 1:   The null and alternative hypotheses are

H_0: Gender and college enrolled in are independent

H_1: Gender and college enrolled in are not independent

Step 2:   We will use $\alpha = .01$

Step 3:   Calculate the test statistic. To do this we will first calculate the expected frequencies (F_{ij}) using Equation 6.1 and place those values in the bivariate frequency distribution as shown in the following table.

<table>
<tr><td rowspan="2" colspan="2"></td><td colspan="4" align="center">COLLEGE</td></tr>
<tr><td align="center">Engineering</td><td align="center">Business</td><td align="center">Education</td><td></td></tr>
<tr><td rowspan="3">G
E
N</td><td rowspan="3">Male</td><td>$f_{11} = 512$</td><td>$f_{12} = 357$</td><td>$f_{13} = 127$</td><td rowspan="3">$f_{1.} = 996$</td></tr>
<tr><td>$F_{11} = 397.85$</td><td>$F_{12} = 315.76$</td><td>$F_{13} = 282.38$</td></tr>
<tr><td></td><td></td><td></td></tr>
<tr><td rowspan="3">D
E
R</td><td rowspan="3">Female</td><td>$f_{21} = 215$</td><td>$f_{22} = 220$</td><td>$f_{23} = 389$</td><td rowspan="3">$f_{2.} = 824$</td></tr>
<tr><td>$F_{21} = 329.15$</td><td>$F_{22} = 261.24$</td><td>$F_{23} = 233.62$</td></tr>
<tr><td></td><td></td><td></td></tr>
<tr><td colspan="2"></td><td>$f_{.1} = 727$</td><td>$f_{.2} = 577$</td><td>$f_{.3} = 516$</td><td>$N = 1820$</td></tr>
</table>

The test statistic is then given by

$$\chi_0^2 = \frac{(512-397.85)^2}{397.85} + \frac{(357-315.76)^2}{315.76} + \; ... \; + \frac{(389-233.62)^2}{233.62}$$

$$= 273.08 \, .$$

Step 4:   The degrees of freedom are $(r - 1)(c - 1) = (2 - 1)(3 - 1) = 2$. Therefore, the area to the right of $\chi_0^2 = 273.08$ in the $\chi^2(2)$ distribution is less than 0.005. This area is shown in the following sketch.

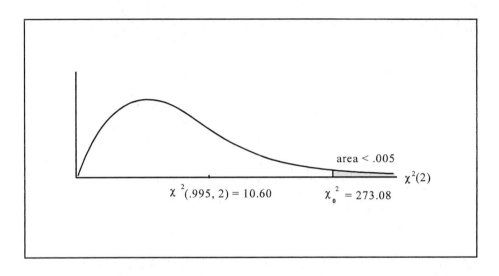

Step 5:   The P value is the area calculated in Step 4, which is much less than 0.005.

Step 6:   Since $P < \alpha$, we conclude H_1. That is, gender and college enrolled in are not independent, and we can be more than 99.5% confident in our conclusion.   ∎

    A test very similar to the test for independence is the test for homogeneity. In the χ^2 test for homogeneity, we test to see if several populations are similar when classified by a single variable. We take a simple random sample of each population and classify those samples by the variable of interest. The hypotheses are given by

H_0: The populations of interest are homogeneous

H_1: The populations of interest are not homogeneous

In the example above suppose that, instead of selecting 2500 students at random from several universities, the experimenter selected 1250 males and 1250 females. The two populations would then be male and female university students.

If sampling the two populations resulted in the data given in Example 6.11, then the same contingency table as shown in Example 6.11 would be formed and, in fact, the same test statistic would be calculated. The test for homogeneity, however, draws conclusions about two or more populations. The sampling scheme is usually the determining factor in how the hypotheses are defined.

6.9 Critical Value Approach to Hypothesis Testing[2]

An alternative but equivalent approach to hypothesis testing uses a number called a critical value. The critical value is a certain percentile of a given distribution, such as the Z distribution. For example, the critical value for the Z distribution at $\alpha = .05$ for a two-sided hypothesis test (e.g., H_0: $\mu_1 = \mu_2$ vs H_1: $\mu_1 \neq \mu_2$) would be $Z(1 - \alpha/2) = Z(.975) = 1.96$. This value is then compared with the standardized test statistic Z_0, which is found in Step 3 of the 6-step procedure. The following decision rule is then used to make the final conclusion:

$$\text{If } \left|Z_0\right| > Z\left(1 - \frac{\alpha}{2}\right), \quad \text{then reject } H_0 \text{ and conclude } H_1 \text{ with at least}$$

$$(1 - \alpha)100\% \text{ confidence.}$$

$$\text{If } \left|Z_0\right| \leq Z\left(1 - \frac{\alpha}{2}\right), \quad \text{then fail to reject } H_0.$$

The steps involved in the critical value approach to hypothesis testing can be generalized as follows:

---

[2]This section was provided by Kaye A. de Ruiz, USAF Academy.

Step 1:   State the hypotheses to be tested.

Step 2:   Determine a planning value for α **and** use this value to obtain the appropriate percentile (or critical value) of the sampling distribution from which the test statistic is drawn (e.g., the Z, t, χ^2, or F distributions).

Step 3:   Compute the test statistic from the sample evidence. The test statistic will be of the form Z_0, t_0, χ_0^2, or F_0, as shown previously in all of the hypothesis tests in this chapter.

Step 4:   Compare the test statistic with the critical value. If the test statistic is more extreme than the critical value, then conclude H_1 with at least $(1 - \alpha)100\%$ confidence. Otherwise, fail to reject H_0 (i.e., conclude H_0).

The P-value approach to hypothesis testing which is used in the previous sections compares a P-value (an area under a curve) with an α value (also an area under a curve) to decide on H_0 or H_1. The critical value approach, on the other hand, compares a test statistic (a value or point on the horizontal axis) with the critical value (another point on the horizontal axis) to make the decision between H_0 or H_1. The methods are equivalent because the areas being compared in the P-value approach are directly related to the two points on the horizontal axis which are used in the critical value approach to hypothesis testing. Figure 6.6 compares these two approaches. In either approach, the test statistic is computed the same way. In this figure, we assume the test statistic, $Z_0 = 2.54$, comes from the Z distribution, and that $\alpha = .05$ in a 2-tail hypothesis test. Note that the test statistic, $Z_0 = 2.54$, is greater than the critical value, $Z(1 - \alpha/2) = Z(.975) = 1.96$. Also note that the P-value, $2(.0055) = .0110$, is less than $\alpha = .05$. Hence, the same conclusion, namely H_1, is obtained. Thus, if the test statistic is more extreme than the critical value, then the P-value will be less than the α value, and vice versa. The P-value is the area associated with the test statistic, and the α value is the area associated with the critical value. The P-value is a table lookup based on the test statistic, whereas the critical value is a table lookup based on the α value.

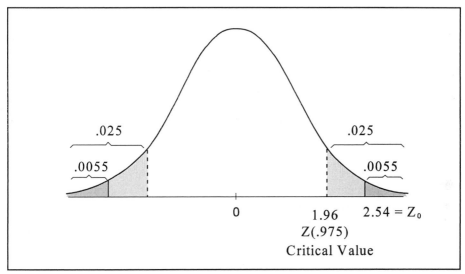

Figure 6.6 Critical Value vs P-Value

If an H_1 conclusion is reached, one can conclude H_1 with at least $(1 - \alpha)$ 100% confidence when using the critical value approach, whereas one can conclude H_1 with $(1 - P)$100% confidence when using the P-value approach. Since almost all software packages produce and display a P-value and since the P-value is a consummation of the evidence (namely the data), it is important for the student to understand what the P-value represents. It denotes the probability of making a Type I error. The α value, on the other hand, is often established by "the seat of the pants" and does not depend on the data at all.

The critical value approach to hypothesis testing has already been mentioned and used in the Rules of Thumb (e.g., see Table 6.2). Recall the ROT that says, "if $|Z_0| \geq 2$, we can be approximately 95% confident in H_1." Here the critical value is 2, and we are comparing the test statistic Z_0 with the critical value of 2. This ROT is designed to bypass the table lookup when $\alpha = .05$ and use the critical value of 2, rather than the exact value of 1.96.

6.10 Problems

6.1   The trailing edge of a turbine blade is to be robotically machined to a target of 6 mm. A simple random sample of 50 blades are inspected and found to have a mean of 5.98 mm with a standard deviation of 0.047 mm. Test at the $\alpha = .01$ level whether the process is centered at the target value.

6.2   An airline executive wishes to know if her airline is on schedule. She collects data on the deviations from actual takeoff time. From a random sample of 35 takeoff deviations she finds the mean to be 4.59 minutes (positive deviations being late) with a standard deviation of 16.32 minutes.

   a)   Is there sufficient evidence to conclude that the mean airline takeoff deviation is not zero at the 0.05 level of significance?
   b)   Suppose the executive took a sample of 100 takeoff deviations and found the same sample mean and standard deviation. What would her conclusion be then?
   c)   Discuss the difference between statistical significance and practical significance as related to this scenario.

6.3   A shell manufacturer claims to have produced a projectile having a mean muzzle velocity of more than 3000 fps. From a random sample of 60 shells he calculates a sample mean of 3012 fps and a standard deviation of 112 fps. Does the data from the sample support his claim?

6.4   A government contractor is required to produce a part that has a mean service life greater than 2000 hours.

   a)   What would the hypotheses be if the government were trying to prove that the mean service life was actually less than 2000 hours?
   b)   What would the hypotheses be if the contractor needed to prove that it was meeting the requirement?

6.5   A large corporation has determined that the nationwide average starting salary for engineers is $28,000. Last year they hired 36 new engineers and paid an average starting salary of $27,500 with a standard deviation of $957.

   a)   With what level of confidence can these new engineers who work for this corporation claim that their average starting salary was less than the national average?

b)   What other factors might be considered in the company's argument that lower wages are justified?

6.6   Solar cells are attached to a panel with a new heat resistant adhesive. The adhesive must provide 2.80 pounds of force in order to hold the cell on the panel. A random sample of 50 cells is tested and the following forces are necessary to break the adhesive.

2.65	2.76	2.82	2.89	2.95
2.67	2.77	2.83	2.90	2.95
2.68	2.77	2.84	2.90	2.97
2.68	2.77	2.84	2.91	2.98
2.70	2.78	2.84	2.92	2.98
2.70	2.78	2.85	2.92	2.99
2.73	2.79	2.88	2.93	3.00
2.74	2.80	2.88	2.93	3.01
2.75	2.81	2.88	2.93	3.02
2.75	2.82	2.89	2.94	3.03

a)   Test at the 0.01 level of significance the hypothesis that the mean adhesive strength is at least 2.80 pounds.
b)   Assuming the data follow a normal distribution, what proportion of the cells would you estimate to be out of tolerance (i.e., less than 2.80 pounds of adhesive strength)?
c)   Discuss your results from parts a and b above. Is a test for the mean appropriate in this case? Why or why not?

6.7   A machining process is established to produce metal wafers 2 cm thick. We wish to conduct a test to see if the process mean is set correctly. Eighteen wafers are sampled and their thicknesses are measured. The results are listed below.

2.005	2.003	2.001
1.999	1.998	2.007
2.008	2.002	2.000
2.001	2.005	2.006
1.997	2.004	2.003
2.005	2.005	2.004

a) What assumptions do we need to make in order to conduct this test?

b) Test at the 0.05 level of significance to see if the process mean is set correctly.

c) Suppose the process variance was known to be 0.00001 cm². Test to see if the sample variance is different. Use $\alpha = .05$.

d) Using the known process variance of 0.00001 cm², test the same hypothesis as in part b.

6.8 A machine that bottles 12 oz soft-drinks is being tested. The manufacturer wants to ensure that the machine puts at least 12 oz into each bottle on the average. A random sample of 24 bottles produces an average volume of 12.07 oz with a standard deviation of 0.11 oz. Assuming that volume is normally distributed, test the manufacturer's hypothesis at the 0.05 level of significance.

6.9 The IRS is interested in knowing the proportion of correctly filled-out tax forms in a given year. They take a sample of 200 forms and find that 52% are filled out correctly. Test the hypothesis that at least 50% of all tax forms are correct. Test at the 0.05 level of significance.

6.10 A medical researcher is testing a new drug to see if it slows the growth of malignant tumors in rats. The control group of 40 rats is given none of the drug and the test group of 40 rats is given a dose every day for two weeks. At the end of the test, the following statistics were calculated.

	Control Rats	Test Rats
Average Growth of Tumor	6.2 grams	4.1 grams
Standard Deviation	3.1 grams	2.8 grams

Conduct the appropriate hypothesis test to determine if the new drug appears to slow the mean rate of tumor growth.

6.11 A tire company believes that their new line of tires can increase gas mileage by at least 2 mpg over mileage achieved by using a competitor's tire. Twelve cars were run for 1000 miles on the competitor's tires and 1000 miles on the new tires. Assume that the difference between tire mileages are normally distributed.

Car	New Tire Mileage	Competitor Mileage	Difference
1	31.1	28.4	2.7
2	25.2	24.3	0.9
3	20.9	19.0	1.9
4	35.2	31.7	3.5
5	27.5	28.2	-0.7
6	30.4	28.9	1.5
7	26.2	23.1	3.1
8	28.5	26.4	2.1
9	33.7	31.5	2.2
10	24.3	21.9	2.4
11	32.6	29.5	3.1
12	37.5	33.2	4.3

a)   Can the tire manufacturer support his claim at the 0.10 level of significance?

b)   Can the manufacturer state that the tires improved gas mileage?

6.12   A research laboratory is testing two versions of a thermocouple used in jet engines. The purpose of the test is to see if the new, more expensive thermocouple (#1) has a longer mean lifetime than the old one (#2). The thermocouples are put on an accelerated test at extremely high temperatures. Thirty of each type of thermocouple are tested. We assume the lifetimes to be normally distributed. The results of the test are as follows:

	Thermocouple # 1	Thermocouple # 2
Average Lifetime (hrs)	24.2	18.4
Variance (hrs^2)	23.3	30.3

a)   Test to see if we can assume equal variances. Provide a bound on the P-value.

b)   Is there enough evidence to support a change to the new thermocouple? Use $\alpha = .05$.

6.13   A subcontractor produces pistons and sleeves for a race car manufacturer. The pistons are 7.5 cm in diameter with a variance of 0.01 cm². The subcontractor suspects that by increasing the percentage of magnesium in the alloy that variability will be reduced. He manufactures 35 pistons and finds that the new variance is 0.0075 cm². Is this significantly better? At what level?

6.14   A marketing company wants to determine if a new product will sell better on the east coast or west coast. Fifty individuals on each coast are randomly selected to test the new product. The following table shows the results of the market test:

Estimated Market Share with New Product	
East coast	18% (i.e., 9 of 50 strongly in favor)
West coast	26% (i.e., 13 of 50 strongly in favor)

a)   The company will first market the product in the area which shows the most potential for high sales. Are the market shares of the two coasts significantly different?

b)   The company feels that the new market share must be at least 15% if the product is to be successful. Test each sample to see if they exceed this level.

6.15   A scientist needs to simulate exponential failure times on a computer. A colleague gives her the set of data below, claiming that it comes from an exponential distribution with parameter $\lambda = 3.0$.

.002	.046	.109	.164	.229	.327	.413	.511	.675	1.012
.005	.046	.109	.165	.233	.329	.415	.515	.683	1.064
.005	.047	.115	.172	.247	.332	.420	.516	.689	1.064
.007	.050	.120	.180	.250	.339	.425	.518	.691	1.077
.012	.052	.124	.184	.260	.351	.433	.518	.714	1.099
.013	.054	.131	.184	.261	.367	.442	.522	.776	1.117
.015	.056	.134	.188	.280	.383	.462	.635	.805	1.257
.018	.070	.136	.216	.296	.397	.472	.640	.815	1.382
.028	.090	.145	.225	.303	.403	.503	.657	.860	2.171
.044	.090	.145	.227	.311	.407	.508	.668	.976	3.107

a)   Construct a frequency distribution of the data.
b)   Conduct a goodness-of-fit test to see if the data are exponential with $\lambda = 3.0$.
c)   Conduct a goodness-of-fit test to see if the data could be fit by any exponential distribution (i.e., you must estimate λ).

6.16   The following frequency distribution represents response times for a fire station in a large city. The times are measured in minutes from the time the alarm sounds until the fire truck arrives at the scene. The shortest response time was 12 seconds (next-door fire) and the longest was 23.6 minutes.

Response Time (minutes)	Frequency
0 to less than 4	11
4 to less than 8	13
8 to less than 12	22
12 to less than 16	17
16 to less than 20	17
20 to less than 24	20

a) Test the hypothesis that the data come from a uniform distribution with parameters a = 0, b = 24.
b) Using the maximum likelihood estimators, test the hypothesis that the data come from a uniform distribution.
c) Using the unbiased estimators, test the hypothesis that the data come from a uniform distribution.
d) Test the hypothesis that the data are normally distributed.

6.17   A company motor pool has 35 Chrysler products and 40 General Motors cars. All are compact vehicles bought at comparable prices. Management wants to know if the make has any effect on the mileage. The results of the study are listed below in a frequency distribution followed by the summary statistics.

Mileage

Make		20-29.9	30-39.9	Over 40	Total
Make	**Chrysler**	12	16	7	35
	GM	14	20	6	40
	Total	26	36	13	75

	Chrysler	General Motors
Average Mileage	33.326 mpg	32.878 mpg
Variance	44.903 mpg^2	37.884 mpg^2

a) Assume that the mileages are normally distributed. Test for equal variances.

b) Test at the 0.02 level of significance the hypothesis that the two makes have the same average mileage.

c) Conduct a test of homogeneity (see test for independence) to see if the two makes have the same distribution of mileage.

d) Test to see if the proportions of cars with mileage in the middle class are the same for both makes.

e) Discuss your results from the previous three tests.

6.18   A national educational lobby would like to know if gender and degree are independent among teachers at universities throughout the nation. They surveyed 1000 university educators nationwide and constructed the following table. (Only 870 responded to the survey.)

	Bachelor's Degree	Master's Degree	Ph.D.
Male	107	245	367
Female	42	73	36

a) Test at the 0.10 level of significance to see if the degree held is independent of gender.

b) Test to see if the proportion of male educators with a Ph.D. is the same as the proportion of females having a Ph.D. Test at the 0.10 level of significance.

6.19   The following frequency distribution lists the number of male jurors serving on a 12-person jury in a random survey of 100 juries in a Midwestern state. Test to see if the data can be fit by a binomial distribution with parameters $p = .5$ and $n = 12$. Use $\alpha = .05$. (Hint: the classes are already pooled to ensure that the expected frequencies are large enough.)

Number of Males on the Jury	Observed Frequencies
0 to 3	12
4 to 5	33
6	20
7 to 8	29
9 to 12	6

6.20 Review the data in Problem 6.6.

a) Create a frequency distribution with six equal-sized classes.
b) Conduct a goodness-of-fit test to see if the data could be fit by a normal distribution.
c) Conduct a goodness-of-fit test to see if the data could be fit by a uniform distribution.

6.21 A student experiment at the U.S. Air Force Academy was conducted to determine if acid rain was present near that area. Acidity of rainfall was measured at three stations on the Academy grounds. Station 1 was high elevation, Station 2 was medium elevation, and Station 3 was low elevation. Measurements are in pH.

	Station	
1	2	3
8.1	7.4	7.6
7.3	7.4	7.4
7.5	7.6	7.5
6.9	7.4	7.4
6.8	7.6	7.7
7.0	7.7	7.1
7.6	6.9	6.8

a) Acid rain is indicated by a pH below 7.0. Do any of the individual stations indicate the presence of acid rain? (use $\alpha = .05$)
b) Are there any differences in the acidity among the stations? (use $\alpha = .05$)

6.22   Two lakes on the Colorado front range were examined in order to determine the effect of urbanization on the amount of dissolved nitrogen and oxygen in the water. Ten subsurface samples were obtained from each lake and the sample statistics are listed below. Assume data are normally distributed.

	"Urbanized" Lake		"Non-urbanized" Lake	
	Oxygen	**Nitrogen**	**Oxygen**	**Nitrogen**
Average:	6.5800	.04307	8.5400	.07520
Std Dev:	.3178	.01071	.2853	.01160

a) Is there a difference in dissolved oxygen?
b) Is there a difference in dissolved nitrogen?

6.23   Two water treatment reservoirs were examined to determine if there was a difference in productivity. Differences in productivity are measured by the presence of four types of micro-organisms. After four samples were taken at each reservoir, the following sample statistics were calculated. Assume data are normally distributed.

	Reservoir #1		Reservoir #2	
	Mean	**Std Dev**	**Mean**	**Std Dev**
Copepoda	75.00	50.00	675.0	394.75
Daphnia	13.75	9.46	12.5	5.77
Mayfly			35.0	19.15
Volvox			17.5	9.52

a) Are there significant differences in the Copepoda and Daphnia populations at the two reservoirs?
b) Are there significant differences in Daphnia and Volvox at Reservoir #2?

6.24   A study was conducted to determine if the species of trout caught in a "riffle" area or a "pool" will differ. Assume data are normally distributed.

	Riffle	Pool
Cutthroat trout	7	15
Brook trout	1	6

Is the type of fish caught independent of the type of area in which it is caught?

6.25   The following are the length of trout caught (in cm) over a two-day period at the U.S. Air Force Academy. Assume data are normally distributed.

Cutthroat Trout (22 fish caught)		Brook Trout (7 fish caught)
22.4	10.0	
14.4	28.1	
8.3	15.2	12.7
9.0	17.5	13.2
11.6	15.7	12.4
19.6	24.0	19.9
13.2	17.6	11.9
20.3	11.8	9.1
15.2	15.9	12.4
16.5	14.8	
11.2	10.5	

a) Construct a 95% confidence interval for the mean size of each type of fish caught.

b) Is there a significant difference in the size of Cutthroat and Brook trout?

Chapter 7

Regression Analysis

7.1 Simple Linear Regression (SLR)

7.1.1 Introduction to SLR

Consider the scenario of measuring yield as our output characteristic (response) of a certain process at four levels of temperature: 70°, 80°, 90°, and 100°. Assume that our objective is to develop a model which will allow us to estimate the response at levels other than those mentioned above and be able to place prediction intervals about these estimates. We also desire a measure of model effectiveness. The technique to be discussed next, **Simple Linear Regression (SLR)**, will assist us in accomplishing these objectives. Assume we have collected three response values or observations from each of the four temperature settings. The data is shown in Table 7.1.

Temperature			
70°	**80°**	**90°**	**100°**
2.3	2.5	3.0	3.3
2.6	2.9	3.1	3.5
2.1	2.4	2.8	3.0

Table 7.1 Three Yields at Each of Four Temperatures

A graph of the data is given in Figure 7.1. The line represents an "eyeball fit" or free-hand regression line. The closeness of all the observations to the line indicates the accuracy of the predicted values of y for any given temperature.

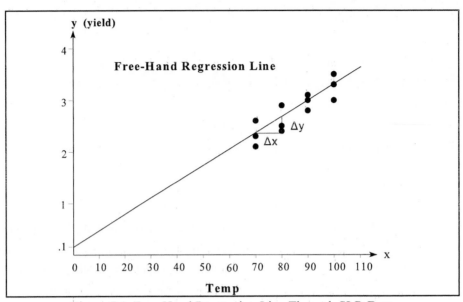

Figure 7.1 Free-Hand Regression Line Through SLR Data

The goal for line placement is to minimize the distance that observations are from the line. Using the slope-intercept formula [$E(y) = \beta_0 + \beta_1 x$],[1] we can estimate "$\beta_0$" graphically as 0.1 by looking at Figure 7.1. The slope, β_1, is estimated by measuring the change in y (Δy) for some specific change in x (Δx), i.e.,

$$\frac{\Delta y}{\Delta x} = \frac{2.7 - 2.3}{80 - 70} = .04.$$

Thus, our "eyeballed" regression line takes the form of $y = .1 + .04x$. Since all observations do not lie exactly on the line, there is obviously some error in our straight line estimate. To incorporate this **error** in the formula for predicting y, given any x value, we use $y = \beta_0 + \beta_1 x + \epsilon$, where ϵ represents the error term which is typically assumed to be distributed normally about 0; the ϵ are assumed

---

[1]$E(y)$ is the expected value of some predicted y given a specific x value. β_0 is the true intercept and β_1 is the true slope.

to have equal variability for all levels of x; and the error values are assumed to be independent.[2]

The relationship or model $y = \beta_0 + \beta_1 x + \epsilon$ is applicable for the **population data** (i.e., the set of *all* possible x and y values). The true regression line imbedded in this model is represented by $E(y) = \beta_0 + \beta_1 x$. Unfortunately, β_0 and β_1 are population parameters which are not known. Thus, the true regression line is not known. Data used in experimentation and process control are almost always sample data (a subset of population data); therefore, we use $\hat{y} = b_0 + b_1 x$ to approximate the true regression line; i.e., $\hat{y}$, b_0, and b_1 estimate $E(y)$, β_0, and β_1, respectively. In addition, $e_i = y_i - \hat{y}_i$ is called the ith residual which estimates ϵ_i. These terms are displayed graphically in Figure 7.2. The subscript "i" as used here is associated with the "i^{th}" observation.

From Figure 7.2, it can also be seen that each observation's deviation from $\bar{y}$ can be partitioned such that

$$(y_i - \bar{y}) = (y_i - \hat{y}_i) + (\hat{y}_i - \bar{y}).$$

Squaring both sides and then summing both sides over all observations (i.e., over i) yields, after some algebra, the following equation:

$$\sum_{i=1}^{n} (y_i - \bar{y})^2 = \sum_{i=1}^{n} (y_i - \hat{y}_i)^2 + \sum_{i=1}^{n} (\hat{y}_i - \bar{y})^2$$

where $SST = \sum_{i=1}^{n} (y_i - \bar{y})^2$ is the numerator of what we have already seen to

be the variance of y. SST is called the sum of squares total (or total sum of squares). The two terms partitioning SST are called the sum of squares due to error, SSE, and the sum of squares due to regression, SSR. SSE and SSR are given by

$$SSE = \sum_{i=1}^{n} (y_i - \hat{y}_i)^2 \quad \text{and} \quad SSR = \sum_{i=1}^{n} (\hat{y}_i - \bar{y})^2,$$

[2]For data that is in strong violation of these assumptions see [Myer 90].

respectively. Thus, in abbreviated notation,

$$\textbf{SST} \; = \; \textbf{SSE} \; + \; \textbf{SSR}$$

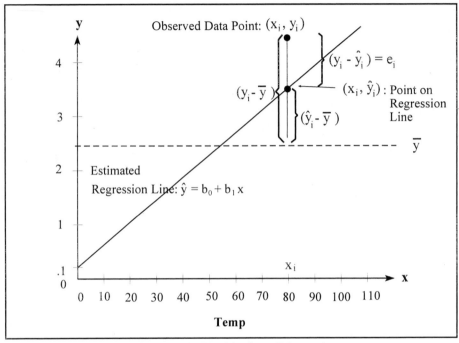

Figure 7.2 Partitioning of $(y_i - \bar{y})$

If we don't use x to predict y, then the best prediction for some future y is $\bar{y}$. The variance of observations about $\bar{y}$ is

$$s_y^2 \; = \; \frac{\sum\limits_{i=1}^{n} (y_i - \bar{y})^2}{n - 1} \; = \; \frac{\text{SST}}{n-1}.$$

When information on x is used to predict y, then $\hat{y} = b_0 + b_1 x$ is the prediction, and the variance of observations about $\hat{y}$ is given by

$$\frac{\sum_{i=1}^{n} (y_i - \hat{y}_i)^2}{n - 2} = \frac{SSE}{n - 2} = MSE.$$

MSE is referred to as the **mean square error** or the error variance which represents the unexplained variation about the regression line.

The objective of Simple Linear Regression (SLR) is to develop a regression line such that MSE is much smaller than s_y^2. The size of the error variance determines how good a fit the regression line will be and it plays a key role in the prediction intervals to be developed later in this chapter.

One mathematical method for finding b_0 and b_1 is the **method of least squares**. The name is derived from the property of minimizing the sum of squared deviations from the line, i.e., minimize Σe_i^2. The appropriate equations are

$$b_1 = \frac{\sum_{i=1}^{n} x_i y_i - \frac{\sum_{i=1}^{n} x_i \sum_{i=1}^{n} y_i}{n}}{\sum_{i=1}^{n} x_i^2 - n\bar{x}^2} \quad \textbf{and} \quad b_0 = \bar{y} - b_1\bar{x}$$

Equation 7.1 Least Squares Estimates for β_1 and β_0 in SLR

For the example data, the calculations are shown in Table 7.2. The estimates from the free-hand regression, namely $b_1 = .04$ and $b_0 = .1$, are fairly close to those of least squares for this example. However, when the data are more spread out and more variables are added, the "eyeball fit" becomes an impossible task.

n	x	y	x^2	xy	y^2
1	70	2.3	4900	161	5.29
2	70	2.6	4900	182	6.76
3	70	2.1	4900	147	4.41
4	80	2.5	6400	200	6.25
5	80	2.9	6400	232	8.41
6	80	2.4	6400	192	5.76
7	90	3.0	8100	270	9.00
8	90	3.1	8100	279	9.61
9	90	2.8	8100	252	7.84
10	100	3.3	10000	330	10.89
11	100	3.5	10000	350	12.25
12	100	3.0	10000	300	9.00
	1020	33.5	88200	2895	95.47

$$\bar{x} = 85 \quad \bar{y} = 2.79$$

$$b_1 = \frac{\sum xy - \dfrac{\sum x \sum y}{n}}{\sum x^2 - n\bar{x}^2} = \frac{2895 - \dfrac{1020(33.5)}{12}}{88200 - 12(85)^2} = \frac{47.5}{1500} = .032$$

$$b_0 = \bar{y} - b_1\bar{x} = 2.79 - 0.032(85) = .07$$

$$\hat{y} = .07 + .032x$$

Table 7.2 Simple Linear Regression Calculations

7.1.2 SLR Example[3]

Consider the 16 data points in Figure 7.3 and draw a straight line that you believe is a "reasonable fit" for the data (i.e., draw your regression line through these data points). The equation for a linear regression line is of the form $\hat{y} = b_0 + b_1x$. The value of the intercept b_0 can be found by extending your

---

[3]Most of this section was provided by Charles Hendrix of Union Carbide Corporation, South Charleston, West Virginia.

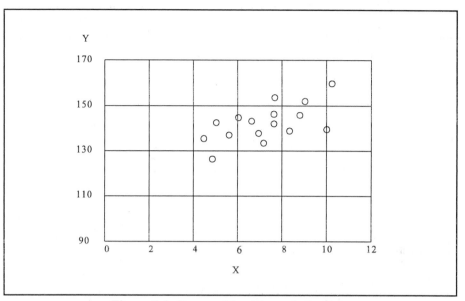

Figure 7.3 Draw a Line Through the Data

straight line to where it intersects the y-(or vertical) axis. This value on the y-axis
is b_0. The value of b_1 is the slope of the line. To find b_1, read the values of y on
the regression line that correspond to x = 10 and x = 0, respectively. Subtract the
second y-value from the first y-value. This difference is called Δy. Δx will be
10 - 0 = 10. Now find $b_1 = \Delta y/\Delta x$. When finished, you have derived the values
of the intercept and slope for the line you drew through the data in Figure 7.3.
Place the values you found for b_0 and b_1 in the following template.

$$\hat{y} = \boxed{} + \boxed{} \; x.$$

$$\quad\quad b_0 \quad\quad\quad\quad b_1$$

You may want to compare your solution with that of a classmate's. If you
do this, you will most likely find that your solution differs from that of the other
person. Comparisons with other people may reveal a diversity of opinions about
where the line should be drawn, and hence a diversity of combinations of values of
b_0 and b_1. More than 5000 people have drawn their lines through these data points,

resulting in about that many different lines. People tend to have strong opinions about drawing lines through data and even stronger opinions about their ability to do this in a fair and unbiased manner.

The (b_0, b_1)-combination found by each person in one class of 23 people is plotted in Figure 7.4. This plot shows not only a diversity of opinions, but also a strong relationship between b_0 and b_1. This relationship stems from the following. If a person drew his or her line steeper than the average of the class, then his or her intercept was lower than the average intercept, and vice versa. You can see this by drawing several alternative lines on Figure 7.3. Vary the slope and as a result, the intercept will vary accordingly.

Now add your own (b_0, b_1)-combination to Figure 7.4. It will almost surely lie along the trend-line formed by the (b_0, b_1)-points generated by those 23 students. If your combination of (b_0, b_1) is very far from this trend-line, chances are good that you made an arithmetic error; double-check on this before going further. This exercise is designed to reveal the astonishing diversity of answers found by highly trained professionals who run experiments, collect and analyze data, and study reports and papers written by others. The pattern seen in Figure 7.4 is representative of our total experience with this classroom experiment.

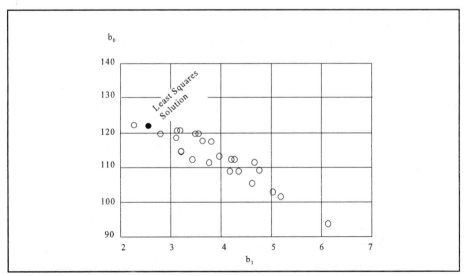

Figure 7.4 Values of b_0 and b_1 Found by 23 Students —
Add Your (b_0, b_1) Point to this Figure

It is natural to ask, "what is the best line?" Or, perhaps, "is there a best line?" While opinions are allowed, we must have some objective criteria for determining the "best" line. The method of least squares is accepted as the standard for fitting equations to data. It is also capable of fitting equations to data that have more than one independent variable (e.g., $\hat{y} = b_0 + b_1x_1 + b_2x_2$). It offers ways of judging how well we have estimated the true model coefficients and, perhaps best of all, the method of least squares has predictable properties.

The **least squares equation** for the data in Figure 7.3 is $\hat{y} = 123.11 + 2.68x$. These coefficients were obtained by using Equation 7.1 with the exact values given in Table 7.3. In Figure 7.4 you will notice that nearly all of the 23 students overestimated the slope, b_1. This is consistent with our experience in the classroom. Your own (b_0, b_1)-combination may have come closer to the least squares combination than those of highly trained professionals in science and engineering. People who work with experimental data almost every day develop some strong opinions about drawing lines through data; they can become a little dogmatic about it. Therefore, we need to use statistical modeling techniques such as Simple Linear Regression and least squares to assist us in properly analyzing sample data.

Some Properties of the Least Squares Equation

If we substitute the 16 known values of x into the equation $\hat{y} = 123.11 + 2.68x$, we will get the corresponding **predicted values** of y. We expect that the predicted values of y, denoted $\hat{y}$, will closely match the observed values of y. The differences between **observed** and **predicted** values of y are called **residuals**.

The first predicted value is $\hat{y} = 123.11 + 2.68(4.6) = 135.44$. The first residual is $136.0 - 135.44 = 0.56$. The complete set of 16 residuals is shown in Table 7.3.

Information about the linear relationship between y and x is in the equation. The residuals contain all of the remaining (or residual) information in the data, that is, information about possible curvature, time trends, key-entry errors, other variables (x's) not in the equation, and random variation. The residuals contain no information about the relationship between y and x; that information is stored in the equation. The residuals are very important; you will be learning more about them later. If our model properly describes the data, then the residuals should be nothing but random variation about the **regression equation** (or, **regression line**) which, in this case, is $\hat{y} = 123.11 + 2.68x$.

If we square each of the residuals and sum those squares, we will obtain

$$(0.56)^2 + (-7.71)^2 + . \; . \; . + (9.54)^2 \; = \; 644.792$$

This value is our estimate of background noise or experimental error and is also what we referred to in Section 7.1.1 as SSE, the sum of squares due to error or the residual sum of squares.

i	x_i	y_i (observed)	$\hat{y}_i$ (predicted)	$e_i = (y_i - \hat{y}_i)$ (residuals)
1	4.6	136	135.44	0.56
2	4.7	128	135.71	-7.71
3	5.0	142	136.51	5.49
4	5.6	137	138.12	-1.12
5	6.0	147	139.19	7.81
6	6.6	144	140.80	3.20
7	6.9	138	141.61	-3.61
8	7.2	132	142.41	-10.41
9	7.7	142	143.75	-1.75
10	7.7	146	143.75	2.25
11	7.7	154	143.75	10.25
12	8.3	139	145.36	-6.36
13	8.9	145	146.97	-1.97
14	9.0	151	147.24	3.76
15	10.0	140	149.92	-9.92
16	10.2	160	150.46	9.54

Table 7.3 Complete Table of Observations and Residuals

If we choose any other (b_0, b_1)-combination, compute the residuals, sum the squares of the residuals, we will find a **residual sum of squares (SSE)** that is larger than the one calculated here for the least squares solution. Try it. Use your (b_0, b_1)-combination found by drawing a line through the data in Figure 7.3. Calculate the predicted values of y, then the residuals, and then sum the squared residuals. The resulting residual sum of squares will be larger than the residual sum of squares obtained above.

The method of least squares minimizes the sum of squares of the residuals.

As an extension to this, let's consider what would happen if we were to draw alternative lines through the data in Figure 7.3. Each such line is represented by a (b_0, b_1)-combination. Table 7.4 shows the residual sum of squares for eight (b_0, b_1)-combinations. Four of these lines are then reproduced in Figure 7.5.

b_0	b_1	**Residual Sum of Squares**
135.00	1.20	769.27
123.11	2.68	644.79
112.00	4.10	750.32
106.00	4.90	893.30
128.00	2.68	1026.49
122.00	4.00	1872.68
118.00	2.20	1838.73
110.00	3.20	2054.01

Table 7.4 Residual Sums of Squares for Eight Different Lines

Those (b_0, b_1)-combinations which lie along the **sum of squares valley** in Figure 7.6 have residual sums of squares which are moderately larger than the minimum. Those which are far off of the valley have much larger sums of squares; they literally represent alternative lines in Figure 7.5 which don't even go through the data. From this we see that while the least squares solution for (b_0, b_1) is accepted as the "best," there are alternative combinations that are almost as good (e.g., {135, 1.20} and {112, 4.10}).

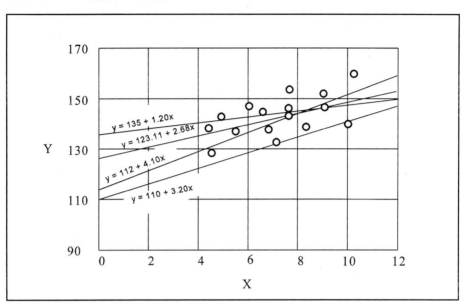

Figure 7.5 Alternative Lines Drawn Through the Data

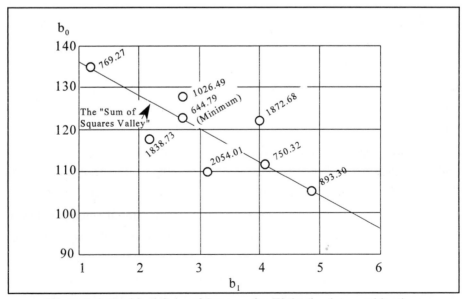

Figure 7.6 Residual Sum of Squares for Eight (b_0, b_1)-combinations

7.1.3 R^2 and ANOVA

A measure of strength of the linear relationship of y with some x is the correlation coefficient,

$$R = \frac{\sum xy - \dfrac{\sum x \sum y}{n}}{\sqrt{\left(\sum x^2 - n\bar{x}^2\right)\left(\sum y^2 - n\bar{y}^2\right)}}$$

For our example (reference Tables 7.1 and 7.2),

$$R = \frac{2895 - \dfrac{1020(33.5)}{12}}{\sqrt{\left[88200 - 12(85)^2\right]\left[95.47 - 12(2.792)^2\right]}} = \frac{47.5}{53.763} = .883$$

The value of R is limited to the interval $[-1,1]$ where -1 indicates perfect negative correlation, 1 indicates a perfect positive correlation and 0 implies no linear relationship between y and x. For SLR, the sign of R will be the same as that for b_1, the slope. The formula for R given above is the formula for the sample correlation coefficient given by Equation 2.13 in Chapter 2. Whenever $|R| > .7$, we say that there is a fairly strong correlation between x and y. Perhaps a more appropriate measure of the relationship between x and y that has a very nice interpretation from a variability point of view is R^2, the coefficient of determination.

The proportion of variability in y which is explained by y's relationship with x is measured by R^2. To understand the meaning of R^2, recall from Section 7.1.1 that SST = SSR + SSE. If we divide both sides of this equation by SST, we obtain

$$1 = \frac{SSR}{SST} + \frac{SSE}{SST}.$$

Since SST represents the sum of squares total, the ratio SSE/SST represents the proportion of total variability represented by the observed points about the regression line and SSR/SST is the proportion of the total variability that has been explained (or removed) by using the regression line instead of the horizontal line $\bar{y}$ to predict response (y) values. Using algebra, it can be shown that

$$\frac{SSR}{SST} = \frac{\sum_{i=1}^{n} (\hat{y}_i - \bar{y})^2}{\sum_{i=1}^{n} (y_i - \bar{y})^2} = R^2$$

and thus

$$R^2 = 1 - \frac{SSE}{SST}.$$

If there is no linear relationship between x and y, then SSE = SST and thus $R^2 = 0$ (see Figure 7.7a). If all the observations fall on the $\hat{y}$ (or regression) line, then SSE = 0 which implies that $R^2 = 1$ (see Figure 7.7b).

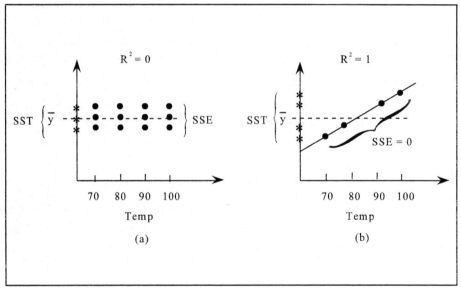

Figure 7.7 Extremes in Strength of Linear Relationship

In the example data from Table 7.1 or 7.2, $R^2 = .780$ indicates that about 78.0% of the variability in y is explained through y's linear relationship with x. The strength of the linear relationship between the two variables is directly related to the amount of variability in y that can be accounted for by the variable x. In this example, the strength of the linear relationship between the two variables is considered to be fairly strong. An $R^2 = .78$ also means that $1 - R^2 = 1 - .78 = .22$, or 22% of the variability in the y values is still unaccounted for. We sometimes call this unexplained or unaccounted variability noise.

Recall that the variance of y without regard to x is

$$s_y^2 = \frac{SST}{n-1} = \frac{\sum y^2 - n\bar{y}^2}{n-1} = \frac{95.47 - 12(2.792)^2}{11} = .1752.$$

We can now solve for SST in the above relationship by computing

$$SST = S_y^2 \cdot (n - 1) = .1752(11) = 1.927.$$

Since we know $SST = 1.927$ and $R^2 = .78$, we can solve for SSR in the relationship $R^2 = SSR/SST$ to get

$$SSR = R^2 \cdot SST = .78 \, (1.927) = 1.503.$$

The last of the three sums of squares, SSE, can be obtained from the previous two via the relationship $SSE = SST - SSR$. We have

$$SSE = SST - SSR = 1.927 - 1.503 = .424.$$

It is customary to formalize the partitioning of the total sums of squares in what is called an Analysis of Variance (ANOVA) table. Table 7.5 gives the format of an SLR ANOVA table.

Source	Sum of Squares (SS)	df	Mean Square (MS)	F_0
Regression	$SSR = R^2 \cdot SST$	1	$SSR/1 = MSR$	MSR/MSE
Error	$SSE = (1 - R^2) \cdot SST$	n-2	$SSE/(n - 2) = MSE$	
Total	$SST = s_y^2 \cdot (n - 1)$	n-1		

Table 7.5   Format for SLR Analysis of Variance (ANOVA) Table

The degrees of freedom (df) for the error term in a regression model are equal to n minus the number of parameters estimated. In SLR, b_0 and b_1 are estimates (obtained from sample data) of the two population parameters β_0 and β_1, respectively. Therefore, $df_E = n - 2$. The degrees of freedom for regression in SLR are the number of parameters estimated minus 1, which is $2 - 1 = 1$.

Since R^2 is computed from sample data, it is only an **estimate** of $\mathbb{R}^2$, the true but unknown strength of the linear relationship between y and x in the entire **population**. We would like to test the value of $\mathbb{R}^2$ to insure that it is significantly different from zero, i.e., we will test the hypotheses:

$$H_0: \; \mathbb{R}^2 = 0$$

$$H_1: \; \mathbb{R}^2 > 0$$

For SLR the hypothesis test for $\mathbb{R}^2$ is equivalent to testing the slope parameter, i.e.,

$$H_0: \; \beta_1 = 0$$

$$H_1: \; \beta_1 \neq 0$$

In either case, concluding H_0 implies that the sample data did not provide overwhelming evidence to indicate a significant linear relationship between y and x. Concluding H_1 indicates the presence of a significant linear relationship with $(1 - P)100\%$ confidence. The Mean Square Error ($MSE = SSE/(n - 2)$) is an estimate of the variance of y at any level of x. Table 7.6 displays the ANOVA table for the regression in Table 7.2. Note that the F-test indicates a significant linear relationship.

Source	SS	df	MS	F_0
Regression	1.503	1	1.5030	
Error	0.424	10	.0424	35.448*
Total	1.927	11		

*   From Appendix F, F(.999; 1, 10) = 21.0 implies P < .001. Thus, we can conclude H_1: $\mathbb{R}^2 \neq 0$ or H_1: $\beta_1 \neq 0$ with at least 99.9% confidence.

Table 7.6 SLR ANOVA Table for Data in Table 7.2

7.1.4 Residual Analysis

In the previous sections we suggested that when an equation is fitted to data, the resulting equation contains the information about the relationship between y and x; and the residuals contain all of the remaining information in the data. This is another way of stating that the residuals contain no information about the linear relationship between y and x.

Returning to our set of 16 data points in Figure 7.3, let us calculate the total sum of squares, $\sum(y - \bar{y})^2$, where $\bar{y}$ is the average value of y. $\bar{y} = 142.5625$, and $\sum(y - \bar{y})^2 = 983.94$. This total sum of squares is the available information in the data. As shown previously, it can be partitioned into two parts: that which is due to the **regression equation** $\hat{y} = 123.11 + 2.68x$ and that which is left in the residuals. It turns out that the sum of squares which is attributed to the regression equation is just the total sum of squares minus the residual sum of squares. We summarize this information in the following ANOVA table.

Source	Sum of Squares (SS)		df
Regression (b_1)	339.15	(SSR)	1
Residuals (Error)	644.79	(SSE)	14
Total	983.94	(SST)	15

We began with a total sum of squares = 983.94. Of this, we used 339.15 to establish the regression equation $\hat{y} = 123.11 + 2.68x$. This is sometimes called the **sum of squares due to regression** (SSR). We literally "spent" this information in order to establish the regression equation. What remains, the residual sum of squares (SSE), is (hopefully) just a manifestation of random variation. In practice, examination of the residuals may reveal patterns (time trends, additional variables not included in the equation, mistakes, etc.) that invalidate the claim that the residuals are just random variation. It is always wise to carefully examine the residuals before accepting the regression equation.

There are typically three types of residual plots that should be examined when conducting regression analysis (simple linear or multiple):

(i)   residuals plotted against predicted values, $\hat{y}_i$.

(ii)   residuals plotted against some x variable.

(iii)   residuals plotted against time sequence of the data.

Four hypothetical results of these types of plots are shown in Figure 7.8 a, b, c, and d. The pattern displayed in Figure 7.8a depicts what we would expect to see if our model adequately describes the data. The patterns in Figure 7.8 b, c, and d indicate that some other information can be gained from the residuals. For example, if we plotted residuals versus $\hat{y}$ or some x and obtained a pattern similar to Figure 7.8b, we most likely have an error in calculations. However, the same plot over time would indicate some type of systematic change in the process which must be investigated. If we plot residuals versus i) $\hat{y}$, ii) some x, or iii) time and see a pattern similar to Figure 7.8c, then i) the $\hat{y}$ plot would indicate increased variance with larger values of y, ii) the x plot would depict a variance shifting variable which can be used to reduce variation, and iii) the time plot would indicate increased variation over time which could be related to tool wear, chemical solution change, etc. Finally, a plot for x similar to Figure 7.8d would indicate the need for a non-linear model.

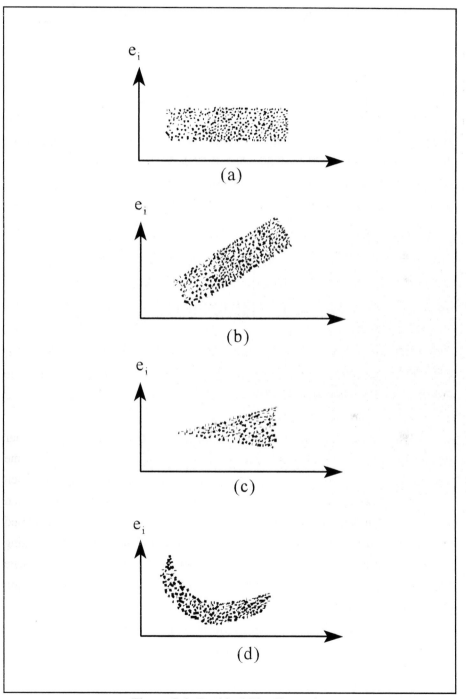

Figure 7.8 Residual Plot Examples

Revisiting the 16-observation data set shown in Figure 7.3 and Table 7.3, we will partition the sums of squares and take a closer look at the residuals. Having begun with n = 16 observations, we consumed one degree of freedom (df) in calculating the average $\bar{y}$. Thus, 15 degrees of freedom remain with the *total sum of squares*, SST = $\sum (y - \bar{y})^2$ = 983.94. The calculation of the regression coefficient b_1 used one degree of freedom. Thus, there are 14 df remaining for the *residual or error sum of squares*, SSE = 644.79. From SST = 983.94 and SSE = 644.79, we have SSR = SST - SSE = 339.15. The ANOVA table can now be completed by dividing each sum of squares by its respective degrees of freedom, as shown here:

Source	Sum of Squares (SS)	df	Mean Square	F-ratio	t-ratio
Regression	339.15	1	MSR=339.15	$F = \dfrac{MSR}{MSE}$	$t = \sqrt{F}$
Residuals (Error)	644.79	14	MSE= 46.05	= 7.36	= 2.71
Total	983.94	15			

The *variance of the residuals* is estimated by MSE = 46.05. If we take the square root of MSE, we get an estimate of the *residual standard deviation*, $s_{e_i} = \sqrt{46.05} = 6.79$. This standard deviation is a proper estimate of the variation we would expect to see if we set x to a fixed value and observed y repeatedly. Note that this *residual standard deviation*, $s_{e_i} = \sqrt{MSE}$, is not quite the same as what we would get if we simply calculated the standard deviation of the 16 residuals in the usual fashion. The reason why they are different is because fitting an equation to data has imposed certain restrictions upon the residuals, leaving just 14 (not 15) degrees of freedom for the residuals.

The following scatter diagrams show plots of the residuals versus x and $\hat{y}$, respectively. The numeric values being plotted can be seen in Table 7.3. These plots are benign, showing no apparent glaring defects in the data, like outliers or quadratic effects.

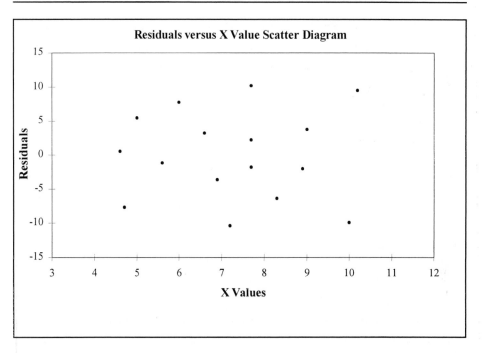

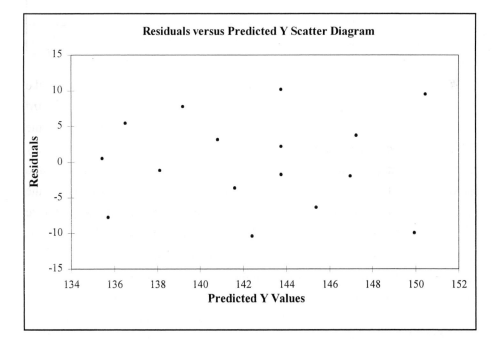

Residuals are often standardized to give the practitioner a quick way of detecting the possible presence of bad data. To standardize a residual e_i, we just divide e_i by $s_{e_i} = \sqrt{MSE}$, to get the standardized residual

$$Z_{e_i} = \frac{e_i}{S_{e_i}}.$$

Under the assumption that the e_i are normally distributed, we would expect roughly 99% of the $|Z_{e_i}|$ to be less than 3. Therefore, any standardized residual value greater than 3 in absolute value should be investigated to see if that data point should be removed from the sample and the data re-analyzed. Various rules exist for removing outliers and more information can be found in [Myer 90]. Common sense must prevail since models based on bad data will confuse us and, similarly, throwing out good data may reduce the sample size below what is desired.

The ANOVA table above also shows the F and t values. The F-ratio is obtained by dividing MSR by MSE. Since the numerator df for F is one, we can take the square root of the F-ratio and obtain the t-ratio. This t-ratio is a t-value from the t distribution (Appendix E) we studied in Chapter 5. Using 14 df, we will find a t-value of 2.71 to be "significant." The t = 2.71 with 14 df and F = 7.36 with (1, 14) df are equivalent, implying that it is unlikely that the slope of the true regression line (β_1) is "zero." Thus, we have confirmed that there really is a linear relationship between y and x.

7.1.5 Prediction

To obtain a point estimate for y, given any x value (x_h), we simply insert the value for x_h in $\hat{y} = b_0 + b_1 x_h$ (Note: whereas the subscript "i" is associated with "observed" values, the subscript "h" is used to denote *any* possible level of x). Point estimates, however, do not include any reference to precision. In order to indicate how precise our estimate is, a **prediction interval** is formed in which the researcher is $(1 - \alpha)100\%$ confident that the true value of y will occur. Since the data is just one sample, it is reasonable to take another sample and compute different values of b_0 and b_1. This implies that there exists sample variability in the slope and intercept of the regression line, in addition to observation variability about the regression line itself. The slope and intercept variability will cause larger deviations in the predicted response (ŷ) as the desired level of x (x_h) moves further away from the center of the domain of x ($\bar{x}$). For these reasons, the prediction limits for the response at any given value of x (labeled x_h) are computed as follows:

$$\hat{y}\big|_{x_h} \pm t\left(1 - \frac{\alpha}{2}, n-2\right) \sqrt{MSE\left(1 + \frac{1}{n} + \frac{(x_h - \bar{x})^2}{\sum (x_i - \bar{x})^2}\right)}$$

While MSE represents the variance of observations about the regression line, the quantity

$$MSE\left[\frac{1}{n} + \frac{(x_h - \bar{x})^2}{\sum (x_i - \bar{x})^2}\right]$$

represents the variance in the average response due to the variability in slope, b_1, and intercept, b_0. Adding the two variances together gives the total variance.

For our temperature example, the prediction intervals (y_L, y_U) based on 95% confidence are displayed in Table 7.7. For example, if the process temperature is set at 80°, we would be 95% confident that any future y value will occur between 2.14 and 3.12.

x	ŷ	y_L	y_U	Δ	
70	2.31	1.79	2.83	1.04	t(.975, 10) = 2.228
*80	2.63	2.14	3.12	.98	MSE = .0445
90	2.95	2.46	3.44	.98	$\sum(x_i - \bar{x})^2 = 1500$
100	3.27	2.75	3.79	1.04	$\bar{x} = 85$

*Computations shown on next page   (Continued on next page)

Table 7.7 Prediction Limits for y Based on 95% Confidence

---

* Using $x_h = 80$,

$$\hat{y}|_{x_h} = .07 + .032(80) = 2.63$$

$$y_L = 2.63 - 2.228 \sqrt{.0445 \left(1 + \frac{1}{12} + \frac{(80-85)^2}{1500} \right)} = 2.14$$

$$y_U = 2.63 + 2.228 \sqrt{.0445 \left(1 + \frac{1}{12} + \frac{(80-85)^2}{1500} \right)} = 3.12$$

Table 7.7 Prediction Limits for y Based on 95% Confidence

Predicting values of y for $x_h < x_{min}$ or $x_h > x_{max}$ is called **extrapolation**, which assumes the same relationship between y and x outside the region of the sampled x as it is within the sampled region. Notice that the 95% prediction interval would be wider outside the interval [70, 100] than it will be inside. This lack of accurate predictability and uncertainty of the relationship between y and x outside the region of sampled x is why extrapolation is not recommended. If we need to predict outside the sampled region for x, we should expand the sampled region. If extrapolation is used, confirmation data should be collected as soon as possible to validate the prediction.

When prediction intervals are too wide for a particular application, there are ways to reduce these intervals. The researcher can

(1)   decrease the confidence level $(1 - \alpha)$; then $t(1 - \frac{\alpha}{2}, n - 2)$ will get smaller;

(2)   increase the sample size; then $1/n$ becomes smaller;

(3)   increase $\sum(x_i - \bar{x})^2$. Note that $\sum(x_i - \bar{x})^2 = \sum x_i^2 - n\bar{x}^2$.

It is not recommended that one alter the confidence level to reduce interval width. In fact, the subject of experimental design discussed later in this text addresses items (2) and (3) as the major areas of interest. For non-experimental data, the primary method of reducing prediction interval widths is increasing sample size. However, in conducting industrial experiments, the sample size is desired to be a minimum, thus putting even more emphasis on the remaining option, i.e., maximizing $\sum(x_i - \bar{x})^2$. This is equivalent to maximizing the variance of x, which is accomplished by placing n/2 observations at each of x_{min} and x_{max}. The problem with sampling at only two levels of x is that we are restricted to linear estimates. For this reason, you will see in the next chapter that center points can be added to 2-level designs to allow for curvature estimation. The idea of maximizing $\sum(x_i - \bar{x})^2$ has another effect in that

$$s^2(b_1) = \frac{MSE}{\sum (x_i - \bar{x})^2}$$

can be minimized. Therefore, the variance of the slope is minimized if $\sum (x_i - \bar{x})^2$ is maximized, creating more power in the F test for H_0: $\mathbb{R}^2 = 0$ or H_0: $\beta_1 = 0$. This implies that if we minimize $s^2(b_1)$, we are more likely to find a real linear relationship if it in fact exists. Once multiple regression is covered, the reader will also see that maximizing $\sum(x_i - \bar{x})^2$ is related to other experimental designs to be discussed later.

7.2 Polynomial Regression and Optimization

Suppose that the example temperature data included 3 more observations taken at x = 50°, which when added to Table 7.1 produces the example data shown in Table 7.8. Graphically, the data are presented in Figure 7.9. This graphic indicates the danger in extrapolation because the function takes on a different shape outside the original sample space of x.

Temperature				
50°	**70°**	**80°**	**90°**	**100°**
3.3	2.3	2.5	3.0	3.3
2.8	2.6	2.9	3.1	3.5
2.9	2.1	2.4	2.8	3.0

Table 7.8 Example Data for Polynomial Regression

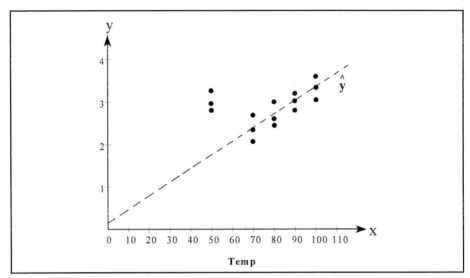

Figure 7.9 Graph of Example Data for Polynomial Regression

As you can see, the linear dashed line is no longer a good fit. In fact, if we try SLR, the result is $b_1 = 0$ which implies $R^2 = 0$. However, the more appropriate thing to do is to try a quadratic term. Staying with SLR, the new model becomes $\hat{y} = b_0 + b_{11}x_1^2$, where the subscript "11" is not read "eleven" but rather "one, one" to indicate that b_{11} is the coefficient of x_1x_1 or x_1^2. Table 7.9 shows the computations for SLR. The low R^2 indicates that something is wrong because we would expect a better fit. What happened? Creating x_1^2 as a new variable and using it in the SLR model above resulted in fitting the points with a parabola whose vertex (lowest point) is on the y-axis. Graphically, we can see that the vertex should be far from $x = 0$.

Computations for $\hat{y} = b_0 + b_{11}x_1^2$				
(x_1^2)	y	$(x_1^2)^2$	y^2	$(x_1^2)y$
2500	3.3	6250000	10.89	8250
2500	2.8	.	.	.
2500	2.9	.	.	.
4900	2.3	.	.	.
4900	2.6	.	.	.
4900	2.1	.	.	.
6400	2.5	.	.	.
6400	2.9	.	.	.
6400	2.4	.	.	.
8100	3.0	.	.	.
8100	3.1	.	.	.
8100	2.8	.	.	.
10000	3.3	.	.	.
10000	3.5	.	.	.
10000	3.0	.	.	.
95700	42.5	710490000	122.61	276810
$\hat{y} = 2.472 + 0.000056(x_1^2)$			$R^2 = .146$	

Table 7.9 SLR Using a Quadratic Term

To properly conduct polynomial regression in the manner in which the problem was stated, we must include both x_1 and x_1^2 in the model. That is, the correct model should be $\hat{y} = b_0 + b_1x_1 + b_{11}x_1^2$. This second order polynomial regression model is similar to a multiple regression problem.   A summary of the computer output for this model is shown in Table 7.10. It indicates that this model is a fairly good fit, i.e., $R^2 = .673$ and the probability of an incorrect model is very small ($P = .001$). Once we have the model, we have a picture of the curve that best fits the data. Figure 7.10 shows us the "best" quadratic fit (parabola) through the data. "Best" here again implies minimizing the sum of the squared deviations of the observed points about the curve, the same criteria as in the SLR scenario. Now that we have the curve, optimizing our process becomes much easier, because we can see which value of x (Temp) will minimize y and where temperature will have to be set to maximize y if that is what may be desired.

Summary of Computer Output for Polynomial Regression

Model: $\hat{y} = 7.96048 - .15371x_1 + .00108x_1^2$ $R^2 = .673$ $P = .001$

Parameter	Estimate	P-value
b_0	7.96048	.00004
b_1	-0.15371	.00087
b_{11}	0.00108	.00059

Table 7.10 Polynomial Regression on Example Data

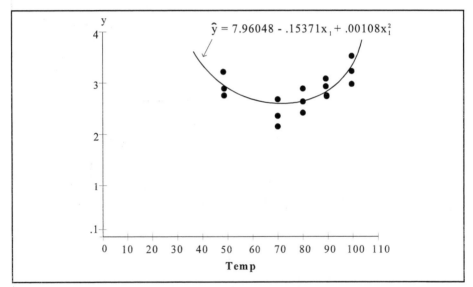

Figure 7.10 Graph of Polynomial Regression

7.3 Multiple Regression

If a variable for pressure, x_2, was also included in our model, the multivariable model would be referred to as a **Multiple Regression Model**. A new set of example data appears in Table 7.11. The initial model to be tested is $y = \beta_0 + \beta_1 x_1 + \beta_2 x_2 + \epsilon$, where the population parameters β_0, β_1 and β_2 will be estimated via the sample data by b_0, b_1 and b_2, respectively. The ϵ are assumed to be normally distributed with mean of 0 and variance of σ^2. We shall use the mean square error, MSE, to estimate σ^2. The estimate for σ^2 is based on the variance of the residuals which have n − p degrees of freedom. It is given by

$$MSE = \frac{\sum_{i=1}^{n}(y_i - \hat{y}_i)^2}{n-p},$$

where p is the number of population parameters estimated by the data. MSE = SSE/(n−p) is referred to as the regression error variance. For our Multiple Regression (MR) example data in Table 7.11, the computer output is displayed in Table 7.12.

MR Data		
y	x_1 = Temp	x_2 = Pressure
3.1	50	100
3.3	50	100
2.6	50	200
2.4	50	200
2.5	70	100
2.6	70	100
3.0	70	200
3.3	70	200
2.4	80	100
2.3	80	100
3.2	80	200
3.5	80	200
2.8	90	100
2.6	90	100
3.1	90	200
3.0	90	200
3.2	100	100
3.4	100	100
2.5	100	200
2.4	100	200

Table 7.11 Example Data for Multiple Regression (MR)

Regression output				
Factor	**Coef**	**Std Error**	**t stat**	**p Value**
Constant	2.695	0.506	5.322	0.000
x1	0.001	0.005	0.108	0.915
x2	0.001	0.002	0.437	0.668
R Squared	0.012		Std Error	0.410
Adj R Sq	-0.043		SS reg	0.034
F Value	0.101		SS resid	2.854

Table 7.12 Multiple Regression Output from SPC KISS
for the model $\hat{y} = b_0 + b_1x_1 + b_2x_2$

This computer output is typical of regression analysis output produced by most statistical software packages. There are various entries in the table, some of which are more important than others. We highlight these important areas and provide Rules of Thumb (ROT) for determining significance. The two adjacent columns labeled "Factor" and "Coef" (short for coefficient) specify the regression model. In this case, $\hat{y} = 2.695 + .001\ (x1) + .001\ (x2)$. Thus, $b_0 = 2.695$, $b_1 = .001$, and $b_2 = .001$. The column labeled "p Value" (or p) gives the significance level for each term in the model. A ROT for p Values is: If $p < .05$, the term is significant and should be included in the model. If $p > .10$, the term is not significant and should be removed from the model. If $.05 \le p \le .10$, p is in a gray zone, where the practitioner usually leaves the term in the model, unless prior knowledge dictates otherwise. In our case, neither x1 nor x2 is deemed a significant predictor of y. The term "R Squared" is the R^2 discussed in previous sections. It represents the proportion of variation in y that is accounted for by the model. In this case, $R^2 = .012$, indicating the model explains or accounts for only 1.2% of the variation; 98.8% of the variability is still unaccounted for. "Adj R Sq" is an adjusted R^2, which adjusts R^2 for the number of terms in the model compared to the number of observations. Section 7.5 addresses this measure. The "F Value" is an overall model F statistic which indicates how well the model will predict y. An F value greater than 6 usually indicates a significant model for prediction purposes. In our case, $F = .101$ indicates the model does not predict well at all. "SS reg" and "SS resid" are the sum of squares due to regression (SSR) and sum of squares due to residuals or error (SSE) discussed in the ANOVA section. "Std Error" or the standard error is $\sqrt{MSE}$, an estimate of σ, the variation in the residuals about the regression line.

The model in Table 7.12 is obviously inadequate for predicting the response, since $R^2 = .012$ and the p-values are not significant. In order to calculate the model's lack of fit, we partition SSE into pure error and lack of fit error. That is, $SSE = SSE_{pure} + SSE_{LF}$. The pure error, SSE_{pure}, is acquired from repeated observations of y at the same levels of x. Since the example data has 2 observations of y for the 10 different combinations of x_1 and x_2, the pure error is estimated by (computations not shown)[4]

$$SSE_{pure} = \sum_{ij=1}^{10} (n_{ij} - 1)s_{ij}^2 = 0.19.$$

$MSE_{pure} = SSE_{pure}/df_{pure} = .19/10 = .019$, where $df_{pure} = \sum(n_{ij} - 1)$. The sum of squares for lack of fit, SSE_{LF}, is calculated using $SSE_{LF} = SSE - SSE_{pure}$. For our example, $SSE_{LF} = 2.854 - .19 = 2.664$ and the degrees of freedom for testing lack of fit are $df_{LF} = df_E - df_{pure}$ where $df_E = n - p$ and p is the number of parameters estimated. Therefore, the example data has a mean square for lack of fit of

$$MSE_{LF} = \frac{SSE_{LF}}{df_{LF}} = \frac{2.664}{17 - 10} = 0.3806.$$

The F test for lack of fit is $F_0 = MSE_{LF}/MSE_{pure} = .3806/.019 = 20.032$. Comparing this to $F(1 - \alpha; df_{LF}, df_{pure}) = F(.95; 7, 10) = 3.14$, we see that the lack of fit is significant. Thus, it is necessary to investigate a better model using interactions and quadratic effects. If we test a 2nd order model, it will appear as

$$\hat{y} = b_0 + b_1x_1 + b_2x_2 + b_{12}x_1x_2 + b_{11}x_1^2$$

---

[4]If the design does not contain any replicated points at designated levels of x, replications can be made at the center point (in our example, this would be 75° and 150 psi) and the SSE_{pure} would then be calculated for this one combination of independent variables. If using replicated center points, the $df_{pure} = n_c - 1$, where n_c is the number of center points.

Notice that we do not estimate a quadratic effect for x_2 because it has only 2 levels. The computer output is shown in Table 7.13.

Regression output				
Factor	**Coef**	**Std Error**	**t stat**	**p Value**
Constant	2.973	2.371	1.254	0.229
x1	-0.004	0.057	-0.079	0.938
x2	0.000	0.009	-0.046	0.964
x1 x2	0.000	0.000	0.137	0.893
x1 x1	0.000	0.000	0.051	0.960
R Squared	0.013		Std Error	0.436
Adj R Sq	-0.172		SS reg	0.038
F Value	0.050		SS resid	2.850

Table 7.13 SPC KISS Output for 2nd Order Regression Model

Once again, R^2 is small (.013) and the p-values are large, indicating an inadequate model. Testing this model for lack of fit, we have $SSE_{LF} = SSE - SSE_{pure} = 2.850 - .19 = 2.66$ (about the same as before). The F test for lack of fit is

$$F_0 = \frac{2.66/(15-10)}{0.19/10} = 28.0$$

which is still significant at the 0.05 level. What's wrong? The problem is that 2nd order models can also have quadratic interaction terms. Therefore, the new model to be tested will include a quadratic interaction of x_1^2 with variable x_2:

$$\hat{y} = b_0 + b_1 x_1 + b_2 x_2 + b_{12} x_1 x_2 + b_{(11)2} x_1^2 x_2 + b_{11} x_1^2$$

The results are shown in Table 7.14. It should be of concern that adding the $x_1^2 x_2$ term increased R^2 to 0.901 which is significant, but the coefficient for $x_1^2 x_2$ is 0.

Regression output				
Factor	**Coef**	**Std Error**	**t stat**	**p Value**
Constant	23.815	2.015	11.818	0.000
x1	-0.596	0.056	-10.656	0.000
x2	-0.139	0.013	-10.934	0.000
x1 x2	0.004	0.000	11.192	0.000
x1 x1	0.004	0.000	10.683	0.000
x1 x1 x2	0.000	0.000	-11.210	0.000
R Squared	0.901		Std Error	0.143
Adj R Sq	0.875		SS reg	2.602
F Value	25.504		SS resid	0.286

Table 7.14 SPC KISS Output for Quadratic Interaction Model

The problem here is that only 3 decimal place accuracy is shown. Formatting the output to 4 or more places will produce non-zero coefficients.

A final note is that the MR prediction limits for a future observation at some independent level h of the k predictor variables, call it $\underline{x}_h = (x_{h1}, x_{h2}, \ldots, x_{hk})$, can be obtained in a manner similar to the SLR case shown in Section 7.1.5. The exact mathematical formulation for finding the upper and lower prediction limits in the MR case, however, becomes unwieldy and requires the use of matrix algebra, which is beyond the scope of this text. A very useful Rule of Thumb for finding meaningful approximations to the exact upper (U) and lower (L) prediction limits is given by

$$\begin{matrix} U \\ L \end{matrix} = \hat{y}|_{\underline{x}_h} \pm \begin{bmatrix} 2 \\ \text{or} \\ 3 \end{bmatrix} \sqrt{MSE},$$

where $\hat{y}|_{\underline{x}_h}$ is the point estimate obtained by evaluating the regression equation at $\underline{x}_h$; $\sqrt{MSE}$ is the standard error estimate of σ obtained from the computer output; and we use 2 or 3 as a multiplier, depending on whether we want a 95% or 99% prediction interval, respectively.

7.4 The Importance of Using Coded Data

Table 7.15 contains data at three equally spaced values for each of the three variables A, B, and C. The corresponding coded variables are designated x_1, x_2, and x_3, respectively. For this example, the actual data for A $= 70, 75, 80$ is coded as $x_1 = -1, 0, 1$, (or $-$, 0, $+$, for short) respectively. Building a second order polynomial model on the actual data results in the multiple regression output displayed in Table 7.16. This output is from DOE KISS, a design of experiments software package. DOE KISS contains the same major outputs as SPC KISS, namely the model, p-values, R^2, etc, but it also contains a "TOL" column.

	Actual Variables			Coded Variables			
Run	A	B	C	x_1	x_2	x_3	$\bar{y}^*$
1	75	60	71	0	$-$	$+$	1332
2	70	60	68	$-$	$-$	0	1370
3	75	60	65	0	$-$	$-$	1386
4	80	60	68	$+$	$-$	0	1377
5	75	65	68	0	0	0	1387
6	80	65	71	$+$	0	$+$	1365
7	80	65	65	$+$	0	$-$	1392
8	75	65	68	0	0	0	1411
9	70	65	71	$-$	0	$+$	1372
10	70	65	65	$-$	0	$-$	1417
11	75	65	68	0	0	0	1397
12	75	70	71	0	$+$	$+$	1394
13	80	70	68	$+$	$+$	0	1405
14	75	70	65	0	$+$	$-$	1410
15	70	70	68	$-$	$+$	0	1399

Table 7.15 Actual and Coded Data for 3 Variables

$^*\bar{y}$ is the average of five data points.

"TOL" is short for tolerance, which measures the degree of independence among the independent variables. Tolerances are numbers between 0 and 1, and the closer to 1 the better. Tolerance is computed as $1 - R_j^2$, where R_j^2 is the squared multiple correlation coefficient obtained by regressing predictor variable x_j on all other predictor (x) variables. If all of the tolerances are 1, then the independent variables are truly independent. That is, they are completely uncorrelated. Tolerance values less than 1 indicate that a phenomenon known as ***multicollinearity*** exists. Tolerances close to zero imply strong multicollinearity conditions which can grossly impact the accuracy of the estimated coefficients. Low tolerances indicate gross dependencies among the independent variables. A Rule of Thumb for adequate tolerances is: ***all tolerances should be greater than 0.5***.

FACTOR	COEF	P(2 TAIL)	TOL	LOW	HIGH	EXPER	ACTIVE
Constant	-934.7130	0.8470					
A	-6.4750	0.8945	0.000				X
B	4.3750	0.9242	0.000				X
C	74.6019	0.4776	0.000				X
AA	-0.0917	0.7151	0.001				X
BB	-0.3317	0.2210	0.001				X
CC	-1.0602	0.1686	0.000				X
AB	-0.0100	0.9667	0.001				X
AC	0.3000	0.4655	0.001				X
BC	0.6333	0.1564	0.001				X
R Sq	0.9040						
Adj R Sq	0.7312						
Std Error	11.3966						
F	5.2322	**PRED Y**					
Sig F	0.0416						

Table 7.16 Quadratic Regression Model Using Actual Data

The low tolerance values in Table 7.16 indicate high multicollinearity which is due to modeling the actual factor levels. High multicollinearity inflates the standard error estimates which results in large P(2-tail) or p-values. Even though R^2 is high (.904), we see nothing significant from the p-values and the coefficients cannot be trusted. The following simple example illustrates the problem which can result from using non-coded data in non-linear models.

Actual Data			Coded Data	
A	A^2	vs	x	x^2
1	1		-1	1
1	1		-1	1
2	4		0	0
2	4		0	0
3	9		1	1
3	9		1	1
$R^2 = 0.979$			$R^2 = 0.0$	

The R^2 values show that A and A^2 are highly intercorrelated, whereas the coded columns, x and x^2, have no intercorrelation, indicating they can be evaluated independently. Coding the data removes the dependencies among the independent variables.

Using the coded data in Table 7.15 to build a 2nd order model, we obtain output as shown in Table 7.17. Notice that the R^2, standard error, and the F values are the same as in the model from actual data (Table 7.16). However, the high tolerance values in Table 7.17 indicate there is little or no multicollinearity. Because of this, the P(2-tail) values are useful in finding which effects are significant, and identifying the important variables becomes much easier. We can interpret the P(2-tail) values as follows: (1 - P)100% is the confidence we have that that term actually belongs in the model. Notice that there is some correlation among the variables, although not nearly as much as before. Specifically, the quadratic terms are slightly intercorrelated. Using the ROT for P values, we see that only the x_2 and x_3 terms are significant. Therefore, if we believe that the true model is $Y = f(x_2, x_3)$, then we must recompute the regression coefficients with only x_2 and x_3 in the model, thereby producing the computer output in Table 7.18. The recomputation is also necessary to find the correct standard error and R^2. Notice that the coefficients for x_2 and x_3 have not changed but the constant term, R^2 and the standard error are different from those indicated in Table 7.17.

FACTOR	COEF	P(2 TAIL)	TOL	LOW	HIGH	EXPER	ACTIVE
Constant	1398.3333	*0.0000*					
X1	-2.3750	0.5812	1.000	70	80		X
X2	17.8750	*0.0068*	1.000	60	70		X
X3	-17.7500	*0.0070*	1.000	65	71		X
X1X2	-0.2500	0.9667	1.000				X
X1X3	4.5000	0.4655	1.000				X
X2X3	9.5000	0.1564	1.000				X
X1X1	-2.2917	0.7151	0.989				X
X2X2	-8.2917	0.2210	0.989				X
X3X3	-9.5417	0.1686	0.989				X
R Sq	0.9040						
Adj R Sq	0.7312						
Std Error	11.3966						
F	5.2322	**PRED Y**					
Sig F	0.0416						

Table 7.17 Quadratic Regression Model Using Coded Data

FACTOR	COEF	P(2 TAIL)	TOL	LOW	HIGH	EXPER	ACTIVE
Constant	1387.6000	*0.0000*					
X2	17.8750	*0.0011*	1.000	60	70		X
X3	-17.7500	*0.0012*	1.000	65	71		X
R Sq	0.7504						
Adj R Sq	0.7088						
Std Error	11.8637						
F	18.0345	**PRED Y**					
Sig F	0.0002						

Table 7.18 Reduced Regression Model Using Only Significant Terms

The resulting prediction equation is:

$$\hat{y} = 1387.600 + 17.875(x_2) - 17.750(x_3)$$

If we wish to predict y for $x_2 = 1$ and $x_3 = 0.5$ (x_1 can be any value since it is not in the model), then the 95% prediction limits using the Rule of Thumb are given by

$$\frac{U}{L} = \hat{y}\big|_{x_h} \pm 2\sqrt{MSE}$$

$$= 1396.6 \pm 23.7$$

$$= (1372.9, \; 1420.3)$$

That is, for the coded values of $x_2 = 1$ and $x_3 = .5$, we can be about 95% confident that an individual observed response will fall somewhere between 1372.9 and 1420.3. To recover the actual values from the coded values, we will use linear interpolation as shown in Chapter 8.

7.5 Crossvalidation

When conducting SLR or MR, researchers typically use R^2 as a measure of the strength of the model. It should be noted that R^2 is dependent on the model, the sample, and the sample size. It is possible to continue to add variables up to k n 1 to produce an R^2 which is monotonically increasing up to 1.0. This results in an overfit condition which inaccurately predicts the strength of the model. To compensate for models with large numbers of terms compared to the number of observations, some software packages such as SPC KISS and DOE KISS will calculate an adjusted R^2 value. The formula for adjusted R^2 is shown below.

$$\text{Adj } R^2 = 1 - \left(\frac{n-1}{n-p} \right)(1 - R^2)$$

where   n = total number of observations and

p = total number of terms in the model (including the constant).

A more accurate measure of model strength is to use half of the sample data to build the model and the other half for crossvalidation, or perhaps to build the model on sample data and obtain a crossvalidated R^2 from confirmation experiments. To get the crossvalidated estimate of R^2, calculate $e_i = y_i - \hat{y}_i$, where the $\hat{y}_i$ are from the regression model built from the original data and the y_i are the observed y values in the crossvalidation sample data. As before, we then compute

$$R^2 = 1 - \frac{\sum e_i^2}{SST_2}$$

where SST_2 is the total sum of squares for the data in the crossvalidation sample. It is also possible to correlate the y_i from the original data with the $\hat{y}_i$ model based on the second data set.

Since splitting a sample into two parts requires a large amount of data, a more efficient measure for estimating model strength is the Relative Press (Rel PRESS) constant. This measure of the strength of a model is calculated through the following steps:

(1)   Compute the first of a series of regression model coefficients by withholding (or setting aside) the first observation (i).

(2)   Use the model obtained in (1) to predict the observation (i) not used in building the model, namely $\hat{y}_{i,-i}$.

(3)   Calculate the press residual $(e_{i,-i}) = (y_i - \hat{y}_{i,-i})$ for observation (i) not in the model.

(4)   Repeat steps (1) through (3) n times, deleting a different observation each time.

(5)   Calculate the Prediction Sum of Squares, referred to as

$$PRESS = \sum_1^n (e_{i,-1})^2 .$$

(6)   Calculate Relative Press as follows:

$$\text{Rel PRESS} = 1 - \frac{\text{PRESS}}{\text{SST}},$$

where SST is the Total Sum of Squares for all n actual values of the response y.

For a more efficient approach to calculating PRESS and more information on Rel PRESS see [Myer 90]. Note that calculating Rel PRESS requires a software capability beyond that of SPC KISS.

One way to reduce overfit is to ensure that for any regression model, n is sufficiently large compared to the number of parameters estimated. A good Rule of Thumb for the final prediction model is for n to be at least 10 times the number of parameters in the model.

7.6 Notes of Caution

When using SLR or MR on "happenstance data," one should be cautious of high correlations, for they may be significant only because a spurious condition unknown to the analyst exists. This problem can be averted by designing proper experiments instead of analyzing non-experimental data. Furthermore, there are times when a researcher has properly designed an experiment but obtains unanticipated results due to improper levels for independent variables. For example, consider the relationship between y and x in Figure 7.11. If the levels of x are a and d, we will find a significant linear relationship between y and x. However, if the range on x is b to c, this relationship will not be discovered.

A less likely, but certainly possible, scenario appears in Figure 7.12. Due to the quadratic relationship, if the sampled levels of x are a and b for a SLR, a non-significant relationship will be found even though a significant quadratic relationship exists. To find a significant linear relationship, the sampled region of x must be [a, c] or [c, b]. Obviously, using a quadratic (or 2nd order) model over [a, b], one could detect a significant quadratic relationship. Thus, experimenters should carefully plan not only which variables to put in the model, but also which levels are necessary to detect anticipated relationships.

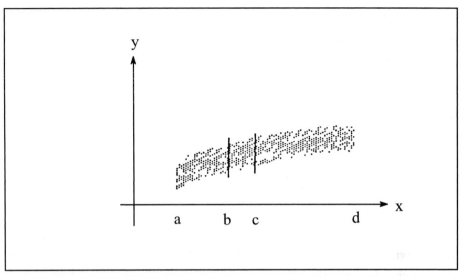

Figure 7.11 Effect of Varying Levels of an Independent Variable

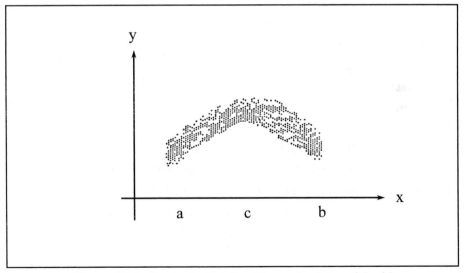

Figure 7.12 Quadratic Relationship on the Interval [a, b]

Finally, it will be briefly stated that outliers in the data can significantly alter regression results. Before any regression is conducted or any other data analysis takes place, one should explore the data distribution for each variable, identify outliers, ascertain their accuracy, and take the appropriate action. One can also use more sophisticated techniques to identify outliers in a multi-variable setting [Myer 90]. The following section provides the practitioner a step-by-step approach to analyzing historical data.

7.7 Historical Data Analysis Strategy

A question commonly asked by someone who is confronted with analyzing some set of data is, "Where do I begin?" This section is provided to give the practitioner a step-by-step approach to analyzing data. This approach applies equally well to data for which the collection procedure/process may not be entirely known. While an efficient and systematic process for collecting data (see Chapter 8, Design of Experiments) typically makes the analysis of the data easier and is certainly encouraged, sometimes it is necessary to analyze data without knowing exactly how the data was collected. We present this approach as a 10-step process. It presupposes a knowledge of the basic tools presented in previous chapters and the regression methodology of this chapter.

Step 1:

- ▸Obtain descriptive statistics on all variables. That is, find the mean, median, max, min, range, variance, and standard deviation, etc.
- ▸Generate histograms and/or stem-and-leaf plots for all variables.
- ▸Generate scatter plots for each pair of variables.
- ▸Detect any outliers in the data set, determine the special cause of any outlier, and remove the cause if possible.

Step 2:

- ▸Delete any legitimate outliers detected in Step 1.
- ▸Go back to Step 1 and redo, verifying that all legitimate outliers, if any, have been removed.

Step 3:

> ▸Check for any intercorrelations between independent variables (i.e., multicollinearity). Do this by examining the plots (from Step 1) and the tolerance values, and by generating the correlation matrix.
> ▸Remove or reduce the multicollinearity by centering the data, standardizing the independent variable data, coding the data (see Section 7.4), removing the variables, etc.

Step 4:

> ▸Build a linear regression model which relates the response (dependent) variable to the independent variables in a linear fashion.
> ▸Examine the regression analysis output for significant predictor variables, tolerances, R^2, etc.
> ▸Generate the residual plots. See Section 7.1.4.

Step 5:

> ▸Examine the residual plots to detect any other outliers that may not have been detected previously.
> ▸Remove any legitimate outliers and go back to Step 4, repeating it with a reduced data set.

Step 6:

> ▸Examine the model for insignificant terms.

Step 7:

> ▸Obtain the best model and examine the residuals from this model. The criteria for the "best" model vary but certainly should include examining MSE and adjusted R^2 values at the least. Common sense and process experience may also help determine which terms to include or exclude.

Step 8:

> ▸Test for adequacy of the model. That is, once again, examine residual
> plots and perform a lack of fit analysis. See Section 7.3.
> ▸Do a crossvalidation study. See Section 7.5.

Step 9:

> ▸Based on evidence from Step 8, add quadratic and/or interaction terms.

Step 10:

> ▸Repeat Steps 6-9 for the quadratic model.

The interested reader is encourage to apply this methodology to the scenario
and data set provided in Problem 7.6 at the end of this chapter. Also, Case
Study 13: Historical Data Analysis of a Plating Process illustrates the use of this
procedure to obtain an optimum model.

7.8 Problems

7.1   The heights(x) and weights(y) of 12 men are given in the following table.

x (in)	y (lbs)
68	135
70	140
71	165
69	155
71	175
72	170
74	190
67	130
69	145
70	158
68	142
72	180

a)   Plot a scatter diagram of the data, with x on the horizontal axis and y on the vertical axis.

b)   Using your "eyes only," draw what you think is the "best fitting" line through the data.   Determine the equation of this line graphically (recall y = mx + b, where m is the slope of the line and b is the y-intercept).

c)   Evaluate $\sum x$, $\sum y$, $\sum x^2$, $\sum y^2$, and $\sum xy$, with all summations being from i=1 to 12. Use Table 7.2 as a guide. Also find $\bar{x}$ and $\bar{y}$.

d)   Find b_0 and b_1 so that $\hat{y} = b_0 + b_1 x$. That is, use least squares to find b_0 and b_1 in the SLR model. How does this regression line compare to your "eye-fit" line?

e)   Evaluate $\sum e_i = \sum (y_i - \hat{y}_i)$. Will this result always be the case?

f)   Find SST = $\sum (y_i - \bar{y})^2$.

g)   Find SSR = $\sum (\hat{y}_i - \bar{y})^2$.

h)   Find SSE = $\sum (y_i - \hat{y}_i)^2$.

i)   Do your findings support the fact that SST = SSR + SSE?

j)   Find the correlation coefficient, R.

k)   Verify that R^2 = SSR/SST. What is the meaning of R^2?

l)   Construct the SLR ANOVA Table for this data. Use Table 7.5 as a guide.

m)   Test the overall regression via an F-test at the 0.01 level of significance. Reference Table 7.6.

n)   Find a 99% prediction interval for the weight of a man selected at random who is 70 inches tall.

7.2   The vehicle weight (in 100 pounds) and gas mileage data (mpg) given in Section 2.2 is reproduced here.

x: Wt (100 lbs)	y: Mileage (mpg)
30	18
28	21
21	32
29	17
24	31
33	14
27	21
35	12
25	23
32	14

Using this data and SPC KISS, find the SLR model that could be used to predict gas mileage from vehicle weight. Analyze this model, as you did the model obtained in problem 7.1, and determine whether it is a useful model. Compare the scatter diagram produced by SPC KISS with Figure 2.21. What are the limitations of this model? Determine a 95% prediction interval for the gas mileage of the vehicle you are now driving (estimate its weight if you don't know the vehicle's weight).

7.3   Miss distance (y) of a certain air-to-ground missile when fired at a ground target is known to be a function of launch range (x_1) and dive angle (x_2). Given the following data from 18 launches, use multiple linear regression to obtain what you think is the "best" possible model. Defend your choice of model.

Miss Distance (y) (ft)	Launch Range (x_1) (ft)	Dive Angle (x_2) (degrees)
9	600	15
8	600	15
10	600	30
12	600	30
15	600	45
14	600	45
6	800	15
9	800	15
13	800	30
16	800	30
15	800	45
12	800	45
10	1000	15
11	1000	15
12	1000	30
11	1000	30
17	1000	45
19	1000	45

7.4   On January 28, 1986, the space shuttle **Challenger** was launched at a temperature of 31°F. The ensuing catastrophe was caused by a combustion gas leak through a joint in one of the booster rockets, which was sealed by a device called an O-ring. The Rogers Commission which investigated the accident concluded that O-rings do not seal properly at low temperatures [*Journal of the American Statistical Association (JASA)*, Vol. 84, No. 408, December 1989, pp 945-957]. The following data are taken from Figure 1.b of the referenced JASA article. It relates launch temperature to the number of O-rings under thermal distress (i.e., O-rings likely to not seal properly) for 24 previous launches.

Temperature (° F)	# of O-rings under thermal distress
x	y
53	3
57	1
58	1
63	1
70	1
70	1
75	2
66	0
67	0
67	0
67	0
68	0
69	0
70	0
70	0
72	0
73	0
75	0
76	0
76	0
78	0
79	0
80	0
81	0

a)   Obtain a scatter plot of the data, with Temp (x) on the horizontal axis and Number of O-rings (y) on the vertical axis.

b)   Determine b_0 and b_1 in the SLR model. That is, find $\hat{y} = b_0 + b_1 x$ and then graph this on your scatter plot in part (a).

c)   Find $\hat{y}|_{x=31}$. That is, predict the number of O-ring failures (i.e., distressed O-rings) for a launch temperature of 31°.

d)   Construct and complete a SLR ANOVA table for this data.

e)   Perform an F test for the hypothesis: $H_0: \beta_1 = 0$
$H_1: \beta_1 \neq 0$.

f)   What is your calculated p-value from the F test?

g)   What do you conclude about your model and the relationship between temperature and O-ring failures?

h)   Construct a 95% prediction interval for the number of O-ring failures for a launch at 31°F.

i)   Using the analysis you have just performed in the steps above, and supposing that 4 O-rings are used in the booster rocket, what are your conclusions in this study?

j)   Build a quadratic model which predicts the number of O-ring failures, given a particular launch temperature. Which model, the quadratic or the SLR model obtained in (b), is the better model? Defend your answer.

7.5   The total number of man-years required for a software development project needs to be estimated for the purpose of bidding on a contract. The following data gives the total number of man-years (y) required for each of 20 previous software development projects accomplished by this company. The independent variables are

x_1 = number of application subprograms developed, and

x_2 = number of software configuration change proposals implemented.

x_1	x_2	y
135	99	52
128	157	58
221	211	207
82	62	95
401	203	346
360	211	244
241	191	215
130	104	112
252	177	195
220	111	54
112	91	48
29	12	39
57	44	31
28	40	57
41	69	20
27	37	33
33	39	19
7	8	6
17	16	7
94	117	56

a)   Find the SLR model using just x_1 as the only regression (independent) variable. Evaluate the adequacy of the model using an F test.

b)   Find the SLR model using just x_2 as the only regression (independent) variable. Evaluate the adequacy of the model using an F test.

c)   Use MLR to obtain the model $\hat{y} = b_0 + b_1 x_1 + b_2 x_2$. Evaluate the adequacy of this model using an F test.

d)   Compare and contrast the three models developed above.

e)   If your company anticipates developing 100 application subprograms (i.e., $x_1 = 100$) and processing 150 configuration change proposals (i.e., $x_2 = 150$) on this project, how many man-years would you recommend that your company bid? Defend your answer.

f)   Investigate the use of a polynomial regression model for prediction purposes. That is, look at squared terms, as well as interaction (e.g., $x_1 x_2$) terms. What is your conclusion?

7.6   The following 29 observations were taken from a blood analysis process where:

x_1 = incubation time,
x_2 = vendor for a particular substance, and
y = rate (Spec: $y < 3$)

Analyze this historical data and develop an appropriate model which explains the relationship between y and the two independent variables x_1 and x_2. Use this model to determine the best settings for x_1 and x_2 that will minimize y.

	X_1	X_2	Y
CASE 1	15.000	1.000	1.250
CASE 2	16.000	1.000	4.550
CASE 3	20.000	2.000	3.900
CASE 4	16.500	2.000	6.800
CASE 5	11.500	2.000	5.100
CASE 6	18.000	2.000	7.100
CASE 7	16.000	2.000	5.300
CASE 8	2.000	2.000	2.600

	X_1	X_2	Y
CASE 9	1.600	2.000	2.600
CASE 10	1.000	2.000	2.100
CASE 11	1.000	2.000	2.600
CASE 12	1.250	2.000	2.900
CASE 13	1.100	2.000	2.900
CASE 14	2.000	2.000	2.500
CASE 15	1.500	2.000	2.500
CASE 16	2.000	2.000	3.000
CASE 17	2.000	2.000	3.000
CASE 18	1.000	2.000	2.800
CASE 19	1.000	2.000	2.800
CASE 20	1.000	2.000	3.700
CASE 21	1.000	2.000	3.700
CASE 22	1.000	2.000	6.600
CASE 23	1.000	2.000	3.100
CASE 24	2.000	2.000	3.400
CASE 25	1.000	2.000	3.900
CASE 26	1.000	1.000	2.700
CASE 27	1.100	1.000	2.500
CASE 28	1.000	1.000	5.900
CASE 29	1.000	1.000	3.700
N OF CASES	29	29	29
MINIMUM	1.000	1.000	1.250
MAXIMUM	20.000	2.000	7.100
MEAN	4.881	1.793	3.638
STANDARD DEV.	6.588	0.412	1.482

Chapter 8

Introduction to Design of Experiments

8.1 What is a Designed Experiment?[1]

In order to properly define a designed experiment, we must have a good understanding of the term **process**. In a general sense, a process is an activity based upon some combination of inputs (factors), such as people, material, equipment, policies, procedures, methods, and environment, which are used together to generate outputs (responses) related to performing a service, producing a product, or completing a task. Graphically, a process would appear as shown in Figure 8.1.

Some examples of processes are shown in Figures 8.2 through 8.7. Obviously, there exist many different kinds of processes; the ones provided are from various applications of designed experiments. In conducting a designed experiment, we will *purposefully make changes to the inputs (or factors) in order to observe corresponding changes in the outputs (or responses)*. The information gained from properly designed experiments can be used to improve performance characteristics, to reduce costs and time associated with product development, design and production, and to build mathematical models which approximate the true relationship between the output(s) and inputs. These mathematical models will contain information on how to optimize the process, how to perform a sensitivity analysis which can be used for tolerance evaluations, and how to reduce variation and possibly make our response robust (insensitive) to factors we are not able to control.

---

[1]Some of this material is taken from *Understanding Industrial Designed Experiments* by S. R. Schmidt and R. G. Launsby (see [Schm 94]).

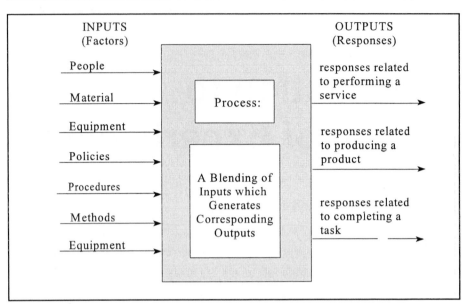

Figure 8.1 Illustration of a Process

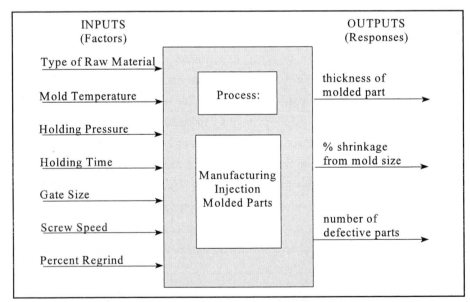

Figure 8.2 Manufacturing Injection Molded Parts

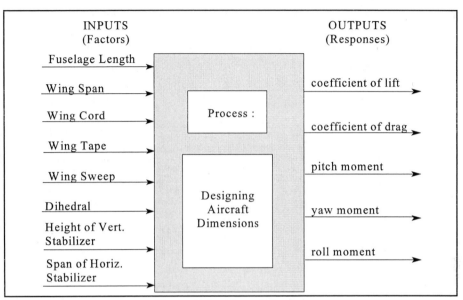

Figure 8.3 Aircraft Design

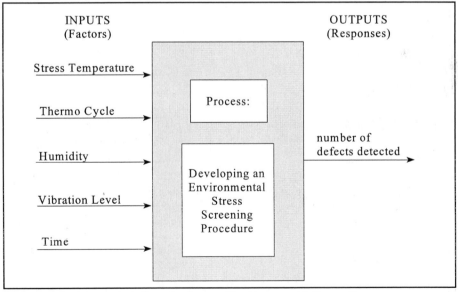

Figure 8.4 Developing an Environmental Stress Screen

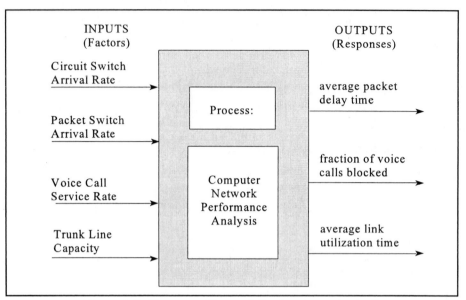

Figure 8.5 Computer Network Performance Analysis

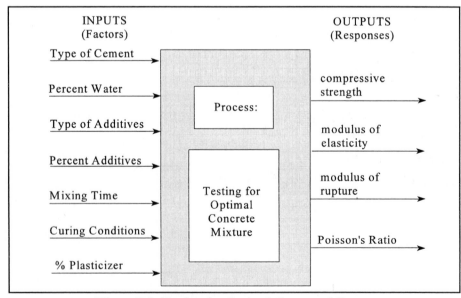

Figure 8.6 Testing for Optimal Concrete Mixture

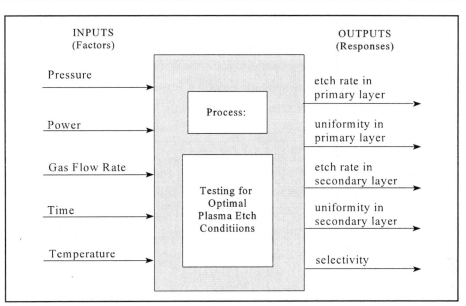

Figure 8.7 Testing for Optimal Etch Conditions

In reality, a process has four categories of variables all of which should have been documented on a cause and effect diagram. These categories are:

(1)   response (output) variables — those variables which are measured to evaluate process and/or product performance.

(2)   controlled variables to be held constant — those variables which have been designated to be held constant during any experimentation. Typically, a standard operating procedure (SOP) is written to ensure consistency of these variables.

(3)   uncontrolled (noise) variables — those variables that cannot be held constant during the experiment and/or later on during production or product usage. These are the variables to which we would like to make the process and/or product robust (insensitive).

(4)   key process variables — those key variables that we intend to vary during an experiment.

Using all four categories of variables, experimenters can represent a process diagram more completely as shown in Figure 8.8.

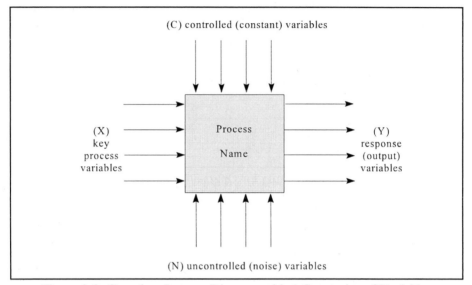

Figure 8.8 Complete Process Diagram with 4 Categories of Variables

8.2 Taguchi Loss Function

In Chapter 1 we discussed the need to adopt a new philosophy of quality by focusing on target values instead of specifications for process output variables. Genichi Taguchi, a Japanese engineer, has developed a loss function which allows the researcher to associate a dollar value to the current state of process/product quality. This dollar amount can be used to identify key processes/products to be improved and to evaluate the amount of improvement. In addition, the loss function will motivate the objectives for a designed experiment [Tagu 87].

The quadratic loss function shown in Figure 8.9 will, in many cases, approximate the true long term loss due to deviations from target. Actual losses result from a combination of scrap, rework, poor performance, lack of customer satisfaction, etc. The measured loss (L) for a single product can be estimated using $L = k(y - T)^2$ where y is the response value, T is the target value, and k is a monetary constant. To determine k we need to estimate the loss for any specific value of y. For example, if the estimated loss for $y = 130$ is \$100.00 then $k = 100 \div (130 - 120)^2 = 1.0$. The loss function $L = 1.0(y - T)^2$ can now be used to estimate the loss associated with other y values.

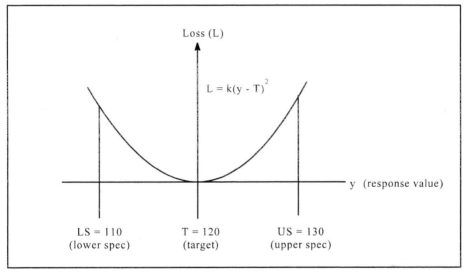

Figure 8.9 Taguchi Loss Function

Another use of the quadratic loss function is to determine the average loss per product. Using expected loss, E(L), to depict average loss, it can be shown that $E(L) = k(\sigma_y^2 + (\bar{y} - T)^2)$ [Schm 94]. The route to reduced loss can now be easily identified, i.e., reduce the variability, σ_y^2, and reduce the deviation of product average from the designated target, $|\bar{y} - T|$. The use of design of experiments (DOE) can assist the researcher in determining which inputs (factors) shift the response average, which shift the response variability, which shift both the average and the variability, and which have no effect on the average or the variability (see Figure 8.10). This type of knowledge of the process allows the researcher to choose the proper input settings to achieve desired targets with minimum variability.

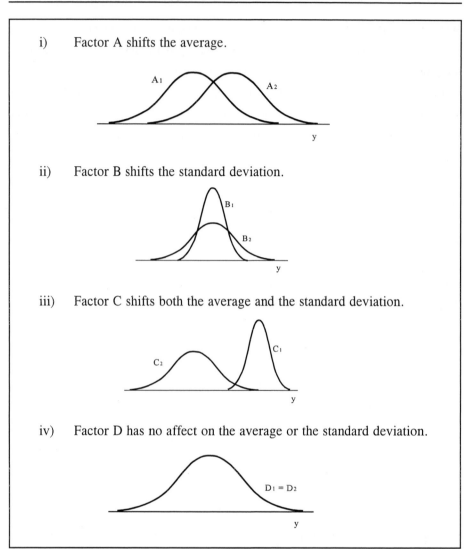

i)   Factor A shifts the average.

ii) Factor B shifts the standard deviation.

iii) Factor C shifts both the average and the standard deviation.

iv) Factor D has no affect on the average or the standard deviation.

Figure 8.10 Graphical Description of Different Types of Factors

The remainder of this chapter will focus on an introduction to designed experiments. For more information on this subject the reader is referred to [Schm 94]. The next three sections address 2-level designs, that is, designs where the input factors are tested at two different settings or levels.

8.3 Full Factorial Designs

A **full factorial** designed experiment consists of testing all possible combinations of factor levels. The total number of different combinations for k factors at 2 testing levels is n = 2^k. For example, if we intend to experiment with 2 factors at 2 testing levels each, there will be a total of n = 2^2 or 4 combinations. The **combinations** to be tested are often displayed in a **test matrix**. See Table 8.1. The advantage of testing the full factorial is that we obtain information on all main factors plus all interactions. The disadvantage is an excessive number of runs, especially when the number of factors (k) is greater than 4.

Example 8.1 Consider a nickel plating process with 2 inputs: plating time and plating solution temperature. The test matrix in Table 8.1 consists of all possible combinations for time (factor A) and temperature (factor B), each at 2 testing levels. The term "run" refers to a particular experimental combination of factor settings. Run number 1 indicates that time is set at 4 seconds and temperature at 16°C for which a response value on thickness (y) is collected.

| | **Factors** | | **Response** |
Run	**Time (A)**	**Temp. (B)**	**thickness (y)**
1	4	16° C	
2	4	32° C	
3	12	16° C	
4	12	32° C	

Table 8.1 Nickel Plating Test Matrix

To conduct the experiment in a way that reduces potential bias from factors not included in the test matrix (e.g., type of raw material, operator, time of day, etc.), we should use a randomization procedure. In other words, the runs should be conducted in a random order. Different randomization strategies exist and can be accomplished through the use of random number tables (see Appendix A). If randomization cannot be achieved (or only partially achieved), emphasis on holding all possible uncontrolled factors at a constant level is even more critical. In Table 8.2 we have displayed 20 observations which were obtained by replicating each test run 5 times. In other words, we plated 5 products for each run and entered a thickness value for each product in Table 8.2.

	Factors		Replicated Response Values						
Run	A	B	y_1	y_2	y_3	y_4	y_5	$\bar{y}$	s
1	4	16	116.1	116.9	112.6	118.7	114.9	115.8	2.278
2	4	32	106.7	107.5	105.9	107.1	106.5	106.7	.607
3	12	16	116.5	115.5	119.2	114.7	118.3	116.8	1.884
4	12	32	123.2	125.1	124.5	124.0	124.7	124.3	.731

$$\bar{\bar{y}} = 115.93$$

Table 8.2 Results of Plating Experiment

Although the actual factor settings shown in Table 8.1 are required for running the experiment, **coded values** are used to set up and conduct the analysis of the experiment. The reason for coding is to avoid confounding information during the model building process and to standardize the units and scaling of all input factors. To properly code the actual factor settings one can use the formula and graph in Figure 8.11.

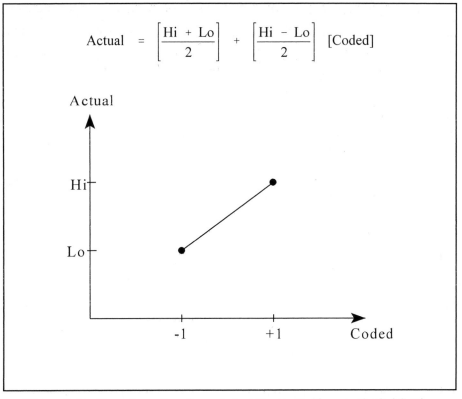

$$\text{Actual} = \left[\frac{\text{Hi} + \text{Lo}}{2}\right] + \left[\frac{\text{Hi} - \text{Lo}}{2}\right] \text{[Coded]}$$

Figure 8.11 Transformation from Actual Factor Settings to Coded Setting
(and Vice Versa)

For example, consider factor B (temperature) = 32 in Table 8.1. The actual value is 32 and using the relationship between actual and coded values shown in Figure 8.11, we have:

$$\text{Actual}_B = \left(\frac{32 + 16}{2}\right) + \left(\frac{32 - 16}{2}\right) \text{(Coded)}$$

$$32 = 24 + 8 \text{ (Coded)}$$

$$\text{Coded} = 1.0.$$

Converting all the actual factor settings in Table 8.1 to coded values produces the coded test matrix in Table 8.3. Basically, the high actual values are coded as (1) and the low actual values are coded as (-1). Figure 8.11 can help transform actual values to coded and vice versa.

Notice that an AB column has been added in Table 8.3. The AB column is used to evaluate the AxB or time x temperature interaction. The values in the AB column are obtained by multiplying the coded values in the A and B columns, row by row. Interaction effects represent a coupling or combined effect of two or more factors. The true interaction effect is represented in the model by a coefficient times the product of the respective factors. The notation AB, A·B, A*B, or AxB all indicate the same thing: the interaction of A and B.

| | Factors & Interactions | | | Replicated Response Values | | | | | | |
Run	A	B	AB	y_1	y_2	y_3	y_4	y_5	$\bar{y}$	s
1	-1	-1	1	116.1	116.9	112.6	118.7	114.9	115.8	2.278
2	-1	1	-1	106.7	107.5	105.9	107.1	106.5	106.7	.607
3	1	-1	-1	116.5	115.5	119.2	114.7	118.3	116.8	1.884
4	1	1	1	123.2	125.1	124.5	124.0	124.7	124.3	.731

Table 8.3   Coded Design Matrix

To determine which terms in the test matrix (A, B, AB) shift the average of the response, we must first evaluate each coded column or term by computing the average response at the low setting (-1) and the average response at the high setting (1). For example, factor A is at its low setting during runs 1 and 2. The average of the 10 response values for these 2 runs is 111.3 (see entry in Table 8.4), which is also the average of the $\bar{y}$'s from runs 1 and 2. The Δ row for Table 8.4 indicates the difference between the response average at the highs (1) and the response average at the lows (-1). The absolute size of these Δ's provides a subjective measure of the relative importance of the terms represented by each column.

Run	\multicolumn{3}{Factors & Interactions} A B AB	\multicolumn{7}{Replicated Response Values}

Run	A	B	AB	y_1	y_2	y_3	y_4	y_5	$\bar{y}$	s
1	−1	−1	+1	116.1	116.9	112.6	118.7	114.9	115.8	2.278
2	−1	+1	−1	106.7	107.5	105.9	107.1	106.5	106.7	.607
3	+1	−1	−1	116.5	115.5	119.2	114.7	118.3	116.8	1.884
4	+1	+1	+1	123.2	125.1	124.5	124.0	124.7	124.3	.731

Factors & Interactions

	A	B	AB	
$\bar{y}$ Avg @ −1	111.3	116.3	111.8	
$\bar{y}$ Avg @ +1	120.6	115.5	120.1	Grand Mean: $\bar{\bar{y}} = 115.93$
Δ	9.3	−0.8	8.3	MSE = 2.409
MSB	432.5	3.2	344.5	
F	179.5	1.3	143.0	F(.95; 1, 16) ≈ 4.54
	*		*	

* these columns have significant effects

Table 8.4 Plating Example Analysis Summary

To evaluate the statistical importance of each of the j terms in the matrix, we can compute an observed F value based on the following formula for 2-level designs.

$$F_j = \frac{\text{Mean Square Between for column } j}{\text{Mean Square Error}} = \frac{MSB_j}{MSE}$$

where

a) $MSB_j = \dfrac{N}{4}(\Delta_j)^2$ and

N =   total number of response values collected in the experiment.

In this example, N = 20 and the MSB for column A is

$$MSB = \frac{20}{4} (9.3)^2 = 432.45.$$

b) $$MSE = \frac{\sum\limits_{r=1}^{n}\left[(n_r - 1)s_r^2\right]}{\sum\limits_{r=1}^{n}(n_r - 1)}$$ where

n = number of combinations (or rows) in the test matrix,

n_r = number of data response values collected in test matrix row r, and

s_r^2 = variance of response values in row r.

For this example,

$$MSE = \frac{(5-1)(2.278)^2 + (5-1)(.607)^2 + (5-1)(1.884)^2 + (5-1)(.731)^2}{(5-1) + (5-1) + (5-1) + (5-1)}$$

= 2.409.

Once the observed F values are computed for each term or column, their magnitudes are compared to a percentile of the F distribution to determine the significance of the term associated with the column. The critical F percentile for 2-level designs is found using $F(1 - \alpha; 1, \Sigma(n_r - 1))$. If α were selected to be .05, the critical F in the plating example would be $F(.95; 1, 16) \approx 4.54$ (see Appendix F)[2]. A summary of the plating example analysis is shown in Table 8.4.

---

[2]As a Rule of Thumb for 2-level designs, an observed $F \geq 4.0$ implies approximately 95% confidence that the term belongs in the model.

Using the significant columns from the experimental results, one can now build an empirical model (similar to a Taylor Series approximation) which relates the output (y) to the coded inputs A and B. The general form for this model is shown here.

$$\hat{y} = \bar{\bar{y}} + \frac{\Delta_A}{2} A + \frac{\Delta_B}{2} B + \frac{\Delta_{AB}}{2} A \cdot B$$

If we use only the terms having significant F_j values in the model, the plating model would appear as

$$\hat{y} = 115.93 + \left(\frac{9.3}{2}\right) A + \left(\frac{8.3}{2}\right) A \cdot B$$

(where A and B can take on values only from the coded scale).

However, the commonly practiced hierarchical rule of modeling dictates that any significant interaction or quadratic term in the model be accompanied by the associated linear terms (whether the linear terms themselves are significant or not). Therefore, the **model** for the plating example is

$$\hat{y} = 115.93 + 4.65A - .4B + 4.15AB.$$

Remember that the input factors represented in the ŷ equation can take on only coded values. This same model could have been generated using multiple regression (see Chapter 7) which for most software packages requires that the data be arranged as shown in Table 8.5. However, using the DOE software package called DOE KISS, one can input the data in the format of either Table 8.3 or Table 8.5. The regression output from DOE KISS is shown in Table 8.6.

Obs #	A	B	AB	y
1	-1	-1	1	116.1
2	-1	1	-1	106.7
3	1	-1	-1	116.5
4	1	1	1	123.2
5	-1	-1	1	116.9
6	-1	1	-1	107.5
7	1	-1	-1	115.5
8	1	1	1	125.1
9	-1	-1	1	112.6
10	-1	1	-1	105.9
11	1	-1	-1	119.2
12	1	1	1	124.5
13	-1	-1	1	118.7
14	-1	1	-1	107.1
15	1	-1	-1	114.7
16	1	1	1	124.0
17	-1	-1	1	114.9
18	-1	1	-1	106.5
19	1	-1	-1	118.3
20	1	1	1	124.7

Table 8.5 Plating Experiment Data in Regression Format

FACTOR	COEF	P(2 TAIL)	TOL	LOW	HIGH	EXPER	ACTIVE
Constant	115.9300	*0.0000*					
A	4.6400	*0.0000*	1	-1	1	-1	X
B	-0.4100	0.2548	1	-1	1	-1	X
AB	4.1400	*0.0000*	1				X
R Sq	0.9527						
Adj R Sq	0.9438						
Std Error	1.5523						
F	107.4449		**PRED Y**				
Sig F	0.0000						

Table 8.6 DOE KISS Regression Output for Experimental Example

We now show how to interpret the regression output in Table 8.6 by addressing the important entries.

(1)   The first two columns, "FACTOR" and "COEF," specify the model. The coefficients in the "COEF" column are the constant ($\overline{\overline{y}}$) and the $\Delta/2$'s shown in the previous analysis. Writing out the model horizontally from these two columns, we get

$$\hat{y} = 115.93 + 4.64A - .41B + 4.14AB,$$

which is the same model, barring slight roundoff differences, we obtained previously. The coefficients from this regression analysis are the more accurate of the two results.

(2)   The "P(2 TAIL)" column specifies the significance level for each of the effects (like A, B, AB). The following Rule of Thumb (ROT) can be used to interpret which effects are important and should be included in the $\hat{y}$ model.

If "P(2 TAIL)" < .05,   the term is significant so include
                              the term in the model.

If "P(2 TAIL)" > .10,   the term is not significant so do
                              not include the term in the model.

If .05 ≤ "P(2 TAIL)" ≤ .10, then the evidence is in a gray zone
                              in which case most experimenters
                              will place the term in the model,
                              barring any prior knowledge that
                              might indicate otherwise.

The "ACTIVE" column indicates via an "x" which terms are currently in the model. The user may remove a term from the model by erasing the "x" (using the space bar) in the appropriate position in this column. The constant term is always assumed to be in the model. Table 8.6 indicates that everything is in the model.

(3)   The "TOL" column specifies the tolerance or degree of independence (i.e., orthogonality) for that term with regard to all of the other input variables/effects. Tolerance is a number between 0 and 1, with numbers closer to 1 being better. A good ROT for tolerances is:

All tolerances should be at least .5.

If any tolerance drops below .5, it indicates the factor or effect is quite highly correlated with the other input factors/effects. These dependencies among the input factors/effects create inflated p-values and loss of confidence in the coefficients. In full factorial designs such as that shown in Table 8.3 or 8.4, all tolerances will be 1.0. This is ideal, allowing the effect of each term to be evaluated independently of the others.

(4)   The "LOW" and "HIGH" columns reflect the factor settings that the user specified when initiating the design in DOE KISS. The "EXPER" indicates the specific settings for the factors at which the model will be evaluated in the event the "Predict Y" button is clicked. Even though coded values appear in these columns in Table 8.6, these three columns allow the user to communicate with the software using actual values, which would no doubt be more convenient than using coded values.

(5)   The "R Sq" value (.9527 in this case) is R^2 and it specifies the proportion of variability in the response values that is accounted for by the model. $1 - R^2$ is the proportion of noise still remaining (or unaccounted for) in the data.

(6)   The "Adj R Sq" value adjusts R^2 for the number of terms in the model compared to the number of observations, as follows.

$$\text{Adj } R^2 = 1 - \left(\frac{n-1}{n-p} \right) (1 - R^2) \quad \text{where}$$

n = total number of observations and

p = total number of terms in the model, including the constant.

(7)   The "Std Error" value is an estimate of the standard deviation of the residuals. That is, "Std Error" $= \sqrt{\text{MSE}}$.

(8)   The "F" statistic of 107.4449 is computed from the Analysis of Variance table as F = MSR/MSE. It indicates the strength of the model for prediction. The "Sig F" gives the significance level for the F statistic which, in this case, is .0000. This is a p-value for

the entire model, meaning the probability of an incorrect model is 0.0000 (to 4 places). A good ROT to use is:

If "SigF" < .05, then the model is a strong model for prediction.

The prediction equation has now provided information on factors shifting the average. What is still needed is information on factors that shift the variability. To determine which factors shift the variability we will use the data in Table 8.3 with the values in the s column as our response values. The actual analysis in Table 8.7 is similar to that discussed with Table 8.4. Note that in Table 8.7, δ is defined as δ = (s Avg @ 1) - (s Avg @ -1). Instead of an F test to analyze $\hat{s}$, we can test columns for significance using the following simple Rule of Thumb for 2-level designs: $\delta \mid \geq \bar{s}$ *indicates that the corresponding term is significant at approximately the* α =.05 *level and should be placed in the* $\hat{s}$ *model.* It is, however, important to remember that a Rule of Thumb is not a razor cut line, but rather a guideline where values close to the Rule of Thumb must be evaluated with lots of common sense.

Run	A	B	AB	s
1	-1	-1	1	2.278
2	-1	1	-1	0.607
3	1	-1	-1	1.884
4	1	1	1	0.731
s Avg @ -1	1.4425	2.081	1.2455	grand s
s Avg @ 1	1.3075	0.669	1.5045	average is
δ	-.1350	-1.412	0.2590	$\bar{s}$ = 1.375

Table 8.7 Plating Analysis for Variance Factors

In general, a model for predicting response standard deviation for 2-level designs is

$$\hat{s} = \bar{s} + \frac{\delta_A}{2}A + \frac{\delta_B}{2}B + \frac{\delta_{AB}}{2}A \cdot B$$

Therefore, the full model for $\hat{s}$ in this case (from Table 8.7) is

$$\hat{s} = 1.375 - \frac{.1350}{2}A - \frac{1.412}{2}B + \frac{.2590}{2}A \cdot B$$

Using the previously stated Rule of Thumb for $\hat{s}$ models, a prediction equation for $\hat{s}$ including only significant terms is

$$\hat{s} = 1.375 - .706B$$

Remember that the $\hat{s}$ equation is for coded values. Therefore, to **minimize** $\hat{s}$, we should set factor B at the coded value of 1 (the high actual setting) which results in a predicted response standard deviation of $\hat{s} = .669$.

Now that we have found the factor settings that will minimize the variance, we must return to the $\hat{y}$ prediction equation to determine the remaining factor settings for achieving a target thickness of 120 units. To accomplish this objective we will follow the steps below.

1)   *Set prediction equation for ŷ equal to the desired target value.*

$$120 = 115.93 + 4.64A - .41B + 4.14AB$$

2)   *Insert factor settings previously determined from variance minimization, i.e., set B = 1.*

$$120 = 115.93 + 4.64A - .41(1) + 4.14A(1)$$

3)   Set all but 1 remaining factor at coded values that correspond to low cost, convenience, etc. Since only A remains, there are no extra factors to set.

4)   Solve the remaining equation for the 1 unset factor.
$$(4.64 + 4.14) A = 120 - 115.93 + .41$$

$$A = \frac{4.48}{8.78} = .51.$$

5)   Decode the coded value in 4) using Figure 8.11.

$$Actual_A = \left(\frac{12 + 4}{2}\right) + \left(\frac{12 - 4}{2}\right) \text{ (Coded)}$$

$$= 8 + 4(.51) = 10.04.$$

Obviously, selecting factor settings to minimize variance, achieve target response values, and minimize cost must include not only the use of the $\hat{s}$ and $\hat{y}$ prediction equations, but a great deal of process knowledge and common sense. Often tradeoffs in objectives must be made. The two prediction equations provide the knowledge needed to make the proper tradeoffs.

For the plating process, the optimal factor settings appear to be 10.04 seconds and 32°C. What is needed at this point are some confirmation runs to verify this conclusion. To **test for confirmation** we would expect all confirmatory response values to fall within $\hat{y} \pm 3\ \hat{s}$. For our example all confirmatory plated parts should have a thickness value between $120 \pm 3(.669)$ or $(117.993, 122.007)$. If conclusions cannot be verified with a reasonable degree of certainty, then more problem solving techniques (see [Schm 94]) and possibly more experiments may be needed. The reason most experiments fail is due to poor experimental discipline and measurement error. It is difficult to salvage anything from our experiment if we have these problems. Therefore, the planning phase of the experiment should insure that we will not have discipline or measurement problems.

Example 8.2 In a metal casting process for manufacturing jet engine turbine blades, the experimentation objective is to determine optimal factor settings to minimize part shrinkage (i.e., difference between mold size and part size) and minimize shrinkage variability. The factors to be tested are mold temperature (A), metal temperature (B), and pour speed (C). The coded test matrix together with the response values and analysis for this experiment are shown in Table 8.8. The A, B, and C columns represent all possible combinations for the three factors at 2 levels each. The interaction columns are obtained by multiplying the respective column elements for each of the main factors, row by row.

Run	Factors and Interactions							Replicated Response Values			$\bar{y}$	s
	A	B	AB	C	AC	BC	ABC	y_1	y_2	y_3		
1	-1	-1	1	-1	1	1	-1	2.22	2.11	2.14	2.16	0.057
2	-1	-1	1	1	-1	-1	1	1.42	1.54	1.05	1.34	0.255
3	-1	1	-1	-1	1	-1	1	2.25	2.31	2.21	2.26	0.050
4	-1	1	-1	1	-1	1	-1	1.00	1.38	1.19	1.19	0.190
5	1	-1	-1	-1	-1	1	1	1.73	1.86	1.79	1.79	0.065
6	1	-1	-1	1	1	-1	-1	2.71	2.45	2.46	2.54	0.147
7	1	1	1	-1	-1	-1	-1	1.84	1.76	1.70	1.77	0.070
8	1	1	1	1	1	1	1	2.27	2.69	2.71	2.56	0.248
$\bar{y}$ Avg @ -1	1.735	1.957	1.945	1.993	1.522	1.975	1.913					
$\bar{y}$ Avg @ 1	2.164	1.942	1.954	1.906	2.377	1.924	1.986	$\bar{\bar{y}}$ = 1.95 $\bar{s}$ = .135				
Δ	0.429	-.015	0.009	-.087	0.855	-.051	0.073					
MSB	1.104	0.001	0.001	0.045	4.386	0.016	0.032	MSE = .02489				
F	$\boxed{44.36}$	0.040	0.040	1.810	$\boxed{176.2}$	0.640	1.281	F(.95;1,16) = 4.49				
s Avg @ -1	0.138	.131	0.113	0.060	0.145	0.130	0.116	$\boxed{}$ indicates significant values				
s Avg @ 1	0.132	.139	0.157	0.210	0.125	0.140	0.154	based on ROT				
Δ	-.006	0.008	0.044	$\boxed{0.150}$	-.020	0.010	0.038					

Table 8.8 ANOVA Analysis for Turbine Blade Casting Example

Based on the information provided in Table 8.8 (ANOVA Analysis), the two equations for modeling the turbine blade casting process are[3]

i) $\hat{s} = .135 + .075C$

ii) $\hat{y} = 1.950 + .2145A + .4275AC - .0435C$

The equivalent (but more accurate) results from the regression analyses in Tables 8.9 and 8.10 are:

$\hat{s} = .1355 + .0748C$

$\hat{y} = 1.9496 + .2146A + .4279AC - .0438C$

FACTOR	COEF	P(2 TAIL)	TOL	LOW	HIGH	EXPER	ACTIVE
Constant	0.1355	Not Avail					
A	-0.0027	Not Avail	1	-1	1	-1	X
B	0.0043	Not Avail	1	-1	1	-1	X
C	0.0748	Not Avail	1	-1	1	-1	X
AB	0.0223	Not Avail	1				X
AC	-0.0097	Not Avail	1				X
BC	0.0046	Not Avail	1				X
ABC	0.0194	Not Avail	1				X
R Sq	1.0000						
Adj R Sq	Not Avail						
Std Error	0.0000						
F	0.0000			PRED S			
Sig F	Not Avail						

Table 8.9   Multiple Regression Output for Std. Dev. ($\hat{s}$)
Model on Table 8.8 Data

---

[3]Many statisticians choose to model ln s (the natural logarithm of s) instead of s because the distribution of ln s is less skewed than the distribution of s. This text will use the simpler $\hat{s}$ model.

FACTOR	COEF	P(2 TAIL)	TOL	LOW	HIGH	EXPER	ACTIVE
Constant	1.9496	*0.0000*					
A	0.2146	*0.0000*	1	-1	1	-1	X
B	-0.0071	0.8289	1	-1	1	-1	X
C	-0.0438	0.1938	1	-1	1	-1	X
AB	0.0046	0.8888	1				X
AC	0.4279	*0.0000*	1				X
BC	-0.0254	0.4422	1				X
ABC	0.0363	0.2776	1				X
R Sq	0.9334						
Adj R Sq	0.9042						
Std Error	0.1580						
F	32.0166			**PRED Y**			
Sig F	0.0000						

Table 8.10 Multiple Regression Output for ŷ Model on Table 8.8 Data

To minimize variability ($\hat{s}$), factor C (pour speed) should be set at the low level (coded level -1). Inserting a value of -1 for C in the ŷ equation results in

$$\hat{y} = 1.9496 + .2146A + .4279(A)(-1) - .0438(-1) \quad \text{or}$$

$$\hat{y} = 1.9934 - .2133A.$$

Thus, minimizing ŷ requires factor A to be set at the high level or coded value of 1. The resulting predicted percent shrinkage for $A_1B_?C_{-1}$ will be

$$\hat{y} = 1.9496 + .2146(1) + .4279(1)(-1) - .0438(-1) = 1.7801.$$

And the predicted standard deviation will be

$$\hat{s} = .1355 + .0748\,(-1) = .0607.$$

Since factor B was not important for shifting the average or the variability, it should be set at an appropriate level to reduce cost.

A more in-depth examination of the ŝ and ŷ equations in Example 8.2 indicates that, by minimizing ŝ first, we have obtained a suboptimal value for ŷ. If we had instead minimized ŷ first, we would have set A at the low level and C at the high level. This type of problem can potentially occur when the ŝ and ŷ equations contain one or more of the same factors. When both ŝ and ŷ cannot be optimized, a **tradeoff** must take place. Usually, the engineer/researcher will weigh the importance of adjusting the mean versus minimizing the variance and use the two equations (ŝ and ŷ) to obtain the best settings.

An alternative tradeoff methodology is to use the loss function discussed in Section 8.2. To compute average loss we will use

$$\bar{L} \; = \; k(\hat{s}^{\,2} + (\hat{y} - T)^{\,2}).$$

For Example 8.2, the target percent shrinkage is zero; therefore,

$$\bar{L} \; = \; k(\hat{s}^{\,2} + \hat{y}^{\,2}).$$

Using this equation, we can construct a table (see Table 8.11) to aid in selecting appropriate factor settings. Thus, the best settings to minimize average loss are $A_{-1}B_?C_1$. The B factor will be set based on cost and/or convenience. Of the different strategies discussed, the best one depends on the objective of the experiment. Similar optimization strategies could include maximizing C_{pk} (see Chapter 9) and/or minimizing the number of defective product.

Factors		ŝ	ŷ	$\bar{L}$
A	C			
−1	−1	.0607	2.2067	k(4.8732)
−1	1	.2103	1.2633	k(1.6402)*
1	−1	.0607	1.7801	k(3.1724)
1	1	.2103	2.5483	k(6.5381)
				* minimum average loss

Table 8.11 Tradeoff Analysis Using Average Loss

The previous examples were full factorial designs for 2 and 3 factors, respectively. The largest reasonable full factorial for k factors each at 2 levels would be for k=4, i.e., a 2^4 or 16 run matrix. Although an example is not presented at this time, you can use Table 8.12 as a template to conduct this type of experiment. The analysis for y and s would be similar to that previously discussed. We highly recommend the use of software like DOE KISS, rather than manual calculations when analyzing data from a designed experiment.

	1	2	3	4	5	6	7	8	9	10	11	12	13	14	15				
Run	A	B	AB	C	AC	BC	ABC	D	AD	BD	ABD	CD	ACD	BCD	ABCD	y_1 y_2 y_3	$\bar{y}$	s
1	-1	-1	1	-1	1	1	-1	-1	1	1	-1	1	-1	-1	1			
2	-1	-1	1	-1	1	1	-1	1	-1	-1	1	-1	1	1	-1			
3	-1	-1	1	1	-1	-1	1	-1	1	1	-1	-1	1	1	-1			
4	-1	-1	1	1	-1	-1	1	1	-1	-1	1	1	-1	-1	1			
5	-1	1	-1	-1	1	-1	1	-1	1	-1	1	1	-1	1	-1			
6	-1	1	-1	-1	1	-1	1	1	-1	1	-1	-1	1	-1	1			
7	-1	1	-1	1	-1	1	-1	-1	1	-1	1	-1	1	-1	1			
8	-1	1	-1	1	-1	1	-1	1	-1	1	-1	1	-1	1	-1			
9	1	-1	-1	-1	-1	1	1	-1	-1	1	1	1	1	-1	-1			
10	1	-1	-1	-1	-1	1	1	1	1	-1	-1	-1	-1	1	1			
11	1	-1	-1	1	1	-1	-1	-1	-1	1	1	-1	-1	1	1			
12	1	-1	-1	1	1	-1	-1	1	1	-1	-1	1	1	-1	-1			
13	1	1	1	-1	-1	-1	-1	-1	-1	-1	-1	1	1	1	1			
14	1	1	1	-1	-1	-1	-1	1	1	1	1	-1	-1	-1	-1			
15	1	1	1	1	1	1	1	-1	-1	-1	-1	-1	-1	-1	-1			
16	1	1	1	1	1	1	1	1	1	1	1	1	1	1	1			

Table 8.12 Full Factorial Design for k=4 Factors (each at 2 levels)

8.4 Fractional Factorial Designs

If we were to test k = 4 factors at 2 levels there would be 2^4 or 16 total possible combinations. The test matrix would include 15 columns for the following terms: A, B, C, D, AB, AC, AD, BC, BD, CD, ABC, ABD, ACD, BCD, and ABCD (see Table 8.12).   In most applications, experimenters seldom find interactions of more than two factors to be significant. Thus, time and resources may be conserved by use of a fraction of the full factorial. For example, if we were to start with k = 3 factors, the test matrix with all interaction columns would appear as shown in Table 8.13. Since there is a low likelihood of a significant 3-factor interaction, we will instead use the ABC column to represent the settings for D, the fourth factor (see Table 8.14). We now have half (8) of the total number of runs we would have had using a $2^4 = 16$ run full factorial. Thus, by "aliasing" factor D with the ABC interaction, we have built a half fraction of the full factorial. We refer to this as a 2^{4-1} design. Letting D = ABC does not mean that the linear effect of factor D is equal to the 3-way interaction effect of factors A, B, and C. It simply means we will use the -1 and +1 settings in the column labeled ABC to generate settings for factor D when we run the experiment. Under the assumption that 3-way interactions are rare, we will conclude that the effect of the D = ABC column reflects only the linear effect for factor D.

If we now look back at the full factorial for 4 factors (see Table 8.12), we will see that Table 8.14 is missing the AD, BD, and CD interactions. If we multiply the -1 and +1 elements in columns C and D, we can see that CD has the same pattern as column AB (see Table 8.15). We could also show that the BD column is aliased with AC and that the AD column is aliased with BC. All of the aliasings generated from the 2^{4-1} fractional factorial in Table 8.14 are presented in Table 8.16. In Table 8.14, the 2^{4-1} notation indicates that we have 4 factors each at 2 levels to be tested in n = 2^{4-1} = 8 runs, which is half of the full factorial. The notation for a fractional factorial design is 2^{k-q}, where k is the total number of factors in the experiment and q is the number of those factors set equal to an interaction column in a full factorial for k - q factors.

Factors and Interactions							
Run	A	B	AB	C	AC	BC	ABC
1	-1	-1	1	-1	1	1	-1
2	-1	-1	1	1	-1	-1	1
3	-1	1	-1	-1	1	-1	1
4	-1	1	-1	1	-1	1	-1
5	1	-1	-1	-1	-1	1	1
6	1	-1	-1	1	1	-1	-1
7	1	1	1	-1	-1	-1	-1
8	1	1	1	1	1	1	1

Table 8.13 Full Factorial for k = 3 Factors

Factors and Interactions							
Run	A	B	AB	C	AC	BC	D
1	-1	-1	1	-1	1	1	-1
2	-1	-1	1	1	-1	-1	1
3	-1	1	-1	-1	1	-1	1
4	-1	1	-1	1	-1	1	-1
5	1	-1	-1	-1	-1	1	1
6	1	-1	-1	1	1	-1	-1
7	1	1	1	-1	-1	-1	-1
8	1	1	1	1	1	1	1

Table 8.14 Fractional Factorial Test Matrix for a 2^{4-1} Design

Run	AB	CD
1	1	1
2	1	1
3	-1	-1
4	-1	-1
5	-1	-1
6	-1	-1
7	1	1
8	1	1

Table 8.15 Aliasing of AB with CD

Test Matrix Term	Aliased Term
A	BCD
B	ACD
C	ABD
D	ABC
AB	CD
AC	BD
AD	BC
BC	AD
BD	AC
CD	AB
ABC	D
ABD	C
ACD	B
BCD	A

Table 8.16 Aliasing Pattern for 4 Factors in 8 Runs

Most experimenters will not be overly concerned about aliasings of 3-way interactions because the likelihood of a significant 3-way interaction is small. However, when there is concern about possible 3-way interactions, we need to run a full factorial most of the time. Aliasings of 2-way interactions, however, should be of more concern. Since the 2^{4-1} design has 2-ways aliased with other 2-ways, this design may not be desirable. For example, the fact that columns AB and CD are identical (aliased) indicates that, if this column is significant, we must use prior knowledge to discern which term is to be included in the model. If our prior knowledge of factors A, B, C, and D does not allow us to rule out any one of the pairs of 2-way interactions depicted in Table 8.16, we should not run this design. Other options would be to do the 2^4 (16 run) full factorial or to hold one of the 4 factors constant and do a 2^3 (8 run) full factorial for just 3 factors. The following example illustrates the use of a fractional factorial, along with a simple graphical analysis approach.

Example 8.3 Assume you desire to improve your car's gas mileage through a 4-factor experiment involving 3 replications of each of the 8 runs. The factors are shown in Table 8.17 and the coded test matrix with response values is shown in Table 8.18.

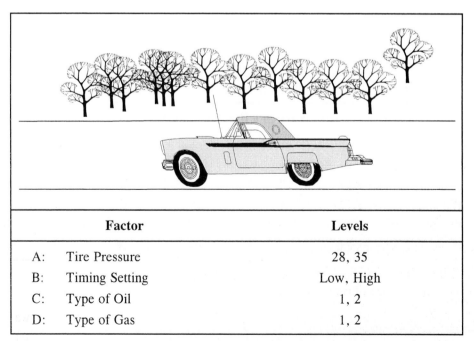

Factor	Levels
A: Tire Pressure	28, 35
B: Timing Setting	Low, High
C: Type of Oil	1, 2
D: Type of Gas	1, 2

Table 8.17 Factors for Gas Mileage Experiment

RUN	A	B	AB CD	C	AC BD	BC AD	D	y_1	y_2	y_3	$\bar{y}$	s
1	-1	-1	1	-1	1	1	-1	22.27	21.12	21.37	21.59	.60
2	-1	-1	1	1	-1	-1	1	14.22	15.40	10.46	13.36	2.58
3	-1	1	-1	-1	1	-1	1	22.49	23.15	22.08	22.57	.54
4	-1	1	-1	1	-1	1	-1	9.96	13.80	11.92	11.89	1.92
5	1	-1	-1	-1	-1	1	1	17.35	18.60	17.97	17.98	.62
6	1	-1	-1	1	1	-1	-1	27.08	24.54	24.57	25.40	1.46
7	1	1	1	-1	-1	-1	-1	18.36	17.63	17.04	17.68	.66
8	1	1	1	1	1	1	1	22.78	26.97	27.14	25.63	2.47

	A	B	AB CD	C	AC BD	BC AD	D	
ȳ Avg @ -1	17.35	19.58	19.46	19.95	15.23	19.75	19.4	
ȳ Avg @ +1	21.67	19.44	19.56	19.07	23.8	19.27	19.88	
Δ	4.32	-.14	.10	-.88	8.57	-.48	.48	
s Avg @ -1	1.41	1.31	1.13	.60	1.44	1.31	1.16	
s Avg @ +1	1.30	1.40	1.58	2.11	1.27	1.40	1.55	
δ	-.11	.09	.45	1.51	-.17	.09	.39	

$$\bar{\bar{y}} = 19.5 \text{ and } \bar{s} = 1.355$$

Table 8.18 Coded Matrix with Responses for Gas Mileage Example

A graphical analysis for each effect in the design matrix is obtained by plotting the average response at the (-1) and (1) levels for each column. After plotting these averages, the dots are connected to form a line. See Figure 8.12 for the complete graphical analysis. For example the large dot, ●, on Figure 8.12 is a result of averaging all the response values when the factor A column is at a $+1$, i.e., the average of runs 5 through 8 or

$$\frac{17.98 + 25.40 + 17.68 + 25.63}{4} = 21.67.$$

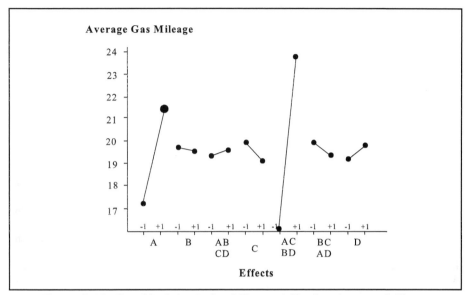

Figure 8.12 Graphical Analysis of Terms Affecting Average Mileage

The steepness of the slope reflects the importance of each effect. From the graphical results in Figure 8.12, it is apparent that the most important effects are factor A and the AC or BD interaction. Since AC is aliased with BD, we must go back to the definition of the factors to determine which interaction is most likely responsible for the steep slope in Figure 8.12. Since it is highly unlikely to have a "tire pressure x type of oil" interaction (AC), it is concluded that the steep slope in Figure 8.12 is due to the BD, or "timing x type of gas," interaction. The BD interaction graph is shown in Figure 8.13. Notice that the large star, ★, is plotted as the average y when B = 1 and D = 1, i.e., the average response values for runs 3 and 8 which is computed as follows:

$$\frac{22.57 + 25.63}{2} = 24.10.$$

The other plotted points are obtained similarly.

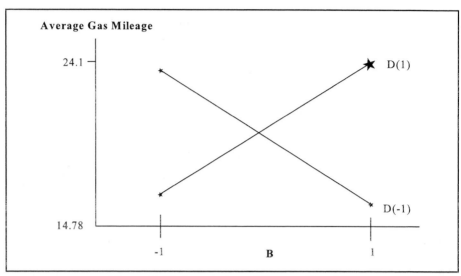

Figure 8.13 Interaction of Factors B and D

A similar graphical analysis is used to determine if any of the factors shift the variability in the gas mileage.   Figure 8.14 indicates that factor C has a big effect on variability.

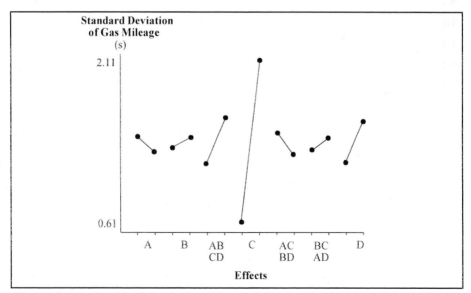

Figure 8.14 Graphical Analysis of Terms Affecting Variability of Gas Mileage

To determine the best factor settings for maximizing gas mileage while minimizing variability, we must return to Figures 8.12, 8.13, and 8.14. Based on Figure 8.12, set factor A at (1). Figure 8.13 indicates that factors B and D should be set at (1). Figure 8.14 indicates that factor C should be set at (−1) to minimize variability. Thus, the final decision on setting the factors is as follows:

Factor	Coded Setting	True Setting
A	1	35 psi
B	1	high timing
C	-1	type 1 oil
D	1	type 2 gas

ANOVA and regression analyses like those in Tables 8.8, 8.9 and 8.10 could also be performed on the gas mileage data (see Problem 8.3).

Example 8.4 To reduce the amount of shrinkage in an injection molded part, an experiment with four factors at 2 levels was performed. Only two interactions, AxC and AxD, were considered worthy of investigation. Therefore, the half fraction was chosen over the full factorial to conserve runs. A sketch of the molding scenario is shown in Figure 8.15. The experimental settings for each factor and the results of the 2^{4-1} fractional factorial design with 4 replications are shown in Table 8.19.

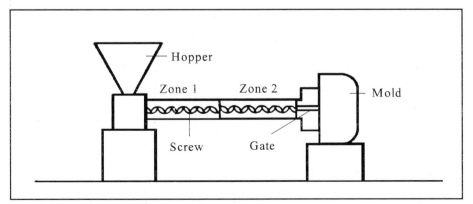

Figure 8.15 Injection Molding Machine

FACTOR	LOW (-)	HIGH (+)
A: Set Time	5	10
B: Zone 1 Temp	170	190
C: Zone 2 Temp	170	190
D: Mold Temp	150	160

	A	B	AB CD	C	AC BD	AD BC	D= ABC						
RUN	1	2	3	4	5	6	7	y_1	y_2	y_3	y_4	$\bar{y}$	s
1	-	-	+	-	+	+	-	6.4	5.5	6	5.2	5.775	0.532
2	-	-	+	+	-	-	+	6.9	5.5	5.7	6.2	6.075	0.624
3	-	+	-	-	+	-	+	6.5	6.3	5.7	4.7	5.8	0.808
4	-	+	-	+	-	+	-	5.9	5.7	6.9	6.6	6.275	0.568
5	+	-	-	-	-	+	+	2.7	3.4	3.5	2.9	3.125	0.386
6	+	-	-	+	+	-	-	4	4.1	4.2	3.8	4.025	0.171
7	+	+	+	-	-	-	-	3.2	3.1	3.6	2.9	3.2	0.294
8	+	+	+	+	+	+	+	4.9	4.9	4.7	4.8	4.825	0.096

$\bar{y}$ Avg @ -1	5.98	4.75	4.80	4.47	4.67	4.77	4.82
$\bar{y}$ Avg @ +1	3.79	5.02	4.97	5.30	5.10	5.00	4.95
Δ	-2.19	.27	.17	.83	.43	.23	.13

$$\bar{\bar{y}} = 4.887 \qquad \bar{s} = .435$$

s Avg @ -1	.633	.428	.483	.505	.468	.474	.391
s Avg @ +1	.237	.441	.386	.365	.402	.395	.475
δ	-.396	.013	-.097	-.14	-.066	-.079	.084

Table 8.19 Matrix and Response Values for Example 8.4

(- implies -1 and + implies +1)

The simplified ŝ and ŷ models, shown below, were obtained from the DOE KISS regression results shown in Tables 8.20 and 8.21, respectively. Note that in using the ROT to detect significant factors for the ŝ model, namely "$|\text{coef}| \geq \left(\dfrac{\text{constant}}{2}\right)$," we see that $|\text{coef}|$ for A is 0.198 which is close to but not at least half of 0.4348. However, ROTs also require judgment and by looking at the size of the coefficients, we see that the coefficient for A is significantly larger than all of the other coefficients. Hence, A was deemed to be significant and was included in the ŝ model. Confirmation runs would indeed verify this decision. Using the ROT for ŷ models, namely "P(2 TAIL) < .05," we see that the only significant effects are from A, C, and AC, which are all included in the ŷ model.

ŝ = .4348 - .198A

ŷ = 4.8875 - 1.0938A + .4125C + .2188AC

FACTOR	COEF	P(2 TAIL)	TOL	LOW	HIGH	EXPER	ACTIVE
Constant	0.4348	Not Avail					
A	-0.1980	Not Avail	1	5	10	5	X
B	0.0067	Not Avail	1	170	190	170	X
C	-0.0703	Not Avail	1	170	190	170	X
D	0.0437	Not Avail	1	150	160	150	X
AB	-0.0485	Not Avail	1				X
AC	-0.0333	Not Avail	1				X
BC	-0.0395	Not Avail	1				X
R Sq	1.0000						
Adj R Sq	Not Avail						
Std Error	0.0000						
F	0.0000			**PRED S**			
Sig F	Not Avail						

Table 8.20 Regression Output for ŝ Model

FACTOR	COEF	P(2 TAIL)	TOL	LOW	HIGH	EXPER	ACTIVE
Constant	4.8875	*0.0000*					
A	-1.0938	*0.0000*	1	5	10	5	X
B	0.1375	0.1256	1	170	190	170	X
C	0.4125	*0.0001*	1	170	190	170	X
D	0.0688	0.4353	1	150	160	150	X
AB	0.0813	0.3577	1				X
AC	0.2188	*0.0186*	1				X
BC	0.1125	0.2065	1				X
R Sq	0.8900						
Adj R Sq	0.8579						
Std Error	0.4901						
F	27.7319			**PRED Y**			
Sig F	0.0000						

Table 8.21 Regression Output for ŷ Model

The next step is to choose the best settings for factors A, B, C, and D to minimize both ŝ and ŷ. Starting with ŝ, the best setting for A is +1 coded (10 actual). This setting for A will result in a predicted

$$\hat{s} = .4348 - .198(1) = .2368.$$

Now to minimize ŷ, with A already set to +1 to minimize ŝ, we have

$$\hat{y} = 4.8875 - 1.0938(1) + .4125C + .2188(1)C = 3.7937 + .6313C$$

which is minimized when C = -1 coded (170 actual). Thus, when C = -1

$$\hat{y} = 3.7937 + .6313(-1) = 3.1624.$$

The next question should be: "Do these predicted $\hat{y}$ and $\hat{s}$ values at the so called optimal settings of $A_{+1}C_{-1}$ produce results that will satisfy the customer?" If the answer is no, we might consider setting A larger than $+1$ coded and/or C less than -1 coded (this is called extrapolating). The cheapest approach would be to reduce C because a larger A would increase cycle time and reduce our throughput. Once the decision on C and A is made, we should predict the ideal $\hat{y}$ and $\hat{s}$ and then return to the process of confirming our predictions. The easiest way to check for confirmation is to build the prediction interval $\hat{y} \pm 3\hat{s}$ which should contain all confirmation response values. If the confirmation runs produce a response value clearly outside the prediction interval, we say that we have "failed to confirm" our models.

Failure to confirm should lead to investigating the following:

(1)   lack of experimental discipline,

(2)   measurement error,

(3)   other errors, like transfer or math errors, and

(4)   too much variability in the y (response) values.

If these four items are not under control prior to experimenting, the best advice is to not experiment. If they can be ruled out as reasons for lack of confirmation, other areas can be investigated. These include something changing between the initial matrix experiment and the confirmation runs, lack of linearity (i.e., presence of a non-linear relationship), aliased terms, too few or too many terms in a model, etc.

∎

If we want to test 5 factors each at 2 levels, the most commonly used design matrix is the 2^{5-1} or 16-run matrix shown in Table 8.22. In this design we alias the fifth factor, E, with the 4-way interaction of ABCD. Notice that 2-way interactions are aliased with 3-way interactions. Since the 2-way interactions are the more likely to be important, we consider only the 2-way terms for inclusion in the model. As previously stated, if 3-way interactions are of concern, we would typically run a full factorial with either 3 or 4 factors while holding the others constant at some level based on prior knowledge, cost, convenience, etc.

RUN	1 A	2 B	3 AB	4 C	5 AC	6 BC	7 ABC DE	8 D	9 AD	10 BD	11 ABD CE	12 CD	13 ACD BE	14 BCD AE	15 E	y_1 y_2 y_3	$\bar{y}$	s
1	-1	-1	1	-1	1	1	-1	-1	1	1	-1	1	-1	-1	1			
2	-1	-1	1	-1	1	1	-1	1	-1	-1	1	-1	1	1	-1			
3	-1	-1	1	1	-1	-1	1	-1	1	1	-1	-1	1	1	-1			
4	-1	-1	1	1	-1	-1	1	1	-1	-1	1	1	-1	-1	1			
5	-1	1	-1	-1	1	-1	1	-1	1	-1	1	1	-1	1	-1			
6	-1	1	-1	-1	1	-1	1	1	-1	1	-1	-1	1	-1	1			
7	-1	1	-1	1	-1	1	-1	-1	1	-1	1	-1	1	-1	1			
8	-1	1	-1	1	-1	1	-1	1	-1	1	-1	1	-1	1	-1			
9	1	-1	-1	-1	-1	1	1	-1	-1	1	1	1	1	-1	-1			
10	1	-1	-1	-1	-1	1	1	1	1	-1	-1	-1	-1	1	1			
11	1	-1	-1	1	1	-1	-1	-1	-1	1	1	-1	-1	1	1			
12	1	-1	-1	1	1	-1	-1	1	1	-1	-1	1	1	-1	-1			
13	1	1	1	-1	-1	-1	-1	-1	-1	-1	-1	1	1	1	1			
14	1	1	1	-1	-1	-1	-1	1	1	1	1	-1	-1	-1	-1			
15	1	1	1	1	1	1	1	-1	-1	-1	-1	-1	-1	-1	-1			
16	1	1	1	1	1	1	1	1	1	1	1	1	1	1	1			
$\bar{y}$ Avg @ -1																		
$\bar{y}$ Avg @ +1																		
Δ																		

s Avg @ -1	
s Avg @ +1	
δ	

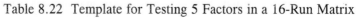
Table 8.22 Template for Testing 5 Factors in a 16-Run Matrix

8.5 Screening Designs

Whenever there are many factors with which to experiment, we often cannot afford to run enough tests to develop a model that includes interactions. As a Rule of Thumb, we recommend that whenever the number of factors (k) to be tested is greater than or equal to 6, we should consider doing a screening experiment first and then follow that with a modeling experiment if needed. In screening experiments we simply desire to partition our factors into the "vital few" and "trivial many." The "vital few" are considered to be the most important factors and if a subsequent modeling experiment is desired, these "vital few" will be tested in the modeling design. The "trivial many" are considered to be the less important factors which will typically be set based on economics, convenience, etc., and held constant during any future experiments on the vital few.

The 2-level design types used for screening are the Taguchi L_8, L_{12}, and L_{16} designs. These designs are also referred to as Plackett-Burman, Latin Square, and/or Hadamard matrix designs of 8, 12, or 16 runs [Schm 94]. Tables 8.23, 8.24, and 8.25 can be used as templates for screening 6 to 15 factors.

				L_8 Design										
Run #	1	2	3	4	5	6	7	y_1	y_2	y_3	y_4	y_5	$\bar{y}$	s
1	-1	-1	-1	-1	-1	-1	-1							
2	-1	-1	-1	+1	+1	+1	+1							
3	-1	+1	+1	-1	-1	+1	+1							
4	-1	+1	+1	+1	+1	-1	-1							
5	+1	-1	+1	-1	+1	-1	+1							
6	+1	-1	+1	+1	-1	+1	-1							
7	+1	+1	-1	-1	+1	+1	-1							
8	+1	+1	-1	+1	-1	-1	+1							
$\bar{y}$ Avg@ -1														
$\bar{y}$ Avg@ +1														
Δ														
s Avg@ -1														
s Avg@ +1														
δ														

Table 8.23 Taguchi L_8 Screening Design

The L_8 design can be used to screen 6 or 7 factors. However, the 2-way interactions are aliased with main effects or other 2-way interactions, so using an L_8 design to screen 6 or 7 factors will not allow us to estimate interaction effects. But recall that screening is not designed to measure interaction effects, only the "vital few" main effects.

Run #	1	2	3	4	5	6	7	8	9	10	11	y_1	y_2	y_3	y_4	$\bar{y}$	s
					L_{12} Design												
1	-1	-1	-1	-1	-1	-1	-1	-1	-1	-1	-1						
2	-1	-1	-1	-1	-1	+1	+1	+1	+1	+1	+1						
3	-1	-1	+1	+1	+1	-1	-1	-1	+1	+1	+1						
4	-1	+1	-1	+1	+1	-1	+1	+1	-1	-1	+1						
5	-1	+1	+1	-1	+1	+1	-1	+1	-1	+1	-1						
6	-1	+1	+1	+1	-1	+1	+1	-1	+1	-1	-1						
7	+1	-1	+1	+1	-1	-1	+1	+1	-1	+1	-1						
8	+1	-1	+1	-1	+1	+1	+1	-1	-1	-1	+1						
9	+1	-1	-1	+1	+1	+1	-1	+1	+1	-1	-1						
10	+1	+1	+1	-1	-1	-1	-1	+1	+1	-1	+1						
11	+1	+1	-1	+1	-1	+1	-1	-1	-1	+1	+1						
12	+1	+1	-1	-1	+1	-1	+1	-1	+1	+1	-1						
$\bar{y}$ Avg @ -1																	
$\bar{y}$ Avg @ +1																	
Δ																	
s Avg @ -1																	
s Avg @ +1																	
δ																	

Table 8.24 Taguchi L_{12} Screening Design

The L_{12} screening design in Table 8.24 can be used to screen up to 11 factors. In this design, however, 2-way interactions are not aliased with main factors; instead, they are partially confounded with all of the other main effects. Because of the partial confounding in the L_{12}, many experimenters prefer this design over the aliasing (or total confounding) that takes place in the L_8. Current research indicates that, under certain circumstances, the L_{12} design may be used to model up to 4 main effects and 3 interaction effects. Reference the software package **DOE KISS with Stealth Regression** which automates this process. It should be remembered that any screening design is sensitive to large numbers of interactions. If the user suspects a large number of interactions, it may be advantageous to run modeling designs on subgroups of the factors that we originally had planned to screen.

The L_{16} design shown in Table 8.25 can be used to screen up to 15 factors. This design will have 2-way interactions aliased with other 2-ways and, if screening more than 8 factors, 2-ways will be aliased with main effects. For any screening design in which not all of the columns are used to test a factor, the columns not assigned a factor are ignored and/or used to estimate error/noise.

Run #	1	2	3	4	5	6	7	8	9	10	11	12	13	14	15	y_1	y_2	y_3	$\bar{y}$	s
							L_{16} Design													
1	-1	-1	-1	-1	-1	-1	-1	-1	-1	-1	-1	-1	-1	-1	-1					
2	-1	-1	-1	-1	-1	-1	-1	+1	+1	+1	+1	+1	+1	+1	+1					
3	-1	-1	-1	+1	+1	+1	+1	-1	-1	-1	-1	+1	+1	+1	+1					
4	-1	-1	-1	+1	+1	+1	+1	+1	+1	+1	+1	-1	-1	-1	-1					
5	-1	+1	+1	-1	-1	+1	+1	-1	-1	+1	+1	-1	-1	+1	+1					
6	-1	+1	+1	-1	-1	+1	+1	+1	+1	-1	-1	+1	+1	-1	-1					
7	-1	+1	+1	+1	+1	-1	-1	-1	-1	+1	+1	+1	+1	-1	-1					
8	-1	+1	+1	+1	+1	-1	-1	+1	+1	-1	-1	-1	-1	+1	+1					
9	+1	-1	+1	-1	+1	-1	+1	-1	+1	-1	+1	-1	+1	-1	+1					
10	+1	-1	+1	-1	+1	-1	+1	+1	-1	+1	-1	+1	-1	+1	-1					
11	+1	-1	+1	+1	-1	+1	-1	-1	+1	-1	+1	+1	-1	+1	-1					
12	+1	-1	+1	+1	-1	+1	-1	+1	-1	+1	-1	-1	+1	-1	+1					
13	+1	+1	-1	-1	+1	+1	-1	-1	+1	+1	-1	-1	+1	+1	-1					
14	+1	+1	-1	-1	+1	+1	-1	+1	-1	-1	+1	+1	-1	-1	+1					
15	+1	+1	-1	+1	-1	-1	+1	-1	+1	+1	-1	+1	-1	-1	+1					
16	+1	+1	-1	+1	-1	-1	+1	+1	-1	-1	+1	-1	+1	+1	-1					
$\bar{y}$ Avg@ -1																				
$\bar{y}$ Avg@ +1																				
Δ																				
s Avg@ -1																				
s Avg@ +1																				
δ																				

Table 8.25 Taguchi L_{16} Screening Design

(For 8 or less factors use columns 1, 2, 4, 7, 8, 11, 13 and 14)

To analyze screening designs, we typically select the factors with the large coefficients or small p-values as the "vital few" and assume the remainder are the "trivial many." When precise models of a process are desired, the screening experiment helps us find the "vital few" which then must be experimented with a full or fractional factorial to obtain a model. The next example illustrates the use of an L_{12} design to screen 7 factors.

Example 8.5 Consider the process of producing the small cylindrical protective mechanism that houses the solid explosive material used to inflate the air bag in an automobile. The performance measure is the diameter of the cylinder and the target value is 800. The following L_{12} screening design with 4 replicates was used to screen 7 factors. See Table 8.26.

Row #	A	B	C	D	E	F	G	y_1	y_2	y_3	y_4	$\bar{y}$	s
1	-1	-1	-1	-1	-1	-1	-1	803.00	800.77	804.64	799.34	801.939	2.347
2	-1	-1	-1	-1	-1	1	1	806.31	804.80	807.19	803.80	805.525	1.515
3	-1	-1	1	1	1	-1	-1	806.89	795.18	797.31	809.94	802.330	7.191
4	-1	1	-1	1	1	-1	1	805.49	795.47	794.50	804.59	800.013	5.831
5	-1	1	1	-1	1	1	-1	802.29	801.69	799.96	802.94	801.720	1.277
6	-1	1	1	1	-1	1	1	811.38	798.87	811.01	800.78	805.511	6.612
7	1	-1	1	1	-1	-1	1	795.73	794.57	801.15	794.03	796.369	3.262
8	1	-1	1	-1	1	1	1	801.36	802.22	798.58	800.09	800.560	1.585
9	1	-1	-1	1	1	1	-1	792.32	799.13	803.69	804.33	799.869	5.538
10	1	1	1	-1	-1	-1	-1	803.23	802.30	798.00	800.21	800.935	2.328
11	1	1	-1	1	-1	1	-1	806.09	801.04	806.97	805.88	804.993	2.677
12	1	1	-1	-1	1	-1	1	799.02	796.58	796.61	800.55	798.191	1.946

Table 8.26 L_{12} Design for Screening 7 Factors

The regression analysis results from DOE KISS are shown in Tables 8.27 and 8.28 for s and y, respectively.

FACTOR	COEF	P(2 TAIL)	TOL	LOW	HIGH	EXPER	ACTIVE
Constant	3.5090	0.0014					
A	-0.6196	0.2358	1	-1	1	-1	X
B	-0.0639	0.8927	1	-1	1	-1	X
C	0.2000	0.6761	1	-1	1	-1	X
D	1.6761	0.0196	1	-1	1	-1	X
E	0.3855	0.4348	1	-1	1	-1	X
F	-0.3083	0.5261	1	-1	1	-1	X
G	-0.0505	0.9150	1	-1	1	-1	X
R Sq	0.8150						
Adj R Sq	0.4914						
Std Error	1.5400						
F	2.5181		PRED S				
Sig F	0.1948						

Table 8.27 Regression Analysis for s Using the L_{12} Design in Table 8.26

The results for s indicate that factor D (p-value $<$ .10) is the critical factor and key to variance reduction. The positive coefficient for D indicates that D should be set low (coded -1) or possibly even extrapolated in the low direction to reduce variability. For any follow-up experimentation, D should be held constant at a low setting.

FACTOR	COEF	P(2 TAIL)	TOL	LOW	HIGH	EXPER	ACTIVE
Constant	801.4963	*0.0000*					
A	-1.3435	*0.0293*	1	-1	1	1	X
B	0.3976	0.5074	1	-1	1	0	X
C	-0.2587	0.6658	1	-1	1	0	X
D	0.0179	0.9762	1	-1	1	0	X
E	-1.0491	0.0853	1	-1	1	0.1	X
F	1.5335	0.0137	1	-1	1	0	X
G	-0.4681	0.4357	1	-1	1	0	X
R Sq	0.2874						
Adj R Sq	0.1627						
Std Error	4.1188						
F	2.3046		PRED Y				
Sig F	0.0451						

Table 8.28 Regression Analysis for y Using the L_{12} Design in Table 8.26

In Table 8.28, we see from the p-values that the factors which significantly affect y are A, E, and F. These are the key variables that we can use to move y to the target of 800. Any further experimentation to detect significant interaction effects and to build a model would involve the factors A, E, and F, while factor D would be held constant at a low setting and factor B, C, and G would be held constant at a level that is most economical.

∎

8.6 Three-Level Designs For Non-Linear (Quadratic) Models

Up to this point the design types presented have been for evaluating linear and interaction effects of factors tested at only 2 levels. It is possible that quantitative factors (time, temp, press, ph, speed, force, etc.) may be related to the output(s) in a non-linear (quadratic) type relationship. To build a non-linear (quadratic) model relating an output to some quantitative factors we must

(1)   test each factor at 3 levels and

(2)   use software to develop the regression model because the $\Delta/2$ technique no longer applies.

Fortunately, a commonly used 3-level design matrix will include what was already presented as a 2-level design plus some center points and some axial points. For example, suppose we have 2 variables to test and they are both quantitative. The 2-level portion of the design would be the full factorial shown as the first 4 runs in Table 8.29. Columns A and B are the only columns needing to be considered at this point. Also included in this design are some center points (see runs 5 and 6 in Table 8.29) and some axial points (see runs 7, 8, 9, and 10 in Table 8.29). The extra run at the center (A = 0, B = 0) is added to improve the statistical properties of the model. The number of replications or sample size of 9 shown in the table is based on the 2-level portion of the design. Sample size Rules of Thumb will be discussed in Section 8.8.

RUN	A	B	A·B	A²	B²	y₁	y₂	y₃	y₄	y₅	y₆	y₇	y₈	y₉	ȳ	s
1	-1	-1	1	1	1											
2	-1	1	-1	1	1											
3	1	-1	-1	1	1											
4	1	1	1	1	1											
5	0	0	0	0	0											
6	0	0	0	0	0											
7	-1	0	0	1	0											
8	1	0	0	1	0											
9	0	-1	0	0	1											
10	0	1	0	0	1											

Table 8.29 Three Level Non-Linear Modeling Design for 2 Quantitative Factors

If we have 3 factors, the matrix would appear as shown in Table 8.30.

RUN	A	B	AB	C	AC	BC	ABC	A²	B²	C²	y₁	y₂	y₃	y₄	y₅	ȳ	s
1	-1	-1	1	-1	1	1	-1	1	1	1							
2	-1	-1	1	1	-1	-1	1	1	1	1							
3	-1	1	-1	-1	1	-1	1	1	1	1							
4	-1	1	-1	1	-1	1	-1	1	1	1							
5	1	-1	-1	-1	-1	1	1	1	1	1							
6	1	-1	-1	1	1	-1	-1	1	1	1							
7	1	1	1	-1	-1	-1	-1	1	1	1							
8	1	1	1	1	1	1	1	1	1	1							
9	0	0	0	0	0	0	0	0	0	0							
10	0	0	0	0	0	0	0	0	0	0							
11	-1	0	0	0	0	0	0	1	0	0							
12	1	0	0	0	0	0	0	1	0	0							
13	0	-1	0	0	0	0	0	0	1	0							
14	0	1	0	0	0	0	0	0	1	0							
15	0	0	0	-1	0	0	0	0	0	1							
16	0	0	0	1	0	0	0	0	0	1							

Table 8.30 Three Level Non-Linear Modeling Design
for 3 Quantitative Factors

The designs shown in Tables 8.29 and 8.30 are a type of Central Composite Design discussed in [Schm 94]. Obviously, to use Table 8.29 or 8.30 we need to use only the main factor columns for running the experiment while randomizing the run order. After the data is collected we will use regression to model the non-linear equations for $\hat{y}$ and $\hat{s}$.

Example 8.6 Assume you are a biochemist who needs to develop a new blood testing assay. The factors you desire to model are pH and concentration of a reagent. The response is signal strength measured by way of an optical reader. The experimental matrix and response values are shown in Table 8.31.

	pH	Conc												
RUN	**A**	**B**	**AB**	**A²**	**B²**	y_1	y_2	y_3	y_4	y_5	y_6	y_7	$\bar{y}$	**s**
1	-1	-1	1	1	1	195	199	195	194	196	200	185	194.9	4.880
2	-1	1	-1	1	1	180	181	182	180	181	178	179	180.1	1.345
3	1	-1	-1	1	1	235	229	240	233	237	235	236	235.0	3.416
4	1	1	1	1	1	222	220	221	222	223	220	221	221.3	1.113
5	0	0	0	0	0	218	219	220	216	218	218	217	218.0	1.291
6	0	0	0	0	0	218	219	218	216	220	218	218	218.1	1.215
7	-1	0	0	1	0	197	199	195	196	197	194	199	196.7	1.890
8	1	0	0	1	0	224	229	226	222	224	227	229	225.9	2.673
9	0	-1	0	0	1	230	231	229	230	230	229	231	230.0	0.816
10	0	1	0	0	1	215	219	210	212	209	223	216	214.9	5.014

Table 8.31 Matrix and Response Values for Example 8.6

The $\hat{s}$ model analysis is summarized in Table 8.32.

FACTOR	COEF	P(2 TAIL)	TOL	LOW	HIGH	EXPER	ACTIVE
Constant	1.6117	0.2800					
A	-0.1523	0.8713	1.000	-1	1	-1	X
B	-0.2732	0.7722	1.000	-1	1	-1	X
AB	0.3078	0.7898	1.000				X
AA	0.3109	0.8368	0.972				X
BB	0.9450	0.5406	0.972				X
R Sq	0.1608						
Adj R Sq	-0.8883						
Std Error	2.1605						
F	0.1532			**PRED S**			
Sig F	0.9679						

Table 8.32 Regression Output for s Using Table 8.31 Data

It is important to note that 3-level designs will have P(2-tail) values for $\hat{s}$ models and the old Rule of Thumb of P(2-tail) $\leq$ .10 applies. In addition, the constant in a 3-level design is not the same as $\bar{s}$.

The $\hat{y}$ model regression analysis is presented in Table 8.33.

FACTOR	COEF	P(2 TAIL)	TOL	LOW	HIGH	EXPER	ACTIVE
Constant	220.3061	*0.0000*					
A	18.4048	*0.0000*	1.000	-1	1	-1	X
B	-7.2619	*0.0000*	1.000	-1	1	-1	X
AB	0.2500	0.7420	1.000				X
AA	-11.2551	*0.0000*	0.972				X
BB	-0.1122	0.9121	0.972				X
R Sq	0.9477						
Adj R Sq	0.9437						
Std Error	4.0010						
F	232.1219			**PRED S**			
Sig F	0.0000						

Table 8.33 Regression Output for $\hat{y}$ Using Table 8.31 Data

Using the results from Tables 8.32 and 8.33 the two reduced models describing our process are:

$$\hat{s} = 1.6117 \text{ (because there are no significant effects)}$$

$$\hat{y} = 220.306 + 18.405A - 7.262B - 11.255A^2$$

The objectives in this experiment are to minimize $\hat{s}$ and maximize $\hat{y}$. Thus, we should set A $=$ +1 and B $=$ -1 for a resulting $\hat{s} = 1.6117$ and $\hat{y} = 234.718$. Obviously, the next step is to confirm the experimental results and also test for lot-to-lot consistency on the reagents, i.e., confirm over multiple lots of raw material.

If we were to test for consistency across lots and find that there exist severe lot differences we now have two choices:

1)   measure different aspects of the lots to find out what is different and then set up a procedure to insure incoming lot consistency

or

2)   conduct the same experiment from Table 8.31 or one similar with replicates from multiple lots to seek a combination of the factors that will provide robustness (insensitivity) to lot-to-lot variability.

The second option has been made popular by the works of Taguchi and is discussed in Section 8.9.

■

The Central Composite Designs (CCDs) are very flexible and efficient for modeling quantitative factors having a non-linear effect on the response. As described here, a CCD design could be implemented in stages:

1)   Conduct the factorial or 2-level portion first and build a linear model using these results.

2)   Use the center points as confirmatory runs.

3)   If we don't confirm, collect data at the axial points and use all of
the data (factorial, center, and axial) to build a non-linear or
quadratic model.

We have shown CCDs for two and three factors. CCDs can also be run for 4, 5,
and sometimes even 6 factors. These designs are available in DOE KISS and the
reader is referred to that software for the design setups.

Another efficient and frequently used 3-level design for modeling
quantitative factors is the Box-Behnken design. Like the CCD, this design also uses
center points. Box-Behnken designs consist of a sequence of embedded 2^2 designs
while holding the other factor(s) at their center points. This structure is apparent
in the 3 and 4-factor Box-Behnken designs displayed in Tables 8.34 and 8.35,
respectively.

Run	A	B	C	y_1 ... y_5	$\bar{y}$	s
1	-	-	0			
2	-	+	0			
3	+	-	0			
4	+	+	0			
5	-	0	-			
6	-	0	+			
7	+	0	-			
8	+	0	+			
9	0	-	-			
10	0	-	+			
11	0	+	-			
12	0	+	+			
13	0	0	0			
14	0	0	0			
15	0	0	0			

Table 8.34 3-Factor Box-Behnken Design

Run	A	B	C	D	y_1 ... y_3	$\bar{y}$	s
1	-	-	0	0			
2	-	+	0	0			
3	+	-	0	0			
4	+	+	0	0			
5	0	0	-	-			
6	0	0	-	+			
7	0	0	+	-			
8	0	0	+	+			
9	0	0	0	0			
10	-	0	0	-			
11	-	0	0	+			
12	+	0	0	-			
13	+	0	0	+			
14	0	-	-	0			
15	0	-	+	0			
16	0	+	-	0			
17	0	+	+	0			
18	0	0	0	0			
19	-	0	-	0			
20	-	0	+	0			
21	+	0	-	0			
22	+	0	+	0			
23	0	-	0	-			
24	0	-	0	+			
25	0	+	0	-			
26	0	+	0	+			
27	0	0	0	0			

Table 8.35 4-Factor Box-Behnken Design

While the CCD and Box-Behnken designs are 3-level modeling designs for quantitative factors, the next section addresses a screening design that can be used with both quantitative and qualitative factors.

8.7 Three-Level Screening Design

The design of choice for screening factors at 3 levels is the Taguchi L_{18} design. As Table 8.36 indicates, this design has 18 runs and 8 columns. The first column has only two levels associated with it, so one factor at 2 levels can be screened along with 7 other factors each at 3 levels. This design, like the L_{12}, has partial confounding among the columns so interaction effects are difficult to estimate.

					L_{18} Design						
Run	1	2	3	4	5	6	7	8	y_1 ...... y_4	$\bar{y}$	s
1	-1	-1	-1	-1	-1	-1	-1	-1			
2	-1	-1	0	0	0	0	0	0			
3	-1	-1	+1	+1	+1	+1	+1	+1			
4	-1	0	-1	-1	0	0	+1	1			
5	-1	0	0	0	+1	+1	-1	-1			
6	-1	0	+1	+1	-1	-1	0	0			
7	-1	+1	-1	0	-1	+1	0	1			
8	-1	+1	0	+1	0	-1	+1	-1			
9	-1	+1	+1	-1	+1	0	-1	0			
10	+1	-1	-1	+1	+1	0	0	-1			
11	+1	-1	0	-1	-1	+1	+1	0			
12	+1	-1	+1	0	0	-1	-1	1			
13	+1	0	-1	0	+1	-1	+1	0			
14	+1	0	0	+1	-1	0	-1	+1			
15	+1	0	+1	-1	0	+1	0	-1			
16	+1	+1	-1	+1	0	+1	-1	0			
17	+1	+1	0	-1	+1	-1	0	+1			
18	+1	+1	+1	0	-1	0	+1	-1			

Table 8.36 Taguchi L_{18} Design

However, the interaction between columns 1 and 2 can be estimated cleanly. Current research indicates that if the interaction and quadratic effects are sparse, the L_{18} can also generate reliable models. This technique is automated in the software ***DOE KISS with Stealth Regression***.

The L_{18} can also be used to evaluate the effect of qualitative factors. Qualitative factors are variables such as machine type, day of the week, site, method, or operator, etc. The L_{18} can evaluate a mix of qualitative and quantitative factors and the results can be easily analyzed by way of graphical plots. The next example illustrates these points.

Example 8.7 Consider the process of a chemical dispense system in which 8 factors are to be evaluated as to their effect on dispense volume (y). The input-process-output (IPO) diagram is shown in Figure 8.16.

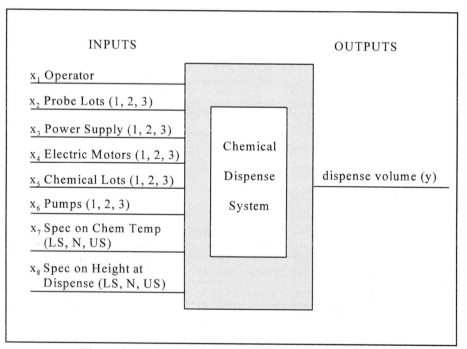

Figure 8.16 Chemical Dispense System IPO Diagram

An L_{18} design with 4 replications was run on the 8 input factors.   Note that $x_1, x_2, \ldots, x_6$ are qualitative factors and that x_7 and x_8 are quantitative factors.   The results of this experiment are displayed graphically in Figure 8.17.

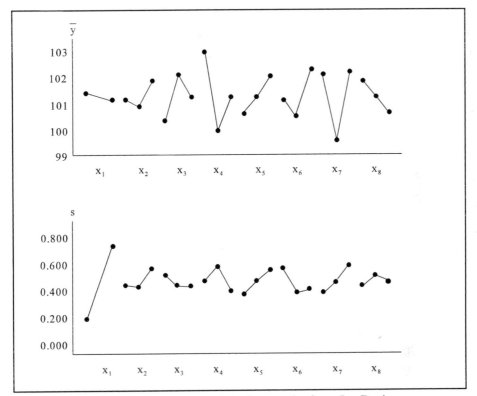

Figure 8.17 Graphical Plots for $\bar{y}$ and s from L_{18} Design

Looking at the results in Figure 8.17, one can quickly grasp which factors have the biggest effect on $\bar{y}$ and on s.   Factors x_4 (Electric Motors) and x_7 (Chemical Temperature) have the greatest impact on shifting the average.   While we cannot say that x_4 has a quadratic effect on y (because x_4 is a qualitative factor), we can say that x_7 (a quantitative factor) does have a quadratic effect on y.   The other quantitative factor, x_8, does not have a quadratic effect on y.   The s plot shows quite clearly that the greatest source of variability is from x_1 (Operator).

8.8 Selecting a Design and a Sample Size

When faced with the need to experiment, we sometimes get caught up in the statistics and forget about everything else that is important for successful experimentation. In order to reduce the amount of time needed to choose a design type and sample size, we have provided the following Rules of Thumb. The sample sizes indicated are for estimating ŝ models with a 95% chance of finding a variance shifting factor, should it exist. If we are only interested in modeling y, a good Rule of Thumb is to do no less than half of what is required to estimate ŝ. For more information on these Rules of Thumb see [Schm 94].

Rules of Thumb for Two-Level Design Choices for the Number of Factors (k) = 2, 3, 4, 5, and greater than 5

a)   k = 2: The only option is a full factorial (See Table 8.4), where $n = 2^2$ = 4 experimental conditions (runs). The ideal sample size per condition (run) is 9. Thus the total resources to conduct the ideal experiment of 2 factors at two levels is 36.

b)   k = 3: The best option is to run a full factorial (see Table 8.8), where n $= 2^3 = 8$ with 5 samples per condition (run). Thus, the total resources required would be 40.

c)   k = 4: Option 1, the full factorial (see Table 8.12), will have n $= 2^4 =$ 16 runs, each with a sample size of 3. Total resources will be 48.

Option 2 is a ½ fraction (see Examples 8.3 and 8.4), where n $= 2^{4-1} = 8$ and the sample size per run is 5. Total resources is 40.

Option 2 will save us 8 resources compared to Option 1. Of the 6 total two-way interactions, Option 1 will evaluate them all and Option 2 will evaluate half of them. Which option to choose will typically depend on the cost per resource and the number of anticipated interactions.

d)   k = 5: Option 1 is a full factorial with n $= 2^5 = 32$ runs, each replicated 2 times for a total of 64 resources.

Option 2 is a ½ fraction (See Table 8.22), where n $= 2^{5-1} = 16$ runs with 3 replications. Total resources will be 48.

In this case, Option 2 is usually the preferred choice because 16 resources are saved and all two-way interactions are still evaluated.

e)   k ≥ 6: In this case, the experimenter will usually conduct a screening experiment followed by a subsequent experiment on the important factors. Two-level design types for screening and ideal sample size are given below for various numbers of factors.

k	Design Type	Sample Size	Comments
6, 7	Taguchi L_8	5	- 2 ways aliased with mains
6 - 11	Taguchi L_{12}	4	- 2 ways partially confounded with mains; sensitive to large numbers of interactions
6 - 15	Taguchi L_{16}	3	- for k ≤ 8, 2 ways unconfounded with mains; - for k ≥ 9, 2 ways are perfectly confounded with mains
12 - 19	Plackett-Burman 20 see [Schm 94]	3	- 2 ways partially confounded with mains

Rules of Thumb for Three-Level Design Choices for the
Number of Factors (k) = 2, 3, 4, 5, and greater than 5

a)   For k ≤ 3 qualitative factors use a full factorial, i.e.,
k = 2 will have n = 9 with 7 replications and

k = 3 will have n = 27 with 3 replications.

b)   For k ≥ 4 qualitative (or some qualitative and some quantitative)
factors use the L_{18} for screening up to 7 factors at 3 levels and
1 factor at 2 levels. The minimum recommended number of
replicates is 4.

c)   For k ≤ 5 quantitative factors use either a Box-Behnken (BB) or
Central Composite Design (CCD) as shown below:

$$k = 2 \; ... \; n_{reps} \geq 9 \; (CCD)$$

$$k = 3 \; ... \; n_{reps} \geq 5 \; (CCD \text{ or } BB)$$

$$k = 4 \; ... \; n_{reps} \geq 3 \; (CCD \text{ or } BB)$$

$$k = 5 \; ... \; n_{reps} \geq 3 \; (CCD)$$

A detailed approach to sample size determination for 2-level designs is
described below (source: David Le Blond).

1.   Let σ =   experimental error.

2.   Let the false detection rate = 0.05 (i.e., α = .05).

3.   Let n =   number of occurrences of ± 1 per column (i.e., the
number of runs in a 2-level design).

To detect a change in ...	... as small as ...	... with a missed detection rate of ... (i.e., a β value)	... run this many reps per trial (n_r)
Mean	$\Delta \sigma$ $(1 < \Delta < 2)$	0.05	$n_r = \dfrac{64}{n\Delta^2}$
		0.10	$n_r = \dfrac{54}{n\Delta^2}$
		0.25	$n_r = \dfrac{36}{n\Delta^2}$
SD	λ fold $(2 < \lambda < 4)$	0.05	$n_r = \dfrac{242}{n\lambda^2} + 1$
		0.10	$n_r = \dfrac{190}{n\lambda^2} + 1$
		0.25	$n_r = \dfrac{128}{n\lambda^2} + 1$

Here is how one might use this table to determine sample size. Suppose we want to know the sample size needed to detect a 1.5σ shift in *average* with a .05 missed detection rate when using a 2-level design of n = 8 runs. Using the table with Δ = 1.5, β = .05, and n = 8, we have

$$n_r = \frac{64}{n\Delta^2} = \frac{64}{8(1.5)^2} = 3.56 \quad \text{or} \quad 4 \text{ replicates.}$$

Similarly, if we want to detect a shift in **standard deviation** of $\dfrac{\sigma_i}{\sigma_j} \geq 2 = \lambda$ with a 0.10 missed detection rate when using a Taguchi L_{12} design (i.e., n = 12), we then use the table with $\lambda = 2$, $\beta = .10$, and n = 12, which results in

$$n_r = \frac{190}{n\lambda^2} + 1 = \frac{190}{12(2)^2} + 1 = 4.96 \quad \text{or} \quad 5 \text{ replicates.}$$

A simplified approach to sample size selection is contained in Table 8.37. This table has the additional advantage of showing how a change in sample size will affect the β (missed detection) risk. This table assumes that the shift in variability to be detected is as small as a factor of $\lambda = e = 2.72$.

Percent Confidence that a term identified as significant, truly does belong in ŝ [ŷ]	Percent chance of finding a significant variance [average] shifting term if one actually exists	Number of Runs in the 2-Level Portion of the Design				
		2	4	8	12	16
		Sample Size per Experimental Condition				
95% ($\alpha = .05$)	40% ($\beta = .60$)	5 [3]	3 [2]	2 [1]	N/A	N/A
95% ($\alpha = .05$)	75% ($\beta = .25$)	9 [5]	5 [3]	3 [2]	2 [1]	2 [1]
95% ($\alpha = .05$)	90% ($\beta = .10$)	13 [7]	7 [4]	4 [2]	3 [2]	N/A
95% ($\alpha = .05$)	95% ($\beta = .05$)	17 [9]	9 [5]	5 [3]	4* [2]	3 [2]
95% ($\alpha = .05$)	99% ($\beta = .01$)	21 [11]	11 [6]	6 [3]	5* [3]	4* [2]

Table 8.37 Simplified Approach to Sample Size Selection

An algebraic expression for calculating the number of replications (n_{reps}) shown in row 4 of Table 8.37 ($\alpha = .05$ and $\beta = .05$) for **2-level** designs is

$$n_{reps} = \left\lceil \frac{32}{n_{runs}} + 1 \right\rceil.$$

A similar algebraic expression for **3-level** designs ($\alpha = .05$ and $\beta = .05$) is

$$n_{reps} = \left\lceil \frac{54}{n_{runs}} + 1 \right\rceil.$$

We summarize this entire section on design choice and sample size selection in Figure 8.18. As stated at the outset of this section, the purpose of this figure is to simplify the statistical aspects of experimentation so that the practitioner can focus on his or her area of expertise, namely the top two boxes in Figure 8.18. The part of the diagram below these boxes is straightforward, allowing the practitioner to bypass the statistical theory and concentrate on those aspects at which he or she can make a significant contribution to the success of the experiment.

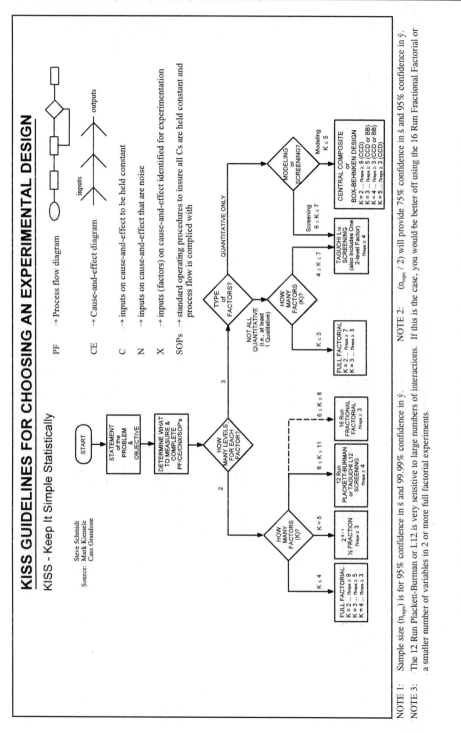

Figure 8.18 KISS Guidelines for Selecting an Experimental Design and Sample Size

NOTE 1:   Sample size (n_{reps}) is for 95% confidence in ŝ and 99.99% confidence in ŷ.   NOTE 2:   (n_{reps} / 2) will provide 75% confidence in ŝ and 95% confidence in ŷ.

NOTE 3:   The 12 Run Plackett-Burman or L12 is very sensitive to large numbers of interactions. If this is the case, you would be better off using the 16 Run Fractional Factorial or a smaller number of variables in 2 or more full factorial experiments.

8.9 Taguchi Approach to Designed Experiments

Genichi Taguchi is a Japanese electrical engineer who studied design of experiments during the 50's and then wrote his own approach to experimentation [Tagu 87]. All of the test matrices used by Taguchi have their origin in classical experimental design developed by R. A. Fisher, D. J. Finney, and C. R. Rao [Box 88]. The primary contributions of Taguchi are the use of the loss function and his development of parameter design. Since the loss function has already been presented (see Section 8.2), the remainder of this section is devoted to an introduction to parameter design.

The objective of parameter design is to determine factor settings that achieve desired response targets (min., max., or specific value) while minimizing the variability due to uncontrolled or noise factors. To illustrate this concept, we return to the gas mileage example. Since weather conditions are outside of the operator's control, it is desirable to find the best settings of factors A, B, C, and D that maximize gas mileage while also making mileage insensitive (as much as possible) to weather conditions. In this case we would make specific replicates correspond to various weather conditions. In other words, the first replicate would all be obtained under rainy conditions (w_1), the second replicate obtained during snowy conditions (w_2), and the third replicate obtained under dry conditions (w_3). Thus, the test matrix would appear as shown in Table 8.38.

	Factors & Interactions							Replicated Responses				
Run	A	B	AB	C	AC	BC	D	w_1	w_2	w_3	$\bar{w}$	s
1	−1	−1	1	−1	1	1	−1					
2	−1	−1	1	1	−1	−1	1					
3	−1	1	−1	−1	1	−1	1					
4	−1	1	−1	1	−1	1	−1					
5	1	−1	−1	−1	−1	1	1					
6	1	−1	−1	1	1	−1	−1					
7	1	1	1	−1	−1	−1	−1					
8	1	1	1	1	1	1	1					

Table 8.38 Replicated Responses Based on Weather Conditions

The only difference between this strategy and that of previous sections is the specific condition of the replicated response. Other examples of noise factors are vendors, lots, environmental variables, etc. After collecting data for the matrix for different levels of the noise factors our objective is to model $\hat{s}$ and hope to find some factor(s) that will reduce the variability. By minimizing the response variability over the noise factor levels, we have found a condition for the process that makes it insensitive (**robust**) to hard-to-control noise variables. For more information on parameter design see [Schm 94].

8.10 Problems

8.1   Given the following settings and results:

A	B	C	y_1	y_2	$\bar{y}$	s
-1	-1	-1	150	170		
-1	-1	1	147	154		
-1	1	-1	144	135		
-1	1	1	154	171		
1	-1	-1	157	153		
1	-1	1	143	148		
1	1	-1	142	139		
1	1	1	160	157		

a)   Complete the remainder of the table.
b)   Find the prediction equations from these results. Include only significant main and 2-factor interactions.
c)   What settings will minimize the response and the variance of the response?

8.2   An engineer needs to improve turbine blade quality by reducing thickness variability around a target value of 3 mm. A brainstorming session identified 3 variables that are likely to affect the thickness. Those variables and their range of values are:

	-1	1
Mold Temperature	3° C	5° C
Pour Time	1 Sec	3 Sec
Vendor	A	B

The engineer has been limited to 8 experimental runs, so he decided to run a full factorial for 3 factors.

Run	A Mold Temp	B Pour Time	C Vendor	Thickness y_1	y_2
1	-1	-1	-1	2.49	2.31
2	-1	-1	1	2.39	2.30
3	-1	1	-1	2.41	2.32
4	-1	1	1	2.27	2.23
5	1	-1	-1	4.11	4.20
6	1	-1	1	4.33	4.26
7	1	1	-1	4.15	4.22
8	1	1	1	4.33	4.19

a)   Do the full analysis on this data so the engineer has a prediction equation. Look for main effects and interactions.

b)   What settings should the engineer use to meet the goal of 3 mm thickness with minimum variability?

8.3   Use the data from Example 8.3 to conduct a complete analysis, i.e., find the significant factors and build equations for $\hat{y}$ and $\hat{s}$. Determine the optimal settings and compare these with the results in the text.

8.4   As a reliability engineer, you have been asked to weed out infancy failures in component-populated printed circuit boards. The four factors of interest are:

		Low	High
A:	Stress Temperature	80° C	125° C
B:	Thermo Cycle Rate	5° C/min	20° C/min
C:	Humidity	15%	95%
D:	g level for a 10 min sinusoid random vibration	3	6

The response is the number of electrical defects per board which contains 1000 bonds.

Given the following design matrix and response data, determine the optimal screening method. The more failures found, the better.

	A	B	AB	C	AC	BC	D	Response		
								y_1	y_2	y_3
1	-1	-1	1	-1	1	1	-1	9	17	12
2	-1	-1	1	1	-1	-1	1	21	37	42
3	-1	1	-1	-1	1	-1	1	29	35	38
4	-1	1	-1	1	-1	1	-1	17	10	15
5	1	-1	-1	-1	-1	1	1	32	41	33
6	1	-1	-1	1	1	-1	-1	21	17	19
7	1	1	1	-1	-1	-1	-1	12	14	18
8	1	1	1	1	1	1	1	33	27	47

8.5   A metal casting process for manufacturing turbine blades has four
controllable factors:

  Metal Temperature
  Mold Temperature
  Pour Speed
  Raw Material

The blades must be 3 mm thick; however, the ambient temperature causes
the blades to expand and contract. The following experiment was run with
ambient temperature in the outer array:

A Metal Temp	B Mold Temp	C Pour Speed	D Raw Material	Low Temp	Med Temp	High Temp
1	1	1	1	3.06	3.14	3.06
1	1	-1	-1	3.01	3.00	3.05
1	-1	1	-1	2.81	2.81	2.80
1	-1	-1	1	2.80	2.88	3.01
-1	1	1	-1	2.62	2.61	2.62
-1	1	-1	1	2.61	2.66	2.72
-1	-1	1	1	2.42	2.43	2.52
-1	-1	-1	-1	2.42	2.43	2.41

a)   Find a combination of settings that will produce the required
thickness and is also insensitive to ambient temperature.

b)   Plot the interactions of:
        A with Ambient Temperature
        B with Ambient Temperature
        C with Ambient Temperature
        D with Ambient Temperature

c)   Use the plots from b) to verify the optimal conditions found in a).

8.6   A company that produces a fuel filter assembly for commercial vehicles was concerned with customer complaints of leaks. Engineers at the company designed an experiment that involved eleven factors set at 2 levels. The response measured the amount of leakage in a fixed amount of time; however, some units did not leak at all. The design and response values are given in the following tables. A response value of 0.00 means "no leak" at all.

Run	A	B	C	D	E	F	G	H	I	J	K
1	-1	-1	-1	-1	-1	-1	-1	-1	-1	-1	-1
2	-1	-1	-1	-1	1	1	1	1	1	-1	1
3	-1	-1	1	1	1	1	1	-1	-1	1	-1
4	-1	1	1	-1	1	-1	-1	-1	1	1	1
5	-1	1	1	1	-1	-1	1	1	-1	-1	1
6	-1	1	-1	1	-1	1	-1	1	1	1	-1
7	1	1	-1	1	1	1	-1	-1	-1	-1	1
8	1	1	-1	-1	1	-1	1	1	-1	1	-1
9	1	1	1	-1	-1	1	1	-1	1	-1	-1
10	1	-1	1	-1	-1	1	-1	1	-1	1	1
11	1	-1	1	1	1	-1	-1	1	1	-1	-1
12	1	-1	-1	1	-1	-1	1	-1	1	1	1

Run	A	D	y_1	y_2	y_3	y_4	y_5	y_6	y_7	y_8	y_9	y_{10}	y_{11}	y_{12}
1	-1	-1	4.11	0.00	0.00	0.00	0.00	0.00	0.00	0.00	0.00	0.00	3.89	0.00
2	-1	-1	4.16	3.18	3.44	3.00	4.30	3.26	3.61	2.45	4.37	2.51	1.79	2.65
3	-1	1	0.00	0.00	1.58	0.00	0.00	0.00	0.00	0.00	0.00	4.05	0.00	0.00
4	-1	-1	0.00	0.00	0.00	0.00	0.00	0.00	0.00	1.84	0.00	0.00	0.00	0.00
5	-1	1	3.29	3.90	2.92	3.73	3.94	2.30	4.16	4.06	4.70	0.00	0.00	0.00
6	-1	1	0.00	5.50	0.00	4.69	0.00	4.89	0.00	0.00	0.00	5.18	0.00	0.00
7	1	1	1.41	0.00	0.00	0.00	0.00	0.00	0.00	0.00	0.00	0.00	0.00	0.00
8	1	-1	0.00	0.00	0.00	0.00	0.00	0.00	0.00	0.00	0.00	0.00	0.00	4.05
9	1	-1	4.45	0.00	0.00	0.00	0.00	0.00	3.48	0.00	0.00	0.00	0.00	0.00
10	1	-1	0.00	0.00	0.00	0.00	0.00	0.00	0.00	0.00	0.00	0.00	0.00	0.00
11	1	1	0.00	0.00	0.00	0.00	0.00	0.00	0.00	0.00	0.00	0.00	0.00	0.00
12	1	1	0.00	0.00	0.00	0.00	0.00	0.00	0.00	0.00	0.00	0.00	0.00	0.00

a)   Factor A at the low level was the old type of seal used by the company. Factor A set at the high level represents a new type of seal. Use the 2-sample test for proportions (Section 6.4.2) to determine if the new seal performs better.

b)   Factor D represents two pressure settings in the assembly process. Use the two sample test for proportions (Section 6.4.2) to determine whether pressure affects leakage.

Chapter 9

Introduction to Statistical Process Control (SPC)

9.1 Introduction[1]

Statistical Process Control (SPC) is a methodology which uses the basic graphical and statistical tools to analyze, control, and reduce variability within a process. SPC uses many of the tools previously presented, such as run charts, control charts, histograms, distributions and confidence intervals. These tools are combined to provide both technical and non-technical workers with powerful methods to participate in the quality improvement process. SPC is commonly associated with manufacturing processes, but the service industries are beginning to successfully apply SPC as well. After all, the statistical methodology doesn't discriminate among the various applications as long as appropriate data is properly collected and the correct method applied.

Recall that a process is defined as some combination of inputs, such as materials, machines, manpower, measurement, environment, methods, and policies, that results in various outputs which are measures of performance. For example, if we are machining a part, a possible set of inputs and outputs is displayed in Figure 9.1.

---

[1] Much of this chapter was provided by Peter Jessup of Ford Motor Company and taken from Ford Motor Company's copyrighted publication, *Continuing Process Control and Process Capability Improvement*, © 1984. All material is reprinted with permission of Ford Motor Company.

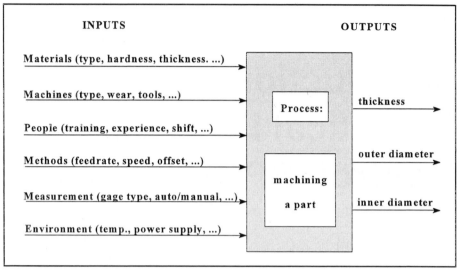

Figure 9.1 Process Example

To systematically track the performance of a process and understand how the process is behaving, we will need to gain knowledge about variability in the critical measurements. Normal procedure usually implies that we track the variability in the critical output(s), but tracking outputs may lead to tracking critical inputs as well. The basic concept of using SPC for process improvement can be applied to *any* type of work where the key measurements exhibit variation. It applies equally well to service as it does to manufacturing. It is unfortunate that the service industry has lagged behind manufacturing in its application of SPC for process improvement. Bob Galvin, former president and CEO of Motorola, once said, *"I wish we would have placed as much emphasis on Six Sigma in the service sector early on as we did in manufacturing. We would have saved an additional $5 billion* [Galv 92]." Some examples where SPC has proven successful are shown in Table 9.1.

The traditional approach to manufacturing has been to depend on production to make the product and on quality control to inspect the final product and screen out items not meeting specifications. In administrative situations, work is often checked and rechecked in efforts to catch errors. Both cases involve a strategy of **detection**. This is wasteful, because it allows time and materials to be invested in products or services that are not always usable. After-the-fact

Work Area	Process	Output(s)
machine shop	machining a part	thickness, outer diameter (O.D.), inner diameter (I.D.), true position, thread depth, cycle time
micro-electronics	manufacturing integrated circuits	oxide or metal thickness, photoresist thickness, resistivity, line width, switching speed, yield, defect rate, cycle time, power consumption
hospital	health care	operating room delays, surgery time, recovery time, number of patients, surgery costs, lab costs, time to process lab results
secretarial	office typing	cycle time, number of errors, number of pages per day, number of phone calls, time to answer a call
customer service	response to complaints	time to return phone calls, number of complaints, time to repair, number of repeat complaints
barber shop	hair styling	number of clients, time to wash, time to dry, time to cut, cycle time, waiting time, number of re-cuts
assembly shop	assemble parts	number of completed parts per day, number of errors, cycle time
pharmaceutical firm	manufacture drugs	weight of tablet, number of defects, cycle time, dimensions of tablets
maintenance shop	scheduled or unscheduled maintenance	time to repair, time between failures, number of failures, number of repeat repairs
engineering	engineering designs	time to complete, number of errors, number of engineering change proposals
software development	design, coding, testing	time to complete, number of bugs, number of design changes, time required to test
training section	employee training	number of trainee complaints in each of several types, amount of required retraining
finance center	processing travel vouchers	time to process vouchers, number of voucher errors

Table 9.1 Examples of Where SPC has Proven Successful

inspection is not economical; it is expensive and unreliable, and the wasteful production has already occurred. Furthermore, inspection alone provides little information as to why the defects occurred and how they can be corrected.

It is much more effective to avoid waste by not producing unusable output in the first place. What is needed is a **prevention** strategy which seems sensible — even obvious — to most people. It is easily captured in such slogans as, "Do it right the first time." However, this kind of cheerleading is not enough. What is required for a viable prevention strategy is an understanding of the elements of a statistical process control system: variation, causes of variation, control charts, process control, and process capability.

9.2 Understanding Variation

"Understanding variation ... is the key to success."
W.E. Deming

"Variability is like a virus. Each process can infect the one it touches, including the management of the process. Infected processes are noisy and produce noisy meetings."
Myron Tribus

In order to effectively use process control measurement data, it is important to understand the concept of variation. No two products or characteristics are exactly alike, because any process contains many sources of variability. Even identical twins are different in many regards. The differences among products may be large, or they may be almost unmeasurably small, but they are always present. The diameter of a machined shaft, for instance, would be susceptible to potential variation from the machine (clearances, bearing wear), tool (strength, rate of wear), material (diameter, hardness), operator (part feed, accuracy of centering), maintenance (lubrication, replacement of worn parts), and environment (temperature, constancy of power supply). For a service example, the time required to process an invoice could vary according to the people performing various steps,

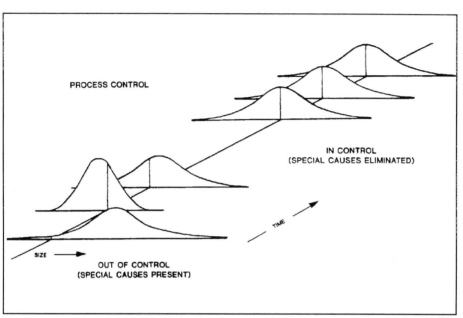

Figure 9.3 Process Control

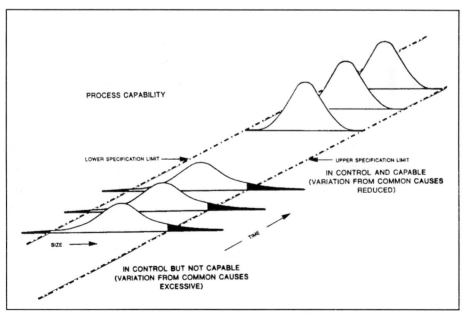

Figure 9.4 Process Capability

The two "c" words — *control* and *capability* — are often confused. The term "process *control*" refers only to the "voice of the process." One needs only data from the process to determine if a process is in control. Specifications or customer requirements are not needed to determine if a process is in control. Nowhere in Figure 9.3 do we see specs or requirements — we see only data from the process. As Figure 9.3 shows, a process is in control if the performance measure we are tracking forms a stable distribution over time. That is, a distribution is stable if the mean and standard deviation are not shifting. This will happen if special causes have been eliminated so that only common causes (natural variation within the system) remain. It is possible for a process to be in control with regard to one performance measure and to not be in control with regard to another performance measure. The tool used to determine if a process is in control is the control chart. The next three sections address the topic of control charts.

Just because a process is in control does not necessarily mean that it is a good process. The goodness of a process is measured by its "process *capability*." Process capability means *comparing* the "voice of the process" with the "voice of the customer." The voice of the customer is given in terms of specifications or requirements. Note that Figure 9.4 shows the voice of the customer as specification limits. Measuring how well a stable distribution (i.e., a process in control) matches up with the specs is the essence of process capability. Four of the most commonly used measures of process capability are defined in Table 9.2, first in words and then mathematically.

Each of these measures is computed and shown graphically in Figure 9.5 for four different processes. In all four cases, the voice of the customer is the same: LSL = 4 and USL = 16 with a target or nominal value of 10. The voice of the process is the distribution shown, which is different in each case. The dpm (defects per million) is the area under the curve outside of the specification limits multiplied by 1,000,000. The σ_{level} is the number of standard deviations (σ's) between the center of the process ($\bar{y}$) and the specification limit nearest the center. C_{pk} is computed in the right-hand margin, and C_p is shown in the left-hand margin. C_{pk}, dpm, and σ_{level} are each dependent on both the center ($\bar{y}$) and the standard deviation (σ) of the distribution, whereas C_p is dependent only on the standard deviation (σ). C_p is therefore considered to be more of a process *potential* measure

dpm	=	defects per million
	=	(proportion of observations outside spec) * 1,000,000

σ_{level} = number of standard deviations between the center of the process and the nearest spec (this is really a Z value)

$$= \text{minimum} \left(\frac{USL - \bar{y}}{\sigma}, \frac{\bar{y} - LSL}{\sigma} \right)$$

C_{pk} = proportion of natural tolerances (3σ) between the center of the process and the nearest specification

$$= \text{minimum} \left(\frac{USL - \bar{y}}{3\sigma}, \frac{\bar{y} - LSL}{3\sigma} \right)$$

$$= \frac{\sigma_{level}}{3}$$

$$C_p = \frac{\text{specification width}}{\text{process width}}$$

$$= \frac{USL - LSL}{6\sigma}$$

Table 9.2 Four Commonly Used Process Capability Measures

than it is a capability measure. C_p and C_{pk} are the same value whenever the process is centered, as in Figure 9.5 (a) and (b); but whenever the process is not centered (e.g., Figure 9.5 (c) and (d)), $C_{pk} < C_p$. Thus, C_p is the limiting value for C_{pk} whenever the standard deviation is fixed. For example, the standard deviation

$$C_p = \frac{USL - LSL}{6\sigma}$$

$$C_{pk} = \min\left(\frac{USL - \bar{y}}{3\sigma} , \frac{\bar{y} - LSL}{3\sigma}\right)$$

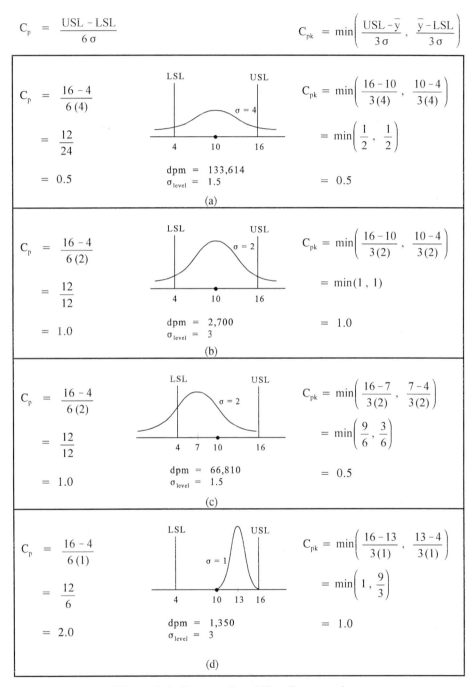

(a)

$$C_p = \frac{16 - 4}{6(4)}$$

$$= \frac{12}{24}$$

$$= 0.5$$

LSL   USL

$\sigma = 4$

4   10   16

dpm = 133,614
σ_{level} = 1.5

$$C_{pk} = \min\left(\frac{16 - 10}{3(4)} , \frac{10 - 4}{3(4)}\right)$$

$$= \min\left(\frac{1}{2} , \frac{1}{2}\right)$$

$$= 0.5$$

(b)

$$C_p = \frac{16 - 4}{6(2)}$$

$$= \frac{12}{12}$$

$$= 1.0$$

LSL   USL

$\sigma = 2$

4   10   16

dpm = 2,700
σ_{level} = 3

$$C_{pk} = \min\left(\frac{16 - 10}{3(2)} , \frac{10 - 4}{3(2)}\right)$$

$$= \min(1 , 1)$$

$$= 1.0$$

(c)

$$C_p = \frac{16 - 4}{6(2)}$$

$$= \frac{12}{12}$$

$$= 1.0$$

LSL   USL

$\sigma = 2$

4 7 10   16

dpm = 66,810
σ_{level} = 1.5

$$C_{pk} = \min\left(\frac{16 - 7}{3(2)} , \frac{7 - 4}{3(2)}\right)$$

$$= \min\left(\frac{9}{6} , \frac{3}{6}\right)$$

$$= 0.5$$

(d)

$$C_p = \frac{16 - 4}{6(1)}$$

$$= \frac{12}{6}$$

$$= 2.0$$

LSL   USL

$\sigma = 1$

4   10 13 16

dpm = 1,350
σ_{level} = 3

$$C_{pk} = \min\left(\frac{16 - 13}{3(1)} , \frac{13 - 4}{3(1)}\right)$$

$$= \min\left(1 , \frac{9}{3}\right)$$

$$= 1.0$$

Figure 9.5 Process Capability Computations

is the same in Figures 9.5 (b) and (c), namely $\sigma = 2$, but the C_{pk}'s are different (as are the dpm's and σ_{level}'s) because Figure 9.5 (c) has a process whose center ($\bar{y}$) is shifted to the left.

Charting (at least one of) these measures over time should be part of a complete scorecard. Lower dpm's are better, while bigger σ_{level}'s, C_{pk}'s, and C_p's are better. However, it should be noted that to improve all four measures simultaneously requires the following:

(1)   center the process, i.e., make $\bar{y}$ halfway between the specs or put $\bar{y}$ on target, and

(2)   reduce the standard deviation.

This is not surprising since the Taguchi Loss Function (see Chapter 8) in the form of

$$\textbf{Average Loss} = \textbf{k} \, (\sigma^2 + (\bar{y} - T)^2)$$

says the same thing: put $\bar{y}$ on target (T) and reduce σ to reduce loss (COPQ).

The Six Sigma (6σ) Quality program first espoused by Motorola stresses the use of measures like C_{pk}, C_p, and dpm to indicate how good a product or process is. As defined by Motorola [Harr 88], the term "6σ quality" means

$$C_p = 2.0, \qquad C_{pk} = 1.5, \qquad \text{and} \qquad dpm = 3.4$$

Definition of a 6σ Capable Process

Table 9.3 provides a glimpse of the relationship between Motorola's $\sigma_{capability}$ index and dpm.

$\sigma_{capability}$	dpm
6.0	3.4
5.5	32
5.0	233
4.5	1,350
4.0	6,210
3.5	22,750
3.0	66,807
2.5	158,655
2.0	308,538

Table 9.3 $\sigma_{capability}$ vs dpm

An example of how one might use this table is as follows: suppose we have found 5 defective invoices out of 1,000 invoices sent out. The ratio 5/1,000 is equivalent to 5,000/1,000,000, i.e., 5,000 dpm. Looking up 5,000 dpm in the table, we see that the corresponding $\sigma_{capability}$ is approximately 4.1 (using "eyeballed" interpolation). A more precise interpolation can be obtained by using the following formula:

$$\sigma_{capability} = .8406 + \sqrt{29.37 - 2.221 * \ln(dpm)}$$

This formula was obtained by applying the regression analysis methodology in Chapter 7 to the data in Table 9.3. Using this formula with 5,000 dpm yields a $\sigma_{capability}$ of 4.07, so our "eyeballed" approximation was quite good. Notice that $\sigma_{capability}$ is a unitless measure which can be used to compare the quality of very dissimilar processes, like an accounting process to a manufacturing process. The conversion process from dpm to $\sigma_{capability}$ assumes an underlying normal distribution.

The meticulously astute reader may have already noticed that Motorola's $\sigma_{capability}$ indexes shown in Table 9.3 are not exactly the same as the σ_{level} defined in Table 9.2. The difference lies in Motorola's definition of 6σ, where one will note that $C_{pk} = 1.5$ and $C_p = 2$. This difference is due to the assumption that a process average may shift and drift long term by as much as 1.5σ without detection. This assumption is quite well founded in the sense that a control chart typically will only detect shifts in average that are greater than 1.5σ. The point is this: Motorola's $\sigma_{capability}$ indexes are based on a 1.5σ shift in average. This means that a Motorola 6σ capable process exhibits the same number of dpm that a stable, centered $4.5\sigma_{level}$ process would exhibit in one tail. Consider Figure 9.5(d) once again. If we view Figure 9.5(d) as a snapshot of an already 1.5σ-shifted distribution, its $\sigma_{level} = 3$ and dpm $= 1350$ as shown, but Table 9.3 tells us that this dpm corresponds to a $\sigma_{capability} = 4.5$. Thus, from a one-tail dpm perspective using the σ_{level} from the worst case shift, $\sigma_{capability} \approx \sigma_{level} + 1.5$.

Before we can legitimately measure process capability, we must first bring the process into a state of statistical control where process performance is stable and predictable. Control charts are the statistical tools used to provide knowledge about whether or not a process is in control, and they are discussed next.

9.4 Control Chart Philosophy and Interpretation

A control chart is just a run chart which includes statistically generated upper and lower control limits. These limits provide the user with bounds on the common cause (or natural) variability of the process output. Dr. Walter Shewhart of Bell Laboratories developed the first control charting procedures in the 1920's. His simple, yet powerful, tool was used to separate common causes of variation from special causes. Since that time, control charts have been used successfully in a wide variety of processes both in the U.S. and other countries, notably Japan.

Several types of control charts have been developed to analyze variables and attribute data. However, according to Shewhart, all control charts have the same basic purpose:

> **to provide evidence of whether a process has been operating in a state of statistical control and to signal the presence of special causes of variation so that corrective action can be taken.**

Process improvement using control charts is an iterative procedure, repeating the fundamental phases of collection, control, and capability. First, data are gathered according to a careful plan; then, these data are used to calculate control limits, which are the basis of interpreting future data. Once a process is in control, the process capability can be measured.

1. Collection: The process is run, and data for the characteristic being studied are gathered and converted to a form that can be plotted on a graph. These data might be the measured values of a dimension of a machined piece, the number of flaws in a bolt of vinyl, railcar transit times, number of bookkeeping errors, etc.

2. Control: Trial control limits are calculated based on data from the output of the process, reflecting the amount of variation that could be expected if only variation from common causes was present. They are drawn on the chart as a guide to analysis. Control limits are *not* specification limits or objectives, but are reflections of the natural variability of the process. The data are then compared with the control limits to see whether the variation is stable and appears to come only from common causes. If special causes of variation are evident, operation of the process is studied to determine what is affecting the process. After actions (usually local) have been taken, further data are collected, control limits are recalculated if necessary, and any additional special causes are studied and corrected.

3. Capability: After all special causes have been corrected and the process is running in statistical control, then process capability can be assessed. If the variation from common causes is excessive, the process will not produce output that consistently meets customer needs. The process itself must be investigated, and management action must be taken to improve the system.

For continuing process improvement, we repeat these three phases. Gather more data as appropriate; work to reduce process variation by operating the process in statistical control; and continually improve its capability.

As a simple example of generating control charts and assessing process capability, consider the gas mileage data in Figure 9.6. As shown, gas mileage varies from week to week. Our concern is whether this variability is natural or due to special causes. The use of a control chart can help us answer this question.

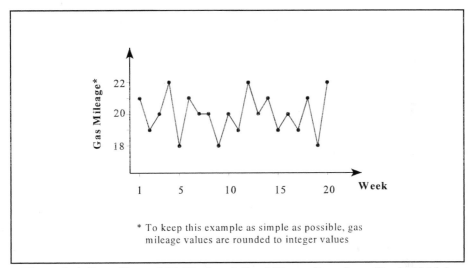

Figure 9.6 Run Chart of 20 Weeks of Gas Mileage Data for a Single Vehicle

The first step in generating a control chart is to construct a run chart such as that shown in Figure 9.6. Normally, 20 - 25 points should be plotted before constructing control limits. Computing the average gas mileage over the 20 weeks will give us a feel for the vehicle's expected gas mileage. This average is found as follows:

$$\bar{y} = \frac{\sum_{i=1}^{20} y_i}{20} = \frac{[21 + 19 + 20 + \cdots + 22]}{20} = \frac{400}{20} = 20.$$

This average will be plotted as the centerline of the control chart.

To obtain a measure of the variability in vehicle gas mileage, we can calculate the variance, s^2, of the 20 mileage values. The equation for s^2 is

$$s^2 = \frac{\sum_{i=1}^{n}(y_i - \bar{y})^2}{n - 1}$$

which can be described as the sum of the squared deviations from the mean, all divided by (n - 1). The deviations from the mean and the variance calculations are displayed in Figure 9.7.

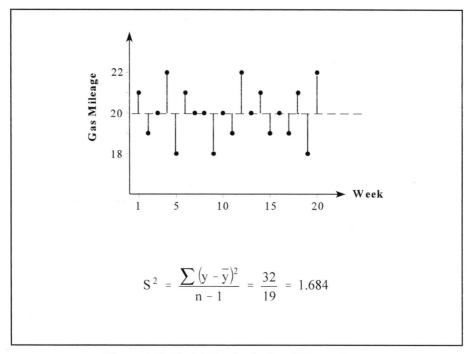

$$S^2 = \frac{\sum(y - \bar{y})^2}{n - 1} = \frac{32}{19} = 1.684$$

Figure 9.7 Variability in the Gas Mileage Data

A common technique which is a simple, yet powerful, way to get a visual feel for the way data is distributed is to draw a histogram. A histogram is a graph which displays how frequently a given outcome occurs. The histogram for our data is shown in Figure 9.8.

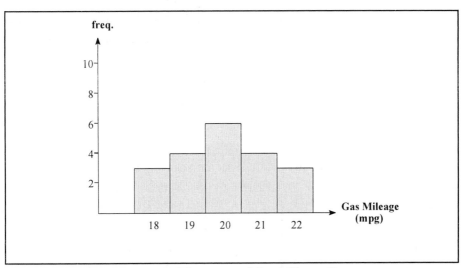

Figure 9.8 Histogram of Gas Mileage Data

If a smooth line were superimposed on the histogram it would appear to be symmetric and bell shaped, much the same as a normal curve. Assuming that the gas mileage values are approximately normally distributed allows us to make use of the empirical rule shown in Figure 9.9. One could validate this assumption of normality by conducting a χ^2 Goodness-of-Fit test as described in Chapter 6.

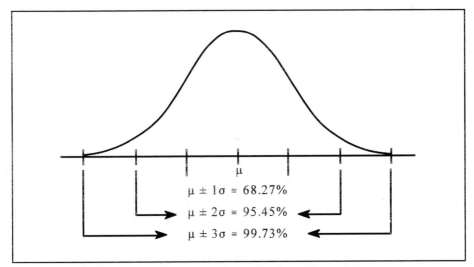

Figure 9.9 Percent of Areas Under a Normal Curve

We can now use the empirical rule to assist us in developing and interpreting the control chart. Most control charts establish the upper and lower control limits at $\pm$ 3 standard deviations from the centerline. For example, $\bar{y} \pm 3s$ = 20 $\pm$ 3(1.3) results in a lower control limit (LCL) of 16.1 and an upper control limit (UCL) of 23.9. See Figure 9.10. The empirical rule states that approximately 99.73% of all gas mileage values for this vehicle should fall within these limits. Therefore, we can use these limits in a control chart (see Figure 9.10) to evaluate current and future performance of this specific vehicle. For instance, if week 21 produced a gas mileage of 23, chances are that nothing special has caused this value; however, a value of 16 is outside our control limits (i.e., outside the natural variability) and indicates that there is an extremely high chance this value was a result of some special cause affecting the operation of the vehicle. Other SPC rules, presented later in this section, are used to detect out-of-control conditions due to trends or shifts in the average or in variability. Figure 9.10 does not exhibit any of these out-of-control conditions. Thus, we would now extend the control limits out in time and watch the process behave for the purpose of detecting any special causes of variation.

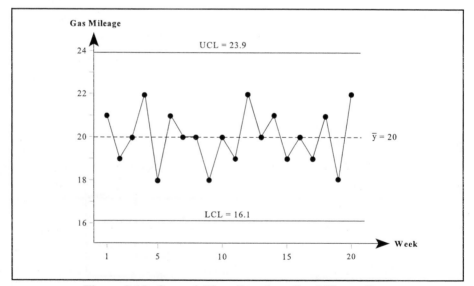

Figure 9.10 Control Chart Based on Gas Mileage Data

If an out-of-control condition such as a mileage reading of 16 is detected, the obvious next step is to determine its cause. To obtain this information, a brainstorming session with the appropriate members should be conducted. The results of such a session might resemble the cause and effect diagram displayed in Figure 9.11. After completing the brainstorming session, we should try to concentrate on those inputs that are most likely to affect gas mileage. We now have a set of input factors which we can refer to each time the mileage control chart detects a significant change.

Suppose our investigation revealed an extremely low tire pressure which we determine to be the cause of the out-of-control mileage value. We would obviously correct the low tire pressure problem and possibly begin to chart the tire pressure on a weekly basis. In this way, we are now controlling a critical input to the process instead of waiting for a substantial change in the process output. Thus, we have entered into a problem prevention mode instead of stagnating in a problem detection mode in which we are detecting poor quality through inspection of output values.

As stated previously, Figure 9.10 shows a stable or predictable process and indicates that the gas mileage process is in control. However, just because a process is in control does not mean that it is necessarily a good process. The process could be predictably bad. That is what measuring process capability is all about. How does our stable, predictable gas mileage process stack up against the customer requirements or specifications? Note that one need not know anything about the process specifications in order to determine if a process is in control. However, to determine process capability one needs to compare the current, stable process with the specifications.

Two of the most commonly used measures of process capability are σ_{level} and C_{pk}. Reference Table 9.2. To determine σ_{level} and C_{pk}, one first uses $\bar{y}$ (the centerline of the "in control" control chart) to estimate the center of the process and the standard deviation (s) to estimate the standard deviation of the overall process (σ). Sometimes the symbol $\hat{\sigma}$ (in this case, $\hat{\sigma} = s$) is used to denote a predicted or estimated value for the unknown σ. For the gas mileage data, $\bar{y} = 20$ and $\hat{\sigma} = 1.3$.

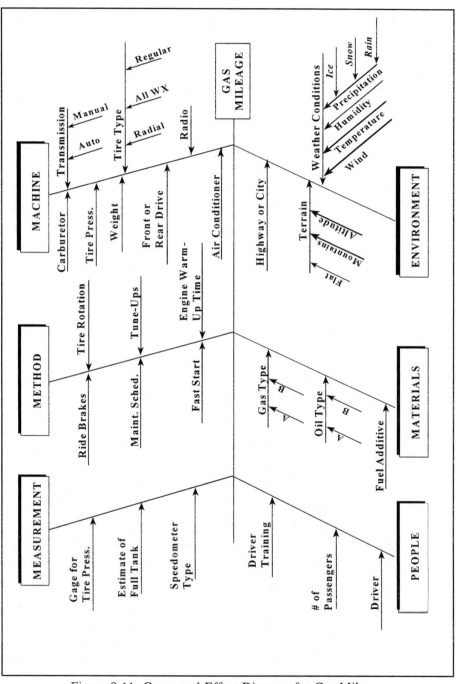

Figure 9.11 Cause and Effect Diagram for Gas Mileage

To determine process capability, we also need to know the specifications. For this make and model vehicle, suppose we are given a lower specification limit (LSL) of 17 mpg and an upper specification limit (USL) of 25 mpg. Not entirely realistic, the USL is used here to illustrate bilateral specs. A sketch of our process along with the superimposed specs (or goalposts) appears in Figure 9.12.

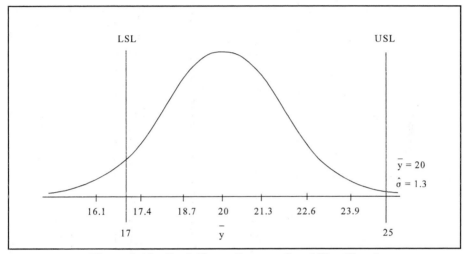

Figure 9.12 Gas Mileage Process Capability Sketch

Recall that σ_{level} is the *number* of σ's we can fit between the center of our process and the *nearest* spec. We can see from the sketch that our process is not centered at exactly the midpoint between the specs (namely, 21). $\bar{y} = 20$ is closer to LSL = 17 than it is to USL = 25. Hence,

$$\sigma_{level} = \frac{\bar{y} - LSL}{\hat{\sigma}} = \frac{20 - 17}{1.3} = 2.31.$$

C_{pk} can be found directly from σ_{level} by

$$C_{pk} = \frac{\sigma_{level}}{3}.$$

In the gas mileage case, $C_{pk} = \dfrac{2.31}{3} = .77$. The C_{pk} capability measure tells us what proportion of the process' natural tolerances falls within spec. Natural tolerances of a process are considered to be $\pm 3\sigma$. That is why σ_{level} is divided by 3 to get $C_{pk} = 0.77 < 1$. A σ_{level} of 3 equates to a C_{pk} of 1, and a σ_{level} of 6 correlates to a C_{pk} of 2, etc.

Obviously, the bigger that σ_{level} or C_{pk} are, the better. Current benchmarks for acceptable capability measures are a σ_{level} of at least 4 or a C_{pk} of at least 1.33. More important than being able to compute these process capability measures is the necessity for the student to understand exactly what must be done to increase the capability. Assuming that the specs or goalposts are fixed, Figure 9.12 clearly indicates that reducing the variation in our processes is the key to increased process capability and continuous process improvement.

Control charts are a simple and effective means to understanding the "voice of the process." They can be used to track any process, in service or manufacturing, from which data is being taken over time. Control charts give the people closest to the process reliable information on when to take action and when not to take action. For example, in the gas mileage control chart in Figure 9.10, it would be wrong to take action on data point #19 (18 mpg) which is nothing more than an instance of random variation. The control chart is telling us that if we go looking for a special cause affecting data point 19, we will more than likely be on a witch-hunt, searching for something that is not there. And witch-hunts usually lead to scapegoats. In statistics, we call this a Type I error. In management, it is referred to as an error of commission, i.e., taking action when action is not warranted. Unfortunately, this is done far too often. Witness our government who often reacts quite vehemently (due to political pressure) to changes in the trade deficit, for example, when the data is showing only random variation — no special causes! Taking action when action is not justified will usually inject more variation into the system than if we left it alone in the first place.

Control charts also give us the knowledge about when we *should* take action. For example, if data point #21 in the gas mileage control chart in Figure 9.10 shows up as 16 mpg, this is sufficient information to justify a search for a special cause. The control chart is giving us the license to search for a special cause because a point such as this (outside the control limits) is not likely to happen by random chance alone. There is a very high likelihood that a special cause influenced this data point. If we do not take action on such a point we will more than likely commit a Type II error. In management, this is called an error of omission, i.e., failing to take action when action is warranted. Control charts help us minimize both Type I and Type II errors.

The generally recommended positions of the control limits on a control chart are at $\pm 3\sigma$ from the mean. Shewhart and others have demonstrated over the last 75 years that the 3σ limits continue to provide remarkable success at eliminating/reducing both overreaction and underreaction (i.e., Type I and Type II errors). If we were to tighten the control limits around the mean, say move them to $\pm 2\sigma$, the result would be to increase the risk of overreacting (a Type I error) and lessen the risk of underreacting (a Type II error). Of course, common sense with regard to the area of application should always be used to help guide us on where to place the control limits. But the 3σ limits are almost always a safe bet.

Points outside of the control limits are not the only indicators that special causes may be affecting the process. There are other symptoms that may occur in a chart that will indicate when a process is changing significantly and when there is more than just random variation present. These symptoms usually involve shifts, trends, or patterns in the data. Although there are others, the seven out-of-control symptoms we present next have been found to be fairly robust to the distribution being plotted on the chart. We will refer to Figure 9.13 as we discuss each of the symptoms. Figure 9.13 shows the control chart divided into zones, along with a normal curve whose areas have been partitioned according to the zones. For the normal curve, each zone represents a standard deviation.

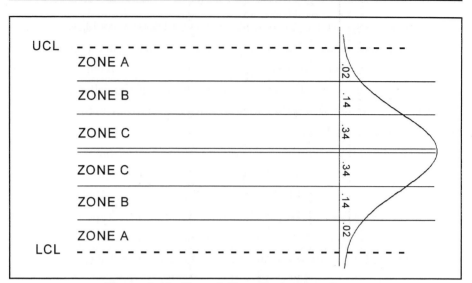

Figure 9.13 Control Chart Broken into Zones

A process is considered to be out-of-control when any one or more of the following symptoms occur.

OUT-OF-CONTROL SYMPTOMS

1.   One or more points are outside the control limits.

This is the one (and, unfortunately, sometimes the only) symptom that many practitioners are aware of that indicates "special cause." We discussed it above in the context of the gas mileage example. If a process is stable and void of special causes, we would seldom see a point like this. That means if we do see it, there is probably a special cause that is producing the outlier.

2. Seven (7) consecutive points are on the same side of the centerline.

A mini-chart showing an example of this symptom is given in Figure 9.14.

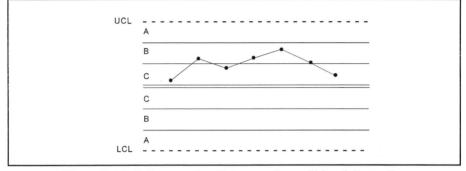

Figure 9.14 7 Consecutive Points on Same Side of Centerline

This chart, as well as the subsequent example charts for the other symptoms, assumes that the UCL and LCL were determined from earlier data. As the control limits were extended out in time, we see this symptom which indicates that the process measure is shifting up. Supposing we have a stable distribution over time, the likelihood of seeing this symptom is remote. In the case of a stable normal distribution (see Figure 9.13), the likelihood of any randomly generated point being above the centerline is 0.5. We have just seen 7 consecutive draws above the centerline. The likelihood of this happening, under the assumption of stability, is $(.5)^7 = .0078$. Thus, the process is very likely to not be stable, i.e., the process average is shifting toward the UCL.

3. Seven (7) consecutive intervals are either entirely increasing or entirely decreasing.

An example of this symptom is shown in the mini-chart in Figure 9.15. This symptom is not likely to be occurring by chance. Rather, the process is showing a trend in the downward direction. This trend indicates the process is changing, not necessarily for the worse. If smaller is better, this trend may indicate the process is improving. But for better or worse, the cause of this symptom needs to be investigated.

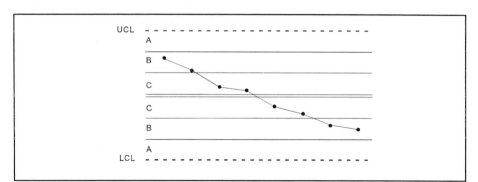

Figure 9.15 7 Consecutive Intervals Downward

4. Two (2) out of three (3) consecutive points are in the same Zone A or beyond.

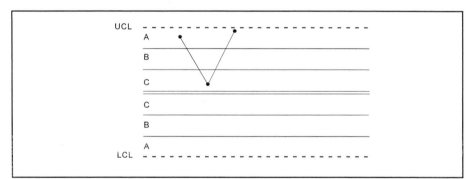

Figure 9.16 2 Out of 3 Consecutive Points in the Same Zone A

Figure 9.16 gives an example of this symptom.   The word "beyond" in the symptom definition refers to the distribution's tail direction for the particular Zone A being addressed. For the upper Zone A, "beyond" means above the UCL and for the lower Zone A, "beyond" means below the LCL. Of course if this happens, then we are also witnessing symptom #1. It is possible to observe several of these symptoms simultaneously. A symptom like this, depending on the kind of chart (see next two sections), could be a preliminary indication of a permanent shift or it could be just a temporary thing, like a change in operator or type of material. Referring to Figure 9.13 again, one can see that in a stable process, points occur much less frequently in a Zone A than they do in a Zone C. Thus, two out of three in the same Zone A is an indication of something "special" happening in the process.

5. Four (4) out of five (5) consecutive points are in the same Zone B or beyond.

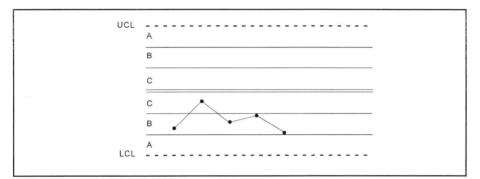

Figure 9.17 4 Out of 5 Consecutive Points in the Same Zone B

Figure 9.17 shows an example of the Zone B symptom which is analogous to the Zone A symptom of Figure 9.16. However, since we expect more points to show in a Zone B than in a Zone A, we need a little more information (like 4 out of 5, instead of 2 out of 3) to make the same conclusion, namely there is a "special" cause affecting the process.

6. Fourteen (14) consecutive points alternate up and down repeatedly.

Figure 9.18 shows an example of this symptom which is sometimes called the "2-stream" symptom. Despite the absence of any of the previous symptoms, this chart does not show random variation. The up/down motion indicates two different distributions (streams) may be operating simultaneously, with the peaks coming from one distribution and the valleys coming from another, as shown in Figure 9.18. The two "streams" could be two different shifts, two different raw materials, two different operators, etc. It could even be the same operator tweaking the system after each data point is recorded (i.e., overreacting). Whatever the case, this variation is not random or natural.

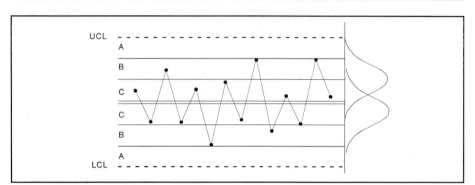

Figure 9.18 14 Consecutive Points Alternate Up and Down Repeatedly

7.   Fourteen (14) consecutive points are in either Zone C.

Figure 9.19 illustrates this symptom which is also known as "hugging the centerline."

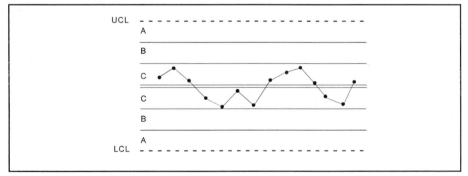

Figure 9.19 14 Consecutive Points in Zones C

Many practitioners, when seeing this symptom, might respond, "This is great. We have reduced the variability." Indeed, that may be the case, especially if we were actively pursuing a variance reduction strategy. However, this symptom is typically associated with some change in the measurement system. The measuring instrument may have lost its ability to discriminate between items (i.e., lost precision) or perhaps an operator is fudging the data. Have you ever heard the statement, "you want .5, we'll give you .5"? In any case, the control limits shown in Figure 9.19 are clearly not the correct limits for the data shown. They are much too wide. If the process has legitimately changed, new control limits will have to be calculated from this new data and the new process evaluated for control.

If a process control chart does not exhibit any of the "significant seven" symptoms shown above, or any other obvious non-random patterns, we say that the process is in control. If this is the case, then process capability can be assessed. Before diving into the construction of a control chart, we must take some preparatory action.

First, the "Steps Prior to Collecting Data" shown in Table 1.3 should be accomplished. That is, establish an environment suitable for action; define the process; determine customer requirements; determine the characteristics to be measured; determine the COPQ and opportunities for improvement; and minimize extraneous variation.

Next, we must define the measurement system. The characteristic must be operationally defined, so that findings can be communicated to all concerned, in ways that have the same meaning today as yesterday. This involves specifying what information is to be gathered, as well as where, how, how often, and under what conditions it is to be gathered. The measurement equipment itself must be predictable for both accuracy and precision — periodic calibration is not enough. Determining how much variability is in the measurement system itself is critical, and a gage capability study (see Section 9.8) may be necessary.

Last, we must select the type of control chart to be used. The definition of the type of characteristic will affect which type of control chart will be used. Control charts are classified into two major categories: control charts for variables data (discussed in Section 9.5) and control charts for attribute data (discussed in Section 9.6). The following flow diagram provides the practitioner with a logical path to follow when deciding which chart to use.

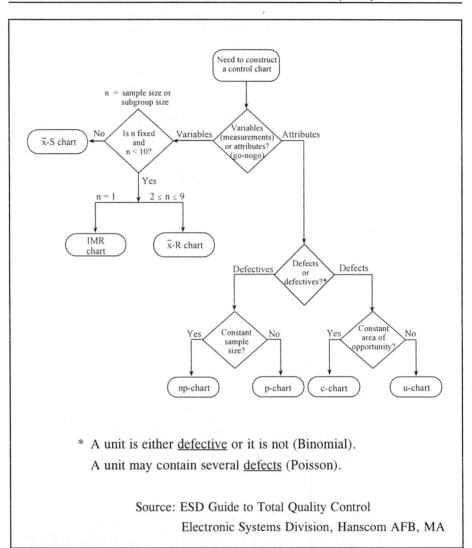

Figure 9.20 Logical Flow for Selecting a Control Chart

Variables data is measurement data that is collected on a continuous scale. Examples of measurement data include time, percentage, temperature, sizes (heights, weights, volumes, diameters), pH, pressure, etc. Even survey data collected on a Likert scale of 1 - 5 is considered variables data. Attribute data is count data associated with defective and/or defect data. When we classify something as defective or not defective, go or no go, good or bad, failed or didn't fail, this is binary or attribute data. When we count a number of defects (or some other category of interest) per interval or area of opportunity, this is also considered attribute data. While variables data is based on the normal distribution and other continuous distributions, attribute data is associated with the binomial and Poisson distributions. The next section addresses control charts for variables data and the following section addresses control charts for attribute data.

9.5 Control Charts for Variables Data

Variables data have certain advantages over attribute data and consequently, variables control charts are typically more powerful than attribute charts. For example, a measurement value (e.g., Flight #454 was 23 minutes late arriving at the gate) contains more information than a simple yes-no response (e.g., Flight #454 was late (vs not late) arriving at the gate). Hence, we will need less measurement data than binary data to acquire the same amount of information. Another advantage of measurement data over attribute data is that with measurement data, the performance of a process can be analyzed even if all the individual values are within the spec limits. This allows continuous improvement to take place, even within the specs.

Control charts for variables or measurement data are almost always prepared and analyzed in pairs — one chart for measuring the variability *between* groups and another chart for measuring the variability *within* a group. The former is also called a chart for *location* and the latter a chart for *dispersion*. The most commonly used pair are the x̄ and R charts. x̄ is the average of a small group of measured values — it is a measure of location. R is the range of values within each small group — it is a measure of dispersion. The small groups are called subgroups or samples, and their size and how they are collected depend on the objectives of the study. The next subsection discusses the x̄ and R charts at length, using an application from the service sector.

If your area of concern is manufacturing and/or the automotive industry, we highly recommend that you refer to the case study by Peter Jessup from Ford Motor Company. It is a very detailed and complete treatment of the construction and interpretation of an x̄ - R chart as used in manufacturing. Those interested in QS 9000 or ISO 9000 certification would also benefit from Jessup's very instructional case study. It can be found in Chapter 12 of this text as Case Study 18: Process Control and Capability in the Automotive Industry.

9.5.1 x̄ and R Charts

To demonstrate the construction of and interpretation for control and capability of an x̄ - R chart, we will consider a computer help desk staff member who wishes to chart his customer survey ratings over time. Since he is concerned with day-to-day variability as well as within day variability, he decides to sample 5 of his customers per day and ask each of these customers to anonymously rate his service on a scale from 1 (extremely dissatisfied) to 6 (extremely satisfied). The complete Likert scale definition is as follows:

Extremely Dissatisfied	Very Dissatisfied	Somewhat Dissatisfied	Somewhat Satisfied	Very Satisfied	Extremely Satisfied
1	2	3	4	5	6

The x̄ - R Chart is recommended whenever we use fixed subgroups (or samples) of size n, where $2 \leq n \leq 9$. The most commonly used size is $2 \leq n \leq 5$. The help desk staffer initially chose to use n = 5, but this could change depending on how the process behaves. He uses a random number generator to select those customers whom he will survey. He will record and chart the overall ratings from his customers in an x̄ - R chart.

He will receive 5 ratings daily; and he will compute the average (x̄) and the range (R) of these 5 ratings and plot them, the average on the x̄-Chart and the range on the R-Chart. Since the Rule of Thumb for chart data collection is to have at least 20 - 25 points plotted before we construct control limits, he decides to collect 4 weeks of data (5 days per week) or a total of 20 days worth of data before he constructs the control limits. Table 9.4 shows the results of his 20-day data collection effort.

	Day 1	Day 2	Day 3	Day 4	Day 5	Day 6	Day 7	Day 8	Day 9	Day 10	Day 11	Day 12	Day 13	Day 14	Day 15	Day 16	Day 17	Day 18	Day 19	Day 20
1	4	4	3	4	3	3	3	3	3	4	3	3	4	5	5	3	2	4	4	4
2	5	4	3	5	2	3	4	5	2	4	4	5	4	4	3	5	4	1	3	3
3	3	5	4	4	4	4	5	4	3	3	4	3	3	5	3	2	2	3	5	6
4	5	3	5	3	5	3	4	2	4	5	3	6	5	4	5	6	3	3	5	5
5	3	2	4	6	6	2	5	3	4	4	3	5	2	3	6	5	5	4	5	4
x̄	4	3.6	3.8	4.4	4	3	4.2	3.4	3.2	4	3.4	4.4	3.6	4.2	4.4	4.2	3.2	3	4.4	4.4
R	2	3	2	3	4	2	2	3	2	2	1	3	3	2	3	4	3	3	2	3

Table 9.4 Raw Data Collection with x̄ and R Calculated
for each Subgroup (Day)

The x̄ and R for each day are then plotted on the x̄ and R Charts, respectively. Figures 9.21a and 9.21b show these two charts.

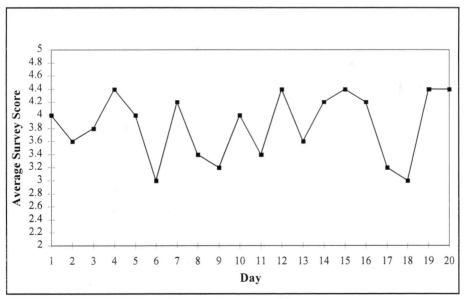

Figure 9.21a x̄ Run Chart

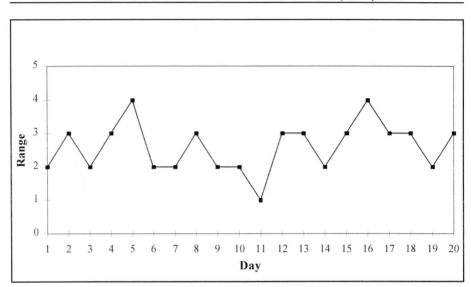

Figure 9.21b   Range Run Chart

These are not yet control charts, only run charts. The x̄-Chart has recorded the day-to-day variability, and the R-Chart denotes the within day variability. Note that the R-Chart displays only 4 different values. A Rule of Thumb for determining if there is sufficient resolution in the measurement system is that there should be at least 5 different values occurring on the R-Chart. That is not the case here, but keep in mind that this is an entirely human measurement system on a scale of only 1 to 6. Hence, the resolution is adequate.

With the initial data collection complete and run charts in hand, the next step is to interpret these charts for control. This is accomplished by constructing the center lines and control limits and then evaluating each chart for out-of-control symptoms.

For the R-Chart, the center line (CL), upper control limit (UCL), and the lower control limit (LCL) are given by

$$CL = \bar{R}$$
$$UCL = D_4 \bar{R}$$
$$LCL = D_3 \bar{R}$$

where $\bar{R}$ = the average of the subgroup range values,

D_3 = a constant dependent on the subgroup size n, and

D_4 = a constant dependent on the subgroup size n.

D_3 and D_4 are Shewhart control chart constants that depend on the subgroup size (n) as shown in the following partial table taken from the complete table in Appendix I.

n	D_4	D_3	A_2	d_2
2	3.27	.00	1.88	1.13
3	2.57	.00	1.02	1.69
4	2.28	.00	.73	2.06
5	2.11	.00	.58	2.33
6	2.00	.00	.48	2.53
7	1.92	.08	.42	2.70
8	1.86	.14	.37	2.85
9	1.82	.18	.34	2.97

Table 9.5 $\bar{x}$ and R Chart Constants for $2 \le n \le 9$
(from Appendix I)

Although Appendix I gives these constants for $n \ge 10$ as well, it is recommended that one use the S-Chart instead of the R-Chart whenever $n \ge 10$. Although not apparent from the formulas, the UCL and LCL are 3 standard deviations above and below the mean. The reader interested in seeing how the constants D_3 and D_4 are developed from 3 standard deviations (and their dependency on n) is referred to Appendix I. For the survey example,

$\bar{R} = 2.6$ and since   n = 5,

UCL $= D_4\bar{R} = 2.11(2.6) = 5.486$ and

LCL $= D_3\bar{R} = 0(2.6) = 0$

The completed R-Chart, which was generated using SPC KISS, is shown in Figure 9.22.

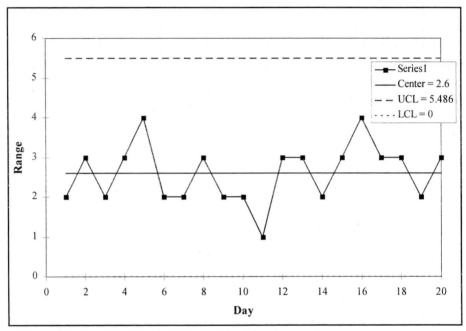

Figure 9.22 R-Chart for Survey Data

For the x̄-Chart, the CL, UCL, and LCL are given by

$$CL = \bar{\bar{x}}$$

$$UCL = \bar{\bar{x}} + A_2 \bar{R}$$

$$LCL = \bar{\bar{x}} - A_2 \bar{R}$$

where

$\bar{\bar{x}}$ = the average of the subgroup averages,

$\bar{R}$ = the average of the subgroup range values, and

A_2 = a constant dependent on the subgroup size n.

Note that $\bar{R}$ has already been computed as the center line in the R-Chart and A_2 can be determined (as a function of n) from Table 9.5 or Appendix I. The x̄ and

R Charts are related via $\bar{R}$. The average within group variability ($\bar{R}$) obtained from the R-Chart is used to determine the control limits in the $\bar{x}$-Chart. The greater the within group variability, the wider the control limits for the between group variability ($\bar{x}$-Chart) will be. While the equations for UCL and LCL in the $\bar{x}$-Chart look more like confidence limit computations than do the equations for UCL and LCL in the R-Chart, it is still not apparent that $A_2\bar{R}$ represents 3 standard deviations in the $\bar{x}$ or "child" distribution (see Chapter 5). The reader interested in seeing the development of $A_2\bar{R}$ as the equivalent to 3 standard deviations is referred to Appendix I. For the survey example,

$$\bar{\bar{x}} = 3.84 \quad \text{and since} \quad n = 5,$$

$$\text{UCL} = \bar{\bar{x}} + A_2\bar{R} = 3.84 + .58(2.6) = 3.84 + 1.508 = 5.348, \text{ and}$$

$$\text{LCL} = \bar{\bar{x}} - A_2\bar{R} = 3.84 - .58(2.6) = 3.84 - 1.508 = 2.332.$$

The completed $\bar{x}$-Chart generated by SPC KISS is shown in Figure 9.23.

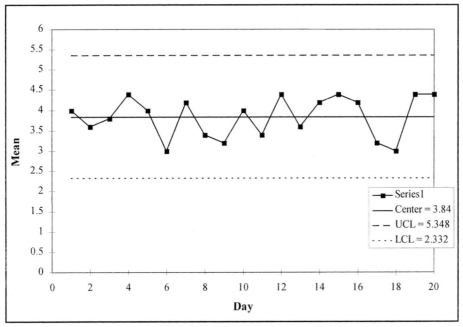

Figure 9.23 $\bar{x}$-Chart for Survey Data

A complete summary of all of the Shewhart constants and the center line and control limit formulas for all of the charts covered in this text is given in Appendix I. The development for the $\bar{x}$ - R Chart constants A_2, D_4, and D_3 is also provided in Appendix I.

Now with the $\bar{x}$ - R Charts in hand, we can evaluate the charts for control. We normally evaluate the R-Chart first because if the R-Chart is out of control, then it is impossible to make conclusions about the $\bar{x}$-Chart. Looking at the R-Chart, we see none of the out-of-control symptoms discussed in Section 9.4. It is always nice to have software like SPC KISS do the checking for out-of-control symptoms, because some of the symptoms (like the 4 out of 5 in Zone B) are difficult to see. Using SPC KISS, we find that there are no out-of-control symptoms in either chart. This does not always happen at the start-up of a chart. The original or so-called "trial" control limits may have to be adjusted due to special causes occurring in the first 20 points. If a special cause is detected in the startup 20 or 25 data points, we remove the affected point(s) and recalculate the control limits without these points. Once the appropriate control limits have been established, we extend those limits out in time and watch the process behave, looking for out-of-control symptoms as the process continues over time. If special causes in the form of outliers, shifts, or trends in the data occur, these causes must be evaluated for removal (if they are detrimental) or they should be made permanent (if they are helpful). If the process average has shifted (seen from the $\bar{x}$-Chart) or the process standard deviation has shifted (seen from the R-Chart), then new control limits need to be established for the new process. In the survey data of this section, it appears that the process is fairly stable, i.e., it is in control, due to the absence of out-of-control symptoms. For this process, we are at the point where we could extend the control limits out in time and watch the process behave. By applying process improvement techniques, we can perhaps generate an out-of-control symptom (like shift the average up) that will lend credibility to the contention we are improving the process.

Because the $\bar{x}$ - R Chart for the survey data is currently in control, we now can calculate process capability. Recall from Section 9.4 that a process in control is not necessarily a good process. The help desk staffer needs to compare the "voice of the process" which is coming from the $\bar{x}$ - R Chart with the "voice of the customer" or specs.   In this survey example, suppose the help desk management team has set a goal of at least 3.5 for all Likert scale survey responses. Thus, 3.5 serves as the lower spec limit (LSL) in the survey process. In this case, there is no upper spec limit (USL) because bigger ratings are better. Having only a one-sided spec is typical of service related processes, because usually our goal is either "bigger is better" (as in the survey responses) or "smaller is better" (as in cycle time, for example).

In conducting a process capability study, we wish to describe the "voice of the process" in terms of the mean (μ) and standard deviation (σ) of the parent (x) population or distribution. What we see in the $\bar{x}$ - R Charts is a picture of the child ($\bar{x}$) distribution. Recall from Chapter 5 that the parent (x) distribution and the child ($\bar{x}$) distribution have the same center, namely μ. Thus $\bar{\bar{x}}$, the center line in the $\bar{x}$ Chart, is a good approximation for μ.

To get an estimate for σ, we can use Shewhart's very useful equation for relating standard deviation to range:

$$\hat{\sigma} = \frac{\overline{R}}{d_2}$$

where $\overline{R}$ is the center line of the R-Chart and d_2 is another Shewhart constant dependent on n. See Appendix I for more details on this relationship. Values for d_2 can be obtained from Table 9.5 or Appendix I. For the survey example, n = 5 so d_2 = 2.33. Using $\bar{\bar{x}}$ = 3.84 and $\overline{R}$ = 2.6 from the $\bar{x}$ and R Charts, we have

$$\hat{\mu} = \bar{\bar{x}} = 3.84$$

$$\hat{\sigma} = \frac{\overline{R}}{d_2} = \frac{2.6}{2.33} = 1.12.$$

Using these estimates for μ and σ, we can construct a process capability chart as shown in Figure 9.24. Since there is only one spec limit, LSL = 3.5, C_{pk} can be calculated as

$$C_{pk} = \frac{\hat{\mu} - LSL}{3\hat{\sigma}} = \frac{3.84 - 3.5}{3(1.12)} = .10$$

SPC KISS was used to generate the chart in Figure 9.24. It shows the process and the LSL on the same axis, as well as C_{pk}, dpm, etc. Obviously, this process leaves considerable room for improvement which can be achieved by shifting the average to the right and reducing the variability.

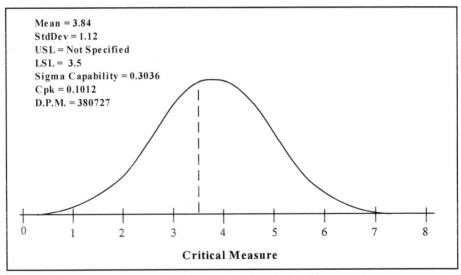

Figure 9.24 Process Capability Chart

While it is instructional to accomplish a control chart manually and we recommend it to master the technique, it should be apparent that computers are much better suited to doing the computations required to construct a control chart. That is why we suggest the use of software like SPC KISS to eliminate the computational drudgery and give the practitioner the time and power to effectively interpret the chart.

9.5.2 x̄ and S Charts

A valid question that commonly arises is, "With today's computing capability, why don't we use the standard deviation, s, instead of the range, R, with the x̄ chart?" It turns out (from theory beyond the scope of this text — see [Mont 91]) that whenever n is small, the range is just as efficient as the standard deviation in estimating the process standard deviation. However, when the sample or subgroup size n gets large enough (n ≥ 10), s is a better estimator than R. Thus, whenever n ≥ 10, we suggest using an x̄ - S chart rather than the traditional x̄ - R chart. We also recommend the x̄ - S chart over the x̄ - R chart whenever the subgroup size n varies from one time increment to the next. For example, if the computer help desk staffer had surveyed 5 customers one day, 8 customers the next day, 6 the next, etc., it would be better to use an x̄ - S chart, than the x̄ - R chart. The following shows how to calculate the center lines and control limits for an x̄ - S chart.

- for each sample or subgroup of size n, calculate x̄ and s, where s is the sample standard deviation

- compute:

$$UCL_{\bar{x}} = \bar{\bar{x}} + A_3\bar{s}$$

$$LCL_{\bar{x}} = \bar{\bar{x}} - A_3\bar{s}$$

$$UCL_S = B_4\bar{s}$$

$$LCL_S = B_3\bar{s}$$

where $\bar{\bar{x}}$ and $\bar{s}$ (the center lines) are the averages of all sample x̄ and s values, respectively. See Appendix I to locate the appropriate values for the constants A_3, B_3, and B_4.

Once the center lines and control limits have been established, the interpretation of these charts for control is exactly as it is for the x̄ - R chart. In determining process capability we use $\hat{\mu} = \bar{\bar{x}}$ and $\hat{\sigma} = \bar{s}$.

9.5.3 Charts for Individual Readings

In some cases it is necessary for process control to be based on individual readings rather than subgroups. This would typically occur when the measurements are expensive (e.g., destructive testing) or when the output at any point in time is relatively homogeneous (e.g., the pH of a chemical solution). Individual readings are also very common in continuous flow kinds of processes or when we get only one reading per time interval. Time series data, like the daily closing price of a particular stock, is an example of individual readings. In cases like these, control charts for individual readings are used to study the process variation. This kind of chart is usually called an IMR (Individuals with Moving Range) chart or an XMR (individual X value with Moving Range) chart. We will use the term "IMR" chart throughout.

Like the x̄ - R chart, the IMR chart is really a pair of charts: one chart for the individual readings and another chart for the ranges. But in the IMR chart, the ranges plotted are the ***differences*** between ***successive*** points on the individuals chart. Unlike the x̄ - R chart in which there are multiple readings (i.e., a subgroup) for each time interval, in an individuals chart there is only one reading per time interval. Thus, to find a range, we must look at successive points on the individuals chart. Hence, the term "moving range" is adopted. The most commonly used procedure is to use ***two*** successive points to find a range value, although using more than two successive points to compute a range value is possible. We recommend using n = 2 successive points in a moving range chart because it minimizes the autocorrelation between range values.

Consider the process of mixing sugar beets with water in a huge vat, where the performance measure is pH, the level of acidity/alkalinity. Since this is a continuous flow type of operation, one pH reading is taken every 15 minutes. These values when plotted form an individuals chart. Suppose the first 25 readings (about 6 hours of data) form the individuals chart shown in Figure 9.25.

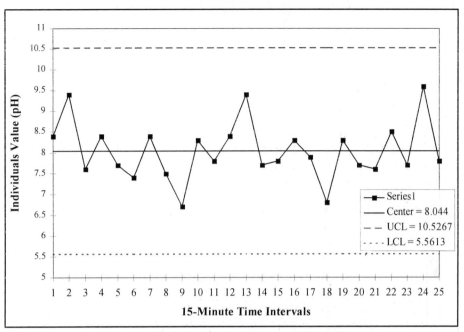

Figure 9.25 Individuals Chart for pH Readings

The moving range chart is obtained from this chart using n = 2, i.e., two successive points are used to find each range value. Since the first reading is 8.4 and the second reading is 9.4, the first range value will be 9.4 - 8.4 = 1.0 and it is plotted in the second time slot in the moving range chart, as shown in Figure 9.26. Note that the first time slot in the moving range chart is empty, i.e., there is no value available for that slot, because we need two readings before we can get our first range value. The second point in the moving range chart (1.8) is found by finding the difference between the second (9.4) and third (7.6) individual readings. SPC KISS was used to generate the IMR charts in Figure 9.25 and 9.26.

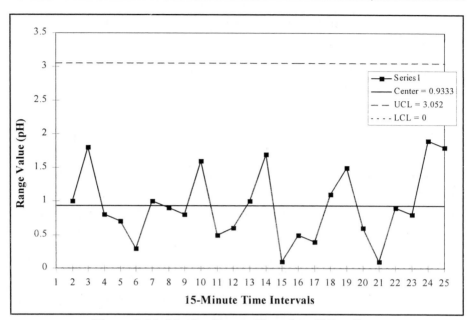

Figure 9.26 Moving Range Chart for pH Readings

The center lines and control limits for the individuals (X) chart and moving range (R) chart are given by

$$CL_x = \bar{x} \qquad\qquad CL_R = \bar{R}$$

$$UCL_x = \bar{x} + E_2 \bar{R} \qquad\qquad UCL_R = D_4 \bar{R}$$

$$LCL_x = \bar{x} - E_2 \bar{R} \qquad\qquad LCL_R = D_3 \bar{R}$$

where $\bar{R}$ = the average of the moving ranges,

$\bar{x}$ = the average of the individual readings, and

D_3, D_4, and E_2 are Shewhart constants dependent on n.

When referencing Appendix I to find the values for D_3, D_4, and E_2, it is important to remember that the subgroup size (n) is the number of successive individual readings used to calculate one range value. In the pH example, n = 2, which is also the default value for n when SPC KISS constructs the moving range chart.

There are several cautions to be exercised when interpreting an IMR chart. First, unlike the x̄-chart where the Central Limit Theorem (revisit Chapter 5) is our ally in the sense that each point plotted is known to be coming from a normal distribution, such is not the case in the IMR chart. Individual readings will tend to vary more than x̄ readings because the individuals come from the parent (X) distribution while the x̄ readings come from the child (X̄) distribution. This means that the individuals chart is not as sensitive to wide fluctuations in the plotted values as an x̄-chart is, i.e., the variability between points in an individuals chart can be expected to be greater than the variability between successive average values in an x̄-chart. One should always keep this in mind when evaluating an IMR chart for out-of-control symptoms.

A second caution is, if the process distribution is not symmetrical, some of the previous symptoms discussed (especially the ones involving zones) may not apply. It is always a good idea to get a histogram of the individual readings to see what kind of distribution we are dealing with. A histogram of the 25 pH readings is shown in Figure 9.27.

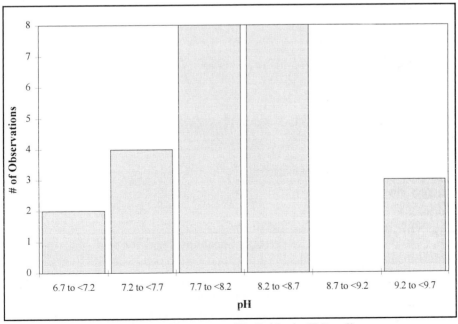

Figure 9.27 Histogram of Individual pH Readings

This histogram does not refute the possibility of an underlying normal distribution, but 25 readings is not a large sample. Recall in the $\bar{x}$ - R chart, we did 20 subgroups of 5 readings each, a total of 100 readings. Thus, it is a good idea to get at least 50 readings on an IMR chart before constructing control limits. Beware of highly non-symmetrical distributions because certain symptoms may signal the presence of special causes when none exist. The IMR charts in Figures 9.25 and 9.26 illustrate a fairly stable process, i.e., the lack of out-of-control symptoms.

When calculating process capability from IMR charts, we will use $\hat{\sigma} = \dfrac{\bar{R}}{d_2}$ as an estimate for σ, just like we did in the $\bar{x}$ - R chart, and we will use $\bar{x}$ as an estimate of μ. The subgroup size (n) for indexing d_2 is the number of successive points used to calculate a range value in the moving range chart. We recommend n = 2 to minimize autocorrelation. If the specifications in the pH process were LSL = 6 and USL = 10 with a targeted pH of 8, then the process capability chart (from SPC KISS) would be as shown in Figure 9.28.

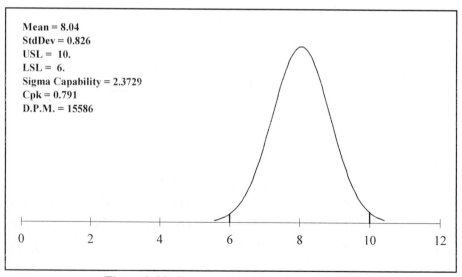

Figure 9.28 Process Capability Chart for pH

This completes the section on variables control charts. A complete summary of notation, tables of Shewhart constants, and equations for these charts is given in Appendix I. The next section addresses attribute charts.

9.6 Control Charts for Attribute Data

While the variables control charts presented in Section 9.5 are based on the normal distribution, the attribute charts presented in this section are based on the binomial and Poisson distributions. Attribute charts track the counts of a number of defectives or defects. If we characterize a unit as being defective or not defective (e.g., a bank deposit slip is either defective or not defective), then we will use the binomial distribution which has the parameters n (sample size) and p (probability any unit is defective). If we examine the same number of units in each subgroup over time (e.g., 200 deposit slips every day), then we may use the np-chart; whereas, if n varies from subgroup to subgroup (e.g., one day we examine 100 deposit slips and the next day we examine 150 deposit slips), then we must use the p-chart. Reference the chart selection flow diagram in Figure 9.20. The np- and p-charts are summarized in Sections 9.6.1 and 9.6.2.

If we are interested in counting a number of occurrences per interval (or area of observation), we then use the Poisson distribution as the basis for charting. For example, suppose we want to chart the number of defects detected on printed wire boards per shift. If we examine the same number of boards on each shift, then we will use the c-chart. If a different number of boards are examined on each shift, then we would use a u-chart to monitor the process. Reference the chart selection flow diagram in Figure 9.20. The details of the c- and u-charts are presented in Sections 9.6.3 and 9.6.4.

9.6.1 np-Charts

If each of the subgroups examined over time has the same fixed size n and if the assumptions of the binomial distribution are satisfied, then the np-chart is the recommended control chart. The following step-by-step procedure outlines the construction and use of the np-chart.

* For each sample subgroup of fixed size n (we label the subgroups i = 1, 2, ..., k), plot the integer X, the number of defective units found in that subgroup.

- Compute $\bar{p} = \dfrac{X_1 + X_2 + \dots X_k}{kn}$, where k is the number of subgroups observed (k is also the number of points plotted), X_i is the number of defective units found in subgroup i (these are the values plotted), and n is the fixed size of each subgroup; $\bar{p}$ is the overall proportion of defective units found over the first k subgroups; and k should be at least 20 before constructing control limits.

- Compute
$$CL_{np} = n\bar{p}$$

$$UCL_{np} = n\bar{p} + 3\sqrt{n\bar{p}\,(1 - \bar{p})}$$

$$LCL_{np} = n\bar{p} - 3\sqrt{n\bar{p}\,(1 - \bar{p})}$$

where $n\bar{p}$ is the average number of defectives.

- Interpret the chart for control using the out-of-control symptoms given in Section 9.4. If special causes are detected and eliminated, remove the affected data points and recompute the center line and control limits. Repeat this process until the chart exhibits control. Then extend the center line and control limits and track the process over time.

- Once the process is in control, we can evaluate process capability by computing dpm = $\bar{p} * 1{,}000{,}000$ and then, if desired, converting dpm to other capability measures such as $\sigma_{capability}$ or C_{pk} by using the conversion principles shown in Section 9.3.

9.6.2 p-Charts

If the subgroup size varies over time and we are tracking the number of defectives in each subgroup, then the p-chart is the control chart of choice. The following steps provide a procedure for the construction and use of a p-chart.

- For each subgroup (we label the subgroups i = 1, 2, . . ., k), plot $p_i = \dfrac{X_i}{n_i}$, where X_i is the number of defective units found in subgroup i

  which has a total sample size of n_i units. p_i is the proportion of defectives in subgroup i.

- Compute $\bar{p} = \dfrac{X_1 + X_2 + \dots X_k}{n_1 + n_2 + \dots n_k}$, the average proportion of

  defective units over all k subgroups.

- Compute $CL_p = \bar{p}$

$$UCL_p = \bar{p} + 3 \sqrt{\dfrac{\bar{p}\,(1 - \bar{p})}{\bar{n}}}$$

$$LCL_p = \bar{p} - 3 \sqrt{\dfrac{\bar{p}\,(1 - \bar{p})}{\bar{n}}}$$

  where ñ = the average sample size = $\dfrac{\sum\limits_{i=1}^{k} n_i}{k}$.

If the subgroup sizes vary widely then one should use n_i instead of ñ in the computation of the control limits. This means that the control limits will change for each data point plotted. This is due to the fact that the standard deviation in a binomial distribution is dependent on the sample size. A Rule of Thumb to determine if n_i, instead of ñ, should be used is as follows: if the smallest subgroup size, n_{min}, is less than 70% of the largest subgroup size, n_{max}, then one should use n_i instead of ñ.

- The interpretation for control and capability is the same as it is for the np-chart. See Section 9.6.1.

The p-chart is used more often than an np-chart because seldom are the subgroup sizes identical. Even though we show the equations used to determine the center line and control limits, we highly recommend the use of software such as SPC KISS to do control charting. SPC KISS was used to generate the p-chart shown in Figure 9.29. This chart tracks on a weekly basis during a 17-week semester the proportion of classes taught by a substitute instructor in the Department of Mathematical Sciences at the USAF Academy.

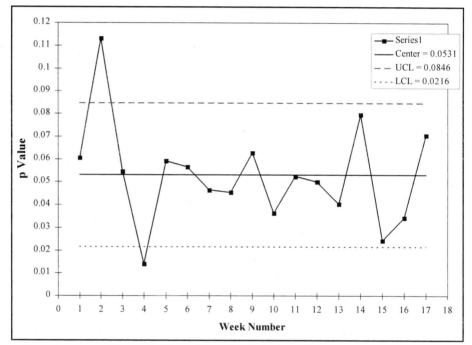

Figure 9.29 p-Chart for Substitute Instructors

The data that generated this chart is shown in Table 9.6. SPC KISS uses only the middle two columns to generate the chart.

Week	Total # of Classes Met	# of Substitute Instructors	p
1	529	32	.060
2	300	34	.113
3	371	20	.054
4	379	5	.013
5	371	22	.059
6	379	22	.058
7	300	14	.047
8	371	17	.046
9	379	24	.063
10	371	13	.035
11	379	20	.053
12	371	18	.049
13	379	15	.040
14	371	32	.086
15	300	7	.023
16	379	12	.032
17	371	28	.075

Table 9.6 Instructor Substitution Data

This example clearly indicates a special cause during Week #2. In this case, the cause is known but cannot be eliminated. Thus, the data point will be included in the calculation of the center line and control limits. There is considerable debate as to whether a point like this (i.e., where the special cause is known) should be eliminated. A Rule of Thumb in control charting is to not remove any points unless the special cause is known ***and*** it can be eliminated. In this example, the special cause (training) is known but it cannot be eliminated and furthermore, it will recur at the beginning of every semester. Hence, it is a legitimate part of the process and the corresponding data points should be retained. There were no other known special causes affecting this data, but again, only 17 points have been plotted. The reader is referred to Case Study 21: Disk Drive Manufacturing p-Chart Analysis for further examples of p-charts.

9.6.3 c-Charts

- For each area of observation (or interval), where we label each interval $i = 1,2,\ldots,k$, plot c_i, the count of the number of occurrences observed in that interval.

- Compute $\bar{c} = \dfrac{c_1 + c_2 + \ldots + c_k}{k}$, the average count of occurrences over all k intervals, where k should be at least 20.

- Compute

$$CL_c = \bar{c}$$

$$UCL_c = \bar{c} + 3\sqrt{\bar{c}}$$

$$LCL_c = \bar{c} - 3\sqrt{\bar{c}}$$

- Interpret the chart for control using the procedures outlined previously.

- Once the process exhibits control, one can estimate process capability by setting $\hat{\mu} = \bar{c}$ and $\hat{\sigma} = \sqrt{\bar{c}}$ and using the definitions of C_{pk} or $\sigma_{capability}$ as shown in Section 9.3.

If the counts are large enough (ROT: $c_i \geq 20$ for all i), or if the Poisson assumptions are not satisfied, it may be advantageous to use an IMR chart. The IMR chart also provides the moving range which can give additional insight into process control. The following example illustrates the use of a c-chart.

A software technical support service monitored the number of abandoned calls daily. This example follows the Poisson template: number of occurrences (abandoned calls) per interval (day). A c-chart is thus the chart of choice, and the first 25 days are plotted in Figure 9.30.

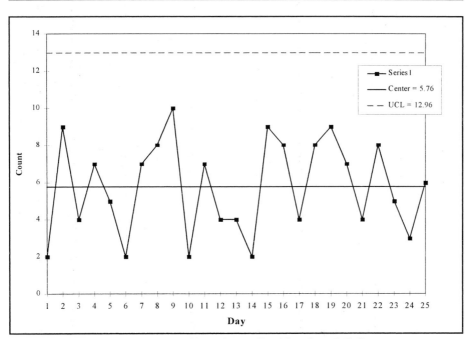

Figure 9.30 c-Chart for Daily Abandoned Calls

This chart does not show any out-of-control symptoms (SPC KISS did the automatic check), so this chart is a good starting point for monitoring the process. The computed LCL would be negative in this case, but since a count cannot be negative, the assumed LCL is 0.

9.6.4 u-Charts

• For each area of observation (we label each of the areas or intervals as $i = 1, 2, \ldots, k$), plot

$$u_i = \frac{c_i}{a_i},$$ where c_i is the count of the number of occurrences

in area a_i. u_i is the ratio of occurrences to the observed area or the number of occurrences per unit area for interval i.

- Compute $\bar{u} = \dfrac{c_1 + c_2 + \ldots c_k}{a_1 + a_2 + \ldots a_k}$, the overall average number of

  occurrences per unit area over all k observations.

- Compute

$$CL_u = \bar{u}$$

$$UCL_u = \bar{u} + 3\sqrt{\dfrac{\bar{u}}{\bar{a}}}$$

$$LCL_u = \bar{u} - 3\sqrt{\dfrac{\bar{u}}{\bar{a}}},$$

where $\bar{a}$ = the average area (or interval) = $\dfrac{\sum\limits_{i=1}^{k} a_i}{k}$.

If the areas of observation (a_i) vary widely (ROT: if $a_{min} < .7\ (a_{max})$, then use a_i instead of $\bar{a}$). This ROT is analogous to the ROT for the p-chart.

The u-chart is used for count data on unequal areas of opportunity. However, if the Poisson assumptions hold (and they should if one is using the u-chart), one could pro-rate the counts based on a constant size area of observation (or interval) and then use the c-chart. In this case, the c_i's may not be integers, but the control limits will not vary from interval to interval like they may in a u-chart, thus making the interpretation of the control chart somewhat easier. Of the four attribute charts discussed in this section, the p-chart and the c-chart are the most important because one can always use a p-chart when one would otherwise use an np-chart and one can use the proportionality assumption of a Poisson process to be able to use a c-chart instead of a u-chart.

9.7 Multi-Vari Charts[5]

Introduction

Multi-Vari Charts are simple yet powerful tools to generate and display process data. Correctly applied, they can reduce the time and resources required to achieve quality improvements in manufacturing processes. In some instances, the information they display can lead directly to improvement opportunities. In other cases, they provide focus to enhance the efficiency of statistical experimentation.

The charts are based on systematic sampling of an existing process, without intervening with how that process operates. The sampling enables the team to categorize sources of process variation as positional, cyclical, or temporal. Coupled with the team's expertise regarding the respective process technology, this information identifies where to focus efforts to reduce variation.

In addition, the data gathered for the chart may be used to estimate short term process capability, using such metrics as σ_{level}, C_{pk}, etc. When coupled with reductions in variation, this feature enables the team to document process improvements.

Description of the Tool

Multi-Vari charts share some features with Shewhart-type control charts. Both display performance data on the vertical axis, with the horizontal axis corresponding to some type of time scale. Both tend to sample subgroups from the process. Both rely to some degree on visual interpretation of graphical data displays. In rare cases, data from a control chart can be adapted to prepare a multi-vari chart.

There the similarities end. Data for multi-vari charts are obtained without interfering with the process, so that naturally occurring variation may be detected. In comparison, a key reason to maintain control charts is to decide whether or not to interrupt the process in some way. While effective use of control charts can

---

[5]This section is provided by Jonathon L. Andell of Andell Associates.

result in gradual, incremental quality improvements in a process, the multi-vari chart is better suited for rapid and substantial improvements.

The premise of multi-vari charts is that a variety of phenomena can cause variation in a process. The resulting variation often may be classified into one of three "families."

Positional variation manifests itself at differing locations simultaneously undergoing the same process. A few examples could be:

- Temperature variations inside a thermal chamber
- Variations in ingredient concentrations through a chemical reactor
- Variations among integrated circuit die, within a wafer
- Cavity-to-cavity variations in a plastics injection mold

Cyclical variation occurs among sequential repetitions of a process over a fairly short time, typically less then fifteen minutes apart. For instance:

- Variations between consecutive batches of a process
- Differences from lot to lot of raw materials
- Changes between consecutive shots of a plastics injection mold

Temporal variation appears over longer periods of time, such as several hours, days, weeks, or even longer.

Please note that batch and lot variation can be either cyclical or temporal. As a rule of thumb, batches or lots that change more often than four times per hour tend to represent cyclical variation. Batches or lots that change less than once hourly more likely apply to temporal variation. Intermediate frequencies need to be considered in context with other, potential sources of variation.

As examples below will show, the principle of multi-vari charts begins with the sampling methodology. For now, though, let us address what the graphical display reveals.

Figures 9.31a, 9.31b, and 9.31c, show multi-vari charts dominated respectively by positional, cyclical, and temporal variation. Consider the common points of the graphs:

- Positional variation appears as the height of the vertical bars. Tops of bars indicate maximum readings, while the bottoms correspond to minimum values.

- Cyclical variation is shown by the lines connecting clusters of vertical bars. The lines intersect the vertical bars at the respective means for those bars.

- Temporal variation is evident by the differences among clusters of vertical bars.

In many instances, the dominant source of variation may exhibit a pattern, such as a trend, a shift, or a cycle. Patterns may serve as clues, to reduce further the potential sources of variation, based on the team's understanding of the possible underlying causes for such variation.

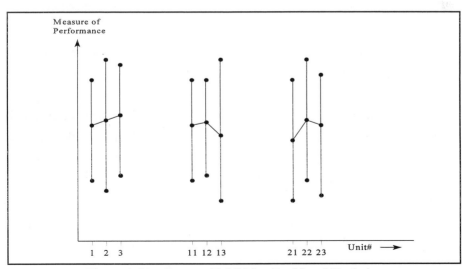

Figure 9.31a Process Exhibiting Positional Variation

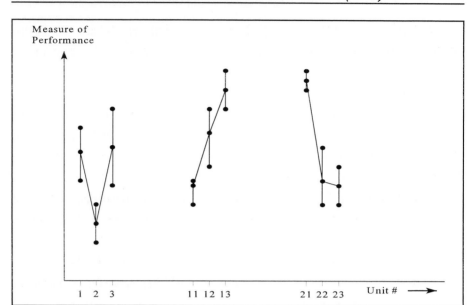

Figure 9.31b Process Exhibiting Cyclical Variation

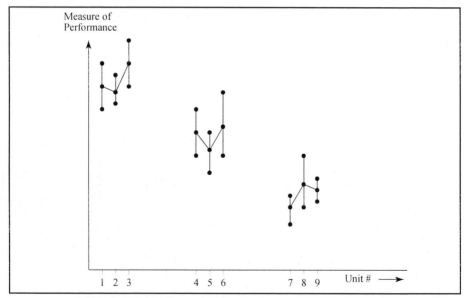

Figure 9.31c Process Exhibiting Temporal Variation

Example 9.1   A conveyorized solvent cleaner removes spent solder flux after components have been soldered onto printed wiring boards (PWB's). When spent solder flux builds up to excessive levels, the solvent must be replaced. Need for new solvent is determined by sampling the bath and performing an analysis in a laboratory.

One operator observed that the frequency of solvent replacement seemed excessively high for the volume of PWB's being cleaned. However, the laboratory procedure was verified as accurately reflecting the sample contents.

The team decided to develop a multi-vari chart, to determine whether an unidentified source of variation was present. A sample could be drawn in roughly thirty seconds. Three consecutive samples were drawn from the bath at starting times of zero, ten, and twenty minutes. Two hours after the test started, the cycle was repeated, and again at four hours. Table 9.7 shows the sampling times and the contamination levels, by volume percent.

Time (min)	Sample # Standard Sampling	Sample # Frugal Sampling	Results
0.0	1	1	1.3
0.5	2	2	1.6
1.0	3	3	1.4
10.0	4	4	1.5
10.5	5	5	1.4
11.0	6	6	1.4
20.0	7	7	1.6
20.5	8	8	1.7
21.0	9	9	1.5
120.0	10	10	2.2
120.5	11	11	2.1
121.0	12		2.3
130.0	13	12	1.9
130.5	14	13	2.2
131.0	15		2.1
140.0	16		2.1
140.5	17		2.3
141.0	18		2.0

Table 9.7 Sampling Times And Results for Example 9.1

(Continued on next page)

Time (min)	Sample # Standard Sampling	Sample # Frugal Sampling	Results
240.0	19	14	2.7
240.5	20	15	2.5
241.0	21		2.8
250.0	22	16	2.6
250.5	23	17	2.9
251.0	24		2.9
260.0	25		2.8
260.5	26		2.7
261.0	27		2.9

Table 9.7 Sampling Times And Results for Example 9.1

Figure 9.32 shows the results of the test. The primary source of variation is temporal, with contamination levels steadily increasing. It is significant to note that sampling was performed after the system was shut down for the day, just as actual samples had been taken in practice. In other words, no new contaminants were entering the system, yet contamination levels appeared to be increasing.

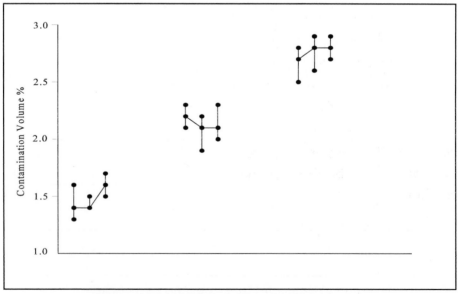

Figure 9.32 Multi-Vari Chart for Example 9.1

The team investigated the sampling technique. Due to interferences in the equipment, samples had to be drawn from the top of the bath. When they observed the laboratory procedure, they learned that the spent solder flux was lighter than the solvent, and would rise. If the operator let the system sit idle overnight, and sampled from the top of the bath, he or she would obtain a specimen with far more contaminant than was representative of the entire system.

Based on those findings, sampling was done with the machine running, solvent circulating, just before being shut down for the day. The resulting, well mixed specimens were far more representative of actual contamination levels in the system. Solvent expenditures were reduced, from $50,000 per month to $50,000 per year. The equipment manufacturer was notified to relay the findings to their other customers, since the solvents used were known to deplete the earth's ozone layer.

Example 9.2 A manufacturer of injection molded plastic components produces drive rollers for cassettes that store computer data. Departures from roundness, measured as Total Indicated Run-Out (TIR), led to unacceptable variation in the speed at which the tape would pass by the read/write heads, which in turn was causing data errors.

The plastics molder needed to solve the problem as quickly as possible, since their customer's line was down until acceptable drive rollers were delivered. Multi-vari sampling was chosen by the team.

Since the molding tool produced eight parts per molding cycle (a molding cycle is known as a shot), positional variation was based upon the cavity in which individual parts were made. Three consecutive shots provided data regarding cyclical variation. Temporal variation was evaluated by sampling three shots each at two and at four hours.

Figure 9.33 shows the multi-vari chart. Note that positional variation (the height of the vertical lines) is the predominant source of variation. Also, note that the mean values for each shot are toward the top half of each vertical line. This led the team to investigate cavity-to-cavity variation more closely, as shown in Figure 9.34.

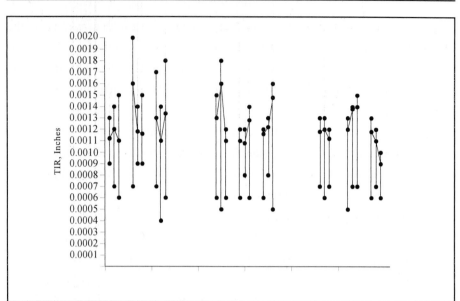

Figure 9.33 Multi-Vari Chart for Example 9.2

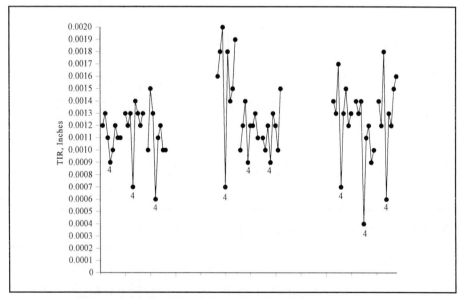

Figure 9.34 Positional Variation Detail for Example 9.2

Here note that cavity # 4 consistently produced parts with lower TIR values than any other cavities. Investigation revealed that cavity #4 had a unique gate configuration. That is, resin was routed to that cavity differently than the others. The tool was pulled from the press, and the other seven cavities were modified to resemble cavity #4.

After the molding tool re-started, performance improved, from an estimated (short-term) C_{pk} of 0.86, to an estimated C_{pk} of 1.84. The customer's line was up the next day.

Application Considerations

Determination of Sampling Scheme

Clearly, the experimental team has some latitude with their sampling scheme. For example, consider a semiconductor fabrication process. In this instance, a silicon wafer contains 100 separate device "dice." Twenty-five wafers sit in a quartz "boat," and three boats are loaded into a diffusion furnace. The diffusion cycle lasts four hours.

Table 9.8 displays how two teams might come up with two different sampling schemes for the same process. There exists no exact way to know whether one approach will reveal variation that another will miss. The best advice is the same as for any other application of statistical methods of gathering information:

1. Invest the time and energy to develop an appropriate response.

2. Use the collective wisdom of the team to make decisions.

Type of Variation	Sample Scheme #1	Sample Scheme #2
Positional	Die-to-die on a wafer	Wafer-to-wafer in a boat
Cyclical	Wafer-to-wafer in a boat	Boat-to-boat in a furnace
Temporal	Batch-to-batch in a furnace	Day-to-day

Table 9.8 Wafer Fab Sampling Options

Economics of Sampling

Every experiment costs money to conduct. There are costs associated with designing the study, taking the sample, measuring, and analyzing the data. Product might have to be quarantined or scrapped while the experiment proceeds. Frequently, the pressure for statistical validity appears to contradict pressure to reduce the time and expense of an experiment. A practical experiment must account for these factors.

There is no sure way to reduce sampling without some risk of missing critical information. However, a few approaches might be tried, under the guidance of an experienced practitioner of the methods:

1.   Reduce the number of positional or cyclical samples at chosen temporal gates. Examine Example 9.1 again. An extra column in Table 9.7 shows how the team could have reduced sampling. At two and four hours, they might take only two positional and two cyclical samples. Figure 9.35 shows the resulting multi-vari chart. In this instance, the conclusion regarding temporal variation would have been unchanged.

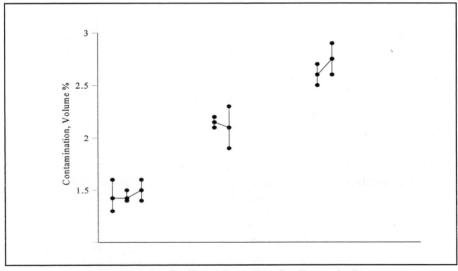

Figure 9.35 Reduced Sampling for Example #1

The risk is that any family of variation can be intermittent, and that reduced sampling might cause the team to miss such variation.

2.   Consider an uneven temporal sampling scheme. That is, after "time-zero," take temporal samples at perhaps one, two, four and eight hours. This improves the chances of "capturing" variation in several time domains.

3.   Often the biggest expense is measurement. In this case, an option is to sample using a more traditional scheme, but use one of the above approaches to reduce the number of measurements taken. The team may retain the extra samples if more data are required later.

Each of the above options assumes certain statistical risks. An experienced practitioner of statistical methods should be consulted to identify those risks, so that the team's decision balances economic and statistical risks appropriately.

Coordination with Other Statistical Tools

Recall that multi-vari charts are designed and intended to capture the greatest possible amount of variation in the time frame of the sample. That attribute can be employed as a means of estimating the overall performance of a process.

An extra benefit of using multi-vari charts occurs when the team achieves a process improvement. In such a case, they have the option of gathering further data and generating a graphical display that demonstrates exactly what they achieved.

For example, consider Example 9.2. Recall that, initially, cavity #4 behaved differently from the rest of the tool. Modification of the other cavities resulted in significant improvements. Figure 9.36a shows the "before" and "after" multi-vari charts. The respective data for before ($C_{pk}=0.86$) and after ($C_{pk}=1.84$) appear as histograms in Figure 9.36b.

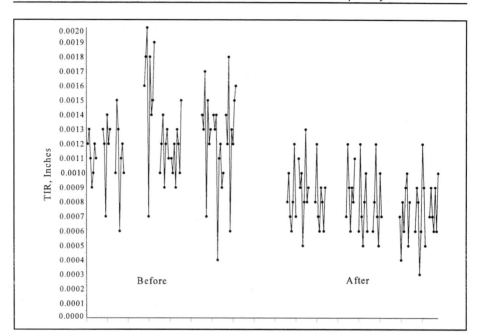

Figure 9.36a Example 9.2 Multi-Vari Charts "Before" and "After"
Improvements

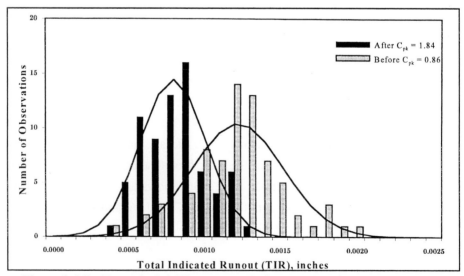

Figure 9.36b Example 9.2 Histograms "Before" and "After" Improvements

At this point, either positional or cyclical variation may dominate the modified process. The team has the option to pursue further improvement, or to proceed to a different process. The decision becomes more managerial than statistical.

One note of caution is warranted. Some types of variation occur over even longer periods of time, such as weeks or months. Obviously, multi-vari charts are impractical tools for characterizing those forms of variation. Therefore, they are regarded as a means for estimating "short term" metrics only.

In perhaps an ideal scenario, multi-vari charts and control charts can complement each other in a comprehensive quality assurance scheme. Here, a team would use multi-vari charts—with statistically designed experiments as appropriate—to identify, and to eliminate or control, existing causes of process variation.

Once those causes are known, control charts could be maintained to keep those causative factors at levels which sustain the ultimate quality of the final process response. In this scenario, defects are prevented "upstream" from the final quality measurement stage. This enables the team to reduce the sampling frequency used to monitor and confirm final process performance.

Figure 9.37 shows a possible process flow chart for one such scenario.

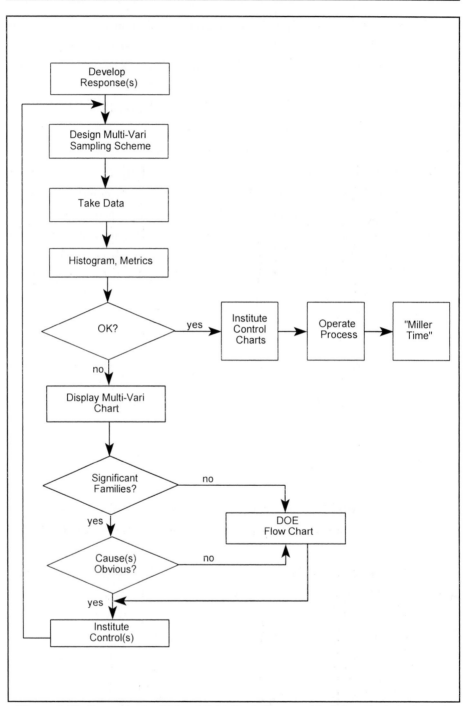

Figure 9.37 Scenario for Using Multi-Vari Charts with Other Statistical Tools

9.8 Gage Capability Analysis

The focus of this section is on the measurement system or instrument (i.e., gage or test equipment) used to measure the quality characteristic of interest. In any process there is variability in the product or service being measured, but there is also variability in the way we measure that product or performance. Although viewing the same match, three different boxing judges rarely score the match the same way. As another example, suppose an operator is responsible for determining the number of defects on a printed wire board (PWB). Assume the operator, unknowingly, was given the same board at three different times during the day and asked to record the number of defects on the board. Perhaps the operator would record the same number every time, but process experience tends to indicate such is not the case. Variability in measurement — in this case, the number of defects — is typically present in every process; and in order to understand and improve our processes we need to be able to quantify and reduce that measurement variability.

In any measuring device or system there are three desired properties:

(1)   **Accuracy** is the ability to produce an average measured value which agrees with the true value or standard being used.

(2)   **Precision** is the ability to repeatedly measure the same product or service and obtain the same results.

(3)   **Stability** is the ability to repeatedly measure the same product or service over time and obtain the same average measured value.

The purpose of a gage or measurement study is to assess how much variation is associated with the measurement system and to compare it to the total process variation.

The relationship between total, product, and measurement variation is given by

$$\sigma^2_{total} = \sigma^2_{product} + \sigma^2_{measurement}$$

The variation in measurement can itself be partitioned as

$$\sigma^2_{\text{measurement}} = \sigma^2_{\text{repeatability}} + \sigma^2_{\text{reproducibility}}$$

where

repeatability   is the variation obtained by the same person using the same instrument on the same product or service for repeated measurements (i.e., the variability **within** operator/device combination), and

reproducibility   is the variation obtained due to differences in people who are taking the measurements (i.e., variability **between** operators).

We demonstrate the concepts and computation of these two components of measurement error by considering two different operators each of whom takes two measurements of each of 20 different products. The products measured are bearings and the measurements are outer diameter in millimeters. The same device is used by both operators. The data collected is shown in Table 9.9.

We reproduce the data in Table 9.9 in the form of x̄ - R charts for each of the operators, as shown in Figure 9.38.

Product #	Operator 1 Measurement				Operator 2 Measurement			
	1	2	$\bar{x}$	R	1	2	$\bar{x}$	R
1	39	37	38.0	2	39	42	40.5	3
2	38	39	38.5	1	41	36	38.5	5
3	44	44	44.0	0	47	45	46.0	2
4	40	41	40.5	1	43	46	44.5	3
5	43	45	44.0	2	48	44	46.0	4
6	50	49	49.5	1	50	49	49.5	1
7	46	46	46.0	0	44	48	46.0	4
8	40	40	40.0	0	39	41	40.0	2
9	41	39	40.0	2	37	39	38.0	2
10	45	46	45.5	1	46	44	45.0	2
11	39	39	39.0	0	43	39	41.0	4
12	43	45	44.0	2	44	45	44.5	1
13	44	43	43.5	1	42	39	40.5	3
14	42	41	41.5	1	43	44	43.5	1
15	41	43	42.0	2	41	39	40.0	2
16	38	39	38.5	1	37	41	39.0	4
17	47	47	47.0	0	47	48	47.5	1
18	41	40	40.5	1	40	41	40.5	1
19	40	41	40.5	1	39	42	40.5	3
20	44	43	43.5	1	44	46	45.0	2
			$\bar{\bar{x}}_1 = 42.3$	$\bar{R}_1 = 1.00$			$\bar{\bar{x}}_2 = 42.8$	$\bar{R}_2 = 2.5$

Table 9.9 Data Collected to Determine Repeatability and Reproducibility (R & R)

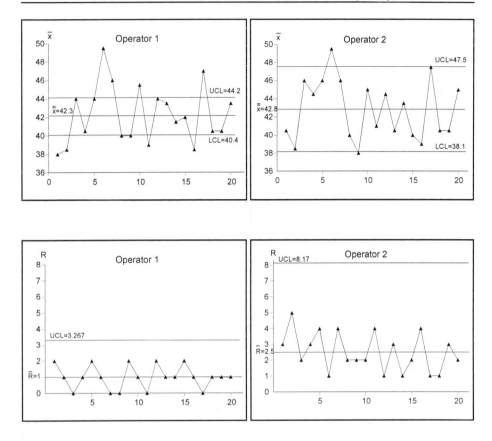

Figure 9.38 x̄ - R Charts for R & R Data

The x̄ - R charts in Figure 9.38 show that while both operators' R-Charts are in control, Operator 2 shows substantially more repeatability variability, i.e., the Operator 2 R-chart has much greater ranges (R̄ = 2.5) than does the Operator 1 R-chart, where R̄ = 1.0. Operator 2 is clearly having more difficulty in making consistent measurements than is Operator 1. The x̄ chart for Operator 1 appears to have at least half of its points beyond the control limits. However, this is not bad because it shows that Operator 1 has the ability to discriminate among the products. Operator 2 does not show that kind of discriminatory power because its control limits are spread apart so far, due to the large R values for Operator 2. Note that this is a somewhat different interpretation of an x̄-chart than the Shewhart interpretation. While the center lines of the x̄-charts are fairly close, indicating decent reproducibility between operators, the center lines of the R-charts are quite different, thus indicating that Operator 2 is exhibiting greater repeatability problems.

To compute the repeatability variation, we find

$$\overline{\overline{R}} = \frac{\overline{R}_1 + \overline{R}_2}{2} = \frac{1 + 2.5}{2} = 1.75$$

and then divide $\overline{\overline{R}}$ by d_2 (using n=2 in Appendix I) to get

$$\sigma_{repeatability} = \frac{\overline{\overline{R}}}{d_2} = \frac{1.75}{1.13} = 1.55.$$

We find the reproducibility variation by first finding

$$R_{\overline{x}} = \overline{x}_{max} - \overline{x}_{min} = 42.8 - 42.3 = .5.$$

Now we divide $R_{\overline{x}}$ by d_2 (again using n=2 in Appendix I) to get

$$\sigma_{reproducibility} = \frac{R_{\overline{x}}}{d_2} = \frac{.5}{1.13} = .44.$$

Thus, reproducibility appears to be less of a problem than repeatability. Combining the variances for repeatability and reproducibility, we have

$$\sigma^2_{measurement} = \sigma^2_{repeatability} + \sigma^2_{reproducibility}$$

$$= (1.55)^2 + (.44)^2$$

$$= 2.60 \text{ and, hence,}$$

$$\sigma_{measurement} = 1.61.$$

From all 80 measurements, we have (computations not shown):

$$\sigma^2_{total} = s^2 = 11.314, \text{ and}$$

$$\sigma_{total} = 3.364.$$

A Rule of Thumb commonly used to determine if the measurement system is capable is to see if the precision-to-total ratio is less than 10%. That is,

$$\frac{\sigma_{measurement}}{\sigma_{total}} \le .10.$$

In our case, $\dfrac{\sigma_{measurement}}{\sigma_{total}} = \dfrac{1.61}{3.364} = .48 > .10$. Hence, our measurement system needs to be examined. In this example, repeatability is by far the greatest contributor to measurement error, and Figure 9.38 indicates that Operator 2 is having considerably more problems with repeatability than is Operator 1. Perhaps a new measuring device and/or training will be required to reduce the variability in measurement.

Another commonly used capability index for a measurement system is the precision-to-tolerance Capability Ratio (CR), which is defined as

$$CR = \frac{6\sigma_{measurement}}{USL - LSL}.$$

The Rule of Thumb for CR is:

if CR $\le$ .10, then adequate measurement system;

if CR $\ge$ .30, then unacceptable measurement system.

9.9 Problems

9.1   An aerospace company machines parts to be used in supersonic aircraft. A critical dimension for one of these parts is its thickness: the target value is 1.04 millimeters with lower and upper specifications of 1.00 and 1.08 millimeters, respectively.   Given the following data, complete an appropriate control chart and answer the stated questions.

Date	6/5	6/5	6/5	6/6	6/6	6/6	6/7	6/7	6/7
Time	1000	1400	1800	1000	1400	1800	1000	1400	1800
	1.04	1.05	1.04	1.07	1.04	1.04	1.06	1.07	1.04
	1.06	1.05	1.06	1.03	1.05	1.05	1.04	1.06	1.03
	1.04	1.06	1.05	1.02	1.01	1.03	1.04	1.06	1.02
	1.03	1.07	1.06	1.05	1.03	1.03	1.05	1.05	1.03

a) Does the process appear to be in control?
b) What is the process capability?
c) What other observations can you make?

9.2   Assume a process in control has provided the following information:

$$n = 5, \quad \overline{\overline{x}} = 2.214 \text{ and } \overline{R} = .029.$$

If the specifications are 2.21 and 2.22, what is C_{pk}?

At what σ level is this process operating?

9.3   A finance center takes daily samples of process vouchers with sample subgroup size $n = 50$. The following data depicts the number of errors over a period of 25 days.

Day	1	2	3	4	5	6	7	8	9	10	11	12	13
Errors	5	4	3	3	7	5	1	7	10	3	2	3	7

Day	14	15	16	17	18	19	20	21	22	23	24	25
Errors	1	4	2	4	0	6	1	8	1	2	3	4

Determine appropriate control limits and comment on the state of process control and process capability.

9.4   The time to repair a downed radar system has been tracked for 30 repairs.
        Using the repair time data below, compute an appropriate control chart and
        evaluate.

Time to repair (in minutes)

Repair #	Time	Repair #	Time	Repair #	Time
1	118	11	116	21	138
2	121	12	117	22	141
3	132	13	126	23	116
4	117	14	117	24	133
5	131	15	127	25	123
6	114	16	113	26	126
7	121	17	119	27	125
8	112	18	121	28	121
9	123	19	128	29	122
10	114	20	145	30	115

9.5   A process is said to be operating in statistical control when common causes
        are the only source of variation.

        a)   True
        b)   False

9.6   _____ causes refer to any factors causing variation that cannot be
        adequately explained by any single distribution of the process output.

9.7   Common causes of variation are usually the responsibility of management
        to correct.

        a)   True
        b)   False

9.8   Assume a process operating in statistical control has provided the following
        information: $n = 8$, $\bar{\bar{x}} = 16.15$, $\bar{R} = 0.19$. If the specification limits are
        15.8 and 16.4, calculate the process capability index C_{pk}.

9.9   In March 1990, some math instructors and cadets visited Microtome
       Precision Inc. of Colorado Springs to help them improve their process
       capability for manufacturing a precision Delrin ring used on high speed
       disk drives. The specifications for the outer diameter of this ring are 1.557
       $\pm$ 0.001 inches, which is a critical dimension. A capability analysis of the
       process before improvement techniques were implemented yielded the
       following statistics from 86 runs: $\bar{x}$ = 1.55708, s = 0.000493. (Assume
       that the outer diameter is normally distributed.)

       a)   Calculate the number of defects per million. (Hint: find the
             probability that a diameter will fall outside the specifications and
             multiply this by one million.)
       b)   Calculate the process capability index, C_{pk}. (Hint: let $\hat{\sigma}$ = s.)

9.10   Which of the following $\bar{x}$ charts illustrate a process in statistical control?

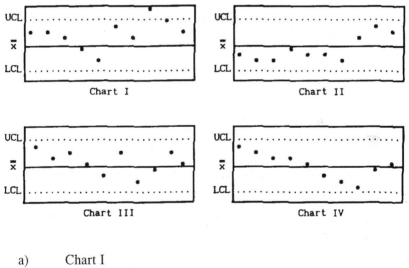

Chart I   Chart II

Chart III   Chart IV

       a)   Chart I
       b)   Chart II
       c)   Chart III
       d)   Chart IV
       e)   At least two of the charts illustrate a process in statistical control.

9.11   A gage capability study was performed on the measurement system used
        to assess wall thickness of a leading manufacturer's 5/8" garden hose. The
        specifications on garden hose wall thickness are 0.145 ± 0.015 inches.
        Each of three operators used the same micrometer and operating
        procedures to measure 10 different product samples twice. The resulting
        measurements are shown in the following table.

Sample	Operator A 1st	Operator A 2nd	Operator B 1st	Operator B 2nd	Operator C 1st	Operator C 2nd
1	.165	.164	.157	.154	.155	.160
2	.157	.154	.152	.153	.159	.154
3	.167	.167	.155	.156	.163	.159
4	.179	.179	.168	.173	.169	.170
5	.157	.156	.159	.157	.156	.162
6	.163	.160	.159	.158	.154	.152
7	.153	.157	.153	.158	.155	.155
8	.155	.155	.158	.156	.153	.155
9	.172	.172	.159	.158	.166	.169
10	.150	.155	.150	.150	.151	.152

$$\overline{\overline{x}}_A = .16185 \qquad \overline{\overline{x}}_B = .15715 \qquad \overline{\overline{x}}_C = .15845$$

$$\overline{R}_A = .0017 \qquad \overline{R}_B = .0021 \qquad \overline{R}_C = .0029$$

a)   Compute $\sigma_{measurement}$ and σ_{total}.
b)   Compute the precision-to-total ratio.
c)   Compute CR.
d)   Is this measurement system capable?

Chapter 10

Introduction to Reliability

10.1 Introduction

Reliability has continued to gain importance over the last few years in both the government and civilian communities. With utmost concern about government spending (in particular, within the Department of Defense), government agencies are trying to purchase systems with higher reliability and lower life cycle costs. These systems range from complete weapon systems such as aircraft or tanks to individual critical components of satellites. As consumers, we are concerned with buying products that last longer and are cheaper to maintain, i.e., have higher reliability. Reliability is also important in the biomedical field. In this case, we define reliability in terms of the "failure" of organisms. Sometimes the organism is the human body. The key is that we want whatever we have to last longer. But how do we measure reliability? How can we compare reliability between two products?

In this chapter we provide an introduction to basic reliability concepts. We begin by defining the terms that are necessary in the discussion of reliability and follow that with a presentation of methods for calculating the reliability of systems.

10.2 Definition of Terms

Reliability is a measure of the likelihood that a product (or system) will operate without failure for a stated period of time (call this time t). Since "probability" is the "measure of likelihood" or "level of certainty" (see Section 3.2) we have dealt with thus far, reliability is considered to be a probability. Therefore, reliability will always be a number between 0 and 1. More formally, we define the reliability of a component or system over time as follows:

The **reliability** of a component or system at time t is given by

$$R(t) = P(T > t),$$

where

      T =   a continuous random variable denoting time to failure. Typically, we will assume that T follows some probability distribution, e.g., normal or exponential, and

      t =   some specific time (e.g., 450 hours).

Definition 10.1 Reliability of a Component at Time t

Thus, R(50), read as "the reliability of the component at 50 hours (or whatever unit of time is being used)," denotes the probability that the component will last past the 50-hour mark.

      Recall that, for a random variable X, we used f(x) to denote the probability density function (pdf) for X. Similarly, we shall use f(t) as the pdf for the time-to-failure random variable T. Sometimes the pdf f(t) is referred to as a **failure function**. The two primary failure functions we shall investigate are the bell-shaped curve (i.e., when T is assumed to be normally distributed) and the exponential curve (i.e., when T is assumed to be exponentially distributed).

      Since reliability is a probability, and a probability is an area under a pdf, reliability is also an area under a pdf. Specifically, it is an area under the failure function f(t). Recall that (see Definition 4.14) a cumulative distribution function (cdf) for a random variable T is defined as $F(t) = P(T \le t)$. Since $R(t) = P(T > t)$, it is clear that $R(t) = 1 - F(t)$, i.e., R(t) and F(t) are complementary probabilities. Figure 10.1 illustrates the scenario for a normal failure function, f(t), at a specific time t_0.

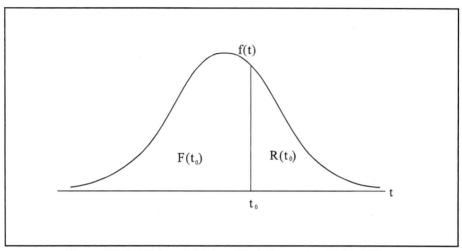

Figure 10.1 $F(t_0)$ and $R(t_0)$ as Complementary Probabilities

While $F(t_0) = P(T \le t_0)$ represents the probability that the component or system will fail prior to time t_0, $R(t_0) = P(T > t_0)$ represents the probability that the component will survive or operate past time t_0. That is, while the cdf represents the area under the curve (or pdf) to the **left** of a specific value, reliability is the area to the **right** of a specific value. If we examine $F(t)$ and $R(t)$ over all possible values of t, we obtain the relationship shown in Figure 10.2.

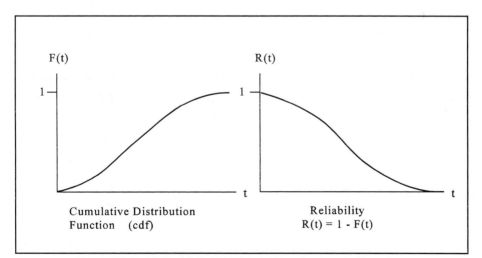

Figure 10.2 CDF and Reliability as Complementary Functions

Another useful term in reliability is the hazard function. It is defined as

$$h(t) = \frac{f(t)}{R(t)}$$

where f(t) is the failure function and R(t) is the reliability function.

Definition 10.2 Hazard Function

The hazard function is interpreted as the "instantaneous failure rate." It is the rate of failure at an instant in time, given that the item has lasted until that time. Consider a new automobile. Immediately after it is manufactured, it has a high failure rate. This initial period of high failures is called the "break-in" period. After the car is sold, the failure rate remains fairly constant for several years. This is referred to as the "useful life" of the system. Of course, we all know that as the car gets older, breakdowns are more frequent and the failure rate increases. This is referred to as the "wearout" period. These phases of the lifetime of an automobile are graphed in Figure 10.3. The curve you see is called a "bath-tub" curve for obvious reasons.

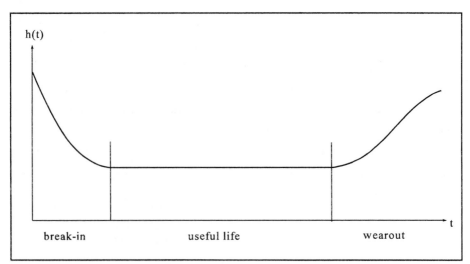

Figure 10.3 Life Cycle Hazard Function or Failure Rate

10.3 Exponential Failure Model

Perhaps the most common mathematical model used in computing component reliability is the exponential model. That is, if the random variable T denoting time to failure is exponentially distributed, then the following properties hold:

$$f(t) \; = \; \lambda e^{-\lambda t}, t \geq 0$$

$$F(t) \; = \; \text{P(failure before time t)} = P(T \leq t) = 1 - e^{-\lambda t}$$

$$R(t) \; = \; \text{P(failure after time t)} = P(T > t) = e^{-\lambda t}$$

$$E(T) = \; \text{mean time between failures} = \text{MTBF} = \; \frac{1}{\lambda}$$

$\lambda \quad = \quad$ average number of failures per time interval, which is the only parameter in the exponential distribution

One important property of an exponential reliability model is that the failure rate or hazard function is constant. Knowing f(t) and R(t), we can compute h(t) as

$$h(t) \; = \; \frac{f(t)}{R(t)} \; = \; \frac{\lambda e^{-\lambda t}}{e^{-\lambda t}} \; = \; \lambda.$$

Thus, an exponential model is often used when modeling the "useful life" period of a system as seen in Figure 10.3. A common application is modeling failures in electrical components. These components seldom have parts that wear out; rather, failures are caused by some Poisson process such as voltage spikes or intermittent current. In fact, any item whose primary failure mode can be modeled by a Poisson process will have an exponential failure function.

Another important property of the exponential distribution is called the "forgetfulness" or "memoryless" property (see Section 4.8). This property says that, no matter how long the item has been working, it is as good as new. We can express this property as

$$P(T > t_1 + t_2 \mid T > t_1) = \frac{P(T > t_1 + t_2 \cap T > t_1)}{P(T > t_1)}$$

$$= \frac{P(T > t_1 + t_2)}{P(T > t_1)} = \frac{R(t_1 + t_2)}{R(t_1)}$$

$$= \frac{e^{-\lambda(t_1 + t_2)}}{e^{-\lambda t_1}} = e^{-\lambda t_2}$$

$$= P(T > t_2)$$

Example 10.1 Suppose data is collected on 40 identical components and they are found to have a mean time between failures (MTBF) of 25 hours. Assume a constant failure rate.

  a. Find the reliability function.

  Since the failure rate is constant, we will use the exponential distribution. Also, the MTBF is 25 hours; therefore, we can estimate the parameter λ by $1/E(T) = 1/25$. Therefore, the reliability function is approximated by $R(t) = e^{-t/25}$.

  b. What is the reliability of the item at 30 hours?

  $R(30) = e^{-30/25} = .3012$

  c. What is the reliability of the item at 70 hours, given that it has already lasted 40 hours?

  Using the forgetfulness property of the exponential distribution, $P(T > 70 \mid T > 40) = P(T > 30) = e^{-30/25} = .3012$.

  ∎

The reliabilities just calculated provide estimated probabilities that given components or systems will not fail until after a given time t. However, just as we sometimes find it valuable to calculate confidence limits for a mean (see Section 5.4), we may want to calculate a confidence limit for reliability. Generally, we are not concerned with upper confidence limits for reliability, only lower limits. If an exponential failure mode is used for R(t), a $(1 - \alpha)100\%$ lower confidence limit for reliability at time t is given by:

$$R_L(t) = \exp\left(\frac{-t\,\hat{\lambda}\,\chi^2(1 - \alpha,\, 2n)}{2n}\right),$$

where

t = time

$\hat{\lambda}$ = estimated average number of failures per time interval = 1 / MTBF,

n = sample size, and

$\chi^2(1 - \alpha, 2n)$ is a Chi-Square table look up, with α = level of significance.

Using this relationship to calculate a 95% lower confidence limit for the reliability at 30 hours of the component given in Example 10.1, we have

$$R_L(30) = \exp\left(\frac{-30\,(1/25)\,(101.9))}{2\,(40)}\right) = e^{-1.5285} = .2169$$

Thus, we can be 95% confident that the true reliability of the component at 30 hours is at least 0.2169.

10.4 Normal Failure Model

Another common failure model is the normal distribution. In this case, probabilities are computed just as they were in Chapters 3 and 4. The parameters μ and σ can be estimated by computing $\bar{t}$ and s from a random sample of failure times.   The hazard function is much more difficult to compute for the normal distribution and will not be presented.

Example 10.2 Suppose we sample 36 machine parts and put them on test until they fail. We record their failure times, construct a histogram, perform a χ^2 Goodness-of-Fit test, and conclude that time to failure is normally distributed. We also calculate $\bar{t}$ = 50 hours and s = 10 hours.

a.   What is the estimated reliability of the parts at 40 hours? Using $\bar{t}$ and s as estimates of μ and σ, we let μ = 50 and σ = 10.

$$\text{Therefore,} \qquad R(40) \quad = \quad P(T > 40) = 1 - P(T \le 40)$$

$$= \quad 1 - P\left(\frac{T - \mu}{\sigma} \le \frac{40 - 50}{10} \right)$$

$$= \quad 1 - P(Z \le -1) = 1 - .1587$$

$$= \quad .8413$$

b.   What is the estimated reliability at 60 hours, given that the system has already lasted 20 hours? Recall that the forgetfulness property applies only to exponential failure rates. Therefore, we will calculate the probability by using the definition of a conditional probability as follows:

$$P(T > 60 \mid T > 20) \quad = \frac{P(T > 60 \cap T > 20)}{P(T > 20)} = \frac{P(T > 60)}{P(T > 20)}$$

$$= \frac{P(Z > 1)}{P(Z > -3)} = \frac{1 - P(Z \le 1)}{P(Z < 3)}$$

$$= \frac{1 - .8413}{.9987} = .1589$$

10.5 Types of Systems

When a system is composed of individual components for which the reliabilities are known, the system reliability can be computed. In this section we look at two types of systems and then at combinations of the two.

10.5.1 Series Systems

In a series system, all of the components must be operating in order for the system to operate. A diagram of a series system with three components is given below.

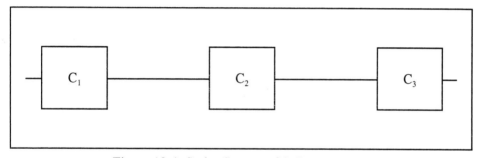

Figure 10.4 Series System of 3 Components

We will assume that each of the components fails independently of the others. If we let T_1, T_2, and T_3 denote the random failure times of the three components and T denote the failure time of the system, then the reliability of the system, R(t), can be computed by

$$R(t) = P(T > t) \quad = P(T_1 > t \cap T_2 > t \cap T_3 > t) \qquad \text{(definition of series system)}$$

$$= P(T_1 > t) \cdot P(T_2 > t) \cdot P(T_3 > t) \qquad \text{(assuming independence)}$$

$$= R_1(t) \cdot R_2(t) \cdot R_3(t)$$

In general, the reliability of n components in series is given by the product of the respective component reliabilities:

$$R(t) = R_1(t) \cdot R_2(t) \cdot R_3(t) \cdots R_n(t)$$

10.5.2 Parallel Systems

In a parallel system, the system fails only if all of the individual components fail. Again, we will assume that the failure times are independent. A diagram of a parallel system is given below.

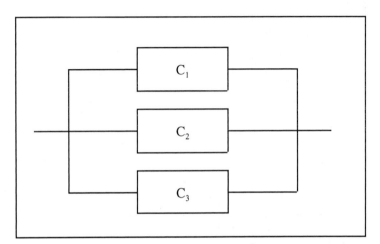

Figure 10.5 Parallel System of 3 Components

We can derive the reliability of this system in a similar manner.

$R(t) = P(T > t) = 1 - P(T \leq t)$ (reliability as the complement of failure)

$= 1 - P(T_1 \leq t \cap T_2 \leq t \cap T_3 \leq t)$ (note: the system fails only if all of the components fail)

$= 1 - [P(T_1 \leq t) \cdot P(T_2 \leq t) \cdot P(T_3 \leq t)]$ (assuming independence)

$= 1 - [(1 - R_1(t)) \cdot (1 - R_2(t)) \cdot (1 - R_3(t))]$ (reliability as the complement of failure)

In general, if we have n components in parallel, the formula becomes

$$R(t) = 1 - [(1 - R_1(t)) \cdot (1 - R_2(t)) \cdot (1 - R_3(t)) \cdots (1 - R_n(t))]$$

10.5.3 Combination Systems

The formulas for series and parallel systems can be combined to calculate the reliabilities of more complex systems. This is best shown in an example.

Example 10.3

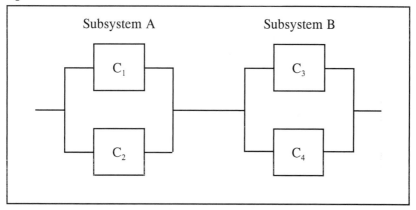

Suppose the reliabilities of the above components are computed at time t = 20 and they are given by $R_1(20)$ = .85, $R_2(20)$ = .90, $R_3(20)$ = .95, and $R_4(20)$ = .88. To calculate the reliability of the system, we first find the reliability of the subsystem containing components 1 and 2 (Subsystem A) and the reliability of the subsystem containing components 3 and 4 (Subsystem B). Each of these subsystems is a parallel system in itself. Thus,

$$R_A(20) = 1 - [(1 - .85)(1 - .90)] = .985$$

$$R_B(20) = 1 - [(1 - .95)(1 - .88)] = .994$$

We notice that the two subsystems are themselves in series so the reliability of the overall system is given by

$$R(20) = R_A(20) \cdot R_B(20) = (.985) \cdot (.994) = .97909$$

10.5.4 Redundancy

Backup or redundant systems are used in many situations. For example, the space shuttle carries redundant computer systems. The booster rocket for the shuttle has a redundant set of o-rings to seal the separate stages. How many redundant systems are required? If we know the individual component reliabilities and assume that the components fail independently of each other and are in a parallel system, we can calculate the number of components that are necessary to achieve a desired overall system reliability.

The system reliability will be given by

$$R(t) \quad = 1 - [(1 - R_1(t)) \cdot (1 - R_2(t)) \cdot (1 - R_3(t)) \cdots (1 - R_n(t))]$$

$$= 1 - [1 - R_1(t)]^n \qquad \text{(since all component reliabilities are the same)}$$

Solving for n in the last equation, we have

$$[1 - R_1(t)]^n \; = \; 1 - R(t)$$

$$n \ln[1 - R_1(t)] \; = \; \ln[1 - R(t)] \quad \text{(taking logs of both sides)}$$

$$n \; = \; \frac{\ln[1 - R(t)]}{\ln[1 - R_1(t)]}.$$

But since n must be an integer, we have

$$n \; = \; \left\lceil \frac{\ln[1 - R(t)]}{\ln[1 - R_1(t)]} \right\rceil,$$

where, if you recall from Chapter 2, $\lceil x \rceil$ denotes the smallest integer greater than or equal to x.

Example 10.4 We desire to operate a system with an overall reliability of 0.99 at 100 hours. If we have components that only provide a reliability of 0.80 at 100 hours, how many components do we need to place in parallel in order to achieve the desired system reliability?

$$n = \left\lceil \frac{\ln[1 - R(100)]}{\ln[1 - R_1(100)]} \right\rceil = \left\lceil \frac{\ln[1 - .99]}{\ln[1 - .80]} \right\rceil = \lceil 2.86 \rceil = 3$$

Therefore, we need $n = 3$ components in parallel to achieve the overall system reliability of 0.99. In fact, our system reliability is

$$R(100) = 1 - [1 - R_1(100)]^3 = 1 - (.20)^3 = 1 - .008 = 0.992$$

Why is this more than the desired system reliability?

■

10.6 Summary

The reliability of a system at time t is the probability that the system lasts beyond the specified time t. We have shown how these probabilities can be calculated using either the normal or exponential distributions. Other models may be used but the same principles apply. Simple system reliabilities may then be calculated using the basic rules of probability.

The models presented in this chapter are quite simple, yet very powerful. They provide a basis for understanding reliability concepts and for performing simple reliability computations. This chapter is intended to be only an introduction to a very diverse and complex subject. If the reader desires a more comprehensive treatment of the subject, there are numerous texts that provide extensive coverage. Among these are [Gros 89], [Mann 74], [Kapu 77], [Bain 78], [Barl 75], and [Barl 65].

10.7 Problems

10.1   An electrical component has a mean time between failures (MTBF) of 200 hours. Compute the reliability of the component at 10, 100, 200, 300, 400, 500 and 600 hours. Graph the probabilities.

10.2   Refer to problem 6.12.

   a)   Calculate the reliability of each of the thermocouples at 20 hours.
   b)   Calculate the reliability of each of the thermocouples at 20 hours given that they have already lasted 10 hours.

10.3   Refer to problem 6.15.

   a)   Using only the 100 (already sorted) failure times, estimate the reliability of this simulated component at t = .10, .20, .40, .60, .80, 1.0, and 2.0 time units.
   b)   Graph the estimated reliability function from the data obtained in part (a).
   c)   Estimate λ (the average failure rate) from the 100 data points and then determine the estimated hazard function.

10.4   Suppose a computer component exhibits a constant failure rate. What is its reliability at its mean time between failures (MTBF)?

10.5   A random sample of 50 microprocessors is put on accelerated test at high temperature. The MTBF is 9.970 hours. Assume a constant failure rate.

   a)   What is the estimated failure rate?
   b)   Find the reliability at 5 hours.
   c)   Find a 95% lower confidence limit for the reliability at 5 hours.
   d)   Find the reliability at 5 hours given that the microprocessor has already lasted 2 hours.
   e)   Find the reliability at 10 hours given that the microprocessor has already lasted 7 hours.

10.6   A government contract calls for a system that uses the microprocessor discussed in the previous problem. If the contract requires an overall reliability of 0.99 at 2 hours, how many processors need to be placed in parallel to achieve this reliability?

10.7   Find the reliability of the following system at t = 10.

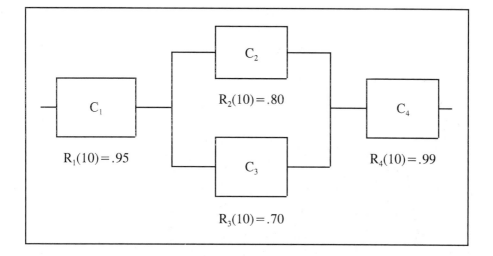

10.8   Below is a cross-section diagram of a booster rocket outer shell at a joint between two stages. The system will fail only if both o-rings fail.

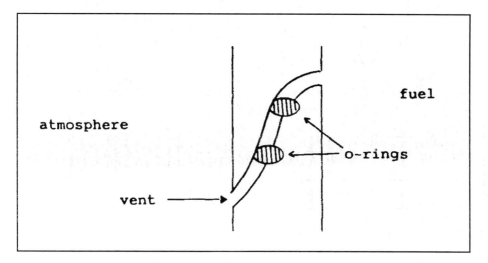

    a)   Do the two o-rings form a series or parallel system?
    b)   Suppose the reliability of each o-ring is 0.95 during the most critical phase of flight. What is the system reliability?

10.9   Suppose components #1, #2, and #3 below have a constant failure rate of 5 failures per 100 hours. Furthermore, assume that the failure rate of component #4 is normally distributed, i.e., $C_4 \sim N(22, 60)$.

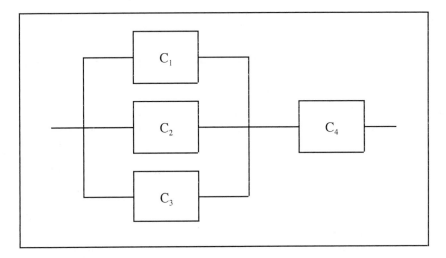

    a)   Find the MTBF for each component.
    b)   Find R(7.2), the system reliability at 7.2 hours.
    c)   How many components must be added to the parallel portion in order to achieve a system reliability of 0.96 at t = 7.2 hours?

10.10   Stress screening is a method for testing components before they are sold. The idea is to subject all components to higher stress (e.g., high temperatures and vibrations) for a specified period of time. This reduces the "break-in" period for the ones that survive.

A computer manufacturer put 24 components through a stress screening program. The following failures were recorded.

Time				Failures
0	to less than	1	hour	3
1	to less than	2	hours	1
2	to less than	4	hours	2
4	to less than	6	hours	3
6	to less than	10	hours	1
10	to less than	24	hours	1

a)   Calculate the MTBF for the components that failed.

b)   Is the exponential distribution a good model for this data? Why or why not?

10.11   Reference Case Study 25: Iranian Hostage Rescue Mission. Using the data in the case study (i.e., λ), determine the number of helicopters that would be needed to provide a 0.9 chance of mission success if the mission length were to somehow be reduced to 9 hours.

10.12   Match the function to its name for the exponential failure model.

_____ failure function, f(t)	a) $1 - e^{-\lambda t}$
_____ hazard function, h(t)	b) λ
_____ reliability function, R(t)	c) $\lambda e^{-\lambda t}$
_____ cumulative probability function for failure time, F(t)	d) $e^{-\lambda t}$

10.13   Suppose a component has a constant failure rate of 0.5 failures per year.

a)   What is the Mean Time Between Failures (MTBF) for this component?

b)   What is the reliability of this component at three years?

c)   What is the probability this component will fail within three years?

d)   What is the reliability of this component at five years, given the component is already three years old?

10.14   A certain electronic component has an exponential failure time with a mean of 50 hours.

a)   What is the instantaneous failure rate of this component?

b)   The reliability of this component at 100 hours is only 0.1353. What is the minimum number of these components that should be placed in parallel if we desire a system reliability of 0.90 at 100 hours?

10.15   Consider the following system:

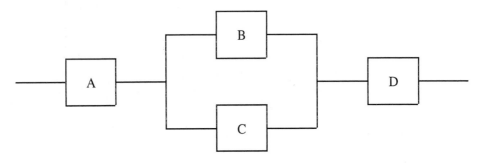

    a)   Component "A" has a constant failure rate with a mean of 200 hours. What is the reliability of Component "A" for 220 hours?

    b)   Suppose the reliability of both the "B" and the "C" components for 220 hours is 0.95 and the reliability of the "D" component is 0.99. What is the entire system reliability for 220 hours? (Use $R_A(t)$ found in part a.)

10.16   An electronic component has a mean life of 500 hours. Given that it has already lasted 450 hours, find the probability that it will last another 100 hours before failing, (i.e., find $P(T > 550 \mid T > 450)$) if:

    a)   The component's life is **exponentially** distributed.

    b)   The component's life if **normally** distributed with a standard deviation of 200 hours.

10.17   A system made up of two components will fail as soon as one of the two components fails. The first component's failure characteristics can be modeled with a normal distribution with a mean of five years and a standard deviation of one year. The second component has a constant failure rate of ten years. What is the reliability of the system at three years if the two components operate independently?

Chapter 11

Quality Function Deployment (QFD)

11.1 What is QFD?

QFD is a systematic process used to integrate customer requirements into every aspect of the design and delivery of products and services. Specifically, QFD is a collection of carefully constructed matrices that are used to facilitate group discussions and decision making. These matrices are constructed using some of the basic quality tools such as affinity diagrams, brainstorming, decision matrices and tree diagrams. QFD can also help utilize other quality techniques such as Design of Experiments (DOE), Statistical Process Control (SPC) and Failure Mode and Effect Analysis (FMEA) in a systematic development approach. When used correctly, QFD can lead to:

- Better understanding of customer requirements

- Increased customer satisfaction

- Reduced time to market and lower development costs

- Structured integration of competitive benchmarking into the design process

This chapter was written by Peter Hofmann, MCI Telecommunications Corp.

- Increased ability to create innovative design solutions

- Enhanced capability to identify those specific design aspects that have the greatest overall impact on customer satisfaction, providing focus for limited development resources

- Better teamwork in cross-functional design teams

- Better documentation of key design decisions

QFD had its origins in the Kobe Shipyard of Mitsubishi Heavy Industries, LTD in the early 1970's. Dr. Yoji Akao developed what he termed "Quality Tables" to help link customer requirements to design targets. By the end of the 1970's, many Japanese companies began using QFD and some, such as Toyota, began to require its use among their key suppliers. In fact, Toyota reported that startup costs for new minivans had been reduced over 60% by using QFD.

QFD was introduced in this country through a 1983 article published by Akao and Kogure that appeared in *Quality Progress*, the journal of the American Society for Quality Control. By the mid 1980's, papers and articles from the American Supplier Institute (ASI) and GOAL/QPC were widely disseminated outlining the mechanics and the benefits of this revolutionary technique. A landmark article, "The House of Quality," was published by Hauser and Clausing in the *Harvard Business Review* in 1988. The House of Quality refers to one of the most widely used QFD matrices and will be discussed at length in a later section of this chapter. Also in 1988, *Quality Progress* published a special QFD edition to highlight some of the recent advancements with this technique. By the early 1990's, numerous QFD books, courses and software aids were widely available for those interested in applying QFD to their organization.

There have been many widely publicized applications of QFD in most all industry sectors. In the automotive industry, Ford (Taurus, Sable, Explorer), GM (Saturn, Camaro) and Chrysler (LHS, Neon) have made extensive use of QFD in their product development processes. There have been publicized examples from Electronics (DEC, Motorola, Texas Instruments), Aerospace (Martin Marietta, McDonnell Douglas, Raytheon, Hughes) as well as organizations as diverse as Bell Labs, Kodak, Proctor & Gamble and the U.S. Air Force. All of these organizations have tailored QFD to help them achieve their own specific business imperatives. While many revolve around the design and introduction of new products and services, there have been several other innovative applications of this technique, including:

- modifying existing products or services
- developing a product test plan
- proposal preparation
- deployment of national security policies
- designing hotel/conference center services

This is just a short list of the innovative ways that QFD can be used to help customer requirements drive organizational activities.

In short, QFD is a structural tool that helps any group create optimal designs that meet customer needs. The cross functional design teams that many organizations are trying to integrate into their business benefit tremendously from the structure and focus that QFD brings to the work environment. The matrices of QFD help facilitate team discussions and keep decisions focused on those things that are important to the customer. This concept of focusing on the "Voice of the Customer" is the very cornerstone of QFD, ensuring that product and service features are chosen to maximize the benefit to the customer. At the same time, QFD helps organizations succeed by ensuring prudent design tradeoffs are made based on competitive benchmarks and process analysis, minimizing costs and overall project risks.

11.2 QFD Approaches

There are two widely taught methods for using QFD. The first method, called the "Four Phase Approach," is illustrated in Figure 11.1.

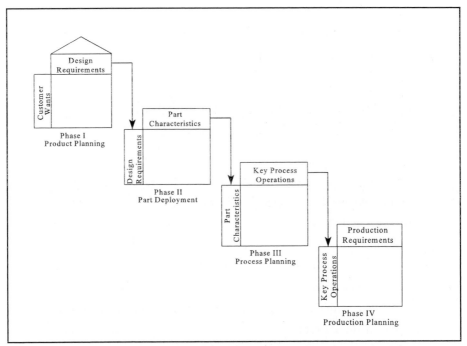

Figure 11.1 The Four Phase Approach

With this method, customer requirements are identified and used to prioritize design requirements during the Product Planning Phase or Phase I. This first matrix, commonly called the "House of Quality," also contains competitive benchmarks, design targets, design tradeoffs and many other key design drivers. After Phase I, the prioritized design requirements are used to identify key part characteristics in the Part Deployment Phase or Phase II. In Phase III, the key parts are used to systematically identify those process operations that are critical to ensuring customer requirements are met. This is also called the Process Planning Phase. Finally, in the Production Planning Phase or Phase IV, specific production requirements are identified to ensure that the key processes from the previous phase are adequately controlled and maintained. This method has been taught extensively by the

American Supplier Institute and is widely used in the automotive industry. It has also been successfully used in service and other industries.

The other widely taught method, called the "Matrix of Matrices," is illustrated in Figure 11.2.

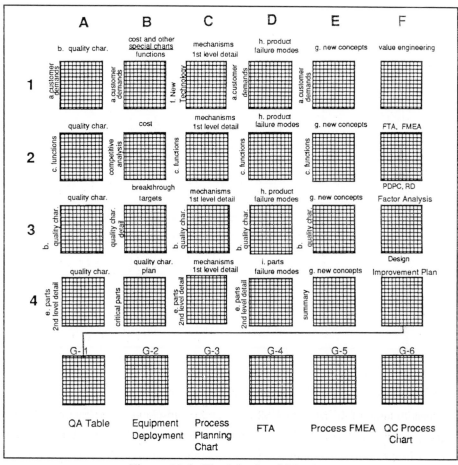

Figure 11.2 The Matrix of Matrices

This method is the outgrowth of twenty years of work by Dr. Akao. It was introduced in this country by Bob King of GOAL/QPC through his book ***Better Designs in Half the Time***. While it appears more confusing at first glance, this method is actually more comprehensive and flexible than the Four Phase Approach.

In fact, all of the matrices of the Four Phase approach are contained in the Matrix of Matrices as illustrated in Figure 11.3, so the two methods are very compatible. For example, the body of the House of Quality (Phase I) is the same as matrix A-1 in the Matrix of Matrices comparing customer requirements against design requirements. Similarly, the roof in Phase I is equivalent to matrix A-3 since both highlight potential interactions between design requirements.

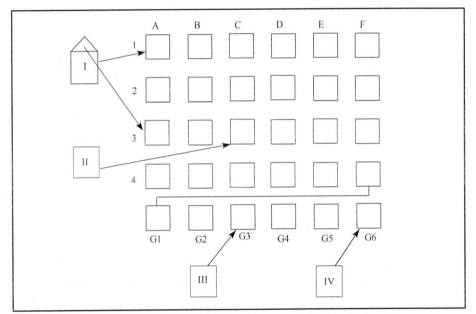

Figure 11.3   Four Phase Method versus Matrix of Matrices

It is best to think of the Matrix of Matrices as a comprehensive toolbox of decision matrices that can help a design team make informed decisions on the details of a project using customer requirements and competitive benchmarks as the primary decision criteria.

It is general practice to use a tailored approach that makes the most sense for the particular application under consideration. In most cases, this almost always involves the use of the House of Quality (HOQ) as the initial starting point in a QFD project. It is also generally accepted that the most benefit is obtained after the project progresses past the first matrix or HOQ, since it is only at that point that detail design decisions, process and failure analyses are performed. Most QFD practitioners agree that since it is very important to progress past the HOQ, teams should not be concerned about using a "right or wrong" matrix. QFD is primarily a decision tool — if the team can use a matrix to facilitate a discussion and to make a design decision, DO IT!

There are several ways to help a team progress through a series of useful matrices. Bob King discusses the use of a tool called the "Voice of the Customer Table" to help identify other applicable matrices for the team to use. Another method that this practitioner has found very successful has been the "Designer's Dozen." The Designer's Dozen is a series of 12 QFD matrices arranged around the Shewhart Cycle (Plan-Do-Check-Act) as illustrated in Figure 11.4.

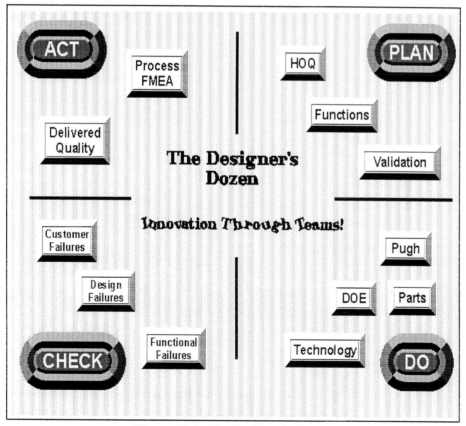

Figure 11.4 The Designer's Dozen

The main benefit derived from this approach is that the QFD team is presented with a "toolbox" of matrices to choose from as well as some suggested sequence to follow. It is not intended that each matrix be used for each project, but the flexibility is there to use what will help the team the most. In addition, most teams are very comfortable working within the familiar PDCA framework. The Designer's Dozen approach to implementing QFD is summarized as follows:

PLAN   The first phase of a project focuses on identifying key requirements and defining broad project goals and targets.

HOQ - The House of Quality or HOQ is the most widely used QFD matrix. It contains comparisons between customer requirements and design requirements, competitive benchmarks and other project planning information.

Functions - This matrix, sometimes referred to as the "Voice of the Engineer," compares functional requirements to design requirements. This matrix provides a golden opportunity to involve technical experts from outside the core QFD team.

Validation - This matrix compares customer requirements to functional requirements. This matrix is used primarily with the "Functions" matrix to assure consistency between customer, design and functional requirements.

DO   The iterative process of developing and evaluating design concepts, performing optimization analyses and developing detailed designs.

Pugh - Developed by Stuart Pugh, the Pugh Concept Selection Matrix is a tool specifically intended to integrate the design creativity of individuals with the refinement and analysis capabilities of a team. After the HOQ, this is one of the most widely used matrices in QFD.

DOE - The Design of Experiments (DOE) Scorecard is used to record and document the various optimization studies that have been conducted on different design concepts.

Parts - In this matrix, the key design requirements are evaluated against the parts of a specific design concept to determine those parts that require additional focus and resources.

Technology - In this matrix, the key parts identified in the previous matrix are evaluated against emerging technologies to highlight those technologies that should be infused into the product or service.

CHECK This is where the potential failure modes of a given design concept are evaluated against the key requirements from the "PLAN" phase. Focus is given to those failure modes that have the greatest overall impact to key requirements.

Customer Failures - In this matrix, those potential product or service failure modes which would be most damaging in the eyes of the customer are identified.

Design Failures - In this matrix, those design requirements which are most susceptible to product or service failures are identified. Potential design weaknesses and key failure modes are identified.

Functional Failures - In this matrix, those potential product or service failure modes that would have the greatest impact on the key functions of the product or service are identified.

ACT   This final phase is where you ensure that a product or service is delivered as planned.

Delivered Quality - This table is used to identify how a product or service will be delivered in accordance with the design requirements.

Process FMEA (Failure Mode and Effect Analysis) - This chart is used to help minimize the potential for the process used for making product/providing service to fail to perform correctly.

11.3 How does QFD Fit into an Organization?

In a traditional (i.e., pre-1990) organization, new product or service development was handled in a process commonly referred to as "over the wall." The marketing and sales organizations defined customer requirements through market research, focus groups, and other accepted forms of market research. This information was then "thrown over the wall" to product development who then interpreted these requirements and translated them to engineering requirements. This refinement happened with little or no interaction with the marketing group or the original customer. These requirements were then thrown over another wall to the manufacturing planners and process engineers who figured out how to build the product, again with little or no interaction with the product designers. Finally, the manufacturing plans and process specs were thrown over the final wall to manufacturing operations who had to finally make some new, lumpy objects.

There are several drawbacks to this approach. The strong functional "stovepipes" that are evident at each wall restrict cross-functional communication instead of enhancing it. For example, people knowledgeable in manufacturing operations are rarely involved in influencing design approaches. Their role was traditionally limited to building what they were told to build. If the design was "unbuildable" for some reason, multiple redesign efforts and accompanying delays were typical at the point where production was scheduled to begin. In fact, it was common to see an emphasis on working out problems after the production launch occurred. With all of the handoffs "over the wall," it was very difficult to relate what customers actually said they wanted to the products that were being produced.

Modern organizations have discovered that cross-functional efforts focused, if not obsessed, with the customer's requirements are crucial to the success of new products and services. Reengineered organizations with collections of horizontal, cross-discipline, customer focused processes are becoming more commonplace with strong functional stovepipes becoming more the exception than the rule. QFD is a valuable tool to assist these cross-functional teams in developing successful new products and services for the following reasons:

- QFD is a team-based tool

- QFD provides a formal structure for documenting and communicating key information internal and external to a team

- Obsession with the Voice of the Customer

- Strong emphasis on systematic planning and minimizing redesign

Finally, most organizations have adopted some company-wide quality improvement efforts. These efforts come under the heading of TQM, CQI, MDQ or some other three-letter acronyms. At a minimum, these efforts involve some type of high performance work teams, increased customer focus throughout the company and a conscious decision to focus on processes in the daily workplace with strong emphasis on management by fact. As stated earlier, QFD uses many of the basic and advanced tools of quality improvement in an integrated framework for developing new products and services. QFD is an excellent complement to these efforts and helps mainstream these quality concepts into the daily work environment.

The Malcolm Baldrige National Quality Award criteria, considered by many to be the most useful blueprint for organizations in search of achieving overall business excellence, contain the following requirements for the Design and Introduction of Products and Services:

"Describe: (1) how customer requirements are translated into product and service design requirements; (2) how product and service design requirements are translated into efficient and effective production/delivery processes, including an appropriate measurement plan; and (3) how all requirements associated with products, services, and production/delivery processes are addressed early in design by all appropriate company units, suppliers, and partners to ensure integration, coordination, and capability."

1996 MBNQA Criteria

These requirements clearly describe a process that is very much aligned with the fundamentals of a QFD approach to developing products and services; translating the voice of the customer into design requirements, tracing these requirements throughout the organization and doing all of these efforts in a cross-functional manner.

11.4 Voice of the Customer

Understanding what customers want or need from a product or service is crucial to the successful design and development of new products and services. Unfortunately, even the simple distinction between wants and needs can become a complex problem to solve. Consider the following situation. I am in possession of a single orange and two of my friends stop by for a visit. My friend Mark tells me he "wants" an orange. On the other hand, my friend Carol tells me she "wants" an orange too. Wanting to best satisfy the requirements of my customers with my limited resources, I cut the orange in half and give half to Mark and half to Carol.

On the surface, it appears that I did an effective job of listening to the voice of my customers and satisfying their needs with my limited resources. Unfortunately, Mark peeled his half, ate the fruit and discarded the rind (although he was still hungry for another half). Carol, on the other hand, peeled her half, discarded the fruit and went to make orange marmalade with the rind (although she had to find more rind at home). Had I taken the time to more fully understand the "needs" of my customers, not just the "wants" that they told me, I could have been more responsive given my limited resources. Fully understanding customer needs or requirements requires dedicated effort to develop an understanding of the purpose and application (eat the fruit of an orange vs. make orange marmalade with the peel) that the customer intends for a given product or service.

There are many models that attempt to quantify this distinction between "spoken" wants and true "needs" of the customer. One of the most widely used was developed by Professor Noritaki Kano. Kano's model defines three types of quality requirements (one-dimensional quality, expected quality and exciting quality) and how the achievement of these requirements affects customer satisfaction. Kano's Model is summarized in Figure 11.5.

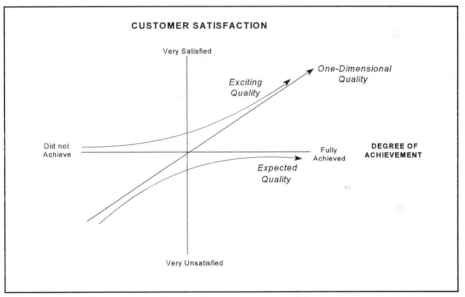

Figure 11.5 Kano's Model

One-dimensional Quality
- Specifically requested items; specifications
- Items from a typical survey (stated wants)
- If present, customer satisfied
- If absent, customer dissatisfied
- Example: "cook my steak medium rare" in a restaurant

Expected Quality

- Not specifically requested, but assumed to be present
- If present, customer neither satisfied nor dissatisfied
- If absent, customer very dissatisfied
- Example: clean silverware in a restaurant

Exciting Quality

- Unknown to the customer
- Most difficult to define and develop
- If present, customer very satisfied
- If absent, customer neither satisfied or dissatisfied
- Example: fine linen or fresh floral arrangements in a restaurant

Clearly, the typical concept of customer requirements is contained within the one-dimensional quality requirements from Kano's Model. Organizations are usually skilled in surveying customers to find out what they want. Unfortunately, surveys don't usually identify what all the customer's true needs are. For example, a restaurant patron won't typically identify clean silverware or non-poisonous food as specific requirements in a survey. Therefore, if these expected requirements are met, the customer is neither satisfied or dissatisfied. However, if these requirements aren't met, the customer will be extremely dissatisfied, probably to the point of never returning. In a similar manner, features that will astound and amaze the customer are rarely articulated beforehand. These exciting quality requirements are usually derived from a detailed analysis of the customer's operating environment for the product or service. The challenge is that not meeting these exciting requirements will not cause the customer to be dissatisfied, but an opportunity to build a strong, positive relationship (and repeat business) will be missed.

The requirements associated with exciting quality and expected quality are most critical to long term success of a product or service since these requirements are closely tied to emotional responses in customers. Safety requirements usually fall into the category of expected quality. Selling an unsafe car, pain reliever or plane ride will usually result in a strong, unfavorable reaction from customers.

However, delivering safe cars, drugs and plane rides doesn't bring any extra points in the eyes of the customer. Similarly, an exceptional environment in a restaurant and good food on airplanes have been identified as exciting requirements. Customers experiencing these services develop strong positive opinions of the service provider and become repeat customers.

These exciting requirements don't have to be as "above and beyond" the normal call of duty as might be implied. The main requirement is to give the customer something they didn't really understand that they needed. For example, an Air Force acquisition organization traditionally operated by finding out the detailed specifications for large radar systems from their users (one-dimensional requirements) and then contracted for the development and production of a single system that satisfied all of the requirements. Unfortunately, there was typically a 5 - 8 year lead time between start and delivery. By working with the user community and understanding the true needs of their "business," the overriding need to have some incremental capability in place quickly that could be modified incrementally over time was revealed. The resulting change in focus for the acquisition organization lead to significant improvements in their levels of customer satisfaction.

How does an organization identify the voice of the customer? Given that there are multiple dimensions, there is no single solution. In addition, these requirements are dynamic. For example, the fresh flowers that were exciting in the restaurant the first five visits will become expected by visit ten. Therefore, assessing customer requirements is a never ending task. There are many methods (special surveys, focus groups, secret shopper, field observations, etc.) for collecting information. The main thing to avoid is becoming dependent on one method. Above all, don't view simple surveys as a "quick fix" to quality problems. It takes a great deal of effort to understand the customer's business or application, effort that can provide huge benefits for developing new products or services.

Finally, when collecting information on customer requirements, don't forget to obtain information about competitors' products. Competitive benchmarks are critical to the success of a QFD project, especially since this information helps the QFD team focus resources and identify design tradeoffs. In addition, the priorities of different customer requirements will also have a significant bearing on the overall focus of the project.

11.5 The House of Quality

Constructing the House of Quality

The initial QFD matrix, normally referred to as the House of Quality (HOQ), is completed by the project team using a group process. The customer requirements are brainstormed and then analyzed using affinity and tree diagrams. These customer requirements, referred to as the "voice of the customer" or the "whats," are then validated, prioritized and benchmarked using direct feedback from customers. Next, design requirements, or the "hows," are brainstormed and analyzed in a similar manner to the process used for customer requirements. Design requirements are defined as those things you would measure to ensure that all customer requirements are met.   Each customer requirement is then systematically compared to each design requirement and the strength of each relationship is determined. After completing a few simple calculations, customer requirements and design requirements are readily prioritized, allowing the project team to focus on those key requirements that will have the greatest positive impact on their customers. The basic inputs and outputs (Figure 11.6) as well as the basic structure of the House of Quality (Figure 11.7) are shown next:

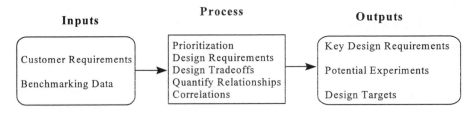

Figure 11.6 IPO Diagram for HOQ

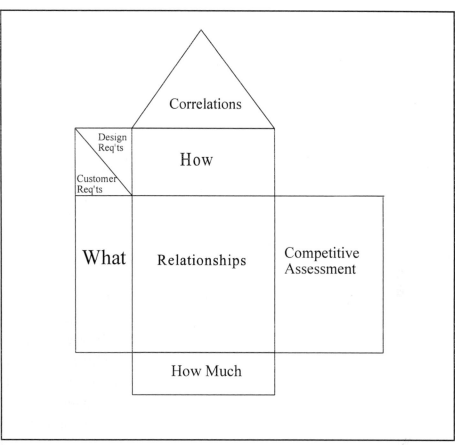

Figure 11.7 House of Quality

The basic process for completing the House of Quality can be summarized with the following questions and the related "room" of the HOQ:

- What does the customer want? (What)

- How well do we and our competitors currently meet these requirements? (Competitive Assessment)

- Where should we focus our efforts for maximum return? (Competitive Assessment)

- What can we measure or control to ensure customer requirements are met? (How)

- How do our proposed design requirements relate to our customer requirements? (Relationships)

- What key design requirements do we need to focus on? (How Much)

- What design tradeoffs need to be analyzed, potentially with designed experiments? (Correlations)

These questions help keep teams focused on the goals associated with completing the HOQ. It should be noted that many teams have quit the QFD process after experiencing burnout from completing the original HOQ. The primary reason for this is a lack of focus. Some teams try to work with matrices with over 100 customer and design requirements. This means at least 10,000 decisions! Most successful teams recognize that QFD is a decision tool and keep the size of the matrices to less than 30 rows and columns.

The following detailed instructions are for constructing a typical HOQ. A blank QFD form (see Figure 11.8) follows this list of instructions. Several completed case studies can be found in Chapter 12.

Identify Customer Requirements

- Obtain customer requirements.
  - See methods under Voice of the Customer section.
  - Obtain priorities and benchmarking data at the same time.

- Organize the verbal data using affinity diagrams and tree analysis.
  - First create the affinity diagram to gain a broad understanding of customer requirements.
  - Use the tree diagram to ensure that the data is complete and covers all logical categories.

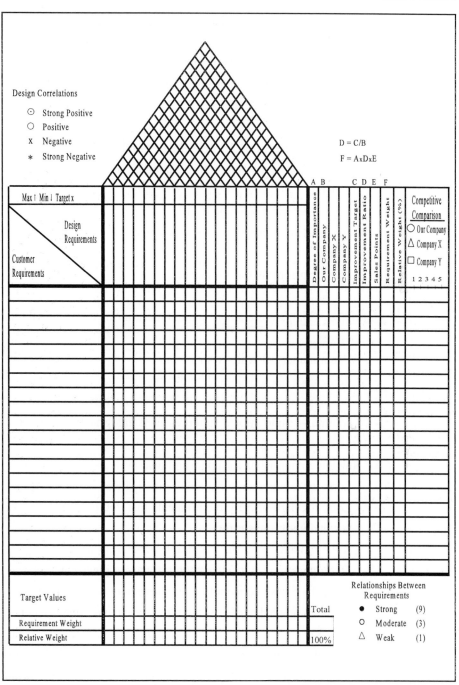

Figure 11.8 House of Quality

- Place the customer requirements in the left column of the HOQ (What).
  - Limit to 25 - 30 requirements at a time.
  - Use the secondary or primary level of the tree diagram as a starting point if necessary.

Perform Prioritizing, Benchmarking, Improvement Goals, Sales Points

- Fill in the "Degree of Importance" column (Column A) for each requirement.
  - Use a numerical scale of 1 - 10.
  - Never use one number more than twice (i.e., maximum of two 10's).
  - Review priorities for consistency.

- Fill in the "Company Ranking" column (column B) for you and your competitors.
  - Use a numerical scale from 1 - 5.
  - Remember, these values should come from customer feedback.
  - For ease of interpretation, fill the "Competitive Comparison" column — same information presented in a graphical format.

- Review the comparisons for opportunities.
  - Look for requirements that aren't satisfied well by any company.
  - Look for high priority requirements that you don't satisfy well but that your competitors do.

- Use these opportunities to identify improvement plans.
  - Use a 1 - 5 scale to identify an improvement target (Column C).
  - Include knowledge of "exciting quality" requirements.

- Compute the improvement ratio (Column D) by dividing the improvement target (Column C) by your current level of performance (Column B).

- Identify sales points (extra emphasis for known key requirements).
    - Use a value of 1.3 or 1.5.
    - Only assign sales points to a few requirements.
    - Be sure to include "exciting quality" requirements as they will not be heavily weighted by customers.

- Determine the requirement weight (Column F) for each customer requirement by multiplying the degree of importance (Column A), improvement ratio (Column D) and sales points (Column E) together.

- Determine the relative weights for each customer requirement.
    - Sum all the customer requirement weights (Column F).
    - Divide each individual weight by the total to obtain the relative weight expressed as a percentage of the total.
    - Think of it as a type of Pareto analysis.

Create Design Requirements

- Brainstorm a list of potential design requirements (How).
    - Perform affinity and tree analysis as with customer requirements.
    - Enlist the support of technical experts in the organization.
    - Focus on things that you would measure or control to ensure a customer requirement is satisfied.
    - Avoid design solutions to maintain flexibility in later matrices.

- Ask someone outside the team to perform a reality check.
    - Confirm design requirements are complete/accurate.
    - Ensure customer requirements are consistent with experience.

Perform Systematic Comparisons

- Compare customer requirements to design requirements.
    - Evaluate all combinations for relationships.
    - Assign a strength (strong, moderate, weak) to those requirements which have relationships.

- There will not be an appreciable relationship between all requirements.

• Determine the requirement weight for each design requirement.
   - Multiply the value of each relationship by the corresponding relative weight for the customer requirement.
   - Sum all of these values for each design requirement (i.e., for each column).

• Determine the relative weight for each design requirement in the same manner used for customer requirements.

Identify Correlations Between Design Requirements (Correlations)
   • Identify discrete "target values" for each design requirement.

   • Identify the movement of the target value (minimize, maximize, target).

   • Identify the correlations between design requirements.
      - There should always be at least one negative correlation.
      - Potential tradeoffs should become apparent.
      - Possible designed experiments should also be identified.

Interpreting the House of Quality
   Once the initial House of Quality is finished, a great deal of information is now available for the design team to use. Some of the information includes prioritized customer requirements, prioritized design requirements and key design targets just to name a few. It is important to critically review the HOQ to ensure that the data about to be used to drive the rest of the design process is valid. Similar to the signals on a control chart that indicate a process is behaving in a non-random manner, there are indications on a QFD matrix that part of the analysis may have been flawed. While these "signals" lack the statistical basis of the signals on control charts, many practitioners agree that the interpretations outlined in the

following matrices should at least be considered as a team progresses past a QFD matrix. Examine the main body of the matrix for the following eight characteristics:

1. blank columns: unnecessary design requirements

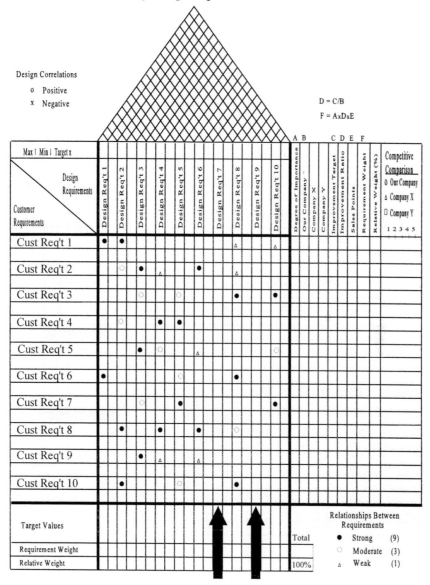

2. blank rows: missed customer requirements

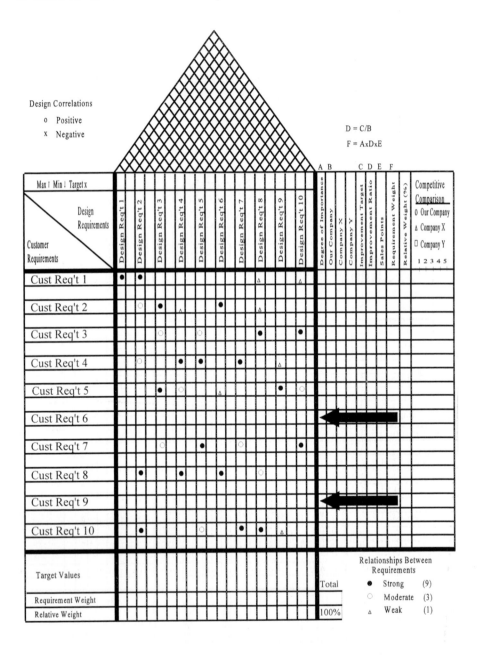

3. rows with no strong relationships: similar to blank rows — missed customer requirements

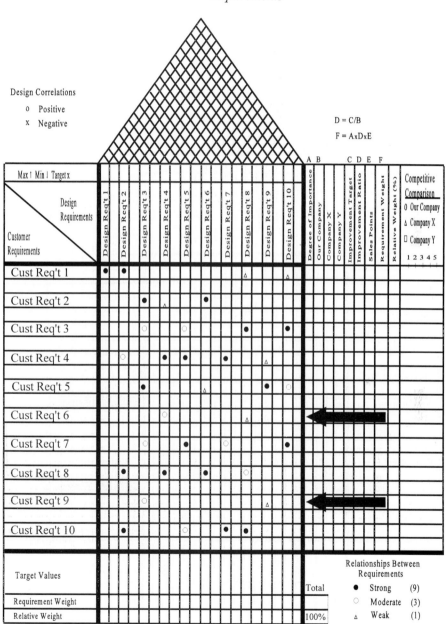

4. identically weighted rows: possible misunderstanding of customer requirements

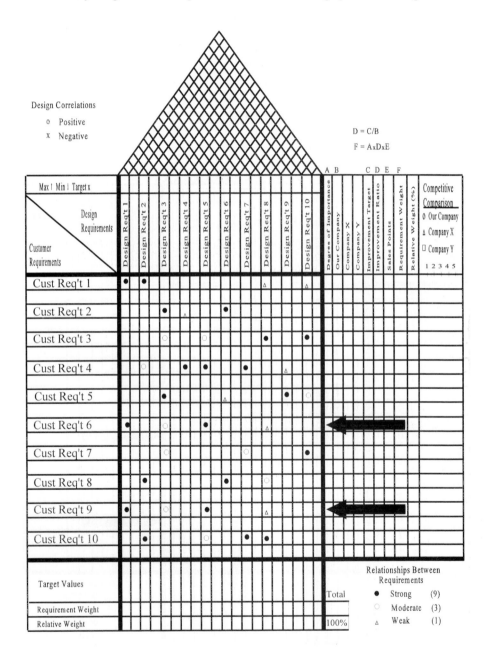

5. strong diagonal pattern: *customer requirements probably contain design solutions*

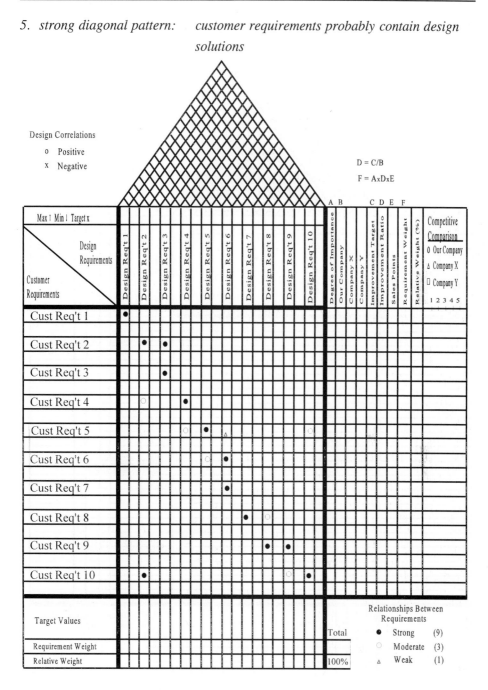

6. complete or nearly complete row: customer requirement involves cost,
reliability, safety or tier problem

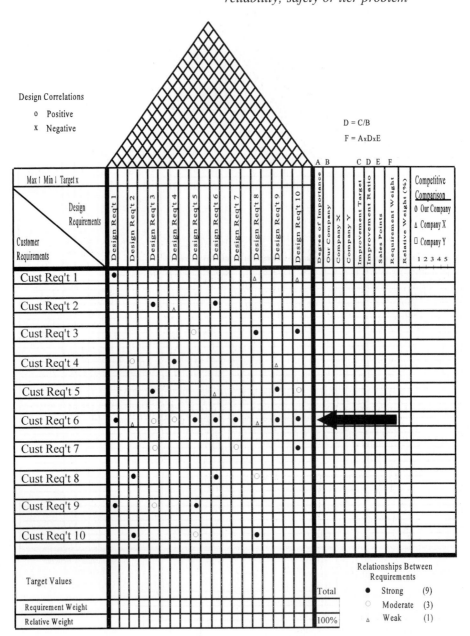

7. *complete or nearly complete column: design requirement involves cost, reliability, safety or tier problem*

Design Correlations

  o Positive

  x Negative

$D = C/B$

$F = A \times D \times E$

Max ↑ Min ↓ Target x — Customer Requirements / Design Requirements	Design Req't 1	Design Req't 2	Design Req't 3	Design Req't 4	Design Req't 5	Design Req't 6	Design Req't 7	Design Req't 8	Design Req't 9	Design Req't 10	A Degree of Importance	B Our Company	Company X	Company Y	C Improvement Target	D Improvement Ratio	E Sales Points	F Requirement Weight	Relative Weight (%)	Competitive Comparison o Our Company △ Company X □ Company Y 1 2 3 4 5
Cust Req't 1	●						●		△		△									
Cust Req't 2		○				●	●	△	○											
Cust Req't 3					○		△	●		●										
Cust Req't 4		○			●		●													
Cust Req't 5			●			△	○													
Cust Req't 6	●					○	●													
Cust Req't 7			○				●			●										
Cust Req't 8				●			●	○												
Cust Req't 9				△		△	●		●											
Cust Req't 10		●				○		○	●											

Target Values											
Requirement Weight											Total
Relative Weight											100%

Relationships Between Requirements

●	Strong	(9)
○	Moderate	(3)
△	Weak	(1)

*8. large number of weak relationships: too much fuzzy weight — clear decisions
will be confounded with noise*

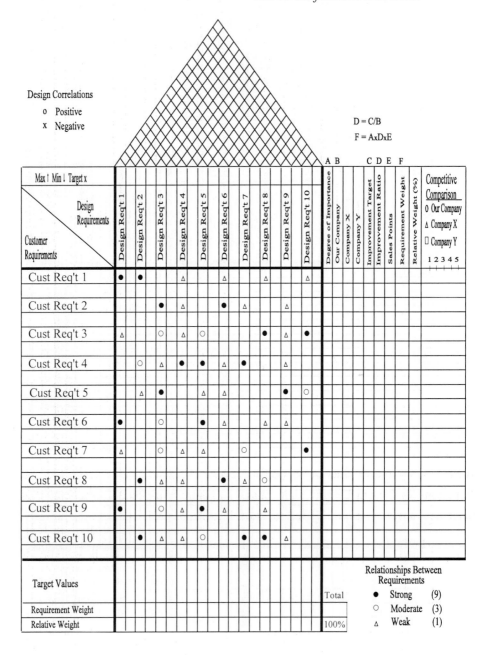

11.6 Other Useful Matrices

As mentioned earlier, the House of Quality is only one of many useful QFD matrices. In this section, a brief summary of the purpose of each of the remaining 11 matrices in the Designer's Dozen (Figure 11.4) is presented along with a high level view of the Inputs and Outputs for each matrix. A summary of the basic processing steps involved with each matrix is also included. Case Study 29: Lemon/Lime Juicer Project gives additional completed examples of these matrices or charts.

Functions - *Functions vs Design Requirements*
Purpose: This chart is used to identify all functions of a product or service that are incorporated into the design, including those that may be unknown to the customer. This is where the engineering "experts" in your organization are asked to contribute to the study. This chart can also help identify unnecessary design requirements.

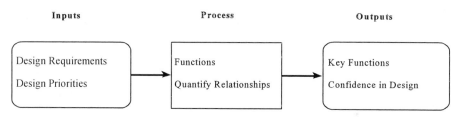

Validation - *Key Customer Requirements vs Functions*
Purpose: This chart is used to quantify the relationships between key customer requirements and product or service functions. It is also used, along with the Functions vs Design Requirements chart, to assure consistency between design priorities, customer priorities and functional priorities.

Pugh - *Evaluation of New Design Concepts*

Purpose: This chart is used to evaluate which design concepts are most promising and least promising. It was developed by Stuart Pugh during his research of the design process. The chart is a tool to help integrate the design creativity of individuals with the refinement and analysis capabilities of a team. Ideally, several iterations of the chart are performed, changing the datum for each. Mr. Pugh's research has shown that this method is very helpful in stimulating creativity in the design process.

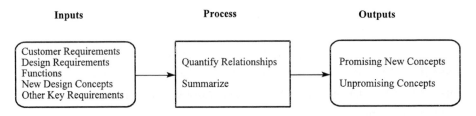

DOE - *Designed Experiments Scorecard*

Purpose: This chart is used to document all of the designed experiments that are to be performed as part of the design process. The experiments are normally run on design requirements that are negatively correlated with each other, thus implying that a tradeoff or optimization is required. These "conflicting" requirements are identified on the roof of the House of Quality.

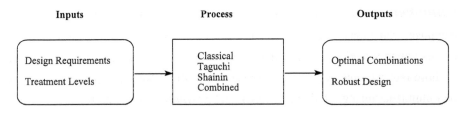

Parts - *Design Requirements vs Parts/Service Details*

Purpose: This chart is used to identify those parts or service details which have the greatest impact on meeting the design requirements for the product or service. It also helps clearly identify how the key design requirements are to be met.

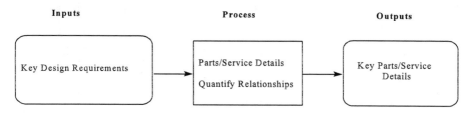

Technology - *New Technology vs Parts/Service Details*

Purpose: This chart is used to identify where new or emerging technologies can be infused into the product or service.

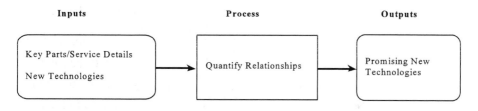

Customer Failures - *Potential Failures vs Customer Requirements*

Purpose: This chart is used to identify those potential product or service failure modes which would be most damaging in the eyes of our customers. The impact of those failures which are highly correlated with the customer requirements should be minimized during the design process, i.e., ensure they do not occur.

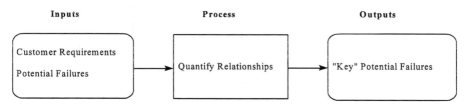

Design Failures - *Potential Failures vs Key Design Requirements*

Purpose: This chart is used to identify those design requirements which are most susceptible to product or service failures. Not only are "key" failure modes identified to help focus failure prevention efforts, but potential shortcomings of the design are also highlighted. This might signal that certain aspects of the design are not optimal and should be reconsidered.

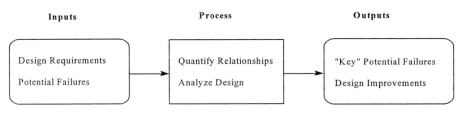

Functional Failures - *Potential Failures vs Key Functions*

Purpose: This chart is used to identify those potential product or service failure modes which would have the greatest negative impact on the functions of the product or service. Failure prevention efforts during design should be prioritized using the list of "key" failure modes identified by this chart to ensure key functions of the product or service are as robust as possible.

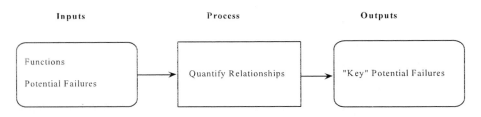

Delivered Quality - *Delivered Quality*

Purpose: This chart is used to identify how the product or service will be delivered in accordance with the design requirements. This is accomplished by identifying the process to be used for each key part or service detail, how that process is "controlled" and any special training that is required.

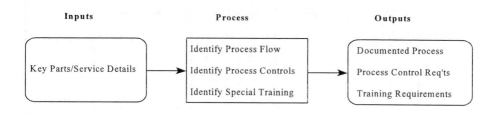

Process FMEA - *Process Failure Mode and Effect Analysis (FMEA)*

Purpose: This chart is used to help minimize the potential for the process used to make product/provide service to fail to perform correctly. This is accomplished by first identifying the process to be used for each key part or service detail. The potential failure modes for each process, the effects of these failure modes and the current methods for failure prevention are then outlined. Finally, an action plan is developed to minimize the effect of the potential failure modes.

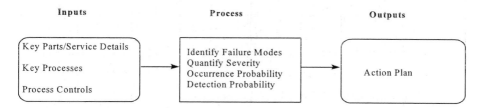

11.7 Points to Remember

- *QFD is used to prioritize requirements.* QFD will not replace your entire design process. QFD will help keep a project team focused on the requirements that are most critical to their customers. It also helps identify those key areas where scarce, discretionary resources should be used to obtain the greatest positive impact to your customers. Most organizations find that QFD is a good complement to their existing development processes.

- *The goal of QFD is to make decisions, not charts.* To succeed, QFD matrices must remain manageable in size — no more than 30x30 elements. Most projects fail when the team focuses on the mechanics of making charts and loses sight of the project objectives. Any team trying to analyze a 50x50 or 100x100 will become more involved with chart making instead of deploying key customer requirements throughout their design process and will burn out.

- *Use a diverse group of people to staff a QFD team*. Engineers, managers, logisticians, contractors and customers all bring invaluable insights to the QFD process. Periodically involve outside members of your organization to perform reality checks throughout the process.

- *Keep the core QFD team to a manageable size*. Staff the core group with 6 - 12 people who are supporters of the QFD process. It is equally important to regularly perform reality checks with functional experts outside the core group, but make sure that the core team can keep the project inertia moving in a positive direction.

- *Go after "low hanging fruit" at first*. Similar to other quality improvement efforts, it is imperative that the project scope be actively managed.   Learn the technique on an important, but limited modification to an existing product or service. Don't jump in and use QFD for the first time on the next "make or break the company" product. Once your organization is familiar with the tool, you can expand to larger projects.

Example 11.1 Design of a Mission Planning System

In response to several complaints from the end user of a mission planning system, a program manager decided to use QFD to more fully integrate the user's needs in a pending major system modification. A mission planning system is an information system used to integrate intelligence, weather and other critical information that the soldiers who both plan and fight can carry out their missions. The program office worked with their user and their contractor to develop the following House of Quality:

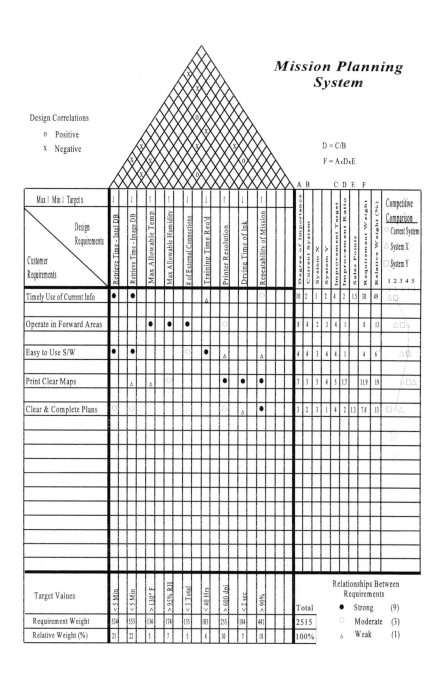

Mission Planning System

Design Correlations
o Positive
x Negative

D = C/B
F = AxDxE

From the HOQ, it is clear that fast access to both intelligence and imagery databases and clear plans that aren't open for interpretation are key requirements for this system. After performing some additional requirements analysis, the team decided to analyze several design options that were readily available. They decided to use a QFD matrix that is based on Pugh's Concept Selection Method. The following was one of several iterations of this method that the design team performed:

Design Options

Key Requirements	1	2	3	4	5		
Timely Use of Current Info		+	+	+	+		
Operate in Forward Areas		+	+	+	+		
Easy to Use S/W	D	s	s	+	s		
Print Clear Maps		s	+	+	+		
Clear & Complete Plans		s	+	+	+		
Retrieve Time - Intel DB		+	+	+	+		
Retrieve Time -Image DB	A	+	+	+	+		
Max Allowable Temp		s	s	+	+		
Max Allowable Humidity		s	s	+	+		
# of External Connections		s	s	+	+		
Training Time Req'd	T	−	−	+	−		
Printer Resolution		s	+	+	+		
Drying Time of Ink		s	−	+	+		
Repeatability of Mission		+	+	+	+		
Create Mission Folders	U	+	+	+	+		
Develop Flight Plans		+	+	+	+		
Analyze Intel		s	+	+	+		
Analyze Imagery		+	+	+	+		
Organize Forms/Output	M	s	+	+	+		
Perform in Adverse Climates		s	s	+	+		
Cost		s	+	−	−		
Reliability		s	−	+	+		
Maintainability		s	−	+	+		
+ 's		8	14	21	19		
− 's		1	4	2	3		

Option 1 -	current system, 386, Windows, 300 dpi, ruggedized commercial, secure 9600 modem
Option 2 -	486, OS/2, 300 dpi, V.42 secure modem, multimedia, ruggedized commercial
Option 3 -	Pentium, Windows, 600 dpi, SATCOM link, multimedia, ruggedized commercial
Option 4 -	Quadra, System 7, 600 dpi, SATCOM link, multimedia, hardened case
Option 5 -	Pentium, OS/2, V.42 secure modem, multimedia, hardened case

The team decided to use these and other QFD matrices as tools to help them work together more effectively as a group. They also decided to use QFD to help identify those critical areas of their program that warrant additional attention beyond the level provided by the current design process.

■

Chapter 12

Case Studies

Case Study 1:   Use of Statistics in Software Development [1]

Introduction

The software industry is littered with the remains of many software companies that had good initial ideas and products, but were unable to support the products due to poor quality. Software has many of the same problems as any mass produced product. The number of problems inherent with software development, usually called "bugs," can be controlled and reduced using probability and statistics. The relative cost of finding and fixing a bug in the last weeks of development compared to finding and fixing the same bug at the beginning of development is 100 to 1. Finding the bug after the system goes operational is even higher in cost. It should also be noted that Japanese programmers average about 75 percent fewer bugs than programmers in the United States. Figure 1 gives a general relationship of the cost for fixing a bug. The information in Figure 1 indicates that the cost of fixing a bug is exponential as the software product moves from the user requirement definition stage to the development stage, and finally to the operational stage. Problems found during the requirement definition and development stages result in unplanned costs and possible impacts to the implementation schedule, but problems found during the operational stage have by far the highest cost to fix, plus lead to intangible costs such as customer dissatisfaction.

---

[1]This case study was provided by Captain Bill Gaught, AF Space Command.

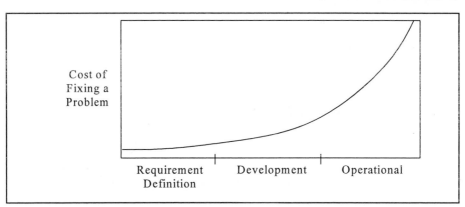

Figure 1. Cost of Fixing a Software Problem

Overview of Software Development Life Cycle

Figure 2 shows an overview of the software development life cycle. The life cycle contains 5 phases: analysis, design, development (coding), testing, and installation/operational.

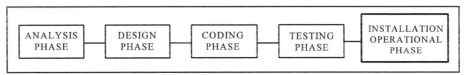

Figure 2. Software Development Life Cycle

The analysis phase is where user requirements are collected and analyzed for technical and financial feasibility. Requirements can be classified as either new capabilities, to include modifying existing functions, or fixes to correct problems with current software. A detailed technical analysis is accomplished to determine if the existing resources (i.e., programmers, operating system, application programming languages, hardware architecture, etc.) can support the requirement, or if new resources must be purchased. A financial feasibility study is conducted after the technical feasibility report has been completed. The purpose of the financial feasibility study is to cost *everything* required to implement and support the requirement. This includes: designing the software using sound software engineering principles, coding the software, testing the software, installation and support costs, and documentation costs. Figure 3 shows many of the inputs and outputs for the analysis phase.

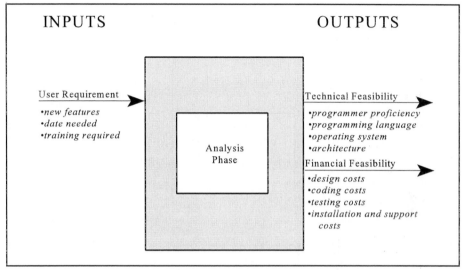

Figure 3. Analysis Phase of Life Cycle

It should be pointed out that the analysis phase sets the foundation for statistical analysis of the life cycle. Since everything about the requirement is considered during analysis, the analysis phase must be very accurate at predicting costs, schedules, and impacts to the overall system. The only way these predictions can be supported is using statistical data and conclusions drawn from sound analysis of the data.

After analysis comes the design phase. The design phase takes the technical and financial feasibility studies and develops a written design document. There are many documents produced during the design phase, but the two most important are the logical and physical design documents. The logical design document contains at a minimum the flow charts, data flow diagrams, descriptions of data structures, security requirements, and performance requirements. It also lays out the flow of the data items to be processed (i.e., customer records, files, etc.). The physical design document describes how the system is to be implemented on the actual hardware. It contains such information as the actual disk drives where files are to be placed and where new terminals, printers, etc., will be installed. Figure 4 shows some of the inputs and outputs of the design phase.

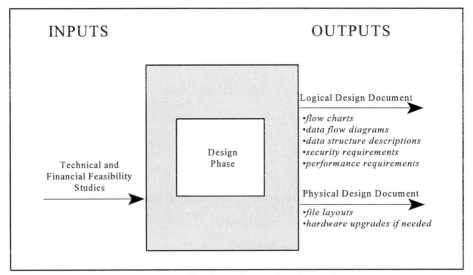

Figure 4. Design Phase of Life Cycle

The coding phase is where the computer programmer takes the logical and physical design documents from the design phase and transfers the design into actual computer source code. The source code conforms to the specifications laid out in the design documents. It should be noted that some testing is accomplished in the coding phase. The computer programmer will perform unit and string testing. Unit testing tests a single program module to ensure the module performs its tasks. Items checked during unit testing are: bounds checking of numerical values, paths through the module, error conditions, etc. The string tests are performed by "stringing" together two or more modules that will interface in the actual program and testing the string as a single unit. This limited amount of testing during the coding phase is necessary before the entire system is tested. Figure 5 shows inputs and outputs from the coding phase.

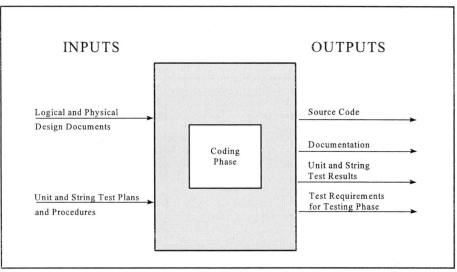

Figure 5. Coding Phase of Life Cycle

The testing phase is used to find errors that slipped past the unit and string tests performed in the coding phase. Some of the tests performed during the testing phase are: integration tests, system tests, regression tests, security tests, and performance tests. The integration tests are nothing more than large string tests consisting of 10 or more modules. System tests are groups of integration tests. Regression testing ensures that the old features still work the same as they did before. Security and performance tests make sure the system meets the specifications in the logical design document. The last test that is accomplished is the acceptance test which occurs at installation time. The acceptance test is performed during the installation/operational phase and is described later. Figure 6 shows the inputs and outputs of the testing phase.

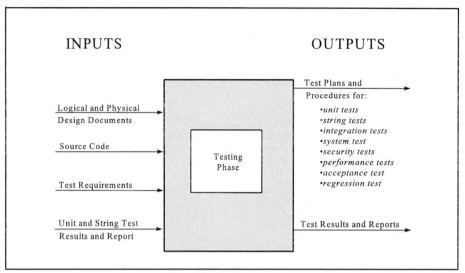

Figure 6. Testing Phase of Life Cycle

The installation/operational phase is the process of taking the completed system and actually putting it into operation. The installation/operational phase is also considered an acceptance period where the user evaluates the system to make sure it meets the original user requirement. An acceptance test is usually conducted at the end of the installation phase. Once the acceptance test is completed and the user is satisfied that the system meets the requirement, then the system is considered operational. At this time the system goes into an operational support phase. The operational support phase ensures the system performs on a daily, weekly, and monthly basis. Figure 7 shows some support requirements (inputs) that must be accomplished during the installation and operation of the system. It also shows benefits (outputs) that a user should expect from the system.

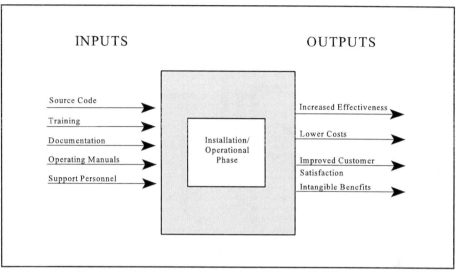

INPUTS   OUTPUTS

Source Code

Training

Documentation

Operating Manuals

Support Personnel

Installation/
Operational
Phase

Increased Effectiveness

Lower Costs

Improved Customer
Satisfaction

Intangible Benefits

Figure 7. Installation/Operational Phase of Life Cycle

How Statistics Can Help

Statistics can be used in every phase of the software life cycle. An analyst working in the analysis phase must estimate the amount of time it takes to design, code, test, and install the system so an accurate financial feasibility study can be done. It is possible to take data from previous development projects and use this to estimate future development efforts. Say, for example, that the analyst took the time cards of three testing personnel for each of the past 4 software projects. Suppose the data looked like Figure 8. The analyst would notice that all testers do not perform the testing task in the same amount of time. Since tester 2 is typically the fastest, it would be a big mistake to estimate the amount of test time using solely the average of tester 2. Better estimators would be the average of all three tester times or use the slowest and fastest testers as worst and best case estimators, respectively.

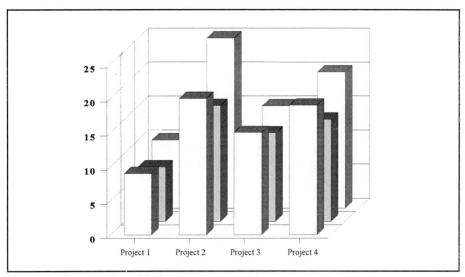

Figure 8. Amount of Test Time for Past Projects

The problems that customers experience with software can be broken down into six primary areas: application software, system software (such as operating system, communication software, etc.), database software, hardware, procedural (customer does something wrong), and other. The personnel working in the design, coding, and testing phases should be aware of the number of problems generated by each over time. Figure 9 is a run chart of application software problems over time. By looking at the chart it is easy to see trends. All six problem areas should be evaluated to determine where the resources can be allocated to minimize problems with the software.

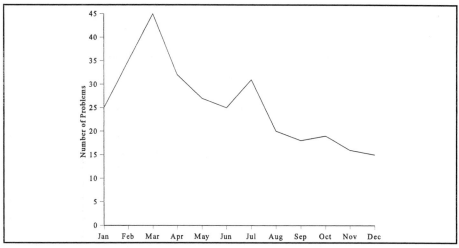

Figure 9. Run Chart of Problems with Application Software

The personnel that support the system during the installation/operational phase are probably the biggest users of statistics. Figure 10 shows some statistics that need to be gathered during the operational phase to ensure the system is performing to the user's requirements.

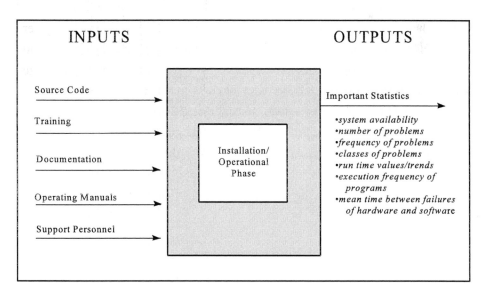

Figure 10. Statistics Required in Installation/Operational Phase

For example, one of the most important statistics is the amount of Central Processing Unit (CPU) time that is being used. The percent of time the CPU is busy is a good estimator of the overall performance of the system. Figure 11 shows a typical run chart for the CPU utilization rate by day.

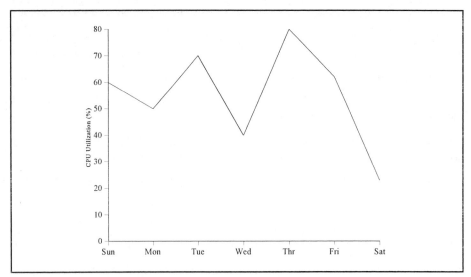

Figure 11. Percent CPU Busy by Day

Conclusion

　　　This case study stresses that the software development life cycle is an excellent application for statistics. With the cost of a typical software error costing approximately $1000 to fix, it becomes obvious that improving the quality of software should be a developer's goal. In addition, the cost of correcting problems increases exponentially as the problem progresses through the life cycle. Problems found early in the analysis phase cost much less to fix than problems found after the system is operational. The use of statistics helps the software developer control and reduce the number of errors, thereby improving the quality of the product.

Case Study 2:   A Generalized Transplantation Process for Patients with End Stage Renal Disease[1]

Approximately 35,000 Americans wait each year for a life enhancing organ. Depending on their blood and antigen types, the organ or organs they need, the region of the country they live in, their urgency of need, and their overall state of health, their wait can last for as long as three to five years. During this waiting period, their quality of life is considerably diminished. Statistics indicate that about ten percent of these patients perish while waiting. In the case of renal transplantation, patients who are fortunate to have suitable living-related donors wait months versus years, and have a 91.5% average first year graft survival rate versus a cadaveric renal transplant whose first year average graft survival is approximately 80.9%.[2] The reason for the higher survival probabilities rests in the fact that better antigen matching occurs along with a shortened waiting time for living-related cases versus the cadaveric graft alternative. The generalized process flow chart in Figure 1 depicts those steps starting when a patient is first informed that end stage renal disease is imminent to when a living-related or cadaveric renal transplantation occurs. From a patient's perspective, the most significant issue is the amount of time required to locate a compatible kidney within one of the eleven transplant regions in the United States. While many of the steps within this process flow are being accomplished with far greater efficiency than just a few years ago, the demand for organs far exceeds the supply. The shortage of life-enhancing organs continues to be an issue of national concern.

---

[1]This case study was provided by Dr. Lee R. Pollock of Electronic Systems Center (ESC) who from personal experience developed this process from a patient's perspective. Lee, who piloted ESC to the Presidential Quality Award in 1994, is a renowned author and presenter. He has authored numerous technical articles in accelerated life cycle testing and more recently has written extensively on the subject of renal transplantation. This example clearly depicts the steps that will need to be anticipated by someone diagnosed with end stage renal disease. Although most doctors would describe these steps verbally, the flow diagram presents a picture that clearly communicates what the patient can expect to go through. Our special thanks to Lee, a true professional, gentleman, and friend.

[2]Data from the United Network for Organ Sharing (UNOS) and the New England Organ Bank.

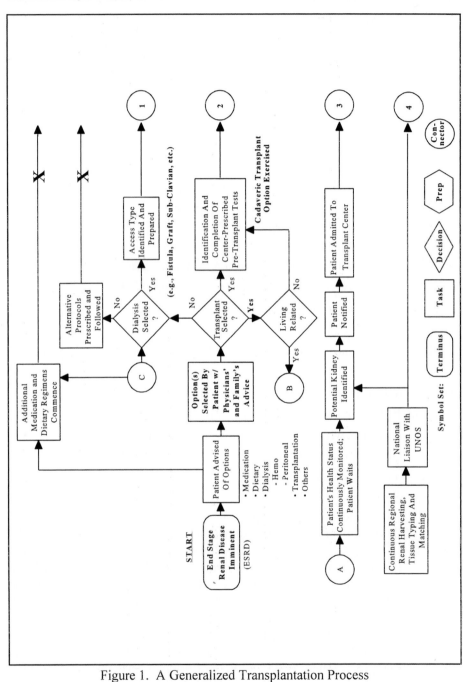

Figure 1. A Generalized Transplantation Process
for Patients with End Stage Renal Disease

(Page 1 of 2)

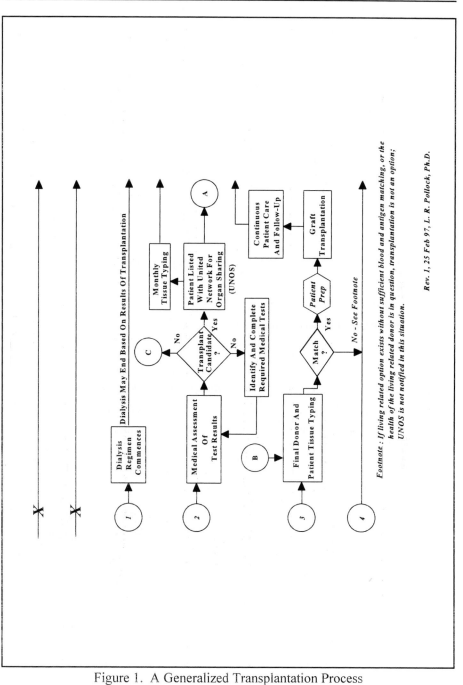

Figure 1. A Generalized Transplantation Process
for Patients with End Stage Renal Disease

Case Study 3: Travel Request and Authorization Process Flow

Problem Statement:

The current time to complete a travel request (T.R.) and authorization process averages 5.76 weeks with an overall 28.3% discrepancy rate. Thus, travelers are currently required to have an undesirable amount of lead time to initiate travel requests.

Objective:

Form a crossfunctional team to i) complete a detailed process flow diagram; ii) identify and remove non-value added activities; iii) evaluate process performance after quality improvement efforts are in place.

Improvement Effort:

The crossfunctional team consisted of 3 traveler representatives, 2 business unit clerks, 2 travel service employees, one travel accounting employee and one corporate level liaison. A process flow diagram depicting the current process appears in Figure 1. After completing the process flow diagram and collecting data, the team determined that of the 28.3% total discrepancies, about 17% were found at the travel services (Step 4) and 11.3% at travel accounting (Step 10). The correction process for discrepancies causes considerable increased cycle time. Therefore, a solution was sought to reduce non-valued activities and to reduce the number of discrepancy corrections. It was suggested that each agency needed increased awareness of the typical errors which cause delays and be provided a means to avoid errors. Thus, a cover letter for each T.R. now contains a list of typical errors and a checklist for the traveler, business unit clerk, travel services employee and travel accounting employee.

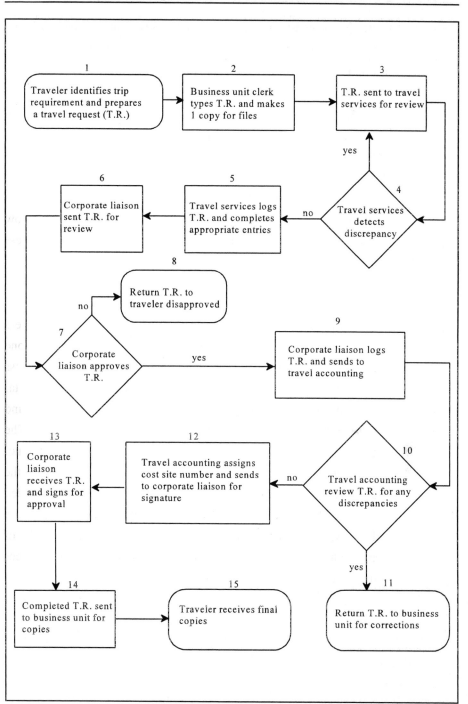

Figure 1. Initial Process Flow Diagram for Travel Request & Authorization

An additional suggestion was to have the business unit managers be the approving authority and coordinate any concerns with the corporate liaison. This process change would free up travel services from working a T.R. which would later be disapproved anyway, empower business unit managers, and remove a bottleneck at the corporate liaison approval step. The revised process flow diagram is found in Figure 2.

Results: Six months after the process flow was implemented the average cycle time was reduced to 2.4 weeks with only a 7% discrepancy rate. The percent of relative improvement was calculated as follows:

$$\frac{(5.76 - 2.4)}{5.76} * 100\% = 57.89\% \text{ improvement for cycle time}$$

$$\frac{(28.3 - 7)}{28.3} * 100\% = 75.26\% \text{ improvement for discrepancies}$$

Most importantly, company waste was removed and now travelers require a more tolerable lead time to process a travel request.

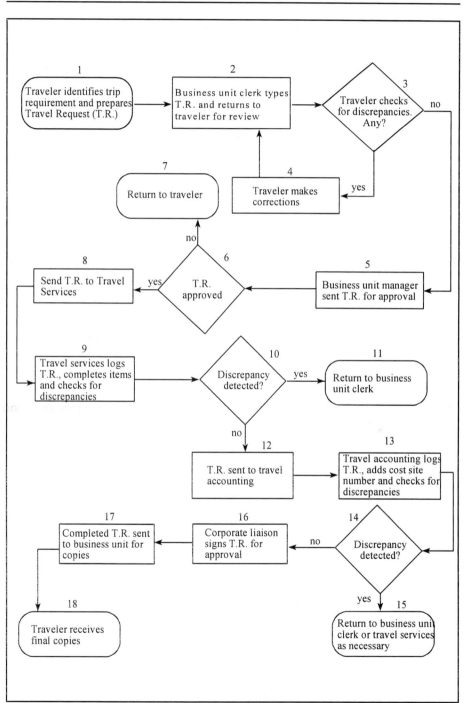

Figure 2. Revised Process Flow Diagram for Travel Request & Authorization

Case Study 4:   Disk Drive Pareto Analysis

Suppose you are the Quality Engineer responsible for a diskette drive that is used in the assembly of a personal computer. You receive a report listing the number of units rejected by the manufacturing process for the first five days of production. This report is shown below. Based on the report you decide that the defect level must be reduced. To understand what is causing the problems with the diskette drives, analyze this report using Pareto Diagrams and recommend what must be done to improve the diskette drive quality.

Thursday, June 8			
Serial Number	**Code**	**Serial Number**	**Code**
12435	410	13684	440
12485	465	13168	841
12438	412	12985	413
12504	410	13764	465
12515	218	13245	410
12486	440	13352	413
12490	410	14233	412
18245	413	14057	410
13057	410	13084	218
13515	465	11997	413
14082	410	15421	841
14215	412	13775	410
13950	841	13241	412
13842	410	14022	410
13425	413	14111	440
13352	410	12864	410
12897	413	12932	410
13405	465	14127	412
13205	410	14274	413
13851	410	21748	465

Disk Drive Defect Report (continued on next page)

Friday, June 9			
Serial Number	**Code**	**Serial Number**	**Code**
14457	841	15014	410
15024	412	14738	465
15047	410	14848	410
16014	218	15727	412
15471	465	14274	413
14877	413	16122	218
15820	410	17154	465
16120	410	16407	410
15830	412	15487	413
14688	410	16124	410
17024	841	16722	465
16748	410	17112	413
16499	218		

Monday, June 12			
Serial Number	**Code**	**Serial Number**	**Code**
18424	410	20114	413
18274	218	19735	465
18345	841	18954	413
17995	413	18342	440
19014	465	18964	218

Disk Drive Defect Report (continued on next page)

Tuesday, June 13			
Serial Number	**Code**	**Serial Number**	**Code**
17642	440	20410	412
20427	413	19758	218
18962	181	19547	465
19276	410	19254	413
20145	412	19028	440
20046	465	20487	841
19993	440		

Wednesday, June 14			
Serial Number	**Code**	**Serial Number**	**Code**
20487	465	21486	413
20867	440	21014	412
20964	413	21017	841
21140	412	20310	465
21842	413	22004	440
20053	465	24081	410

Disk Drive Defect Report

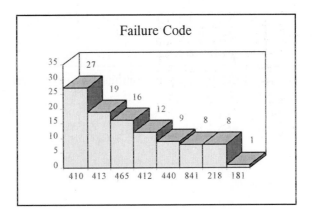

Failure Code

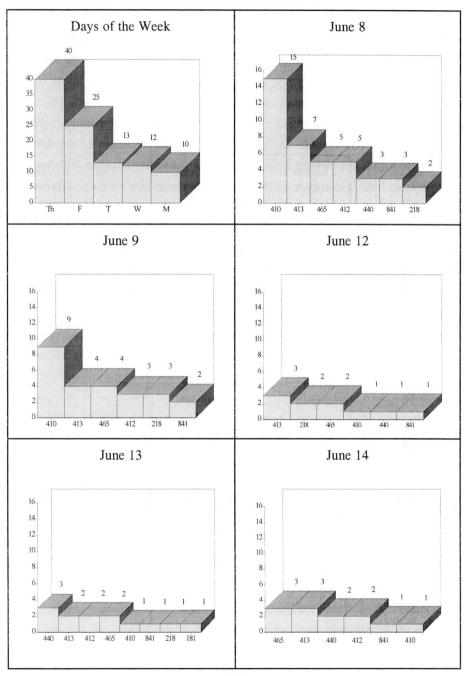

Pareto Diagrams

Analysis:

* The majority of the failures occurred on June 8. What was different about the first day's production that caused so many more failures than the other days?

* The majority of the failures were 410 error codes. In addition, most of these were concentrated in the first days of production. By the third day the frequency of this error was the same as the others. What was happening at the beginning of the week that caused so many 410 errors?

* After analyzing the units that had failed with the 410 error code, it was discovered that the diskette drive design allowed a manufacturing worker to unknowingly damage the unit. As the workers became more experienced at installing this unit they made this mistake less frequently. This reduced the number of 410 errors. Since the product could still be damaged by an inexperienced person, a simple design change was made to prevent the damage from occurring.

Case Study 5: Electro-Mechanical Assembly Histogram Analysis

The electro-mechanical assembly shown below will be built for use in manufacturing hard disk drives. Specifications A, B, and C must be maintained by this manufacturing process. These specifications follow:

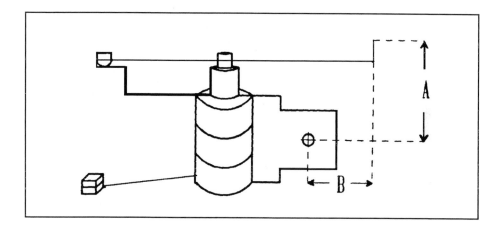

Specifications

A	11.5 ± 1.0 mm
B	8.5 ± 2.0 mm
C	2.5 m-amps max, where
C =	current required to move the solenoid

To evaluate this manufacturing process, data is recorded on the first 60 units built. Use a histogram analysis to develop an opinion of the process used in building this assembly.

Serial	A	B	C	Serial	A	B	C
1	11.5	8.15	1.9	31	11.4	8.31	2.4
2	11.8	7.98	2.0	32	12.0	8.47	1.9
3	11.7	8.04	2.1	33	11.7	8.74	2.3
4	11.1	7.93	1.9	34	11.5	8.30	2.0
5	11.4	8.15	1.8	35	11.8	6.91	2.6
6	11.4	7.57	1.6	36	11.8	8.73	1.8
7	11.8	8.58	0.9	37	11.8	8.37	2.1
8	11.4	7.88	0.7	38	12.1	8.37	2.4
9	11.4	7.76	2.6	39	11.5	7.39	2.4
10	11.5	8.33	1.3	40	11.5	7.51	2.3
11	11.2	8.19	1.3	41	11.4	8.38	1.4
12	11.4	6.91	1.2	42	11.7	7.31	1.2
13	11.5	7.68	2.4	43	11.8	8.74	2.0
14	12.0	7.57	2.2	44	11.8	7.99	1.7
15	11.4	9.02	2.1	45	12.1	7.92	1.9
16	11.7	8.38	2.0	46	11.5	7.42	1.9
17	12.0	7.84	2.1	47	11.8	8.59	1.7
18	12.0	7.74	2.4	48	11.5	8.16	3.7
19	11.5	8.62	1.1	49	11.8	7.11	2.3
20	11.2	8.32	2.0	50	11.7	7.89	2.1
21	11.2	8.46	1.8	51	11.5	8.88	2.0
22	11.4	7.68	2.1	52	11.5	7.70	1.2
23	11.5	8.72	2.5	53	11.4	7.56	1.4
24	11.2	8.42	2.0	54	11.7	8.96	1.4
25	12.1	8.51	1.7	55	11.1	8.60	2.3
26	11.8	8.85	1.8	56	12.0	7.29	1.7
27	11.5	7.99	1.1	57	11.5	8.18	1.3
28	11.4	8.12	1.8	58	11.5	7.67	1.5
29	11.2	8.38	1.5	59	11.4	8.68	1.1
30	11.5	8.81	1.6	60	11.4	9.11	1.2

SPECIFICATION A:

11.1	xx	2
11.2	xxxxx	5
11.3		
11.4	xxxxxxxxxxxx	13
11.5	xxxxxxxxxxxxxxx	16
11.6		
11.7	xxxxxx	6
11.8	xxxxxxxxxx	10
11.9		
12.0	xxxxx	5
12.1	xxx	3

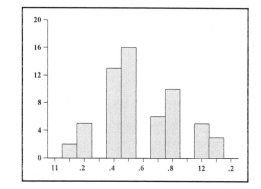

SPECIFICATION B

6.755-7.005	xx	2
7.005-7.255	x	1
7.255-7.505	xxxx	4
7.505-7.755	xxxxxxxxx	9
7.755-8.005	xxxxxxxxx	9
8.005-8.255	xxxxxxx	7
8.255-8.505	xxxxxxxxxxxx	12
8.505-8.755	xxxxxxxxxx	10
8.755-9.005	xxxx	4
9.005-9.255	xx	2

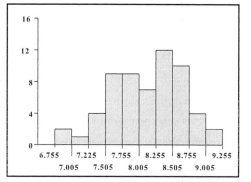

SPECIFICATION C

0.6-0.7	x	1
0.8-0.9	x	1
1.0-1.1	xxx	3
1.2-1.3	xxxxxxx	7
1.4-1.5	xxxxx	5
1.6-1.7	xxxxxx	6
1.8-1.9	xxxxxxxxxx	10
2.0-2.1	xxxxxxxxxxxxx	13
2.2-2.3	xxxxx	5
2.4-2.5	xxxxxx	6
2.6-2.7	xx	2
3.7	x	1

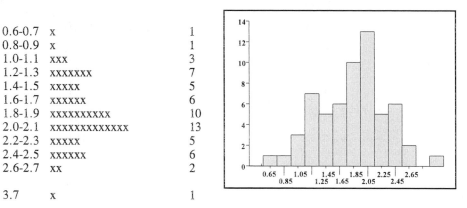

Analysis:

Specification A:

Good normal distribution. Process is capable of meeting specification.

Specification B:

Good normal distribution; however, the process is not located in the center of the specification. Process will have problems meeting the 6.5 lower specification.

Specification C:

Good normal distribution. The process is not capable of meeting the specification. Also one reading is suspected to be an error, since it is so extreme. This unit needs to undergo more testing to determine if the unit has an intermittent failure mechanism or if a data recording error has occurred.

In this situation the supplier needed to build the units with B at the low side of the specification in order to be able to move the solenoid with the required amount of current. As the histograms show, the design was not able to meet Specification C even with the B dimension shifted toward the low end. A product redesign is needed in this situation.

Case Study 6:   Using a Stem and Leaf Plot to Select a Long Distance Calling Card Carrier[1]

Four long distance calling card carriers were soliciting business from Air Academy Associates. Their billing rates are shown below, but this data alone seemed inadequate for making a decision to go with a certain carrier.

	A	B	C	D
Costs	.29/min	.21/min	.25/min	.21/min
Increments	1 min	6 sec	6 sec	6 sec
Surcharge	0	.60	.80	.65

Using two months of calling data from Air Academy Associates along with the above rate structure, we evaluated the carriers and identified the carrier with lowest cost. Two months worth of long distance calling card data was collected and displayed in a stem and leaf plot as shown on the next page. For each row of the stem and leaf we found the midpoint and number of calls. Using the midpoint as a representative time for each row we found the carrier cost for that time from the table and multiplied it by the number of calls to obtain a total cost for that particular carrier for that row. For example, consider line one: the cost for carrier A is $.29 for a 0.2 minute call because the minimum increment is 1 minute. Thus, 58 calls at $.29 each is $16.82. After completing costs for each line we summed the line costs to obtain a total cost for each carrier (see accompanying cost table). Our company used this information to choose the minimum-cost carrier which is the carrier whose rate structure best matched our calling structure, namely, Carrier A.

---

[1]Written by Vicki Schmidt, Air Academy Associates, Colorado Springs, CO

MINIMUM IS: 0.10 (This is the length in minutes of shortest call)
LOWER HINGE IS: 0.60 (This is the 25th percentile value)
MEDIAN IS: 1.80 (This is the 50th percentile value or median)
UPPER HINGE IS: 6.50 (This is the 75th percentile)
MAXIMUM IS: 37.70 (This is the length in minutes of longest call)

Stem and Leaf Plot for Long Distance Carrier Data (Output from Mystat)

Number of Calls	Mid Point	Line		
58	.2	1	0	11111111111111111111111112222222222222233333333334444444444444
59	.6	2	0 H	555555555555555556666666666666666666677777777777777788888888
31	1.1	3	1	0000000000000001112222233344444
15	1.7	4	1 M	555566677788999
17	2.2	5	2	00000112222223344
9	2.7	6	2	556678889
6	3.25	7	3	012334
9	3.7	8	3	566777889
11	4.3	9	4	00113333444
8	4.65	10	4	55667788
3	5.2	11	5	224
9	5.7	12	5	566777889
10	6.3	13	6	1223333344
5	6.8	14	6 H	55899
5	7.8	15	7	57899
9	8.2	16	8	011222333
8	8.65	17	8	56667789
3	9.0	18	9	003
1	10.1	19	10	1
1	10.8	20	10	8
1	11.4	21	11	4
3	11.6	22	11	567
1	12.0	23	12	0
2	12.6	24	12	66
4	13.25	25	13	2234
3	13.8	26	13	689
3	14.1	27	14	111
2	14.7	28	14	77
1	15.2	29	15	2
2	15.55	30	15	47
3	16.4	31	16	449
4	17.45	32	17	2458
4	18.2	33	18	2226
1	19.9	34	19	9
2	20.7	35	20	59
1	22.1	36	22	1
3	23.4	37	23	247
2	24.45	38	24	09
2	26.65	39	26	49
1	27.7	40	27	7
1	31.6	41	31	6
1	32.1	42	32	1
1	33.0	43	33	0
1	37.7	44	37	7

Line No.	Carrier			
	A	B	C	D
1	16.82	37.12	49.30	40.02
2	17.11	43.07	56.05	46.02
3	17.98	25.73	33.33	27.28
4	8.70	14.40	18.38	15.11
5	14.79	18.05	22.95	18.90
6	7.83	10.50	13.28	10.95
7	6.96	7.70	9.68	8.00
8	10.44	12.39	15.53	12.84
9	15.95	16.53	20.63	17.08
10	11.60	12.61	15.70	13.01
11	5.22	5.08	6.30	5.29
12	15.66	16.17	20.03	16.62
13	20.30	19.23	23.75	19.73
14	10.15	10.14	12.50	10.39
15	11.60	11.19	13.75	11.44
16	23.49	20.90	25.65	21.35
17	20.88	19.33	23.70	19.73
18	7.83	7.47	9.15	7.62
19	3.19	2.72	3.33	2.77
20	3.19	2.87	3.50	2.92
21	3.48	2.99	3.65	3.04
22	10.44	9.11	11.10	9.26
23	3.48	3.12	3.80	3.17
24	7.54	6.49	7.90	6.59
25	16.24	13.53	16.45	13.73
26	12.18	10.49	12.75	10.64
27	13.05	10.68	12.98	10.83
28	8.70	7.37	8.95	7.47
29	4.64	3.79	4.60	3.84
30	9.28	7.73	9.38	7.83
31	14.79	12.45	15.08	12.60
32	20.88	17.06	20.65	17.26
33	22.04	17.69	21.40	17.89
34	5.80	4.78	5.78	4.83
35	12.18	9.89	11.95	9.99
36	6.67	5.24	6.33	5.29
37	20.88	16.54	19.95	16.69
38	14.50	11.47	13.83	11.57
39	15.66	12.39	14.93	12.49
40	8.12	6.42	7.73	6.47
41	9.28	7.24	8.70	7.29
42	9.57	7.34	8.83	7.39
43	9.57	7.53	9.05	7.58
44	11.02	8.52	10.23	8.57
Total Costs	519.68	533.06	662.49	549.38

Line Costs and Total Cost for Each Carrier

Case Study 7: A Simple Measurement System Study

The variability in any process arises from a variety of sources: machine, operator, materials, environment, methodology, and measurement, to name a few. In any data collection effort, it is important to understand that there may be variability in the measurement system itself. Understanding and quantifying this measurement error is an important aspect that is often overlooked when one is charting the performance of a process. This case study addresses the simplest of all measurement capability studies. Its purpose is to give the practitioner an idea of what a measurement (or gage) capability study is and how it can be accomplished. A more rigorous approach using a more complicated scenario is presented in Chapter 9.

A quality improvement team involved in developing a statistical process control system is interested in assessing the measurement capability of an instrument used to measure the thickness of a printed wire board. To assess the ability of the instrument to repeatedly measure thickness of a board and obtain the same results, a process operator decides to take one board and measure it at the same position 30 times. The operator takes the 30 measurements randomly throughout the day, always using the same board, the same instrument, and measuring at the same place on the board every time. The following histogram shows how the 30 measurements are dispersed. The unit of measurement is mils. We have found $\bar{y}$ = 249.7 and s = 1.664 (calculations not shown).

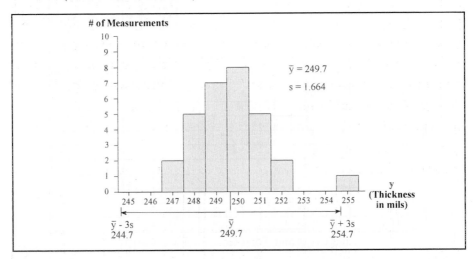

The figure also shows the points three standard deviations (3s) to the left and right of center (ȳ). While the histogram shows how the 30 measurements are dispersed and that the one measurement of 255 is an outlier (i.e., beyond 3s from ȳ), a run chart shows the order in which the measurements were taken and illustrates any trends. The run chart, along with the centerline and upper and lower control limits, is shown as follows.

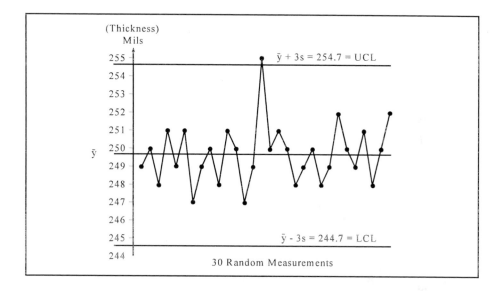

Except for the one point which exceeds the upper control limit (UCL), the measurements tend to fall within a 5-mil range. The outlying point, which is evident in both charts, needs to be investigated further because such a point (or value) is not likely to have occurred as part of the natural variation in the measurement system. Thus, the cause for this large value should be identified and removed from the measurement system. This control chart describes the variability in measurements obtained by the same person using the same instrument on the same product for repeated measurements. This is sometimes called repeatability variation or the variability within an operator/device combination. Chapter 9 addresses the situation when we measure different products with different operator/device combinations.

Case Study 8: Probability Not Always Intuitive

The following question was posed to Marilyn Vos Savant, an obviously intelligent individual who writes the "Ask Marilyn" column in the weekly *Parade Magazine*:

"Suppose you're on a game show, and you're given a choice of three doors. Behind one is a car; behind the others, goats. You pick a door — say, No. 1 — and the host, who knows what's behind the doors, opens another door — say, No. 3 — which has a goat. He then says to you, 'Do you want to pick door No. 2?' Is it to your advantage to switch your choice?" [Para 91]

Vos Savant's response was, "Yes, you should switch. The first door has a 1/3 chance of winning, but the second door has a 2/3 chance ..." [Para 91]. This response created what one might consider a national furor, especially among mathematicians and those who consider themselves knowledgeable in the area of probability. The controversy created by Ms. Vos Savant's correct response is evidence that probability does not always side with one's intuition. Of the many variations on this problem, some of which have been published, none has captured the vulnerability of the intuitive mind to probability quite as well as this one has. The reader interested in further pursuing some issues regarding the counter-intuitive nature of probability is referred to the article "Probability Blindness: Neither Rational nor Capricious," by Massimo Piatteli-Palmarini [Piat 91]. The following "simulation" should help convince the reader that Vos Savant's reply was correct.

Rules of the Game

Step 1:   Player picks one of three doors. In this "simulation," the player always picks Door 2. However, the same results would be obtained if any other door was chosen and the strategy carried out. A circled entry indicates the player's original choice.

Step 2:   The host opens a door that differs from the player's choice and behind which is a goat. An "X" through an entry shows this door. NOTE: the host knows where the car is, so such a door always exists. When does the host, under these rules, have a choice of which door to open?

Step 3:   The host asks the player if she wants to switch from the originally selected (still unopened) door to the other unopened door.

An arrow indicates when the player switches.

	Door 1	Door 2	Door 3	Strategy/Result	
				Stay	Switch
Case 1	Car	Goat	Goat	Lose	Win
Case 2	Goat	Car	Goat	Win	Lose
Case 3	Goat	Goat	Car	Lose	Win
	Probability of Winning the Car =			1/3	2/3

Case Study 9: Probability and the Law

Reference the Collins Case presented in Section 1.1 as an application of statistics in the legal profession. The following presents the mathematical basis upon which the California Supreme Court overruled the lower court conviction of the Collins'.

Judge Raymond Sullivan of the California Supreme Court wrote the court's opinion overruling the lower court conviction of Janet and Michael Collins and cited three primary reasons, any of which would have been sufficient to reverse the lower court.

The first difficulty was that the professor had failed to give any demographic evidence for his probabilities (e.g., how do we know the probability of a yellow car is 1/10?). The figures given were, in the court's view, little more than wild conjecture and certainly not grounds for convicting a couple on circumstantial evidence.

The second objection was a mathematical one. To use the multiplication rule in probability theory requires that the various events be statistically independent, that is, the events have nothing to do with each other. Certainly having a beard and a mustache are not statistically independent, nor are being blond and having a ponytail, etc. Hence, even if the demographic probabilities were accurate, which they probably weren't, the true probability of finding such a couple was certainly greater than the stated 1 in 12,000,000.

The third point was the most interesting. Upon appeal, the defense used the following application of the binomial distribution together with conditional probability to show that a "reasonable doubt" was indeed still present. If we define the random variable X to be

$$X = \text{number of couples that match the Collins' description.}$$

Let the parameters n and p be as follows:

n = total number of couples who could have been in the area and possibly committed the crime, and

p = 1/12,000,000, the prosecutor's original estimate.

Using the binomial pmf,

$$P(X = k) = \binom{n}{k} p^k (1-p)^{n-k} , \quad k = 0,1,\dots ,n,$$

we have $P(X = 0) = q^n$, where $q = 1 - p$,

$$P(X = 1) = npq^{n-1} \text{ and}$$

$$P(X \geq 1) = 1 - P(X = 0) = 1 - q^n.$$

It follows that

$$P(X > 1) = 1 - P(X \leq 1) = 1 - P(X = 0) - P(X = 1)$$

$$= 1 - q^n - npq^{n-1}.$$

Now, we consider the conditional probability

$$P(X > 1 \mid X \geq 1) = \frac{P(X > 1)}{P(X \geq 1)}$$

$$= \frac{1 - q^n - npq^{n-1}}{1 - q^n},$$

which represents the probability that there is **more** than one couple matching the description, given that there is at least one couple. If we use the convenient but certainly plausible value of $n = 12{,}000{,}000$, we find that

$$P(X > 1 \mid X \geq 1) = \frac{1 - .3679 - .3679}{1 - .3679}$$

$$= \frac{.2642}{.6321} = .418 .$$

The reader is encouraged to use other values of n to see what effect it has on the conditional probability. One could also use the Poisson pmf, where $\lambda = np$, to ease the computational effort. The bottom line here, however, is that given one such couple exists, the likelihood of another such couple existing is much greater than the prosecution's original probability of 1/12,000,000.

Case Study 10:   Relationship Between Spousal Abuse and Spousal Murder

Perhaps no other event in modern America has precipitated as much public attention to the perception of "there are lies, damned lies, and statistics" as the recent O. J. Simpson murder trial did. In particular, the question of whether Judge Ito should admit the evidence of spousal abuse was strongly debated. The following Bayesian probability analysis uses recent statistics to shed some light on the relationship between spousal abuse and spousal murder. First, for the notation:

S =   "woman murdered by a current or former *spouse/mate*"

$\bar{S}$ =   "woman murdered by someone *other than a current or former spouse/mate*"

A =   "murdered woman had previously been *abused*"

Note that the universal set is the set of murdered women. Now we investigate some probabilities based on fairly current statistics. First,

$P(S)$ = probability that a murdered woman was murdered by a current or former spouse/mate =.29 [Crim 92]

Note that $P(S) = .29$ is a prior probability. This is a statistic not difficult to obtain because most murders are solved. Crime data [Crim 92] indicates that between 1992 and 1995 inclusive, about 29% of all murdered women were known to have been murdered by a current or former husband or boyfriend. Surprisingly enough, this means that, without any other evidence whatsoever, the likelihood of a former spouse or boyfriend committing the murder is almost 30%. In [Ders 95], Alan Dershowitz, a member of the Simpson defense team, claims that one cannot predict if a spouse abuser will murder his spouse because more than 99.9% of all spouse abusers have not murdered their spouses. Predicting a priori is not what we are attempting to do here. What we are investigating is: does the additional information of a murdered woman having been abused alter the prior probability of P(S) = .29? Bayesian analysis can help us answer this question.

What we are searching for is $P(S|A)$, the probability that the spouse/mate committed the murder given the additional information that the murdered woman was abused. Is $P(S|A)$ significantly different from P(S) = .29?   From Bayes' rule (see Equation 3.1) we have

$$P(S|A) = \frac{P(S \cap A)}{P(A)} = \frac{P(S \cap A)}{P(S \cap A) + P(\overline{S} \cap A)}$$

$$= \frac{P(A|S)\, P(S)}{P(A|S)\, P(S) + P(A|\overline{S})\, P(\overline{S})}$$

From the known prior probability P(S) = .29, we also know the complementary probability $P(\overline{S})$ = .71. Assuming that spousal abuse against women murdered by someone other than a former or current spouse/mate, namely $P(A|\overline{S})$, is not likely to be significantly different from spousal abuse in the general population, which is about 0.07 [Comm 93], we have $P(A|\overline{S})$ = .07. The only remaining unknown on the righthand side is $P(A|S)$, the probability that a woman murdered by a former or current spouse/mate was abused. Data on this is also available. Campbell [Camp 94] says that approximately two-thirds of those murdered by intimate partners or ex-partners have been physically abused before they were killed. Thus, we have $P(A|S)$ = .67 and we solve for $P(S|A)$ as follows.

$$P(S|A) = \frac{(.67)(.29)}{(.67)(.29) + (.07)(.71)} = .80$$

What we have just seen is that

(1)   without any other evidence, female homicide data suggests there is a 29% chance that a murdered woman was murdered by a current or former spouse/mate, i.e., $P(S) = .29$, and

(2)   if a murdered woman was also previously abused, the likelihood the murder was committed by a current or former spouse/mate increases from 29% to 80%, certainly a significant difference. Again, this assumes no additional information or evidence other than the murdered woman had been previously abused.

Case Study 11:   Estimating Number of Hospital Beds

An Intensive Care Unit (ICU) planning committee wanted to estimate the number of beds that would be needed in their hospital's new ICU.   Their investigation of historical ICU data indicated that

1)   the arrival rate of (ICU) patients was on the average 2 per day (i.e., 2 per 24-hour period), and

2)   the average length of stay of a patient in the ICU was 60 hours.

This planning team wanted to be able to meet the customer (patient) demand of needing an ICU 99% of the time.   That is, if a doctor's diagnosis and subsequent treatment requires an ICU bed, the hospital wanted to meet that demand at least 99% of the time, and no more than 1% of the time did they want to be forced into sending a patient to another hospital.

Since ICU patient arrivals are Poisson distributed, a simple application of the Poisson distribution will allow the team to estimate the number of beds required to satisfy the 99% criterion stated above.   They proceeded as follows:

Let X $=$ # of ICU patient arrivals per 60-hour period, and

λ $=$ 2 per 24-hour period

$=$ 5 per 60-hour period.

If n $=$ the number of beds, then $P(X > n)$ represents the probability that there would be more patients arriving in a 60-hour period than there are beds to accommodate those patients.   They wanted to determine the smallest value of n for which $P(X > n) < .01$.   Using the computer program SPC KISS (or appendix C) with $\lambda = 5$, the following table was generated.

n	P(X > n)	n	P(X > n)
1	.9596	9	.0318
2	.8753	10	.0137
3	.7350	11	.0055
4	.5595	12	.0020
5	.3840	13	.0007
6	.2378	14	.0002
7	.1334	15	.0001
8	.0681	16	.0000

This table shows that n = 11 is the minimum number of beds required to satisfy the 99% criterion. It should be clear that this table of values could aid the planning team in performing tradeoff analyses such as how much each additional bed buys them. The next case study presents another example of how probability distributions can be used to help solve resource allocation problems.

Case Study 12:   Antigen Testing Resource Allocation

The ability of a pharmaceutical laboratory to complete validations of antigen testing depends on:

1)   the number of operators available to perform validations,

2)   the length of time it takes to perform a validation, and

3)   the rate and length of interrupts which prevent an operator from completing validations.

The question to be answered is, "how many operators will be needed if, say, 9 validations are to be completed in a 6-month period?" To answer this question, we will need data for items 2 and 3 above. Since there is variability in every process, we would like to accurately estimate the number of operators required. Obviously, we don't want to assign 20 operators when only 3 are needed. Conversely, if we assign a certain number of operators to the task of validating antigens, we would like some level of confidence that this number will suffice. In particular, we will accept at most a 5% risk that the number we assign to this process will not be enough.

There are several ways to go about solving this problem. Simulation is one approach, but the time and cost needed to build a simulation model may be prohibitive. We present an alternative method that is quick and easy to apply. It will give the practitioner an immediate estimate as well as a method for tradeoff analysis. It involves the direct application of the Poisson distribution.

Historical data indicates that it takes on the average about 6 weeks for an operator to complete a validation, provided there are no interrupts. Unfortunately, interrupts do occur, and the interrupt data collected indicates that the interrupts are Poisson distributed with $\lambda = 3$ interrupts per week. Furthermore, an interrupt takes an operator away from the validation process for an average of one week. We assume operators work 4 weeks per month.

Letting Y_n = the number of validations completed in a 6-month (or 24-week) period by n operators,

we seek the smallest value of n for which $P(Y_n \geq 9) \geq .95$.

We also let   X = the number of interrupts per 24-week period. Then

λ = 3 interrupts per week

= 72 interrupts per 24-week period

Since Y_n depends on X, we now establish the relationship between Y_n and X to be:

$$Y_n = \frac{\text{total time (in weeks) spent by n operators on validation work}}{6 \text{ weeks per validation}}$$

$$Y_n = \frac{\text{total work effort in weeks (includ. interrupts) } - \text{ interrupt time}}{6 \text{ weeks per validation}}$$

$$Y_n = \frac{24n - 1X}{6}$$

To find $P(Y_n \geq 9)$, we substitute $\frac{24n - X}{6}$ for Y_n and find an equivalent probability statement involving X, as follows:

$$P(Y_n \geq 9) \quad = \quad P\left(\frac{24n - X}{6} \geq 9 \right)$$

$$= \quad P(24n - X \geq 54)$$

$$= \quad P(24n - 54 \geq X)$$

$$= \quad P(X \leq 24n - 54).$$

Using the software package SPC KISS, we compute a table of probabilities for varying values of n.

n	$P(Y_n \geq 9) = P(X \leq 24n - 54)$
1	0.0000
2	0.0000
3	0.0000
4	0.0001
5	0.2621
6	0.9828
7	1.0000

Note: X is Poisson with $\lambda = 72$

The smallest value of n that will give us at least a 0.95 probability of completing 9 validations in a 6-month period is n = 6. Based on this criterion, we should therefore assign 6 operators to the antigen validation process. This case study illustrates the use of the Poisson distribution as a tool for estimating the number of resources to be allocated to a particular process.

Case Study 13: Historical Data Analysis of a Plating Process

A customer of a nickel plated product had complained about product performance, specifically that plating thickness was not consistent. In discussions with the customer, the producer indicated that they had some historical data on the process. This data consisted of the following variables at various settings with the associated plating thickness.

Variables	Typical Range
x_1 = Time	3 - 15 seconds
x_2 = Bath Temperature	12° - 38° C
x_3 = Percent Nickel	8 - 20 %
x_4 = Vendor	1 or 2
x_5 = Phosphorus Content	25 - 60 mg
y = Plate Thickness	5 - 90 mils

The producer assigned an analyst to determine whether there were any relationships between the variables and plate thickness. The 45 observations in Table 1 are the raw data given to the analyst, and a summary of the analysis follows.

COLUMN #	1	2	3	4	5		
ROW #	Time	Bath Temp	% Nickel	Vendor	Phos		Y
1	5	15	9	1	29		49.5
2	6	13	11	1	33		47.5
3	7	14	13	1	28		47.4
4	6	17	20	2	51		55.6
5	4	18	16	2	53		28.2
6	7	13	15	2	55		44.8
7	6	30	8	2	28		21.4

(Continued on next page)

Table 1. Historical Data (45 Observations)

COLUMN #	1	2	3	4	5		
ROW #	Time	Bath Temp	% Nickel	Vendor	Phos		Y
8	5	32	11	2	30		16.5
9	3	34	13	2	35		8.2
10	5	33	19	1	55		18.4
11	5	31	17	1	51		17.9
12	4	30	18	1	54		19.4
13	14	18	9	2	34		65.4
14	11	15	12	2	27		79.4
15	10	17	11	2	34		72.6
16	13	14	20	1	51		74.3
17	14	19	16	1	54		70.1
18	15	17	19	1	56		78.4
19	11	31	8	1	27		71.2
20	12	33	11	1	34		68.7
21	14	37	13	2	31		93.1
22	11	32	12	1	28		66.4
23	12	30	17	2	56		74.6
24	14	29	18	2	52		70.6
25	13	28	20	2	57		68.4
26	9	26	16	1	43		50.4
27	10	23	13	2	46		78.6
28	8	24	14	1	40		49.7
29	8	24	15	2	43		53.8
30	11	28	12	1	46		73.1
31	5	23	13	1	43		33.9
32	7	35	15	1	40		38.5
33	9	14	16	1	46		59.1
34	21	17	17	1	54		48.8
35	8	24	19	1	57		50.2
36	7	23	11	1	32		51.3

(Continued on next page)

Table 1. Historical Data (45 Observations)

COLUMN #	1	2	3	4	5		
ROW #	Time	Bath Temp	% Nickel	Vendor	Phos		Y
37	11	26	16	2	44		66.8
38	5	24	15	2	43		27.6
39	9	31	14	2	37		61.2
40	7	19	12	2	38		54.4
41	10	24	19	2	58		62.4
42	7	27	8	2	34		58.3
43	6	31	20	1	52		38.3
44	14	24	12	2	29		66.9
45	3	35	15	1	42		17.6

Table 1. Historical Data (45 Observations)

The first step in the analysis was to look at scatterplots of the input variables. The scatterplots shown in Figures 1 through 10 are the pairwise plots of the 5 input variables. We see in Figures 1, 2, 3, and 4 that one data point was collected where time was at a value of 21 minutes (observation # 34). This point is far from the other data points. Points such as this are called influential points because their presence is highly influential on the resulting model. Also notice the strong linear relationship between phosphorus content and % Nickel in Figure 9. This correlation among the input variables causes problems when trying to determine which variables are significant.

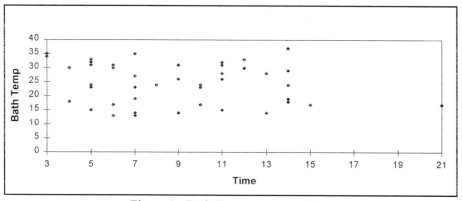

Figure 1. Bath Temperature vs Time

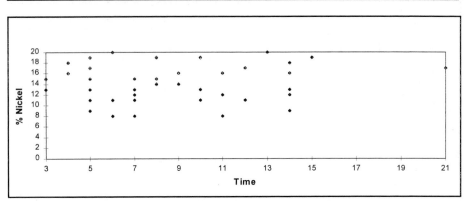

Figure 2. % Nickel vs Time

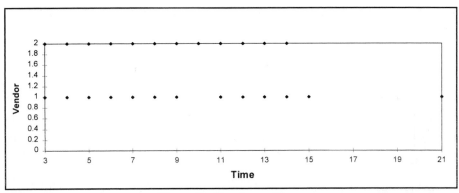

Figure 3. Vendor vs Time

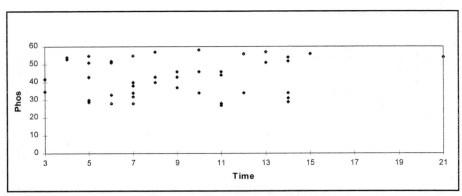

Figure 4. Phosphorus vs Time

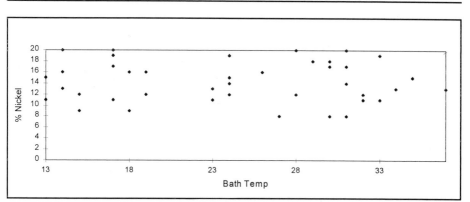

Figure 5. % Nickel vs Bath Temperature

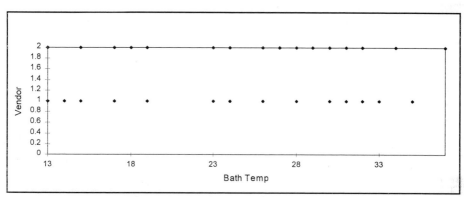

Figure 6. Vendor vs Bath Temperature

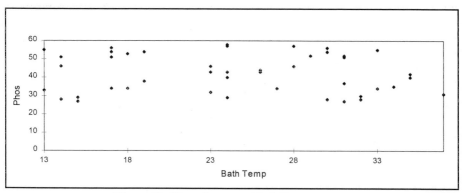

Figure 7. Phosphorus vs Bath Temperature

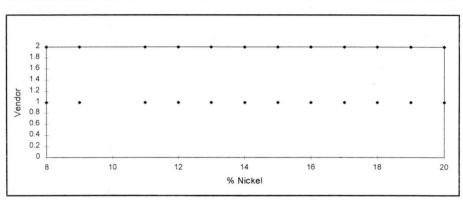

Figure 8. Vendor vs % Nickel

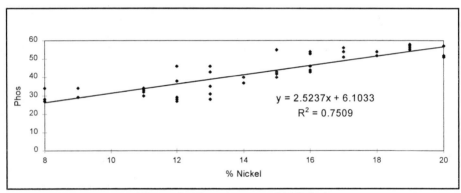

Figure 9. Phosphorus vs % Nickel

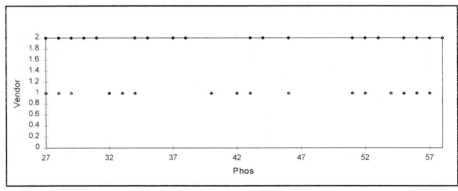

Figure 10. Vendor vs Phosphorus

When we look at the scatterplots of the input variables with the response variable (plate thickness) we notice several things (see Figures 11 through 15):

1. There seems to be a quadratic relationship between time and plate thickness. (See Figure 11.)

2. Observation number 34 is seen to be an outlier in Figure 11.

3. The other plots are relatively unremarkable.   If important relationships exist between these inputs and the response, they may be through interactions.

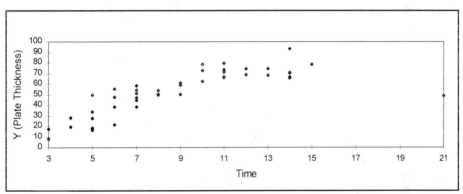

Figure 11. Y (Plate Thickness) vs Time

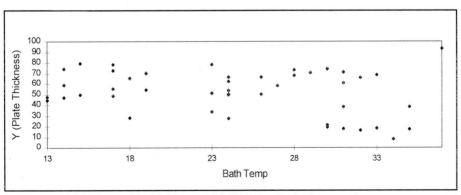

Figure 12. Y(Plate Thickness) vs Bath Temperature

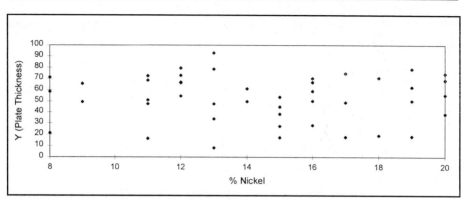

Figure 13. Y(Plate Thickness) vs % Nickel

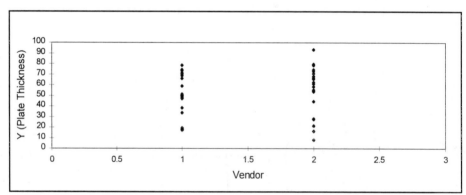

Figure 14. Y(Plate Thickness) vs Vendor

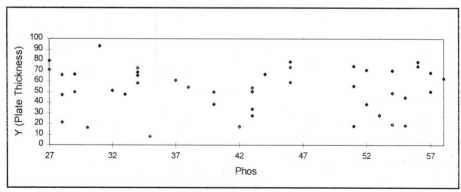

Figure 15. Y (Plate Thickness) vs Phosphorus

Since observation number 34 was identified as both an influential point and an outlier in terms of the response variable, it was eliminated. Outliers and influential points are usually the result of special causes affecting the process. Points such as this should be removed prior to model building.

The remaining 44 observations were used to build a regression model that considered linear effects, linear interactions, and quadratic effects. The x (input) variables were standardized using the following formula:

$$\text{standardized value} = \frac{x - \bar{x}}{s}$$

for each variable. The means and standard deviations are given in Table 2.

	Time	Bath Temp	% Nickel	Vendor	Phosphorus
$\bar{x}$	8.93	24.49	14.40	1.49	42.44
Std Dev (s)	3.87	7.06	3.53	0.51	10.29
Max	21.00	37.00	20.00	2.00	58.00
Midpoint	12.00	25.00	14.00	1.50	42.50
Min	3.00	13.00	8.00	1.00	27.00

Table 2. Summary Statistics for the Independent Variables

The output in Table 3 reflects the results of the full standardized model. The low tolerances (TOL column) indicate a high degree of correlation among the input variables. If the analyst were to eliminate all nonsignificant variables at once, the resulting p-values might change drastically. Therefore, the analyst took the approach to remove higher order nonsignificant terms in a stepwise manner.

FACTOR		COEF	P(2 TAIL)	TOL	LOW	HIGH	EXPER	ACTIVE
	Constant	55.6297	*0.0000*					
A	Time	18.8689	*0.0000*	0.591			8.6591	X
B	Bath Temp	-1.9580	0.2704	0.553			24.6590	X
C	% Nickel	-0.6535	0.8458	0.151			14.3410	X
D	Vendor	0.5764	0.6885	0.825			1.5000	X
E	Phos	0.4310	0.9042	0.133			42.1820	X
	AB	2.4610	0.3583	0.217				X
	AC	2.1959	0.5663	0.116				X
	AD	0.3151	0.9039	0.259				X
	AE	-1.4523	0.7524	0.075				X
	BC	-0.6758	0.8586	0.133				X
	BD	-0.5934	0.8047	0.302				X
	BE	1.2714	0.6937	0.157				X
	CD	4.6615	0.1899	0.144				X
	CE	-12.2078	0.2483	0.027				X
	DE	-3.3692	0.2862	0.179				X
	AA	-4.4334	*0.0290*	0.530				X
	BB	2.1076	0.2872	0.558				X
	CC	7.6645	0.1921	0.049				X
	EE	2.7769	0.6041	0.102				X
	R Sq	0.9098						
	Adj R Sq	0.8384						
	Std Error	8.4595						
	F	12.7436		**PRED Y**				
	Sig F	0.0000						

Table 3. Full Standardized Model

Since A (Time) and AA (Time^2) are important, the analyst certainly wants to keep these terms in the model. The decision was made to eliminate BB, CC, and EE (the quadratic effects of Bath Temp, % Nickel, and Phosphorus, respectively). There is no quadratic effect estimated for D (Vendor) because there were only 2 possible levels for D. The resulting model is shown in Table 4.

FACTOR	COEF	P(2 TAIL)	TOL LOW	HIGH	EXPER	ACTIVE
Constant	57.6935	*0.0000*				
A Time	18.1577	*0.0000*	0.678		8.6591	X
B Bath Temp	-3.1537	*0.0643*	0.654		24.6590	X
C % Nickel	-0.4577	0.8831	0.184		14.3410	X
D Vendor	0.2554	0.8557	0.904		1.50000	X
E Phos	-0.5538	0.8644	0.169		42.1820	X
AB	3.8965	0.1426	0.237			X
AC	-0.0463	0.9901	0.127			X
AD	1.6191	0.5305	0.280			X
AE	1.2718	0.7769	0.083			X
BC	-2.8228	0.4201	0.165			X
BD	-1.3652	0.5730	0.312			X
BE	1.2854	0.6714	0.187			X
CD	1.8089	0.5747	0.179			X
CE	-1.3389	0.4737	0.873			X
DE	-2.0906	0.4988	0.193			X
AA	-3.6700	*0.0604*	0.578			X
R Sq	0.8935					
Adj R Sq	0.8303					
Std Error	8.6684					
F	14.1536	**PRED Y**				
Sig F	0.0000					

Table 4. Reduced Model after Having Eliminated B^2, C^2, and E^2

The results in Table 4 show that AA has become only borderline important now but Bath Temp (B) has increased in importance. Recalling that % Nickel (C) was highly correlated with Phosphorus (E), % Nickel (C) will also be eliminated along with all interactions involving C and all interactions not involving A or B. The resulting model is shown in Table 5.

FACTOR	COEF	P(2 TAIL)	TOL LOW	HIGH	EXPER	ACTIVE
Constant	56.7209	*0.0000*				
A Time	18.5508	*0.0000*	0.817		8.6591	X
B Bath Temp	-2.9233	*0.0481*	0.751		24.6590	X
D Vendor	0.2339	0.8553	0.940		14.3410	X
E Phos	-1.0242	0.4530	0.837		1.50000	X
AB	3.5353	*0.0180*	0.681		42.1820	X
AD	0.5368	0.7920	0.389			X
AE	0.5446	0.7624	0.447			X
BD	-0.6706	0.7078	0.495			X
BE	-0.2398	0.8996	0.411			X
AA	-3.8574	*0.0211*	0.696			X
R Sq	0.8865					
Adj R Sq	0.8521					
Std Error	8.0924					
F	25.7823		**PRED Y**			
Sig F	0.0000					

Table 5. Model After Eliminating C and All Other Interactions
Involving C or Not Involving A or B

As the p-values change, we notice that AB is becoming more important. This changing p-value is typical when variables are correlated among each other. However, the situation is improving. Notice that the tolerances are increasing; there are fewer "small" tolerance values. The next step is to eliminate all nonsignificant effects. This completely reduced model is shown in Table 6.

FINAL STANDARDIZED MODEL

	FACTOR	COEF	P(2 TAIL)	TOL	LOW	HIGH	EXPER	ACTIVE
	Constant	56.9969	*0.0000*					
A	Time	18.6037	*0.0000*	0.891			8.6591	X
B	Bath Temp	-2.6916	*0.0268*	0.962			24.6590	X
	AB	3.5941	*0.0041*	0.854				X
	AA	-4.0184	*0.0059*	0.804				X
	R Sq	0.8841						
	Adj R Sq	0.8722						
	Std Error	7.5236						
	F	74.3631			**PRED Y**			
	Sig F	0.0000						

Table 6. Completely Reduced Standardized Model with Only Significant Effects

The analyst now has a relatively "good" model. All of the tolerances are high and the p-values are all significant. However, when we write the model as given below we recall that the variables were standardized. That means the model is good for predicting Y as long as the values we insert for Time and Bath Temperature are standardized.

$$Y = 57.0 + 18.6*Time - 2.69*Bath\ Temp + 3.59*(Time*Bath\ Temp) - 4.02*(Time^2)$$

Most analysts prefer to communicate with their models via the *actual* values vis-a-vis the standardized values. The next step then, was to perform regression analysis on the original data using only the terms found to be important from the standardized regression. This model is shown in Table 7.

NON-STANDARDIZED MODEL

	FACTOR	COEF	P(2 TAIL)	TOL	LOW	HIGH	EXPER	ACTIVE
	Constant	25.8457	*0.1350*					
A	Time	7.6195	*0.0101*	0.014			8.6591	X
B	Bath Temp	-1.6642	*0.0007*	0.130			24.6590	X
	AB	0.1481	*0.0041*	0.051				X
	AA	-0.3389	*0.0059*	0.025				X
	R Sq	0.8841						
	Adj R Sq	0.8722						
	Std Error	7.5236						
	F	74.3631		**PRED Y**				
	Sig F	0.0000						

Table 7. Final Model for Actual Values

This model gave the analyst a model with coefficients that can be used to predict thickness when given *actual* Times and Bath Temperatures.

$$Y = 25.85 + 7.62*\text{Time} - 1.66*\text{Bath Temp} + 0.15*(\text{Time}*\text{Bath Temp}) - 0.34*(\text{Temp}^2)$$

With this information, the company was able to use time and bath temperature settings that gave more consistent results for various plating thicknesses.

Case Study 14: Operation of an Activated Sludge System[1]

The purpose of this case study is to demonstrate how experimental design can be used in environmental engineering. Specifically, we show how the operation of an activated sludge reactor can be better understood and optimized.

The Activated Sludge System

The activated sludge process is a biological treatment system used at many wastewater treatment plants. Briefly, biological organisms (termed sludge) within the reactor are used to aerobically convert incoming waste (influent) into additional biomass or innocuous carbon dioxide and water. The activated sludge reactor is followed immediately by a settling tank, called a secondary clarifier, where the liquid and solids (sludge) are separated. The sludge from the clarifier is then either wasted, or recycled back to the reactor to maintain an acceptable biomass population. Figure 1 shows a schematic of a typical completely-mixed activated sludge system.

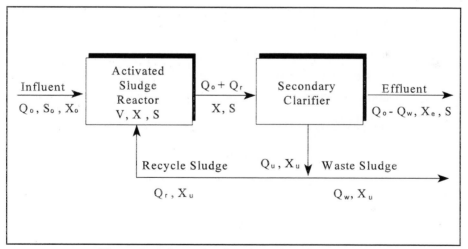

Figure 1. Schematic of Activated Sludge System

[1]This case study was provided by James Brickell and Kenneth Knox, Department of Civil Engineering, USAF Academy.

The parameters shown in Figure 1 are defined as follows:

Q_o, Q_u, Q_w, Q_r = flow rate in influent, clarifier underflow, waste, and recycle flows, respectively (m^3/d)

S_o, S = soluble food concentration (measured as BOD $)_5$ in the influent and reactor, respectively (mg/L)

X_o, X, X_e, X_u = biomass concentration in influent, reactor, effluent, and clarifier underflow, respectively (mg/L)

V = reactor volume (m^3)

Performance of an activated sludge system is measured using the biochemical oxygen demand, BOD. The BOD is the amount of oxygen required to stabilize the decomposable matter in a water using aerobic biochemical action. In general, a lower BOD is indicative of a higher quality water. A typical municipal wastewater would have a BOD_5 concentration (i.e., BOD measured after 5 days at 20°C) of 110 – 400 mg/L. The EPA effluent standards for wastewater, on an average monthly basis, are 30 mg/L BOD_5 [Peav 85].

A mass balance analysis of the system described in Figure 1, using the Monod equation to describe the rate of bacterial growth, yields the following analytical relationship:

$$y = (S_o - S) = \frac{\left(1 + \dfrac{k_d V X}{Q_w X_u}\right)}{\left(\dfrac{Q_o Y}{Q_w X_u}\right)} \qquad \text{[Equation 1]}$$

where:

$y = (S_o - S)$ = amount of BOD removed by the system (the response variable)

k_d = biological growth rate constant (d^{-1})

$$Y \quad = \quad \text{fraction of food (S) converted to biomass (X), or}$$

$$Y \quad = \quad \frac{\text{kg X produced per day}}{\text{kg S consumed per day}}$$

Experimental Design

To demonstrate application of experimental design, a simulated activated sludge system was conceived. Key parameters of the system were then varied between expected ranges, and the results compiled.

The parameters which affect the treatment efficiency for removing BOD $(S_0 - S)$ from wastewater can be gleaned from the analytical equation describing the process (Equation 1). Metcalf and Eddy [Metc 79] provide typical ranges for these parameters for an average municipal activated sludge system:

X, reactor biomass concentration	3000 – 6000 mg/L
X_u, clarifier biomass concentration	8000 – 12000 mg/L
k_d, biological growth rate constant	0.040 – 0.075 d^{-1}
Y, fraction of food to biomass	0.40 – 0.80 kg/kg

The system that will be used for this analysis originated from an example problem (10-1) presented in [Metc 79], where the reactor volume, V, is 4700 m^3, and the influent flow rate, Q_0, is 21,600 m^3/d. Based on typical wasting rates, the waste sludge flow rate for this system was calculated to be 78.5 – 940.0 m^3/d.

Initially, a series of experiments was designed using the five variables (X, X_u, k_d, Y, and Q_w). A two-level design was used, with the low (−1) level at the lower limit of the typical range, and the high (1) level at the higher limit for each variable. A full-factorial design would have required 32 (or 2^5) runs, which may be excessive for an actual plant to perform. Therefore, a quarter-factorial design was used, requiring 8 (or 2^3) runs. Aliasing and the results of each run (based on Equation 1) are shown in the design matrix given in Table 1. The bars over the variables indicate these are linearly "coded" variables, with the value of −1 at the lower limit, 1 at the higher limit, and 0 at the midrange value. To evaluate the usefulness of the derived Taylor approximation of Equation 1, a series of 25 runs was made with the five variables allowed to randomly vary between their typical upper and lower limits. The predicted response was then compared with the "correct" response from Equation 1, the results of which are shown in Figure 2. The least squares regression line of the results has a respectable coefficient of determination (R^2) of 0.905.

An inspection of the prediction equation shows that the most important variables for controlling removal of BOD from wastewater seem to be Q_w and Y. A Pareto diagram, Figure 3, more clearly demonstrates the contribution to BOD removal of each variable. In this figure, X_u and k_d also seem to contribute significantly to the observed responses, while X and the tested interactions (XQ_w and X_uQ_w) appear less consequential.

Quarter-Factorial Design Matrix							
Run	X	X_u	Q_w	XX_u	$k_d =$ XQ_w	X_uQ_w	$Y=$ XX_uQ_w	Response (mg/L)
1	1	1	1	1	1	1	1	775
2	1	1	-1	1	-1	-1	-1	354
3	1	-1	1	-1	1	-1	-1	1001
4	1	-1	-1	-1	-1	1	1	102
5	-1	1	1	-1	-1	1	-1	1371
6	-1	1	-1	-1	1	-1	1	87
7	-1	-1	1	1	-1	-1	1	496
8	-1	-1	-1	1	1	1	-1	195
Avg_{+1}	558	647	911	455	515	611	365	$\bar{y}=548$
Avg_{-1}	537	448	184	640	581	485	730	
Δ	21	198	726	-185	-66	126	-365	
Δ/2	10	99	363	-93	-33	63	-183	

Prediction Equation (First-order Taylor Series Approximation):

$$\hat{y} = 548 + 10\bar{X} + 99\bar{X}_u + 363\bar{Q}_w - 93\bar{k}_d - 33\bar{X}\bar{Q}_w + 63\bar{X}_u\bar{Q}_w - 183\bar{Y}$$

Table 1. Quarter-Factorial Design Matrix and Analysis

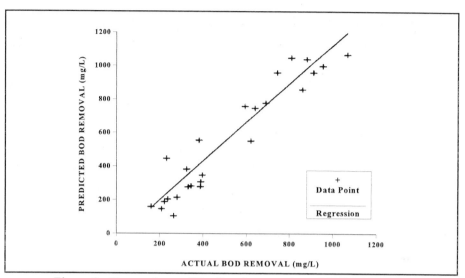

Figure 2. Comparison of Predicted versus Actual BOD Removal
Rates from the Quarter-Factorial Design Model

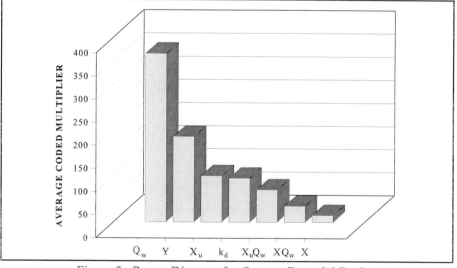

Figure 3. Pareto Diagram for Quarter-Factorial Design

Untested interactions and aliasing could materially impact the prediction.
Aliasing cannot be evaluated without further testing. The untested interactions,
however, were analyzed using two-factor interaction graphs (Figure 4a - e) between

the most significant main effects variables (Q_w, Y, X_u, and k_d). From Figure 4 the following observations can be made:

- Since the interaction graphs are not parallel, significant interaction exists between Q_w and Y (Figure 4a), Q_w and k_d (Figure 4b), and between Y and k_d (Figure 4d).

- Little to no interaction exists between Y and X_u (Figure 4c), and between X_u and k_d (Figure 4e).

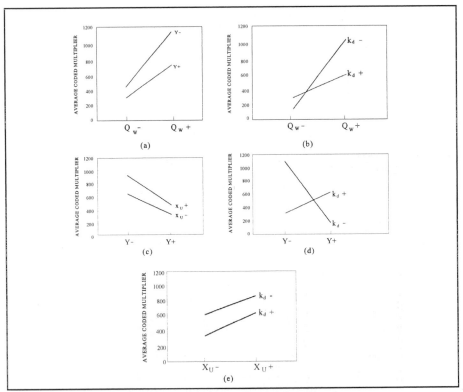

Figure 4. Two-Factor Interaction Graphs

Even though it may be impractical using a "real" activated sludge system, a two-level full factorial model was tested on the computer simulator using all five variables (32 runs) in order to more rigorously evaluate the designed experiments approach. Figure 5 shows that the prediction equation does a good job of modeling the actual BOD removal equation, with an R^2 over 0.99. The corresponding Pareto Diagram, Figure 6, shows that with minor exceptions the activated sludge system accurately identified the key variables using the designed experiments approach with only eight runs.

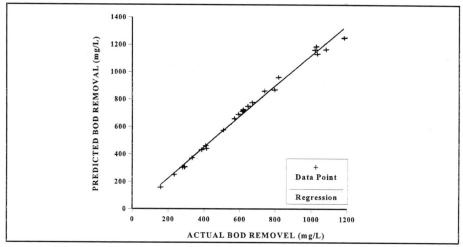

Figure 5. Comparison of Predicted versus Actual BOD Removal
Rates from the Full-Factorial Design Model

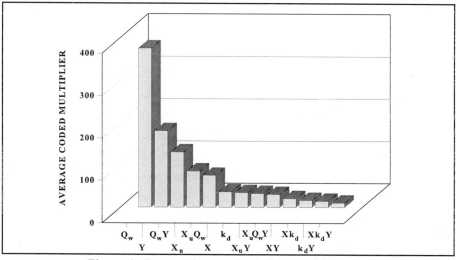

Figure 6. Pareto Diagram for Full-Factorial Design

Operational Evaluation of Results

Based on the results of the quarter-factorial design, to maximize BOD removal from wastewater one should maximize Q_w and X_u, while minimizing k_d and Y. X and the tested interactions are of lesser importance. The interactions between Q_w and Y, Q_w and k_d, and between Y and k_d appear to be significant, but further tests would be required to quantify these relationships.

Operationally, the only variables over which an operator has direct control (by turning a valve, for instance) are X, X_u, and Q_w. k_d and Y are biological rate constants which can only be controlled by relatively extreme measures such as changing the reactor water temperature or changing the type of microorganism used to stabilize the waste products. Therefore, a reasonable operational recommendation based on the results of this study would be to maximize the flow rate of waste sludge (Q_w) and the clarifier biomass concentration (X_u) in order to maximize removal of BOD from wastewater. If the biological growth rate constant, k_d, proved to be unstable for some reason, Figure 4b suggests that adjusting Q_w to approximately 275 m^3/d (the intersection of the two lines, accounting for the "coded" variable) would make the activated sludge system robust to k_d. The power of the designed experiments approach is demonstrated here by noting that these conclusions are certainly not obvious from a simple inspection of Equation 1.

An actual activated sludge process is extremely complex, however, and other considerations would come into play before undertaking changes to the system. One consideration is cost. For example, increasing Q_w will naturally increase the costs of disposing the waste sludge. Operational problems may also arise, such as the sludge in the clarifier may form nitrogen gas and float if the residence time in the settler becomes excessive while attempting to increase X_u.

This case study has attempted to demonstrate the power of the designed experiments approach. Using some simple statistical techniques coupled with engineering judgment, complex systems can be economically modeled, and their operation greatly simplified. As a result, adjustments and modifications to the system can be made, and the consequences of these changes predicted with some degree of confidence.

Case Study 15: Wave Solder Process[1]

Introduction

The problem of electrical shorts resulting from solder bridging is a persistent problem in many wave solder processes. This particular problem accounts for a disproportionate number of test-related nonconformances associated with the product of investigative interest within this study. Based on a hierarchical Pareto analysis, it was determined that the problem of solder bridging should be further investigated relative to the Q2 pin location (RF transistor) on printed wiring board PWB 123. A subsequent designed experiment was used to evaluate the effects of: (1) component lead length, (2) solder flux specific gravity, and (3) wave solder machine chain speed in relation to solder deposition characteristics. An illustration of PWB 123 is illustrated in Figure 1.

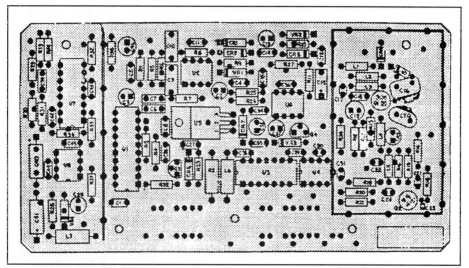

Figure 1. Printed Wiring Board P123

[1]This case study was provided by Dr. Mikel Harry, Research Director at the Six Sigma Institute, Motorola University.

Hierarchical Pareto Analysis

The hierarchical Pareto analysis begins with an investigation to determine which project exerted the greatest amount of leverage in relation to quality problems experienced by manufacturing. Figure 2 illustrates the average defects per million operations over a 12-month period broken down by project. Figure 3 displays the total number of boards processed by each project over the same interval. From this data it was determined that Project A exerted an undue amount of influence in relation to quality problems and operating costs; consequently, this project was selected for further investigation.

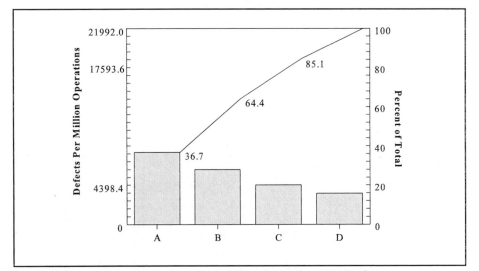

Figure 2. Defects Per Million Operations By Project
During the Criterion Time Interval

The next portion of the study involved analyzing the various defect categories on Project A. From this analysis (see Figure 4) it was determined that solder shorts (SS) accounted for a disproportionate amount of the total number of defects identified during the previously mentioned time interval. The total number of shorts was further decomposed by stratifying board type (see Figure 5). It was then concluded that PWB 123 accounted for the largest proportion of shorts. Note that the axis containing the various board type categories has been coded for ease of graphical notation. Let it suffice to say that category A is representative of PWB 123.

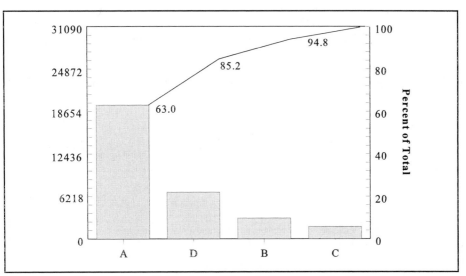

Figure 3. Total Number of Boards Processed By Project
During The Criterion Time Interval

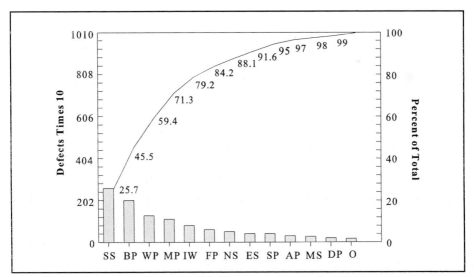

Figure 4. Percent of Total Defects By Defect Category
Relative to Project A Over the Criterion Time Interval

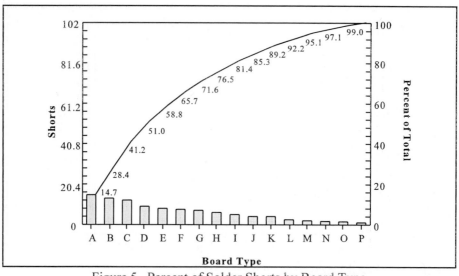

Figure 5. Percent of Solder Shorts by Board Type
Relative to Project A Over the Criterion Time Interval

The next step of the hierarchical Pareto analysis was to determine which functional component, if any, accounted for a disproportionate number of solder shorts on PWB 123. Figure 6 indicates that the Q2 component is the major problem. Note that this component (Q2) is a four lead radio frequency transistor. This particular transistor is utilized on several other PWBs commonly used; consequently, research results may have broad implications for other projects within the organization.

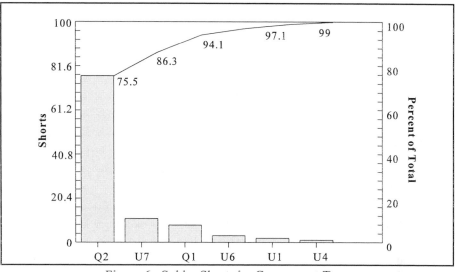

Figure 6. Solder Shorts by Component Type
Relative to PWB 123 Over the Criterion Time Interval

Statement of The Problem

As previously pointed out, a short on a PWB is considered to be one type of functional nonconformance to specification. This particular type of defect can result from an electrical condition known as a "solder bridge." In the case of an electrical component, a bridge condition results when a solder deposit extends from one lead of the component to another. A bridge condition is not restricted to a single component. Such a condition can be displayed in other forms; e.g., lead to lead, pad to pad, track to track, and lead to track. For this investigation, the use of the term "solder bridge" is restricted to electrical shorts resulting from a solder bridge condition between two or more leads on a given component. Figure 7 displays two distinct types of components mounted on a PWB via through-hole technology. This figure should aid the reader in understanding the concept of solder bridging.

The problem under investigation is solder bridging between component leads (pins) on PWBs using through-hole technology. More specifically, the problem centers on excessive electrical shorts resulting from solder bridging between two or more of the four leads associated with the Q2 component on

PWB 123. Refer to Figure 8 for a general conceptual orientation as to the pin location of the Q2 component.

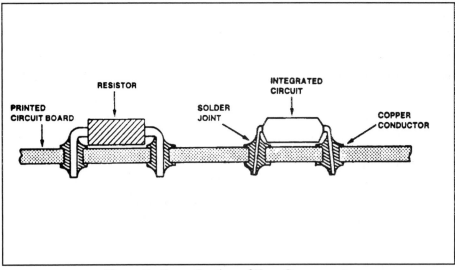

Figure 7. Cross-Section of Two Components
Mounted via Through-Hole Technology

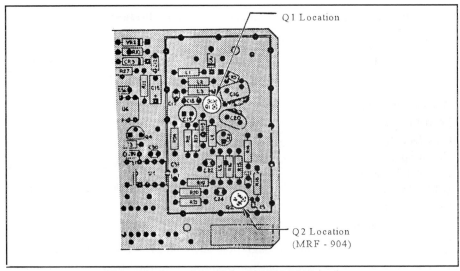

Figure 8. Location of the MRF-904 Transistor
(Q2 Pin Location) on PWB 123

Engineering Research Questions

The investigation of solder bridging at the Q2 pin location was initiated after the following engineering questions were established: Question 1 – What manufacturing variables, if any, account for a disproportionate amount of the variability in solder bridging at the Q2 component location on PWB 123? Question 2 – If the major source of variability in solder bridging is not related to the manufacturing process, can the process compensate for the observable variation until the true source of variation can be isolated and controlled? Question 3 – If the major source of variability in solder bridging at the Q2 pin location is related to the manufacturing process, can the related variables be controlled to rectify the stated problem without inducing an undesirable change in solder deposition characteristics at other pin locations on the specified PWB?

Note that the condition of solder bridging was believed to be a symptom of a more intrinsic problem. Based on this belief, several theories were advanced to offer a plausible explanation of the effect (solder bridging). The theories were rank ordered according to likelihood. Once ranked, the theories were surveyed a second time to assess structural integrity. The theory which was recognized to be most probable with regard to the problem centered around the wave solder operation. In essence, this theory hypothesized that the variables presented in Table 1 accounted for most of the variation in solder bridging at the Q2 pin location.

Variable Description	Rank Order	Experimental	
		Yes	No
Distance Between Pads	1		●
Lead Length	2	●	
Flux Density	3	●	
Chain Speed	4	●	
Pre-heat Temperature	5	●	
Solder Mask	6		●
Contamination	7		●
Wave Configuration	8		●
Humidity	9		●
Orientation	10		●

Table 1. Hypothesized "Vital Few" Variables Associated
with the Primary Theory of Experimental Concern

Review of Related Literature And Research

This section discusses the review of related literature and research. The intent of the review process is to surface information directly related to the aims of the study; e.g., previous experiments, important experimental factors, considerations, etc. In this particular instance several sources of information were obtained from a computer search. Each of the sources were reviewed for applicability to the problem articulated within "Statement of the Problem" of this study.

In relation to the basic soldering process, Woodgate [Wood 83] stated: *"Soldering is a very simple process. The only things necessary to produce a perfect joint are the following:*

(1)   Solderable parts, correctly configured,

(2)   The correct temperature for the correct time,

(3)   The right composition of flux and solder."

In spite of this simplicity, soldering sometimes seems to be one of the major problem areas of the electronics industry and a mixture of science, art, and black magic. The answer, of course, is that the conditions above are often difficult to achieve, especially the solderability requirement. In addition, in the machine soldering of PWBs, the board design can play an important part in the efficiency of the soldering process. Too often these factors are either ignored or forgotten, and they cannot be compensated for later on, except by the expenditure of unnecessary labor in inspection and touch-up. In relation to the problem presented in this study, Woodgate further stated, *"Where the assemblies have components of widely varying thermal capacity mounted close together, the joints of small thermal capacity will be soldered while those of large thermal capacity will not have reached soldering temperature. In many cases, the solder will have heated and joined those parts that are actually in the solder, but the large thermal mass will stop the remainder of the metal in the joint from reaching the temperature at which the solder will melt and fill the joint. The solution is to slow down the conveyer and increase the preheat in an attempt to bring the entire joint up to the necessary temperature. In extreme cases there is a very real possibility that the smaller parts may be overheated, and if the offending components are very large it may prove impossible to increase the temperature sufficiently. This problem is usually seen with such items as heat sinks, bus bars, power diodes and transistors, and similar items where the component itself has to dissipate more than the normal power. In extreme cases it may be necessary to arrange for the component to be mounted in some other way, with low thermal capacity leads into the PWB. Remember that there is a limit to the thermal differences that can be tolerated on any one assembly that is to be machine soldered.*

The basic assumption is that soldering is a cheap process, and if there is a problem it only means putting on another touch-up operator. This, of course, is a total fallacy brought about by the incorrect calculation of the true cost of faulty soldering. Soldering is a low-cost process, but faulty soldering is one of the most costly processes in the electronics industry. When the cost of making soldered joints is considered, the evaluation must stretch further than the actual soldering operation itself. The cost of assuring solderable parts, the cost of any touch-up and inspection, and, of course, the ultimate expenditure of putting right any field failures must be included in the final figure."

A review of the other sources of related literature and research revealed that the following material, process, and design variables play an important role in solder deposition characteristics.

Material Factors
1. Flux Activity·
2. Through Hole Plating
3. Dirty Components/Boards·
4. Contaminated Solder·
5. Pre-tinned Leads

Design Factors
1. Component Thermal Capacity·
2. Component Mounting
3. Board flatness·
4. Pad Spacing

Process Factors
1. Wave Height·
2. Wave Shape·
3. Solder Temperature·
4. Preheat Temperature·
5. Conveyer Angle·
6. Flux Quantity
7. Conveyer Alignment·

Some of these factors influence either excessive or insufficient solder conditions but many influence the entire range of solder joint conditions from no solder to a solder bridge condition. The asterisk notation ($\cdot$) indicates that the factor was identified through the literature searches. A lack of notation ($\cdot$) indicates that no specific reference was found to classify the effect of this factor on solder joint conditions. Based on the nature of the problem, related theories, and the review of related literature and research, it was determined that the most productive and reliable means for ascertaining answers valid to the previous mentioned research questions was through experimental manipulation of the manufacturing variables defined in Table 1.

Experimental Objectives

Based on prevailing manufacturing circumstances and availability of organizational resources, it was determined that the variables ranked 2, 3, and 4, respectively, as identified in Table 1, should be utilized as experimental factors for an in-depth exploration of the problem.   Given this decision, the following experimental objectives were established:

Objective 1:   Determine the main and interactive effects of the Q2 component lead length, specific gravity of the solder flux, and chain speed of the wave solder machine.

Objective 2:   Account for the majority of the variability in solder deposition about the leads of the Q2 component by manipulation of the experimental factors defined in Objective 1.

Objective 3:   For those variables which prove to be statistically significant, as well as practically significant, establish optimum performance levels.

The design matrix used to examine the effects of lead length, flux density and chain speed is a two-level, three-factor, full factorial.   The response measure (solder deposition) was evaluated on a five (5) point qualitative scale. Figures 9 and 10 illustrate this particular measurement scale (often referred to as a "Likert" scale) and its associated interval interpretations as related to this study.

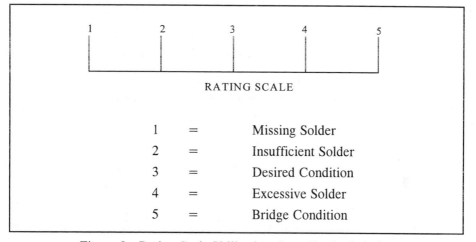

RATING SCALE

1	=	Missing Solder
2	=	Insufficient Solder
3	=	Desired Condition
4	=	Excessive Solder
5	=	Bridge Condition

Figure 9. Rating Scale Utilized to Quantify the Relative
Degree of Solder Deposition

The evaluation criteria was, in part, based on the guidelines presented in MIL-S-45743E. To better control for possible measurement/rating error, several inspectors knowledgeable with the response variable evaluation criteria were selected for comparative purposes. Each inspector was asked to rate several solder joints independent of the other inspectors' judgments. The results of the assessment supported the validity and reliability of the aforementioned measurement scale.

The Sample

To make an estimate of the true state of affairs in the real world, it is often necessary to select a representative sample from the general population. Based on this sample, such estimates as central tendency (mean, median, and mode) and variability (variance and standard deviation) are made. If the sample treatment is not representative of the population, subsequent data analyses will necessarily yield incorrect estimates and, consequently, tend to promote faulty inferences.

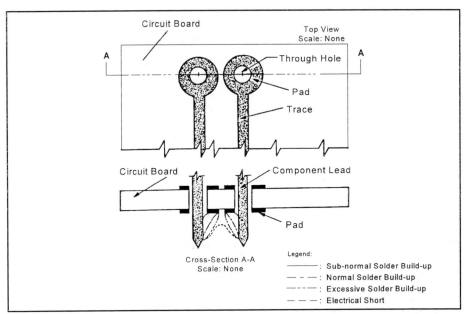

Figure 10. Graphical Rendition of the Five-Point Measurement Scale
Applied to the Determination of Solder Deposition Characteristics

To decrease the likelihood of confounded experimental effects due to initial differences in PWBs and components, all sample PWBs and components were randomly selected from the stock room. The PWBs used in the experiment were selected via random numbers. Furthermore, the components related to the experimental conditions were randomly assigned to the PWBs designated as the experimental group. In essence, the process of random assignment ensured that any biases which may have been present in the PWBs or related components were randomly distributed throughout the experiment. As a final precaution, the order for submission of the samples to the wave solder operation was randomly determined to confound any effects of process variability or process related nuisance variables.

Experimental Findings

The full factorial and the data from the 4 Q2 leads plus the remaining 16 leads is displayed in Table 2. The factors and levels were as follows:

Factor A: Component lead length at the Q2 Pin location (RF transistor)

low level (−)	=	.020 inches
high level (+)	=	.090 inches

Factor B: Specific Gravity of Solder Flux

low level (−)	=	.828 specific gravity
high level (+)	=	.838 specific gravity

Factor C: Wave Solder Chain Speed

low level (−)	=	2 ft/min
high level (+)	=	4 ft/min

		Run							
		1	**2**	**3**	**4**	**5**	**6**	**7**	**8**
	A	-1	-1	-1	-1	+1	+1	+1	+1
Factors	B	-1	-1	+1	+1	-1	-1	+1	+1
	C	-1	+1	-1	+1	-1	+1	-1	+1
Q2 Pin	y_1	4	3	5	3	5	4	5	5
	y_2	1	3	2	2	3	4	2	5
	y_3	2	3	2	3	3	3	5	5
	y_4	2	4	2	2	5	4	5	5
Other Pins	y_5	4	3	1	2	3	4	3	5
	y_6	2	3	1	2	2	3	4	4
	y_7	1	3	2	3	2	3	1	5
	y_8	2	3	4	3	1	4	3	5
	y_9	2	3	4	2	2	4	2	4
	y_{10}	5	4	2	2	5	3	2	4
	y_{11}	4	3	3	2	5	3	4	5
	y_{12}	3	3	3	3	3	3	5	4
	y_{13}	5	3	5	3	3	3	5	5
	y_{14}	2	4	1	2	1	4	3	4
	y_{15}	1	4	4	2	2	4	1	4
	y_{16}	4	3	3	3	4	4	5	5
	y_{17}	2	3	1	2	4	4	5	5
	y_{18}	5	3	2	3	5	4	3	4
	y_{19}	2	3	2	3	2	4	1	4
	y_{20}	1	3	2	3	2	3	5	4
	$\bar{y}$	2.7	3.2	2.55	2.5	3.1	3.6	3.45	4.55
	s	1.42	0.41	1.28	0.51	1.37	0.50	1.54	0.51

Table 2. Full Factorial Design and Response Data from 20 leads

Note: (The data above is not the actual Motorola data. The actual data was altered
to protect the proprietary nature of the results. Also notice a slightly different
layout because of the large number of replicates.)

The models resulting from the data analysis are:

(1) $\hat{y}$ = 3.469 + .781A (For the Q2 Pins only)

(2) $\hat{y}$ = 3.206 + .469A + .269AB + .256C + .056B (For all 20 Pins)

(3) $\hat{s}$ = .87379 - .0592C (For the Q2 Pins only)

(4) $\hat{s}$ = .943 - .4500C (For all 20 pins)

The optimal settings and predictions are displayed in Table 3.

Optimal Settings

A (Lead Length)	= .034 inches
B (Specific Gravity)	= .838 specific gravity
C (Chain Speed)	= 4 ft/min

Optimal Predictions

$\hat{y}$ (for Q2 pins)	= 3.00
$\hat{y}$ (for all 20 pins)	= 3.08
$\hat{s}$ (for Q2 pins)	= .81659
$\hat{s}$ (for all 20 pins)	= .493

Table 3. Optimal Settings and Predictions

Conclusions

Although new settings were found which improved the process, the standard deviation continues to be fairly high. Some of this variability may be a result of the measurement system. An action item will be to work on measurement to insure it is accurate and precise. The confirmation phase (not reported in this study) will attempt to validate the settings and predictions in Table 3. In addition, a confirmation test will also be run with C at a higher setting than 4 ft/min in an attempt to further reduce the $\hat{s}$ predictions. Assuming $\hat{s}$ can be further reduced by an extrapolation of factor C, new A and B settings will be found to readjust $\hat{y}$ to approximately 3.0.

Case Study 16: Reducing Radar Cross Sections Through Design of Experiments[1]

Introduction

Resistive-Card (R-Cards) are used to reduce the radar cross section (RCS) of various vehicles. The R-card manufacturing process revolves around robotic spraying of different layers of conductive inks onto various substrates. The desired end result is to create a continuous resistive gradient across the width of the card. Figure 1 shows the manufacturing setup for this process.

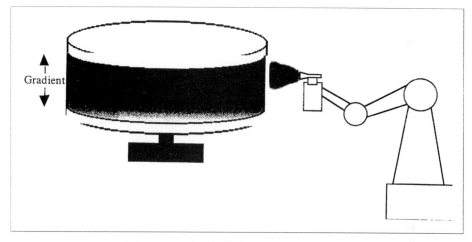

Figure 1. R-Card Manufacturing Setup

The end points of the resistive gradient are typically < 10 ohms/sq. and >2500 ohms/sq. The gradient is specified in 2 to 3 mm steps across the width of the card with a tolerance at each station. Typical card widths vary from 55 to 300 mm and lengths up to 6 meters. The ultimate goal of the Quality Improvement Program was to produce a complete R-card in one spray cycle, with no rework. The major benefits would be labor and material savings. A Design of Experiments (DOE) was

---

[1]This case study was provided by Al Memmolo and John Stinson of McDonnell Douglas Technologies Incorporated.

identified as the methodology for achieving our goal. Figure 2 shows a typical card and specification limits.

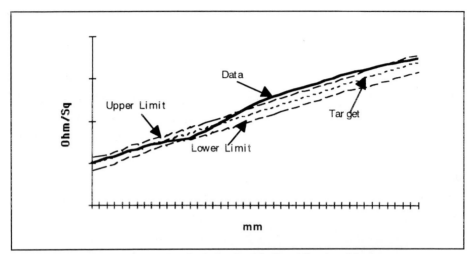

Figure 2. A Typical Card with Specification Limits

Background

During the spring of 1994 the R-card process was being transitioned from R&D to a production process. McDonnell Douglas Technologies Inc. (MDTI) had a contract to produce several R-cards to the same specification. Prior to this contract each R-card was produced individually as a custom card. The process was intentionally flexible and adaptable to meet complex R&D designs. However, this process did not lend itself to the repetitive nature of a production contract. The initial R-cards produced did not meet the required specifications, resulting in a 0% yield. The customer had to purchase R-cards on waiver to meet their schedule. The process of spraying R-cards was out of control.

A process assessment and investigation was initiated. A cross functional team, consisting of Materials and Processes, Manufacturing, Quality, Engineering, and Management, was formed to solve this problem. Preliminary examination of the process revealed that the Standard Operating Procedures (SOPs) were not well defined and the measurement system was not capable. Also, the key process variables needed to be identified and controlled, while reducing the overall standard deviation of the spraying process.

Method

The method used to bring the R-card process in control involved three distinct areas: improving the Standard Operating Procedures (SOPs), understanding and improving the measurement system, and performing a designed experiment.

Standard Operating Procedures

The SOP definition began with the creation of a flow diagram of the manufacturing process. The flow diagrams shown in Figures 3a and 3b indicated which areas needed attention. The first item was material variation. After controlled and detailed experimentation it was discovered that a key ingredient of the ink was being depleted during the spraying process. To limit the effect of this problem it was decided to reduce the batch size. It was also discovered that the room temperature and humidity variation was out of control. Facilities improvements, environmental monitoring, and detailed operational guidelines helped eliminate this problem. Other SOPs were incorporated into the process, such as spray track definitions, controlled procedural planning (work instructions) and R-card marking guidelines. These improvements provided a starting point for establishing process control.

Measurement System Improvements

A process flow diagram of the measurement system for testing R-cards was generated. In-process inspection was being performed on different measurement equipment than the final inspection. Initial correlation studies revealed an unacceptable correlation coefficient. Hence, the faster, less accurate in-process test equipment was eliminated for the more accurate final test equipment. The remaining test equipment was instituted for both the in-process and final inspection, providing 100% correlation. The second major improvement was to quantify the measurement error of the testing procedure. Preliminary evaluation showed the final inspection measurement error to be approximately 50% of the tolerance of the R-card specification in certain resistivity regions. The measurement equipment capability was determined and the results indicated that the equipment met the accuracy requirements as established by the manufacturer. Therefore, it was concluded that the two possible remaining sources of measurement error were R-card fixturing and data reduction.

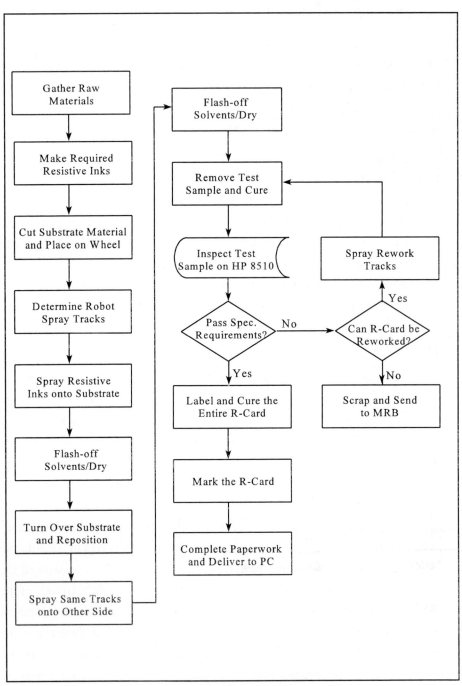

Figure 3a. R-Card Fabrication Process Flow

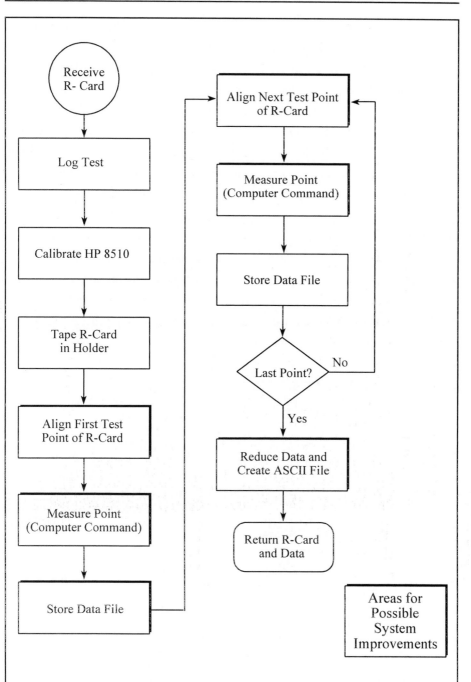

Figure 3b. R-Card Testing Procedure

The fixturing problem revolved around positioning the R-card during the test procedure. A translational stage was incorporated in the test system, which guaranteed that the R-card was positioned correctly every time. The data reduction centered around transferring the test equipment measurements into a usable form for the customer. Major software restructuring removed the possibility of human error by automating the data reduction process. The test equipment was programmed to produce data output which was acceptable to the customer. Testing on the new system showed the measurement error was reduced to less than 10% of the tolerance.

DOE Setup

In order to identify the key variables and quantify the variation in the R-card manufacturing process, a designed experiment was proposed. The team decided the output of the DOE should be an understanding and reduction of the point to point variation across the R-card. The brainstorming process yielded the Cause and Effect diagram shown in Figure 4.

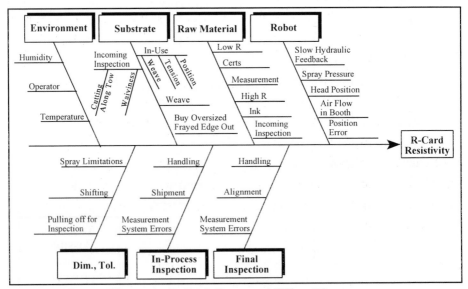

Figure 4. R-Card Cause and Effect Diagram

Seven factors were selected for the experiment. The funding limit for this experiment allowed for an 8 run design (L8) with 4 replications. The experiment layout is shown below in Table 1. However, the experimental runs were randomized to eliminate human bias.

Run	Room Temp	Defects	Fan Adjust.	Ink Feed	Cure Temp	Cloth Tension	Cure Time	y_1	y_2	y_3	y_4	$\bar{y}$	s
1	70°F ±2	No	Yes	Low	High	Taut	Low						
2	70°F ±2	No	Yes	High	Low	Loose	High						
3	70°F ±2	Yes	No	Low	High	Loose	High						
4	70°F ±2	Yes	No	High	Low	Taut	Low						
5	78°F ±2	No	No	Low	Low	Taut	High						
6	78°F ±2	No	No	High	High	Loose	Low						
7	78°F ±2	Yes	Yes	Low	Low	Loose	Low						
8	78°F ±2	Yes	Yes	High	High	Taut	High						

Table 1. 8-Run Experiment Layout

Experimental Analysis

The experimental analysis was challenging because of the vast amount of data the DOE generated. Each card was measured at 40 different places across the card and each point had a unique target value and tolerance. Several measures of goodness were explored, such as linearity and total variation. Linearity did not have enough resolution to evaluate the process. Total R-card variation did not describe the process with respect to the distinct resistivity zones of the R-card. The evaluation method used was point-to-point variation across the entire card. This method produced a linear standard deviation prediction equation for each point of the R-card. The key factors of this experiment and their optimal settings are listed in Table 2.

Cloth Defects:	No (Lo)
Fan Adjustment:	Yes (Hi)
Cure Temperature:	Lo
Cloth Tension:	Taut (Hi)
Cure Time:	Hi

Table 2. Optimal Settings for Key Factors

The size of the effect for each factor in the prediction equation varied across the card, but at no time did the sign of the factor change. Thus, the low setting for cloth defects decreased the standard deviation across the entire card. The predicted standard deviation results and upper and lower limits are shown in the Figure 5. It should be noted that due to the complexity of the analysis, mean shifting variables were not evaluated. The team was only looking for variables which would reduce the standard deviation.

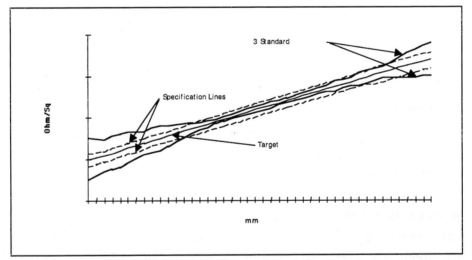

Figure 5. Predicted Standard Deviation Results

Results

The optimal parameter settings from the designed experiment were incorporated in the production process when manufacturing resumed. The resulting production yields improved, but not to satisfactory levels. Although the DOE established the key variables, it did not adequately reduce the process variation.   R-cards could not be produced consistently within the customer's specifications.

Continuous Improvements

The next challenge was to reduce equipment and process variation using the DOE settings as a baseline. A systematic plan of reducing and eliminating possible causes of equipment and process variation was undertaken. New spray pumps were purchased, equipment was placed on a regular cleaning schedule, operational procedures were further defined and ink batching was regimented. Equipment modifications were incorporated, such as limit switches to provided accurate starting and stopping of the spray. Detailed spray track templates were created to pinpoint where rework sprays were needed. Rework sprays were incorporated into original spray tracks to provide one spray cycle capability. Reproducibility of spray patterns was enhanced by using a two station cleaning operation for inks between pump transfers.

The six month effort was effective, the yield improved dramatically and R-cards were produced within the customer's specification. One of the production programs produced 70% of the R-cards in one spray cycle, thus allowing the program to be completed ahead of schedule and under budget.

Conclusions

The reduction of R-card variability was accomplished through the use of experimental design, critical thinking, and other quality tools. The reduction in processing variation allowed a successful transition form R&D to manufacturing.

Completing a DOE was instrumental in the reduction of variability, but on its own it did not improve the manufacturing capability to desired levels. In this case, establishing the SOPs, correcting the measurement system, and systematically eliminating and minimizing processing variation were essential to reducing manufacturing variability.

Acknowledgments

The authors would like to thank the following team members for their contributions: Bob Brenner, Jon Cook, Lynne Crisafi, Bill Doane, Gerry Fay, Bob French, Bryan Imber, Kary Jacobson, Carl Mentzer, Carlos Portugal, Bruce Tracy, Dave Wood, Ed Yarbrough and Dave Zorger.

Case Study 17:   Process Development for Bonding Titanium to Cobalt Chrome[1]

Background

The most widely used metals for orthopedic implants are Titanium (and its alloys), and Cobalt Chrome. Each has distinct advantages. Titanium is preferred as bone interface while Cobalt Chrome is preferred as a wear surface. For quite some time the implants were fixed to the bone using bone cement. An improvement on that would be to fix the implant directly to bone without the use of any cement. This led to the development of porous coated implants.

Porous coatings are applied to the implants in suitable locations to promote bone ingrowth directly on/into the implant, thereby achieving biological fixation. These porous coatings are applied using various methods: sintering, plasma spraying, pressure diffusion bonding, for example. Traditionally, the porous coating material has been the same as the implant material. So Titanium porous coating would be applied to Titanium (or Titanium alloy) implants. Recently the need was felt to combine the two metals so that the best properties of each could be tapped. The requirement therefore was Titanium as porous coating on a Cobalt Chrome implant.

Many variables influence the attachment of porous coatings to a substrate; the cleanliness of the mating surfaces, roughness, particle size of the coating, pressures involved, time of sintering, and temperature of sintering are some of the considerations involved. The most critical dependent variable was identified as the bond strength between the porous coating and the substrate.

Development

In the development of an orthopedic implant manufacturing process, over 25 variables were identified to fuse Titanium and Cobalt Chrome. The company was on a short fuse since market demand was heating up for such a product and competitors were pursuing similar approaches. Technologically, fusing these two materials successfully was considered to be a remote possibility since there exists

---

[1]Contributed by Mr. Rai Chowdhary, Intermedics Orthopedics Engineer.

a low melting eutectic between Titanium and Cobalt that could interfere with the sintering process. Indeed, there was evidence that other researchers had attempted to fuse these materials and ended up with a pool of molten metal in their furnaces.

Considering the number of variables and the pressure to demonstrate continuous rather than discrete progress towards this (so called unachievable) goal, it was decided to take smaller steps in the development process. The simple-to-conduct L8 design was chosen; further, to minimize the risk of facing lost/uncertain knowledge, confounding was kept low by not exceeding 4 variables in any of the experimental designs. The idea was to identify one or two variables through each L8 design initially, by screening all the 25+ variables, and then run successive L8 designs to narrow down the choices, until we had the (absolutely) critical variables for the process identified.

The following table illustrates one of the designs, with representative high and low values, that was used in the program. A total of 7 initial L8 designs, followed by 5 additional ones, were conducted before the critical variables were identified. Once this was accomplished, 15 additional confirmation runs were conducted using the critical variables only, to gain confidence in the process. During these runs, three different operators were chosen at random to determine if the process had any significant operator dependence. The success of this study has resulted in a patented process of bonding Titanium and Cobalt Chrome with a return on investment of over $60 million annually.

An L8 Design Used in the Development Process*

Settings	Grit Size	Air Pressure (psi)	Grit Flow gms/min	Distance, Part to Nozzle (inches)
Low	24	35	10	2
High	40	65	30	4

*Note: All values in the table are representative values.
          Owing to the proprietary nature of the process, actuals and results cannot be disclosed.
          This process has been patented thru Intermedics Orthopedics, Austin, Texas.

Case Study 18: Process Control and Capability in the Automotive Industry[1]

This very instructive style case study gives a detailed step-by-step procedure to show how an x̄ - R chart is developed and used to bring a process in control. Process capability is then computed.

Step 1. Gather Data

An x̄ and an R chart are developed from measurements of a particular characteristic of the process output. These data are reported in small subgroups of constant size, usually containing from 2 to 5 consecutive pieces, with subgroups taken periodically (e.g., once every 15 minutes, twice per shift, etc.). A data gathering plan must be developed and used as the basis for collecting, recording, and plotting the data on a chart.

1.a.   **Select the size, frequency, and number of subgroups** (see Figure 1)

- **Subgroup Size** - The first key step in variables control charting is the determination of "rational subgroups" — they will determine the effectiveness and efficiency of the control chart that uses them.

    The subgroups should be chosen so the opportunities for variation among the units within a subgroup are small. If the variation within a subgroup represents the piece-to-piece variability over a very short period of time, then any unusual variation between subgroups would reflect changes in the process that should be investigated for appropriate actions.

---

[1]Contributed by Peter Jessup of Ford Motor Company and taken from Ford Motor Company's copyrighted publication, *Continuing Process Control and Process Capability Improvement*, ©1984. All material is reprinted with permission from Ford Motor Company.

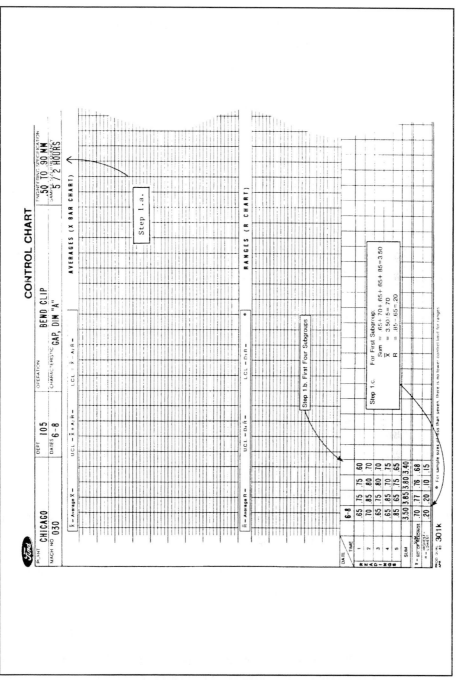

Figure 1. x̄ and R Charts: Steps 1.a., 1.b., and 1.c.

For an initial study of a process, the subgroups could typically consist of 4 or 5 consecutively-produced pieces representing only a single tool, head, die cavity, etc. (i.e., a single process stream). The intention is that the pieces within each subgroup would all be produced under very similar production conditions over a very short time interval. Hence, variation within each subgroup would primarily reflect common causes. When these conditions are not met, the resulting control chart may not effectively discriminate special causes of variation, or it may exhibit some unusual patterns. Sample sizes must remain constant for all subgroups.

- **Subgroup Frequency** - The goal is to detect changes in the process over time. Subgroups should be collected often enough, and at appropriate times, that they can reflect the potential opportunities for change. Such potential causes of change might include shift patterns of relief operators, warmup trend, material lots, etc.

  During an initial process study, the subgroups themselves are often taken consecutively or at short intervals, to detect whether the process can shift or show other instability over brief time periods. As the process demonstrates stability (or as process improvements are made), the time period between subgroups can be increased. Subgroup frequencies for ongoing production monitoring could be twice per shift, hourly, or some other reasonable rate.

- **Number of Subgroups** - The number of subgroups should satisfy two criteria. From a practical standpoint, enough subgroups should be gathered to insure that the major sources of variation have had an opportunity to appear. From a statistical standpoint, 25 or more subgroups containing about 100 or more individual readings give a good test for stability and, if stable, good estimates of the process location and spread.

1.b.   **Set Up Control Charts and Record Raw Data**

x̄ and R charts are normally drawn with the x̄ chart above the R chart, and a data block at the bottom. The values of x̄ and R will be the vertical scales, while the sequence of subgroups through time will be the horizontal scale. The data values and the plot points for the range and average should be aligned vertically. Also, an area should be included to note any unusual circumstances or corrections which occurred during the period. Enter the individual raw values and the identification for each subgroup (see Figure 1).

1.c.   **Calculate the Average (x̄) and Range (R) of Each Subgroup**

The characteristics to be plotted are the sample average (x̄) and the sample range (R) for each subgroup; these reflect the overall process average and its variability, respectively.

For each subgroup, calculate:

$$\overline{X} = \frac{X_1 + X_2 + \ldots + X_n}{n}$$

$$R = X_{highest} - X_{lowest}$$

where the x_1, x_2, . . . are individual values within the subgroup and n is the subgroup sample size.

1.d.   **Select Scales for the Control Charts**

The vertical scales for the two charts are for measured values of x̄ and R, respectively. Some general guidelines for determining the scales may be helpful, although they may have to be modified in particular circumstances. For the x̄ chart, the difference between the highest and lowest values on the scales should be at least 2 times the difference between the highest and lowest subgroup averages (x̄). For the R chart, values should extend from a lower value of zero to an upper value about 2 times the largest range (R) encountered during the initial period.

1.e.   **Plot the Averages and Ranges on the Control Charts**

Plot the averages and ranges on their respective charts. Connect the points with lines to help visualize patterns and trends (see Figure 2).

Briefly scan the plotted points to see if they look reasonable. If any points are substantially higher or lower than the others, confirm that the calculations and plots are correct. Make sure that the plotted points for the corresponding $\bar{x}$ and R are vertically in line.

Step 2. Calculate Control Limits

Control limits for the range chart are developed first, then those for the $\bar{x}$ chart. The calculations for the control limits for variables charts use constants which appear as letters in the formulas that follow. These constant values differ according to the subgroup size (n) and are shown in brief tables accompanying the respective formulas. See Appendix I for a complete set of control chart constants.

2.a.   **Calculate the Average Process Range ($\bar{R}$) and the Overall Process Average ($\bar{\bar{x}}$)**

For the study period, calculate:

$$\bar{R} = \frac{R_1 + R_2 + \dots + R_k}{k}$$

$$\bar{\bar{x}} = \frac{\bar{x}_1 + \bar{x}_2 + \dots + \bar{x}_k}{k}$$

where k is the number of subgroups, R_1 and $\bar{x}_1$ are the range and average of the first subgroup, R_2 and $\bar{x}_2$ are from the second subgroup, etc. See Figure 3.

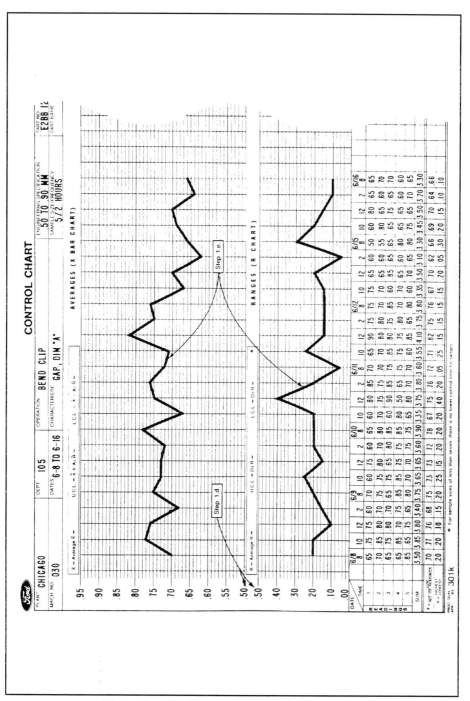

Figure 2. x̄ and R Charts: Steps 1.d. and 1.e.

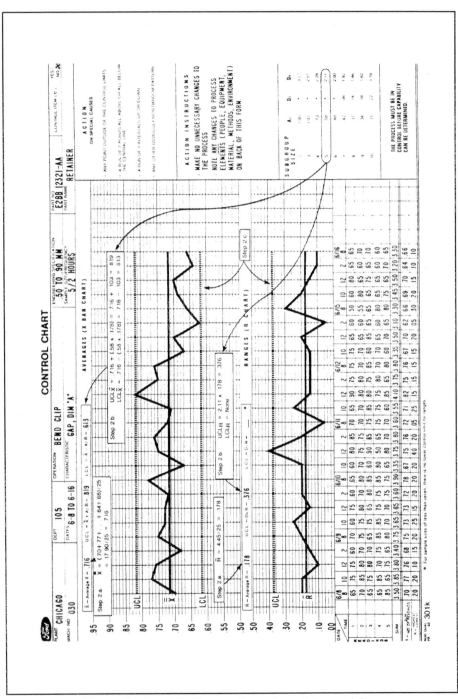

Figure 3. x̄ and R Charts: Steps 2.a., 2.b., and 2.c.

2.b.   **Calculate the Control Limits**

Control limits are calculated to show the extent by which the subgroup averages and ranges would vary if only common causes of variation were present. They are based on the subgroup sample size and the amount of within-subgroup variability reflected in the ranges. Calculate the upper and lower control limits for ranges and for averages:

$$UCL_R = D_4 \bar{R} \qquad\qquad UCL_{\bar{x}} = \bar{\bar{x}} + A_2 \bar{R}$$

$$LCL_R = D_3 \bar{R} \qquad\qquad LCL_{\bar{x}} = \bar{\bar{x}} - A_2 \bar{R}$$

where D_4, D_3, and A_2 are constants varying by sample size, and the sample sizes range from 2 to 10 as shown in the following partial table taken from Appendix I:

n:	2	3	4	5	6	7	8	9	10
D_4:	3.27	2.57	2.28	2.11	2.00	1.92	1.86	1.82	1.78
D_3:	*	*	*	*	*	.08	.14	.18	.22
A_2:	1.88	1.02	.73	.58	.48	.42	.37	.34	.31

*For sample sizes below 7, the LCL_R would technically be a negative number; in those cases we assume the lower control limit to be zero since a negative range value is impossible.

2.c.   **Draw Lines for $\bar{\bar{x}}$, $\bar{R}$, and the Control Limits on the charts**

Draw the average range ($\bar{R}$) and process average ($\bar{\bar{x}}$) as solid horizontal lines, the control limits (UCL_R, LCL_R, $UCL_{\bar{x}}$, $LCL_{\bar{x}}$) as dashed horizontal lines; and label the lines. During the initial study phase, these are considered trial control limits. See Figure 3.

Step 3. Interpret for Process Control

The control limits can be interpreted as follows: If the process piece-to-piece variability and the process average were to remain constant at their present levels (as estimated by $\bar{R}$ and $\bar{\bar{x}}$, respectively), the individual subgroup ranges (R) and averages ($\bar{x}$) would vary by chance alone. Seldom would they go beyond the control limits. Based on the limits calculated above (i.e., the measurements are normally distributed), we would expect less than 1% of the ranges and only 0.27% of the averages to exceed the control limits. Likewise, there would be no obvious trends or patterns in the data beyond that which is due to chance. The objective of control chart analysis is to identify any evidence that points to the process variability or the process average not operating at a constant level — that one or both are out of statistical control — and to take appropriate action. The R and $\bar{x}$ charts are analyzed separately, but comparison of patterns between the two charts may sometimes give added insight into special causes affecting the process.

3.a.   **Analyze the Data Plots on the Range Chart**

Since the ability to interpret either the subgroup ranges or subgroup averages depends on the estimate of piece-to-piece variability, the R chart is analyzed first. The data points are compared with the control limits to detect points out of control or unusual patterns or trends (see Figure 4).

1.   **Points Beyond the Control Limits** - The presence of one or more points beyond either control limit is primary evidence of non-control at that point. Since points beyond the control limits would be very rare if only variation from common causes were present, we presume that a special cause has accounted for the extreme value. Therefore, any point beyond a control limit is a signal for immediate analysis of the operation for special causes. Mark any data points that are beyond the control limits for further investigation and corrective action (see Step 3.b.).

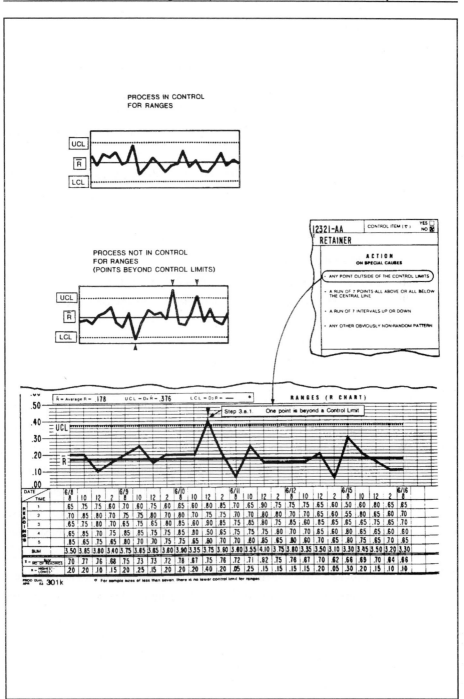

Figure 4. x̄ and R Charts: Step 3.a.1.

A point above the upper control limit for ranges is generally a sign that:
—   The control limit or plot point has been miscalculated or misplotted, or
—   The piece-to-piece variability or the spread of the distribution has increased (i.e., worsened), either at that one point in time or as part of a trend, or
—   The measurement system has changed (e.g., a different inspector or gage).

A point below the lower control limit (for sample sizes of 7 or more) is generally a sign that:
—   The control limit or plot point is in error, or
—   The spread of the distribution has decreased (i.e., become better), or
—   The measurement system has changed (including editing or alteration of the data).

Patterns or Trends Within the Control Limits - The presence of unusual patterns or trends, even when all ranges are within the control limits, can be evidence of non-control or change in process spread during the period of the pattern or trend. This could give the first warning of unfavorable conditions which should be corrected even before points are seen beyond the control limits. Conversely, certain patterns or trends could be favorable and should be studied for possible permanent improvement of the process. Comparison of patterns between the range and average charts may give added insight (see Figure 5).

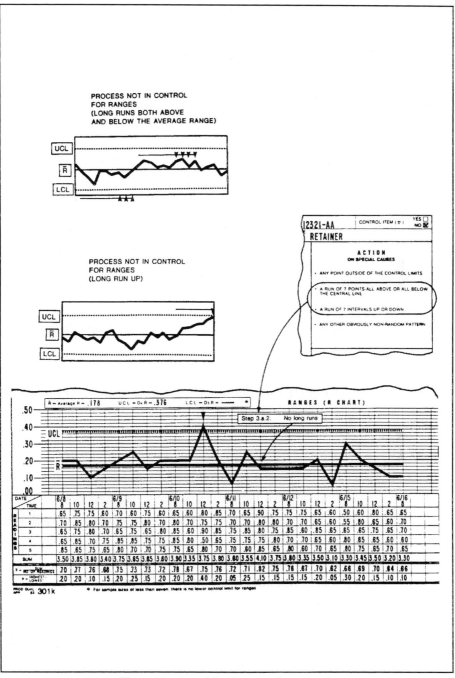

Figure 5. x̄ and R Charts: Step 3.a.2.

2. **Runs** - The following are signs that a process shift or trend has begun:
   - 7 points in a row on one side of the average, or
   - 7 intervals in a row that are consistently increasing (equal to or greater than the preceding points) or consistently decreasing.

   Mark the point that prompts the decision; it may be helpful to extend a reference line back to the beginning of the run. Analysis should consider the approximate time at which it appears that the trend or shift first began.

   A run above the average range or a run up signifies:
   - Greater spread in the output values, which could be from an irregular cause (such as equipment malfunction) or from a shift in one of the process elements (e.g., a new, less uniform raw material lot). These are usually troubles that need correction; or
   - A change in the measurement system (e.g., new inspector or gage).

   A run below the average range or a run down signifies:
   - Smaller spread in output values, which is usually a good condition that should be studied for wider application, or
   - A change in the measurement system, which could mask real performance changes.

3. **Obvious Nonrandom Patterns** - In addition to the presence of points beyond control limits or long runs, other distinct patterns may appear in the data that give clues to special causes (see Figure 6). Care should be taken not to over-interpret the data, since even random (i.e., common cause) data can sometimes give the illusion of nonrandomness. Examples of nonrandom patterns

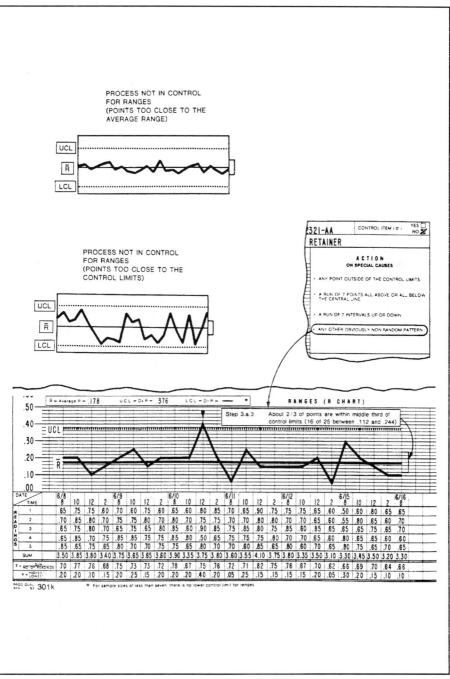

Figure 6. x̄ and R Charts: Step 3.a.3.

could be obvious trends (even though they did not satisfy the runs tests), cycles, the overall spread of data points within the control limits, or even relationships among values within subgroups (e.g., the first reading might always be the highest). One test for the overall spread of subgroup data points is described below:

> Distance of points from $\bar{R}$: About 2/3 of the data points should lie within the middle third of the region between the control limits; about 1/3 of the points should be in the outer two-thirds of the region.

> If substantially more than 2/3 of the data points lie close to $\bar{R}$ (for 25 subgroups, if over 90% are in the middle third of the control limit region), check to see whether:

> — The control limits or plot points have been miscalculated or misplotted, or
> — The process or the sampling method are stratified; that is, each subgroup systematically contains measurements from two or more process streams that have very different process averages (e.g., one piece from each of several spindles), or
> — The data have been edited (i.e., subgroups with ranges that deviated much from the average have been altered or removed).

> If substantially fewer than 2/3 of the data points lie close to $\bar{R}$ (for 25 subgroups, if 40% or fewer are in the middle third), check to see whether:
> — The control limits or plot points have been miscalculated or misplotted, or

- The process or the sampling method cause successive subgroups to contain measurements from two or more process streams that have dramatically different variability (e.g., mixed lots of input materials).

If several process streams are present, they should be identified and tracked separately.

3.b.   Find and Correct Special Causes (Range Chart)
For each indication of a special cause in the range data, conduct an analysis of the process operation to determine the cause; correct that condition, and prevent it from recurring. The control chart itself should be a useful guide in problem analysis, suggesting when the condition began and how long it continued.

Timeliness is important in problem analysis, both in terms of minimizing the production of nonconforming output, and in terms of having fresh evidence for diagnosis. For instance, the appearance of a single point beyond the control limits is reason to begin an immediate analysis of the process. See Figure 7.

It should be emphasized that problem-solving is often the most difficult and time-consuming step. Statistical input from the control chart can be an appropriate starting point, but other simple tools such as Pareto charts, cause-and-effect diagrams, or other graphical analysis can be helpful. Ultimately, however, the explanations for behavior lie within the process and the people who are involved with it. Thoroughness, patience, insight, and understanding will be required to develop actions that will measurably improve performance.

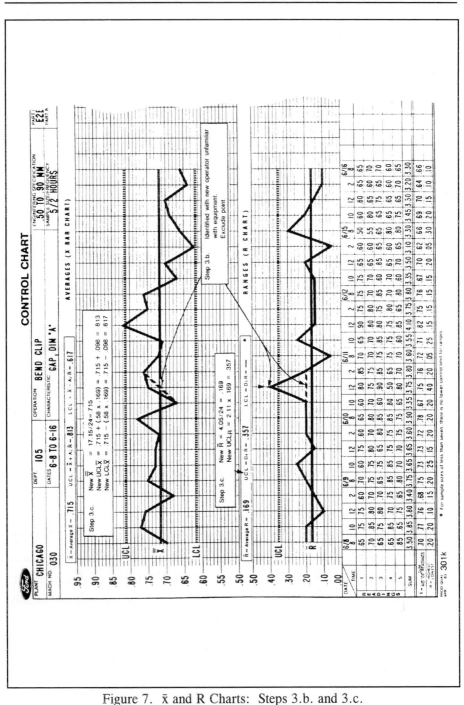

Figure 7. x̄ and R Charts: Steps 3.b. and 3.c.

3.c.   **Recalculate Control Limits (Range Chart)**

When conducting an initial process study or a reassessment of process capability, the control limits should be recalculated to exclude the effects of out-of-control periods for which process causes have been found and corrected. Exclude all subgroups affected by the special causes that have been corrected, then recalculate and plot the new average range ($\overline{R}$) and control limits. Confirm that all range points show control when compared to the new limits, repeating the identification/correction/recalculation sequence if necessary.

If any subgroups were dropped from the R chart because of identified special causes, they should also be excluded from the $\overline{x}$ chart. The revised $\overline{R}$ and $\overline{\overline{x}}$ should be used to recalculate the trial control limits for averages, $\overline{\overline{x}} \pm \overline{R}\,A_2$.

Note: The exclusion of subgroups representing unstable conditions is not just "throwing away bad data." Rather, by excluding the points affected by known special causes, we have a better estimate of the background level of variation due to common causes. This, in turn, gives the most appropriate basis for the control limits used to detect future occurrences of special causes of variation. See Figure 7.

3.d.   **Analyze the Data Plots on the Averages Chart**

When the ranges are in statistical control, the process spread is considered to be stable. The averages can then be analyzed to see if the process location is changing over time. If the averages are in statistical control, they reflect only the common-cause variation of the system. If the averages are not in control, some special causes of variation are making the process location unstable.

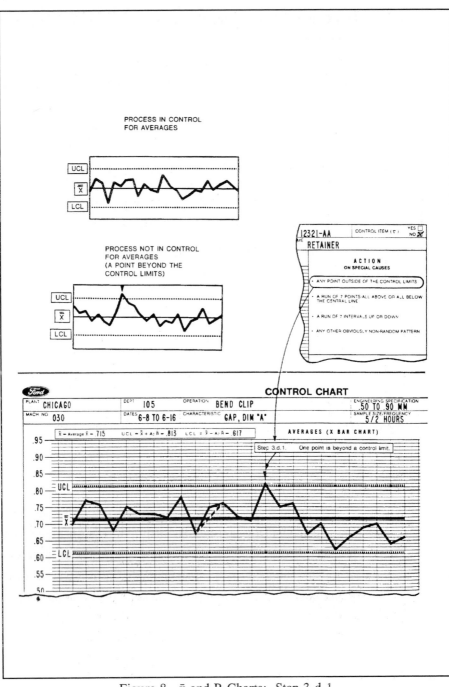

Figure 8. x̄ and R Charts: Step 3.d.1.

1. **Points Beyond the Control Limits** - The presence of one or more points beyond either control limit is primary evidence of the presence of special causes at that point. It is the signal for immediate analysis of the operation. Mark such data points on the chart. See Figure 8.

   A point beyond either control limit is generally a sign that:
   —   The control limit or plot point is in error, or
   —   The process has shifted, either at that one point in time (possibly an isolated incident) or as part of a trend, or
   —   The measurement system has changed (e.g., different gage or inspector).

   Patterns or Trends Within the Control Limits - The presence of unusual patterns or trends can be evidence of non-control or change in capability during the period of the pattern or trend. Comparison of patterns between the range and average charts may be helpful.

2. **Runs** - The following are signs that a process shift or trend has begun:
   —   7 points in a row on one side of the average, or
   —   7 intervals in a row that are consistently increasing or decreasing (see Figure 9).

   Mark the point that prompts the decision; it may help to extend a reference line to the point at which the run began. Analysis should consider the time at which it appears that the trend or shift first began.

   A run relative to the process average is generally a sign that:
   —   The process average has changed — and may still be changing, or
   —   The measurement system has changed.

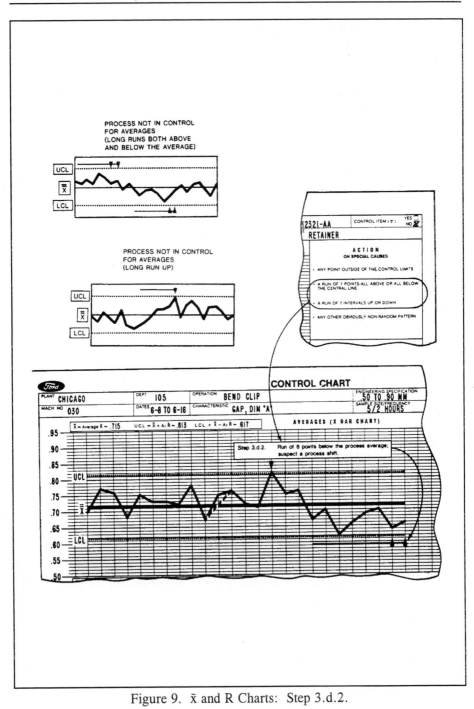

Figure 9. x̄ and R Charts: Step 3.d.2.

3.   **Obvious Nonrandom Patterns** - Other distinct patterns may also
        indicate the presence of special causes of variation, although care
        must be taken not to over-interpret the data. Among these patterns
        are trends, cycles, unusual spread of points within the control
        limits, and relationships among values within subgroups. One test
        for unusual spread is given below:

        Distance of points from the process average: About 2/3 of the
        data points should lie within the middle third of the region between
        the control limits; about 1/3 of the points will be in the outer two-
        thirds of the region; about 1/20 will lie relatively close to the
        control limits (in the outer third of the region). See Figure 10.

        If substantially more than 2/3 of the $\bar{x}$'s lie close to the process
        average (for 25 subgroups, if over 90% are in the middle third of
        the control limit region), check to see whether:
        —   The control limits or plot points have been miscalculated,
                misplotted, or incorrectly calculated, or
        —   The process or the sampling method are stratified; that is,
                each subgroup contains measurements from two or more
                process streams that have different averages, or
        —   The data have been edited.

        If substantially fewer than 2/3 of the data points lie close to the
        process average (for 25 subgroups, if 40% or fewer are in the
        middle third), check to see whether:
        —   The control limits or plot points have been miscalculated
                or misplotted, or

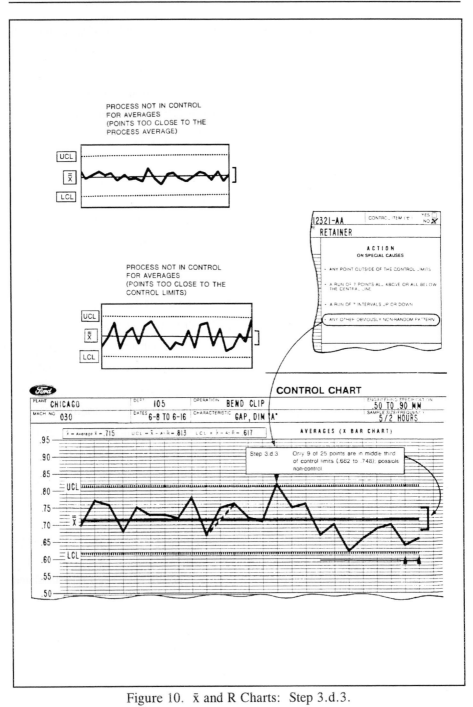

Figure 10. x̄ and R Charts: Step 3.d.3.

— The process or the sampling method caused successive subgroups to contain measurements from two or more very different process streams (this can be the result of over-control of an adjustable process, where process changes are made in response to random fluctuations in the process data).

If several process streams are present, they should be identified and tracked separately.

3.e. Find and Correct Special Causes (Averages Chart)

For each indication of an out-of-control condition in the average data, conduct an analysis of the process operation to determine the reason for the special cause. Correct that condition and prevent it from recurring. Use the chart data as a guide to when problem conditions began and how long they continued. Timeliness in analysis is important, both for diagnosis and to minimize nonconforming output. See Figure 11.

Problem-solving techniques such as Pareto analysis and cause-and-effect analysis can help.

3.f. Recalculate Control Limits (Averages Chart)

When conducting an initial process study or a reassessment of process capability, exclude any out-of-control points for which special causes have been found; recalculate and plot the process average and control limits. Confirm that all data points show control when compared to the new limits, repeating the identification/correction/recalculation sequence if necessary. See Figure 11.

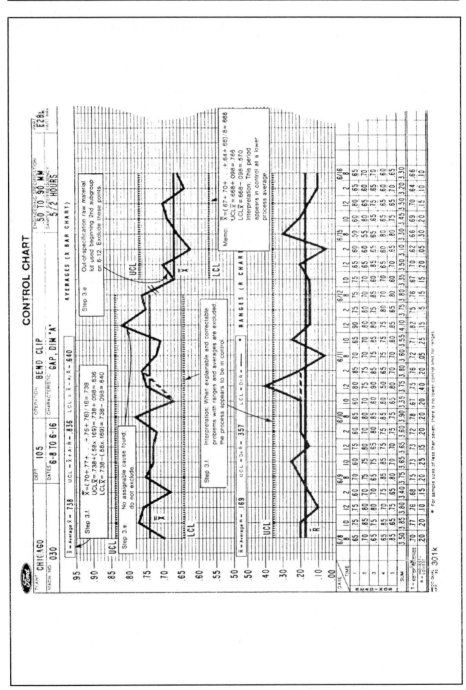

Figure 11. x̄ and R Charts: Steps 3.e. and 3.f.

Note: The preceding discussions were intended to give a functional introduction to control chart analysis. There are, however, other considerations that can be useful to the analyst. One of the most important is the reminder that, even with processes that are in statistical control, as more data are reviewed, the likelihood of false out-of-control situations increases. While it is wise to investigate all signaled events as possible evidence of special causes, it should be recognized that they may have been caused by chance and that there may be no determinable underlying local process problem. If no clear evidence of a process problem is found, any "corrective" action will probably serve to increase, rather than decrease, the total variability in the process output.

3.g.   **Extend Control Limits for Ongoing Control**

When the initial (or historical) data are consistently contained within the trial control limits, extend the limits to cover future periods. These limits would be used for ongoing control of the process, with the operator and local supervision responding to signs of out-of-control conditions on either the x̄ or R chart with prompt action. See Figure 12.

A change in the subgroup sample size would affect the expected average range and the control limits for both ranges and averages. This situation could occur, for instance, if it was decided to take smaller samples more frequently, so as to detect large process shifts more quickly without increasing the total number of pieces sampled per day. For more detail on adjusting center lines and control limits for a new subgroup sample size, see [Ford 85].

Limits for ongoing control should only be extended to cover about 20-25 future subgroups. As long as the process remains in control at constant levels for both averages and ranges, the ongoing limits can be extended for additional periods. If, however, there is evidence that the process average or range have changed (in either direction), control limits should be recalculated based on current performance.

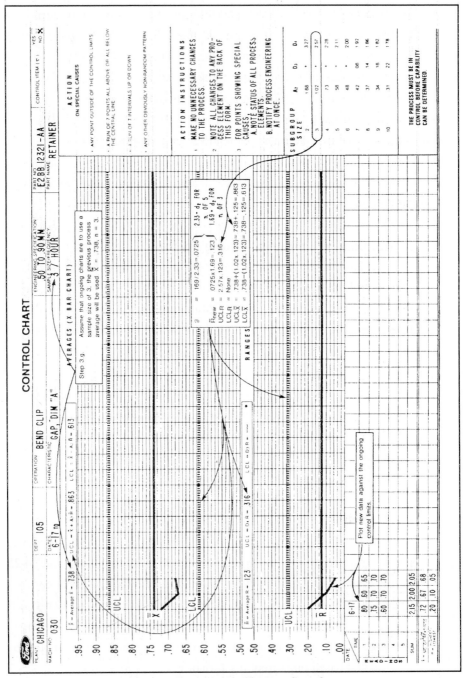

Figure 12. x̄ and R Charts: Step 3.g.

3.h. Summary of Out-of-Control Conditions

Assume that we divide the control chart into the zones indicated in Figure 13. The process is considered to be out of control when any one or more of the following conditions occur:

- one or more points are outside the control limits
- 7 consecutive points are on one side of the centerline
- 7 consecutive increasing or decreasing intervals
- 2 out of 3 consecutive points are in a specific Zone A or beyond
- 4 out of 5 consecutive points are in a specific Zone B or beyond
- 14 consecutive points that alternate up and down
- 14 consecutive points in either Zone C

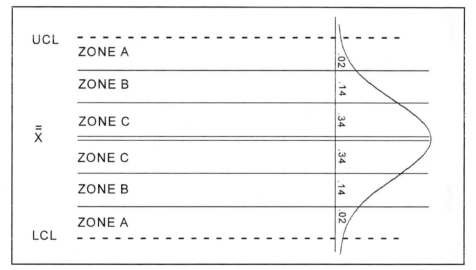

Figure 13. Control Chart Broken into Zones

Step 4. Interpret for Process Capability

Having determined that a process is in statistical control, the question still remains whether the process is capable — i.e., does its output meet customer needs? To understand and improve the capability of a process, an important shift in thinking must occur: capability reflects variation from common causes, and management action on the system is almost always required for capability improvement.

Assessment of process capability begins after control issues in both the $\bar{x}$ and R charts have been resolved (special causes identified, analyzed, corrected and prevented from reoccurring), and the ongoing control charts reflect a process that is in statistical control, preferably for 25 or more subgroups. In general, the distribution of the process output is compared with the engineering specifications to see whether these specifications can consistently be met (see Figure 14).

There are many techniques for assessing the capability of a process that is in statistical control. Some assume that the process output follows the bell-shaped normal distribution. If it is not known whether the distribution is normal, a test for normality should be made (such as reviewing a histogram, conducting a χ^2 Goodness-of-Fit test, or plotting on normal probability paper). If non-normality is suspected or confirmed, more flexible techniques should be used (such as computerized curve-fitting or graphical analysis). When the distribution shape is normal, the technique described below can be used. It involves only simple calculations based on data from the control chart. The process average $\bar{\bar{x}}$ is used as the location (or center) of the distribution. The process standard deviation, a measure of the spread, is estimated from a simple formula involving the average range $\bar{R}$.

Note: Any capability analysis technique, no matter how precise it appears, can give only approximate results. This happens because

(1)   there is always some sampling variation,

(2)   no process is ever "fully" in statistical control, and

(3)   no actual output "exactly" follows the normal distribution (or any other simple distribution).

Final results should always be used with caution and interpreted conservatively.

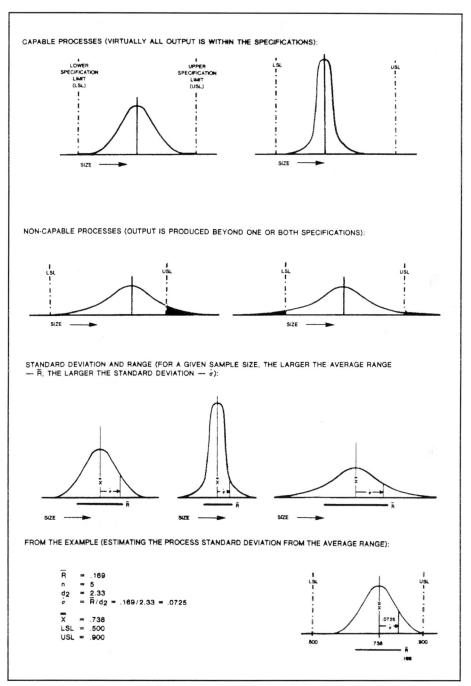

Figure 14. Process Capability

4.a.   Calculate the Process Standard Deviation

Since the within-subgroup process variability is reflected in the subgroup ranges, the estimate of the process standard deviation, ô (read "sigma hat"), can be based on the average range ($\bar{R}$). We calculate ô as (see Figure 14)

$$\hat{\sigma} = \frac{\bar{R}}{d_2}$$

where $\bar{R}$ is the average of the subgroup ranges (for periods with the ranges in control) and d_2 is a constant varying by sample size, as shown in the partial table below:

n:	2	3	4	5	6	7	8	9	10
d_2:	1.13	1.69	2.06	2.33	2.53	2.70	2.85	2.97	3.08

This estimate of the process standard deviation (ô) can be used in evaluating the process capability, as long as both the ranges and averages are in statistical control.

4.b.   Calculate the Process Capability

Capability can be described in terms of the number of standard deviation units, Z, that the specification limits are from the process average. Drawing a diagram that shows the distribution curve, $\bar{\bar{x}}$, ô, the specification limits, and the Z values will be helpful.

●   For a unilateral (one-sided) tolerance, calculate:

$$Z_{min} = Z = \frac{|SL - \bar{\bar{x}}|}{\hat{\sigma}}$$

where SL = specification limit, $\bar{\bar{x}}$ = measured process average, and ô = estimated process standard deviation.

- For bilateral (two-sided) tolerances, calculate:

$$Z_{USL} = \frac{USL - \overline{\overline{x}}}{\hat{\sigma}} \qquad\qquad Z_{LSL} = \frac{LSL - \overline{\overline{x}}}{\hat{\sigma}}$$

and let Z_{min} = Minimum of $|Z_{USL}|$ or $|Z_{LSL}|$ where USL and LSL are the upper and lower specification limits. Z_{min} is also referred to as σ capability or σ level.

Z values can be used with a table of the standardized normal distribution (Appendix D) to estimate the proportion of output that will fall beyond any specification (an approximate value, assuming that the process is in statistical control and is normally distributed). See Figure 15.

The Z_{min} value can also be converted to the capability index, C_{pk}, which is defined as:

$$C_{pk} = Z_{min}/3$$

A process with $Z_{min} = 3$, which could be described as having $\overline{\overline{x}} \pm 3\sigma$ capability, would have a capability index of $C_{pk} = 1.00$. If $Z_{min} = 4$, then the process would have $\overline{\overline{x}} \pm 4\sigma$ capability and $C_{pk} = 1.33$. Obviously, larger C_{pk} values are better than smaller values because a greater proportion of items produced by the process will fall within specification.

4.c. Evaluate the Process Capability

At this point, the process has been brought into statistical control and its capability has been described in terms of Z or C_{pk}. The next step is to evaluate the process capability in terms of meeting customer requirements. The fundamental goal is never-ending improvement in process performance. In the near-term, however, priorities must be set as to which

From the example:

$$\bar{\bar{X}} = .738$$
$$\hat{\sigma} = .0725$$
$$USL = .900$$
$$LSL = .500$$

- Since this process has bilateral tolerances:

$$Z_{USL} = \frac{USL - \bar{\bar{X}}}{\hat{\sigma}} = \frac{.900 - .738}{.0725} = \frac{.162}{.0725} = 2.23$$

$$Z_{LSL} = \frac{LSL - \bar{\bar{X}}}{\hat{\sigma}} = \frac{.500 - .738}{.0725} = \frac{-.238}{.0725} = -3.28$$

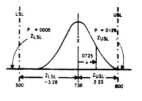

$$Z_{min} = 2.23$$

The proportions out of specification would be:

$$P_{Z_{USL}} = .0129$$
$$P_{Z_{LSL}} = .0005$$
$$P_{total} = .0134 \text{ (about 1.3\%)}$$

The Capability Index would be:

$$C_{pk} = \frac{Z_{min}}{3} = \frac{2.23}{3} = .74$$

- If this process could be adjusted toward the center of the specification, the proportion of parts falling beyond either or both specification limits might be reduced, even with no change in $\hat{\sigma}$. For instance, if confirmed with control charts that $\bar{\bar{X}}_{new} = .700$ (centered), then:

$$Z_{USL} = \frac{USL - \bar{\bar{X}}_{new}}{\hat{\sigma}} = \frac{.900 - .700}{.0725} = \frac{.200}{.0725} = 2.76$$

$$Z_{LSL} = \frac{LSL - \bar{\bar{X}}_{new}}{\hat{\sigma}} = \frac{.500 - .700}{.0725} = \frac{-.200}{.0725} = -2.76$$

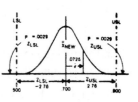

The total proportion out of specification would be:

$$P_{Z_{USL}} + P_{Z_{LSL}} = .0029 + .0029 = .0058 \text{ (about .6\%)}$$

The Capability Index would be:

$$C_{pk} = \frac{Z_{min}}{3} = \frac{2.76}{3} = .92$$

Figure 15. Calculating Process Capability

processes should receive attention first. This is essentially an economic decision. The circumstances vary from case to case, depending on the nature of the particular process in question and the performance of other processes which might also be candidates for immediate improvement action. Whether in response to a capability criterion that has not been met, or to the continuing need for improvement of cost and quality performance even beyond minimum capability requirements, the action required is the same:

Improve the process performance by reducing the variation that comes from common causes.

This means taking management action to improve the system.

4.d.   **Improve the Process Capability**

The problems causing unacceptable process capability are usually common causes. Actions must be directed toward the system — the underlying process factors which account for the process variability, such as machine performance, consistency of input materials, the basic methods by which the process operates, training methods, or the working environment. As a general rule, these system-related causes of process non-capability are beyond the abilities of operators or their local supervision to correct. Instead, they require management intervention to make basic changes, allocate resources, and provide the coordination needed to improve the process performance. Attempts to correct the system with short-range local actions will be unsuccessful. However, the use of more advanced methods of process analysis, including statistical techniques such as designed experiments, may be necessary to achieve significant reductions. See Chapter 8 for a discussion of experimental design techniques.

4.e.   **Chart and Analyze the Revised Process**

When systematic process actions have been taken, their effects should be apparent in the control chart, especially in terms of reduced ranges. The charts become a way of verifying the effectiveness of the action.

As the process change is implemented, the control chart should be monitored carefully. This change period can be disruptive to operations, potentially causing new control problems that could obscure the effect of the system change.

After any instabilities of the change period have been resolved, the new process capability should be assessed and used as the basis of new control limits for future operations.

Case Study 19: Pesticide Mix Interaction Effect[1]

Pesticides that by themselves have been linked to breast cancer and male birth defects are up to 1,000 times more potent when combined, according to a study. A federal environmental official called the finding "astonishing" and said that if confirmed in other labs, it could force a revolution in the way that environmental effects of chemicals are measured.

The study centered on endosulfan dieldrin, toxaphene and chlordane, all pesticide chemicals that are known to switch on a gene that makes estrogen in animals. Estrogen is a hormone that controls formation of female organs. A surplus of the hormone has been linked to breast cancer and to malformation of male sex organs.

By themselves, the pesticides have only a weak effect on the estrogen gene, said John A. McLachlan of Tulane University, leader of a team that tested the chemicals. "If you test them individually, you could almost conclude that they were non-estrogenic, almost inconsequential," he said. "But when we put them in combination, their potency jumped up 500-to 1,000-fold." McLachlan said it was expected that combinations of the chemicals would be additive, that is, the effects of two chemicals together would equal the sum of the effects of the chemicals alone. "Instead of one plus one equaling two, we found in some cases that one plus one equals a thousand," he said. The study is to be published today (June 7, 1996) in the journal Science.

"These findings are astonishing," said Dr. Lynn Goldman, chief of the Environmental Protection Agency's Office of Prevention, Pesticides and Toxic Substances. "The policy implications are enormous about how we screen environmental chemicals for estrogen effects. It is a high priority for us to address the implications of this."

---

[1]This case study is an article taken from *The Arizona Republic* entitled "Pesticide mix called riskier than one alone: linked to breast cancer, birth defects" (June 7, 1996). It is a prime example of the lack of knowledge of interaction effects. To learn more about interaction effects and how proper testing can detect them, refer to Chapter 8, Design of Experiments.

The EPA monitors testing of environmental chemicals one (i.e., varying one factor) at a time, Goldman said, and the agency now must consider how to test for effects of chemicals that might combine in the environment. "We test the ingredients that go into the soup individually," Goldman said. "The combination (or interaction) effect is a very, very new issue for us," she said.

Goldman said the McLachlan study will have to be verified in other labs, including tests that screen the effects of chemical combinations on laboratory animals. "It might not be as simple in whole animals as it is in cell lines," she said. Other scientists also said the work will have to be double-checked by other researchers. But endocrinologist Wade Welshons of the University of Missouri told Science, "It's a very important red flag."

Author's Note:

> The bottom line is that testing factors through one at a time experimentation fails to evaluate combined or interaction effects. Today's most powerful methods of experimentation include full factorial, fractional factorial, Plackett-Burman, etc., designs (discussed in Chapter 8) that will allow interactions of factors to take place and then be evaluated.

Case Study 20: Paint Shop Statistical Process Control[1]

In early 1986, Boeing Wichita began to realize the value and potential benefits of statistical process control (SPC) in improving process performance and needed a pilot project from which to learn. The first project to be targeted in the vast array of applications for SPC was the painting of BMS 10-11 Type I green primer.

Problems indicative of thick paint, such as runs, sags, orange peel, inclusions, etc., had been identified by Quality Assurance. Using eddy current devices to measure paint thickness on aluminum, it was discovered that the BMS 10-11 green primer was two to three times thicker than the maximum thickness of the tolerances (see Figure 1a, Feb 1987). This was not a pleasant discovery, but it provided an ideal opportunity for a pilot project using SPC techniques.

The problems caused by thick paint fall into two categories: weight and rework. The rework portion occurs mainly in assembly where paint that is too thick tends to chip easily when a fastener is installed, thus exposing bare metal to the elements and possible corrosion. The assembly mechanics must then spend their time touching up all bare spots by hand; this becomes quite time-consuming. Of course, primer that is too thin can leave the metal unprotected, as well.

The weight portion of the problem comes from the actual weight of the excess paint. The particular tolerance the painters are working with calls for the paint to be applied in a coat of .0003 to .0008 inches, a thin coat by anyone's standards. To give an idea of how thin that is, consider that a human hair is typically .003 inches in diameter, ten times the minimum. Paint that thin doesn't seem like it would add very much weight to an airplane, but when the size of a 747 is taken into consideration with and knowing that each one of the parts that make up the airplane gets painted at least once, the weight adds up in a hurry.

After preliminary sampling and data collection, the statistician recommended a two-phase program designed to bring primer thickness into tolerance.

---

[1]This case study was provided by George A. Trudeau of Boeing Military Aircraft Company, Wichita, KS.

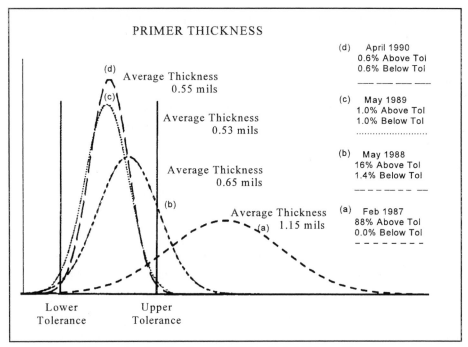

Figure 1. Application of SPC to Paint Primer Process

The first phase   consisted of establishing control charts to plot paint thickness. In the beginning, inspectors from the Quality Assurance organization monitored the process and made the results available to the painters. As the painters adjusted to having their process monitored, the portion of out-of-tolerance parts went down significantly. It seemed that all of the process improvement available had been attained, and yet the process was operating at about 16% above tolerance (see Figure 1b, May 1988). It was felt that the tolerance might have to be changed to reflect the difference between simple and complex parts.

At this point the painters were taught how to use the eddy current devices to measure the paint thickness and plot their own charts. The painters learned that by monitoring the charts, they are afforded the opportunity to know how well they are doing and correct their process **before** a rejection tag is written instead of after. Process improvement from this effort was sufficient to negate the requirement to change the tolerance, and the painters began to assume ownership of their process (see Figure 1c, May 1989). Based on their process knowledge and aptitude, certain shop employees were trained in control chart methodology. As a result, these employees have taken over the implementation portion of the control charts, and are now developing ways to benefit from these methods in other types and colors of paint.

Phase two focused on further improving the controllability of the process by examining some of the important variables that affect the process, such as fluid pressure, air pressure, type of paint system, temperature, painter's experience level, vendor of the paint, etc. Results have led to further experimentation, installation of in-line air pressure gages so the painter can better control his process, and replacement of airless paint systems with air-assisted pressure pots.

Weight and paint savings for military and commercial programs in just one shop for just one type of paint amounted to some $5 million. With our in-house performance improved, it was time to give some attention to vendors' paint processes. A cross-functional task team developed a training package and summary of the in-house project for presentation to suppliers. The selected suppliers provide about 80% of primed parts coming from outside vendors. The goal of the training was to make these suppliers aware of advantages of this process improvement to both themselves and Boeing and to encourage use of the same tools in their respective factories. Results from this effort have been mixed.

Improvements made in the paint process (reducing variability and paint thickness) have resulted not only in weight and dollar savings, but in far-reaching intangible benefits as well. The painters now have a way to measure the quality of their work, which instills pride in their workmanship. This pride spills over into other paints besides BMS 10-11 Type I green primer. Painters often are asked to paint other colors and types of paint, and they bring the skills they have learned with them to the other areas.

Since the start-up of the paint project in January of 1986, the variation in the process has steadily improved and continues to do so (see Figure 1d, Apr 1990). The SPC tools used in this project are now beginning to be used in other paint shops and on other types of paint. With an impressive track record like this, the future for SPC looks bright.

Case Study 21: Disk Drive Manufacturing p-chart Analysis

To approve a disk drive manufacturing process, the process must be capable of producing a consistent quality product. To determine if the process is stable, the product is built for one month after a conditional approval is granted. During this time data is recorded on the number of units that are not passing several key operations. The data is shown in the following table.

Date	Units Built	Base Assembly	Adjustments	Cover
May 01	4750	9	70	30
02	5020	11	78	44
May 05	3790	8	65	31
06	4320	10	42	29
07	4070	7	82	28
08	4650	9	76	38
09	6210	15	102	65
May 12	5210	13	135	40
13	3210	7	145	20
14	6340	13	130	45
15	5870	15	186	38
16	4680	10	192	47
May 19	5030	8	75	42
20	5140	11	58	15
21	5290	8	101	48
22	4870	9	115	38
23	3890	5	150	39

May 26	5210	12	142	27
27	5420	10	136	41
28	3550	5	92	34
29	4240	3	97	13
30	5410	10	136	42

Use a p-chart to evaluate whether the quality of products produced by this line is stable enough to grant approval. Once approved, the manufacturing operation will be run in a production mode rather than in an engineering mode.

ANALYSIS

BASE - Reference Figure 1.   Variations in failure rates are within acceptable limits, although the failure rate seems to have improved during the last half of the p-chart. This small improvement is probably due to worker experience gained during the initial weeks of operation, but could be investigated more thoroughly if resources permit. Approve the process.

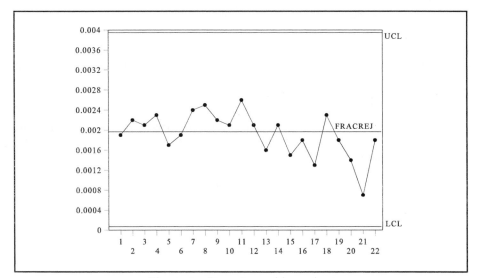

Figure 1. Base Assembly p-chart

ADJUSTMENTS -   Reference Figure 2. Daily failure rates are out of control. The engineer evaluating this process should look at the detail failure modes for the adjustment process. When the exact cause of these large failure rate variations is found, the cause should be eliminated. Withdraw the conditional approval for the adjustment process.

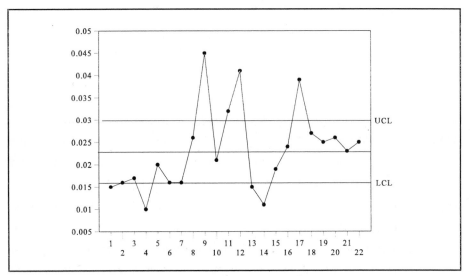

Figure 2. Adjustments p-chart

COVER ASM -   Reference Figure 3. Several points are out of control on the low side and these points should be explained. Withhold approval of the process until the low points have been investigated. There are two possible explanations for this chart and both require action.

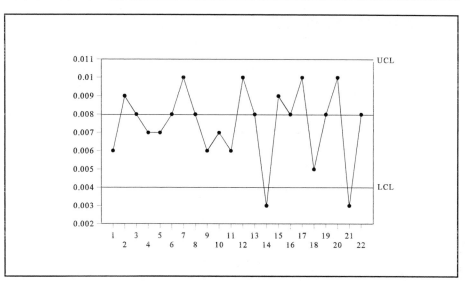

Figure 3. Cover Assembly p-chart

First, the subjective defect categories at this type of final inspection could be judged differently depending on who is doing the checking. If this is the case, training is needed.

Secondly, some assembly workers may be aware of problems that allow them to do a significantly better job. In this case, that information needs to be passed on to all assemblers so the total production can be improved.

Case Study 22: Nuclear Power Plant Process

A nuclear power plant uses deuterium oxide (heavy water) to moderate the neutrons produced during the fission process. The heat released by fission is removed from the reactor core by circulating heavy water through the core. The power plant under study uses four pumps to circulate the heavy water throughout the core. To study and control the water usage, four meters record the water usage by each of the four pumps, respectively, at 30-minute intervals. Each pump is designed to carry its fair share of the load. A stable, predictable pumping process is necessary to avoid catastrophic reactor performance. The data from each of the four pumps for 25 consecutive time intervals is shown in Table 1.

The analysis of this data begins by first generating the R-chart and examining it for any possible out-of-control symptoms. Reference Figure 1, in which we note that the range of subgroup 6 is beyond the upper control limit. We also note that there are 7 consecutive points below the center line, beginning with subgroup 12. An investigation to find special causes for these out-of-control symptoms reveals that at 0850 (just prior to subgroup 6) the electrical power was cut off for several minutes in the vicinity of Station # 1. An operator had cut the power to work on a fuse box and did not realize that this box also controlled the power to the pump at Station # 1. Proper training of the operator should remove this special source of variation in water usage.

This incident seems to be the cause of the lack of control observed at subgroup 6. The investigation does not reveal any special source of variation for the 7 consecutive points below the center line. While this is certainly an unlikely pattern, such occurrences will occasionally be observed in a control charted process without being able to identify a special cause for them. Failure to identify a special cause of variation should not cause panic or disrupt the analysis. False alarms can and do happen.

Time	Subgroup Number	1	2	3	4	x̄	R
0630	1	624	629	624	626	625.75	5
0700	2	621	630	625	624	625.00	9
0730	3	622	624	621	627	623.50	6
0800	4	625	621	625	627	624.50	6
0830	5	629	628	626	630	628.25	4
0900	6	621	629	628	633	627.75	12
0930	7	626	630	629	625	627.50	5
1000	8	630	631	628	627	629.00	4
1030	9	624	626	630	624	626.00	6
1100	10	621	626	628	626	625.25	7
1130	11	628	627	625	622	625.50	6
1200	12	626	624	624	625	624.75	2
1230	13	625	626	623	624	624.50	3
1300	14	630	632	631	631	631.00	2
1330	15	631	633	629	630	630.75	4
1400	16	631	634	633	629	631.75	5
1430	17	632	631	630	631	631.00	2
1500	18	622	622	627	626	624.25	5
1530	19	622	627	625	630	626.00	8
1600	20	627	626	624	625	625.50	3
1630	21	625	622	622	624	623.25	3
1700	22	629	623	627	623	625.50	6
1730	23	622	629	623	627	625.25	7
1800	24	625	627	625	627	626.00	2
1830	25	630	627	621	625	625.75	9

Table 1. Heavy Water Usage for Each of Four Pumps

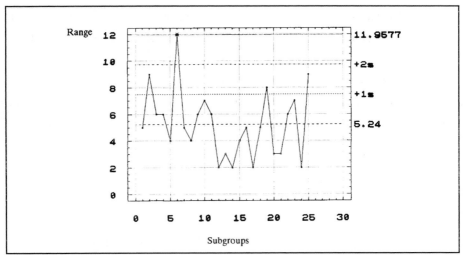

Figure 1. R-chart for Heavy Water Usage

When the sources of special variation are found and eliminated, then the subgroups which were generated from those sources should be removed from the data. In this case, subgroup 6 is removed from the data set. With the operators properly trained, the conditions producing this phenomenon should be removed. Data for subgroups 12-18 are left in place because no special cause was identified. Since data has been removed, a new R-chart is generated. Reference Figure 2.

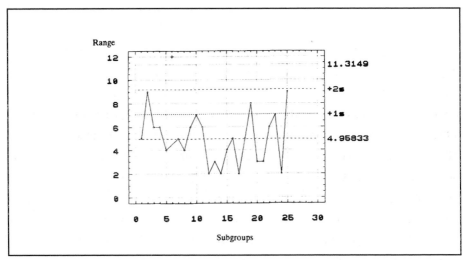

Figure 2. R-chart with Subgroup 6 Removed

Because the center line on the new R-chart has been shifted, subgroup number 18 is no longer the seventh consecutive point below the center line. There are no other indications of a lack of control in the data. Sometimes changing the center line and the control limits unveils other indications of a lack of control, but such is not the case here. Hence, the x̄ portion of the chart may now be constructed using the average range from the now stable R-chart. Reference Figure 3.

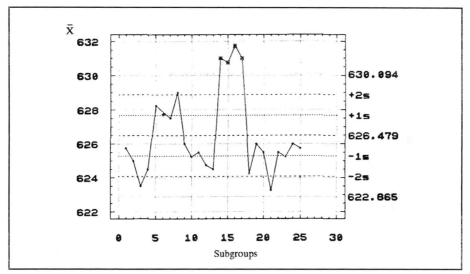

Figure 3. x̄-chart Without Subgroup 6

Analyzing this chart, we note four points (subgroups 14-17) beyond the upper control limit. These occurred from 1300 to 1430. Also, subgroups 18-25 represent 8 consecutive points below the center line. The investigation of this possible lack of control reveals that shortly after 1240 (just prior to subgroup 14) a thermostat monitoring the temperature of the heavy water was replaced as a matter of routine maintenance. A check on the thermostat shows that the thermostat was not properly calibrated, thus leading to an increase in the amount of fresh heavy water pumped into the reactor core. A further check on the lot of thermostats from which this one was selected indicates that the entire lot was improperly calibrated. This leads to a requirement that the thermostat vendor chart his calibration process and demonstrate a stable, predictable process.

As sometimes happens, charting one process can lead to the charting of another process. Such was the case here. No special sources were found for the series of points (18-25) below the center line. However, since removal of these four points will shift the center line downward, little effort was made to search for any other special sources of variation. Thus, we delete subgroups 14-17, recompute, and analyze the resulting R-and x̄-charts in that order once again. As Figure 4 indicates, the new R-chart does not reveal any out-of-control symptoms.

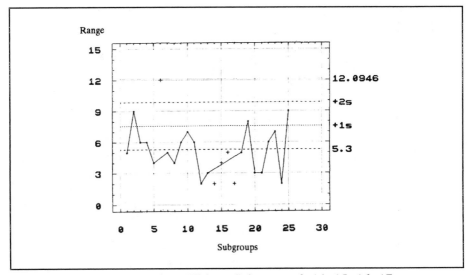

Figure 4. R-chart Without Subgroups 6, 14, 15, 16, 17

Thus, we can examine the x̄-chart shown in Figure 5. Ignoring the missing data point at subgroup 6, it appears that point # 8 is the second of three consecutive points beyond +2s, thereby indicating yet another possible lack of control. In this case, however, since we have already deleted 5 of the original 25 data points, it would probably be advisable to let the process run a while longer. After all, the two-out-of-three condition spans a deleted point anyway. Many times a lack-of-control condition is only a temporary effect that can be attributed to the removal of some of the data. Hence, a conservative approach would be to allow the process to continue to run and demonstrate any evidence of the presence of special causes of variation, if there are any.

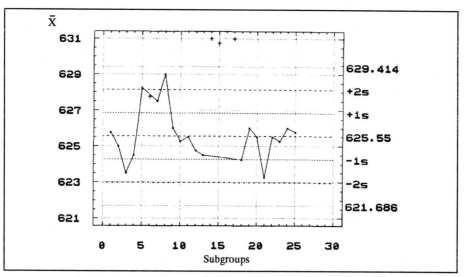

Figure 5. x̄-chart Without Subgroups 6, 14, 15, 16, 17

Case Study 23: SPC in the White Collar Environment[1]

The manager of an accounts payable department in a large corporation faced a serious crisis. This department was responsible for receiving, validating, and paying all the invoices the corporation received from its vendors. Recently, the manager had received numerous calls from vendors complaining that their invoices were not being paid within thirty days of receipt in accordance with the terms of their contracts. The manager was also under pressure from the corporate comptroller not to pay any invoices until they were within ten days of their due date.

Faced with this set of seemingly conflicting demands, the manager decided to take a closer look at the accounts payable process. Five invoices were randomly selected every day for two weeks. The number of days required to receive, validate, and pay these invoices are summarized in the following table:

Day	Inv 1	Inv 2	Inv 3	Inv 4	Inv 5	Average
1	21	18	28	36	18	24.2
2	16	35	18	39	20	25.6
3	28	26	31	11	22	23.6
4	26	25	21	41	14	25.4
5	19	11	18	27	13	17.6
6	44	18	32	36	50	36.0
7	16	26	37	23	25	25.4
8	25	25	23	50	36	31.8
9	10	32	35	46	41	32.8
10	22	33	31	46	34	33.2

The manager then performed the type of analysis that was common in this corporation and reached the following conclusions:

[1]Contributed by Peter Hofmann from MCI Telecommunications Corp.

- On average, the comptroller's policy was being met 90% of the time (9 of 10 days).
- On average, 60% of vendor invoices were being paid on time (6 of 10 days).

The manager then realized that this type of analysis and interpretation did not provide any true insights into how consistent or stable the accounts payable process was. This type of analysis was also flawed since it took only half of the requirements of the process under consideration at a time. When both sets of requirements are considered simultaneously (i.e., payment occurs between 20 and 30 days), the poor performance of this process is very evident. When viewing daily averages, only 50% of the time (5 of 10 days) were both requirements satisfied simultaneously. If individual invoices are examined, only about one third (17 of 50) of those sampled conform to both sets of requirements on the process.

What is the variability of the process? The manager remembered from a recent conference that the key to improving quality centered on understanding, measuring, and reducing the variability of the process. The measures discussed above dealt with average trends and comparisons that put the process performance in the best light possible based on the audience receiving the information. There were no statistically sound measures of how stable the process was and how capable it was of meeting the two sets of requirements placed on it.

In the notes from the conference, the manager found a "how to" guide for implementing SPC. After clearly identifying the process, a control chart was constructed to see how stable the process was. Using the sample data collected above, the following statistics and constants were obtained:

Day	x̄	R	Other
1	24.2	18	$\bar{\bar{x}} = 27.56$
2	25.6	23	$\bar{R} = 24.4$
3	23.6	20	n = 5
4	25.4	27	$A_2 = 0.58$
5	17.6	16	$D_3 = 0$
6	36	32	$D_4 = 2.11$
7	25.4	21	LSL = 20 Days
8	31.8	27	USL = 30 Days
9	32.8	36	$d_2 = 2.33$
10	33.2	24	

Note: A_2, D_3, D_4, and d_2 were obtained from Appendix I, *Control Chart Summary*

The following x̄ - R chart was constructed using the above information:

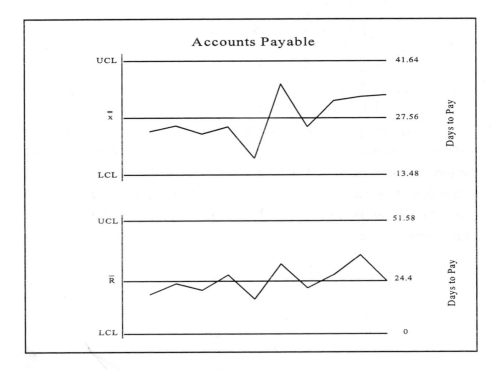

The x̄ - R Chart does not reveal any obvious "out-of-control" symptoms. However, the performance of this process still needed to be compared to the requirements that were placed on it to see how "capable" it was in meeting those requirements. To accomplish this, a process capability analysis was performed as follows:

First, a picture of the process as it currently exists is generated. Figure 1 is obtained by using $\overline{\overline{x}}$ = 27.56 (from the x̄ chart) and $\hat{\sigma}$ = $\overline{R}/d_2$ = 24.4/2.33 = 10.5 (where $\overline{R}$ = 24.4 comes from the R chart) as the mean and standard deviation, respectively, of the normal curve.

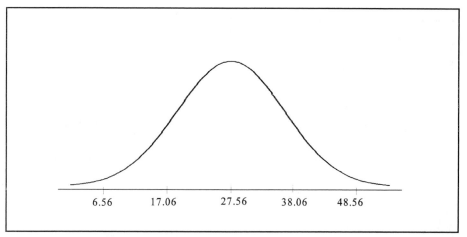

6.56	17.06	27.56	38.06	48.56

Figure 1. Picture of Accounts Payable Process

Next, the process requirements, namely the upper specification limit (USL = 30) and the lower specification limit (LSL = 20), are superimposed upon the process representation as shown in Figure 2. The shaded portion indicates the proportion of payments that do not meet the requirements. It can be shown (computations not provided) that 64% of the payments do not satisfy the requirements.

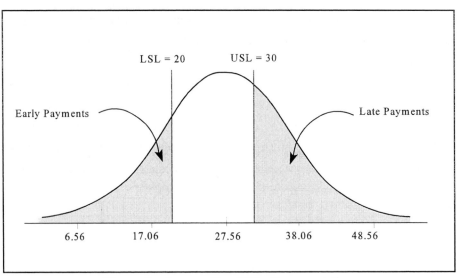

Figure 2. Process Capability Representation

Sometimes this fractional area is presented in terms of defects per million (dpm). In this case it would be $.64 \times 10^6$ or 640,000 dpm. Other capability measures commonly used are σ-capability which, in this case, is

$$\sigma_{capability} = \frac{USL - \overline{\overline{x}}}{\hat{\sigma}} = \frac{30 - 27.56}{10.5} = .23$$

$$\text{and } C_{pk} = \frac{\sigma_{capability}}{3} = .08.$$

At this point, the manager realized that this process was in more trouble than the earlier analysis of average daily performance had indicated. This process needed a fundamental change. To accomplish this change, a meeting was convened with several key members from the accounts payable section.

The manager began the meeting by showing Figures 1 and 2. "You all have known that there were several problems with our accounts payable process. The bad news is that it is much worse than we thought. Invoices paid in less than 20 days violate our corporate fiscal policy, displeasing our comptroller. Invoices paid in more than 30 days violate our contractual agreements with our valued suppliers, causing them great dissatisfaction and fiscal strain. As you can see, we are not doing a good job of satisfying either constraint. The good news is we know what the problem is: variability. We need to reduce the variability of this process so that we can meet the needs of our customers. Your job for the next quarter is to work as a team to identify changes to our process that will reduce the overall variability. Meanwhile, I will present these two charts to both our comptroller and our key suppliers to let them see exactly how we are doing. I will also solicit their assistance in our variability reduction effort.   Remember, challenge any policy or procedure. Our goal is to satisfy our customers' requirements 100% of the time."

Case Study 24: Ice Cream, Engineering, and Statistics[1]

If you're an average American, you single handedly consume many gallons of ice cream each year and the biggest dilemma you have is choosing between "mint chocolate chip" or "rainbow sherbet" — and whether to have it in a dish, a waffle or a sugar cone. How that ice cream gets produced and packaged into neat containers hasn't given most of us a moment's thought.

Cher Nicholas recently gave the subject a lot of thought, however, and the Friendly's Ice Cream company is predicting that the work done by this recent alumna of the Mechanical and Industrial Engineering Department at the University of Massachusetts will save the firm half a million dollars in the next twelve months alone.

The story all began in the spring of 1995 when Nicholas was looking for a project to complete requirements for her master's degree. No stranger to the world of manufacturing — Nicholas had worked in industry for ten years after graduating from UMass with a B.S. in Electrical Engineering in 1981 — the up-and-coming specialist in Statistical Process Control was eager to find a placement that would let her exercise her skills, applying the new mathematical tools being used in the field.

Ronald J. Gathro, MS IE/IOR '90 senior industrial engineer at Friendly's main plant in Wilbraham, Massachusetts was looking for a graduate student who could design some Statistical Process Control files for an existing software package in his automated manufacturing line.

"We had used graduate students for similar projects in the past, felt it was a pretty straight-forward task that would save the rest of us some time, and we really weren't looking for anything more exciting than that," he recalls.

---

[1]This case study is taken, with permission, from an article by Deborah C. Parker entitled "She's got the scoop on statistical process control" which was printed in the *Engineering News*, College of Engineering, University of Massachusetts, Amherst, Volume 4., No. 1, Fall 1996. Our special thanks to Susan Mattei of UMass and to Cher Nicholas and Friendly's Ice Cream for their timely participation.

Scrutinizing the Process

Nicholas began her research, starting out by scrutinizing every step of the ice cream mixing, packaging, and cooling process. There was no question about the quality of the product being packed into the containers. The federal government creates and enforces stringent guidelines for this dairy-based treat and Friendly's has its own standards which are higher than the government's. In fact, ingredients and proper food handling procedures are checked so thoroughly, and so frequently, that the company never has to dispose of product that has "gone bad" for one reason or another.

What they did have to do, almost every day, however, was toss out or break down and re-process the ice cream, sold to consumers in half-gallon block containers, that came off the production line either over-full or under-full.

"If the flavor of the day was vanilla, they were in luck, because it could be re-run and used to make more vanilla, or some more complicated variation on the theme," Nicholas explains. "But if it was mint chocolate chip, say, the components couldn't be separated. It was lost product, and in today's highly competitive marketplace, manufacturers can't afford to produce defective product. It eliminates any profit margin they may have."

Nicholas became intrigued as to why these "over-fills" and "under-fills" occurred, particularly when fairly exact amounts of base ingredients went into each specific recipe — and the mixtures were sent through the same production machines by the same people, every day.

She started at the door of the Receiving Department and began to ask every operator why he or she did each procedure the way that they did — and the answers she received were the kind one would get from the most experienced workers at most successful companies across the nation. Things like, "Why did I just do that? Well, when this machine does this, which it does sometimes on hot days, you have to turn the knob a smidge to the right."

What operators could not tell her was the mathematical equivalent of a "smidge" — or what the quality of the weather outside the plant had to do with too much ice cream overflowing the wax containers inside the plant. Using traditional engineering practices, the challenge of figuring all those variables could have taken hundreds of "hands on" hours to figure out.

Luckily, Nicholas had better tools to work with. Using Statistical Process Control, Design of Experiments and Response Surface Methodology — a sequential, mathematical approach toward achieving optimum results, whatever those might be — Nicholas sat down at a computer and came up with answers that frankly astonished Gathro and staff, and launched the corporation on a path that is unequaled in the industry.

The Next Step

Applying SPC equations to the problem was the first step of the task, Nicholas explains, because that measured the parameters of variations that were occurring. But SPC couldn't explain why they were occurring, and that's where her Design of Experiment came into play. Instead of identifying and testing one variable at a time — such as temperature of the refrigerated mix, or the pressure of the air that, mixed with fluid, creates the high-volume, semi-solid end product — Nicholas used the Design of Experiment approach to assess all of the variables, and in relation to the others, all at once.

Brainstorming and interacting with the Friendly's team, Nicholas identified five variables and ten possible interactions to explore. To minimize costs and efficiently study the problem, she used a classical L-16 orthogonal design — one which required running only 16 different experiments. (In the traditional approach, there would have been two to the 15th power possible experiments required.)

Nicholas was ultimately able to document that air pressure — controlled by the knob that was being turned a "smidge" on occasion — was the culprit. She could have stopped right there and gotten an "A" on her evaluation, as far as her master's project supervisor, Quality Engineering Professor Lawrence Seiford, was concerned. But in her usual thorough fashion, Nicholas elected to take the project one step further. Using Response Surface Methodology, she was able to build a mathematical model describing the interactions between the critical variables and determine the optimal air pressure setting.

It is a discovery that saved the Wilbraham plant a quarter of a million dollars in lost product that fiscal year alone.

Success Beyond Reduced Product Costs

"Cher's original work provided Friendly's with the basis for forming a Total Quality Management team to implement her recommendations, and to do so," Gathro stresses, "without benefit of capital dollars. We improved process capability, reduced process variability, and reduced product costs by $146,000 between September, 1995 and June, 1996. Friendly's expects to realize an additional, incremental savings of $500,000 annually by expanding Cher's work to other production lines and by anticipated implementation of her Response Surface Methodology work."

Nicholas is proud, of course, but even today she speaks with enormous respect for the staff at Friendly's, from Gathro right down to the production line workers.

"What made this project so successful was the attitude I encountered every step of the way," she notes. "In many companies, a graduate student would be perceived and treated as an intruder, even an inconvenience. At Friendly's, the operators treated me as a professional and wanted to know anything I could tell them. They were proud of what they were doing, couldn't figure out what was going wrong, and were eager and willing to implement my recommendations. It was the team approach that made this effort a success."

Gathro says Nicholas is far too modest. "We have sponsored undergraduate and graduate projects in the past and were quite pleased with their performance. But none approached the level of professionalism, enthusiasm and sheer skill that Cher brought to this project. From here on in, we will be looking at graduate students in a whole new light."

In fact, Friendly's was so pleased, they sent high-level company officials to attend Nicholas's Graduate Seminar — and those officials came loaded with "product sample" for everyone who showed up. "It was a memorable ending for a memorable project," Nicholas concludes with a grin. "Usually you're lucky if you get thirty people at these things. And mine? After they heard about the ice cream, well, it's fair to say the room was packed."

Case Study 25: Iranian Hostage Rescue Mission[1]

During the spring of 1980, a joint task force consisting of eight RH-53 Sea Stallion helicopters was launched from the U.S.S. Nimitz in an attempt to rescue the 44 U.S. hostages held in Teheran [Lars 85]. The mission was scrubbed after three of the eight helicopters suffered either electrical or hydraulic malfunctions. Six helicopters would have been needed to get all of the hostages out. The aborted mission turned to tragedy when one of the helicopters crashed into a C-130 at the desert landing site. An interesting question to consider is: could we have predicted beforehand whether eight helicopters was a sufficient number of helicopters to send if six were needed to successfully complete the mission? Some basic principles of probability and reliability provide some revealing information about the mission.

The reliability of the RH-53 Sea Stallion helicopter can be estimated from the maintenance records. By using some basic descriptive statistics (e.g., histograms, frequency distributions) on the maintenance data, we could estimate the mean time between failures (MTBF), as well as individual failure times (T) for this type of helicopter. From this summary data, it could be ascertained that the time to failure (T) was exponentially distributed with $\lambda = 1/\text{MTBF}$. A χ^2 Goodness-of-Fit test could be performed to validate the use of the exponential model. In fact, this was actually done after the operation. The analysis revealed that T was indeed exponential with a MTBF of 20 hours. That means that $\lambda = 1/20$. Knowing that the rescue mission will take a total of 14 hours of flying time, one can compute the probability that one RH-53 will successfully complete the mission. That is, we compute $R(14) = P(T > 14)$, where T is exponential with $\lambda = 1/20$.

$$R(14) = e^{-(1/20)14} = .4966 \approx .5$$

Once the probability of a successful mission for one helicopter is calculated, we can now use the binomial distribution to calculate the probability that six or more helicopters will successfully complete their missions. Thus, if the random

---

[1]Most of the data in this case study comes from Buddy Wood's film on "Application of the Binomial," USAF Academy, 1982.

variable X represents the number of successful helicopters out of the 8 sent, and we assume independence between helicopters, X is a binomial random variable with parameters n = 8 and p = .5.   We then calculate

$$P(X \geq 6) \; = \; P(X = 6) + P(X = 7) + P(X = 8)$$

$$= \; .1094 \; + \; .0312 \; + \; .0039$$

$$= \; .1445,$$

which is not a pleasant postscript!   The real question is, how many helicopters should we have sent? This question can be better analyzed by constructing a chart which reveals the probability of mission success (i.e., $P(X \geq 6)$) as a function of the number of helicopters sent. This chart is shown in Figure 1.

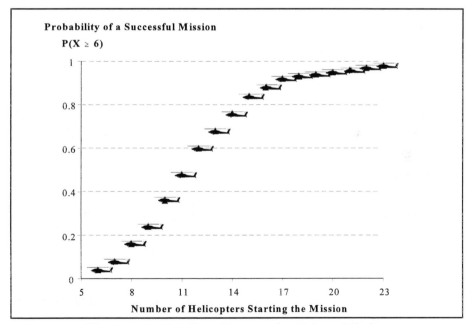

Figure 1. Probability of Success for 14-hour Mission

The reality of the chart indicates that the mission may have been doomed, even without the catastrophe in the desert. Based on the reliability of the helicopters, it would have taken 15 helicopters to have a 0.8 chance of a successful mission. Granted, there are other factors that enter into a decision of how many helicopters to send on such a mission. Certainly, as the number of helicopters increases, the probability of detection also rises. However, the point of this case study is that very basic statistical tools can provide powerful, revealing information to a decision maker, and if one is to be a wise decision maker, one should use all of the available tools.

Case Study 26: Satellite Servicing System[1]

An excellent example of the power of QFD to focus a multifunctional team was demonstrated in support of a Satellite Servicing System Proposal effort performed at Martin Marietta Astronautics Group. A multifunctional team used the Quality Function Deployment methodology to identify all critical system and subsystem design characteristics of a Satellite Servicing System (SSS) to support the development of an integrated technical approach. Major functions of the SSS (pictorially illustrated in Figure 1) were to perform on-orbit autonomous rendezvous and docking, supervised autonomous orbital replacement unit exchange, and supervised autonomous fluid transfer with an experimental space vehicle. In addition, the SSS was required to simulate proximity operations for the future Space Station Freedom.

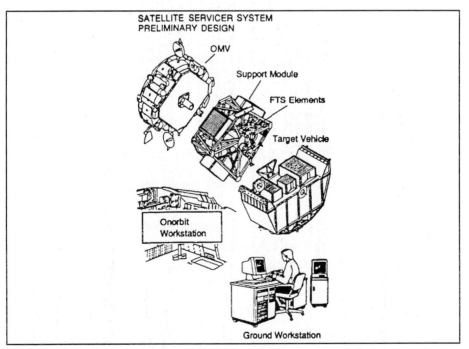

Figure 1. An Illustration of the Satellite Servicing System

---

[1]This case study was provided by Barbara Bicknell.

Training and completion of the House of Quality matrix was completed over a period of approximately 7 days. Over 140 design issues were identified and prioritized. These design issues were integrated into the technical approach. Not only did the use of QFD result in a strong proposal win, but the proposal was completed a week early with no questions prior to best and final awards. In retrospect, the team became empowered through the process and committed to the final product. Each member walked away with a thorough knowledge of how the program goals and mission objectives affected the design element and each of the system components.

Lessons Learned

The lessons learned from the Satellite Servicing program are similar to those experienced by many other successful teams. The following is only a partial list of those lessons:

1.   Expert facilitation is required and critical to the success of initial pilot projects. The facilitator must be able to assist the group in the definition of the problem, formulate the approach and guide the team in the right direction in an environment where the number of options is unlimited. The facilitator must keep the group from drifting in directions guided by personal biases.

2.   Each QFD matrix should be tailored to the needs of the work environment.

3.   In large-scale efforts, QFD matrices can quickly become too large to control. Identifying priorities is essential so that each matrix in a series can be scaled down by focusing on priorities only.

4.   Communication is more important than mathematical elegance.

5.   Program managers may find it difficult not to dominate, interfere or lead discussions.

6. Time is needed to overcome a sense of vulnerability between team members.

7. Success is directly proportional to the listening skills of group members.

8. Differences in terminology will lead individuals to different end results. Provide definitions of terms as appropriate.

9. Individuals have a tendency to leave the process, causing the group to rehash old discussions which result in the loss of valuable time. Set meeting schedules for full participation.

10. Observers cannot resist the temptation to participate, thinking they are being helpful. The group loses valuable time explaining the basis for their opinions or unknown information to the observer. Ultimately, the group can become resentful. Minimize interference for best team progress.

11. Large teams are difficult to focus. A team size of 10 to 15 is recommended.

12. Adequate time should be allowed for information preparation. Conceptual design by committee requires maturity of thought.

Case Study 27: Design of a Mail Order Catalog Using QFD[1]

The president of a medium size clothing company was displeased with the performance of the catalog operations of the company. When the company was founded 15 years ago, it was strictly a catalog house. Since the company began franchising stores 8 years ago, the volume of catalog sales has been steadily falling. While overall company sales have been increasing, the president was very concerned that they were letting one of their key markets slip away. This was especially disturbing since their profit margins on catalog sales were twice that of their retail outlets.

The president asked the company's senior managers to try to find out why the catalog business was declining so steadily. They reported back that with the increased focus on growing the retail business, the mail order business had received very little attention. One of their customers' main complaints was that the catalog was too big and too hard to use. In addition, with the increased emphasis on the environment in the last few years, customers viewed most catalogs as unwanted junk mail that was choking their landfills. However, many of their customers, especially working couples who have very little time for shopping, still preferred mail order shopping.

A team was commissioned by the president to design a new catalog and revamp the mail order business. The team was cross functional, including members from marketing, customer service, shipping, accounting, and even some key suppliers. The team was also trained to use Quality Function Deployment to help them inject the "voice of the customer" into their design efforts.

The first task the team faced was to accurately find out what their present and potential mail order customers wanted. They conducted phone interviews, face to face interviews in their retail outlets, interviews with their customer service representatives and even convened several customer focus groups. The result of these efforts was the following list of desirable, customer defined "characteristics" for a mail order catalog:

---

[1]Contributed by Peter Hofmann, MCI Telecommunications Corp.

Accept Major Credit Cards
Clear Order Form
Clear Colors
Not too Large
Clear Exchange Terms
Lots of Pictures
Accurate Descriptions
Uncomplicated Shipping Rates

Easy to Use Size Tables
Good Index
Environmentally Sound
800# for Ordering
Clear Warranty Terms
True Colors
Complete Descriptions

Next the team had to perform an initial analysis on the descriptive data they had collected. Using the above list, they created the following Affinity Diagram that outlined the characteristics of a Great Clothing Catalog:

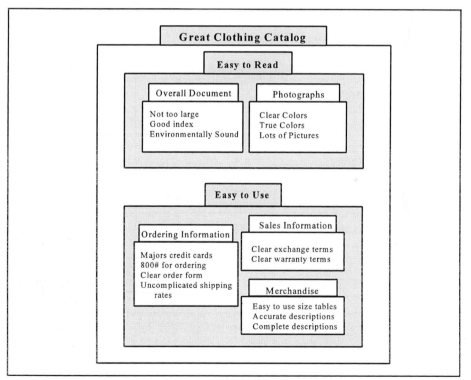

Figure 1. Affinity Diagram

After having several people outside the team perform a "reality check" on their diagram, i.e., ensure it was consistent with actual experience, the team then went back to their customers. They needed their customers to assign a priority to each characteristic as well as assess how effective their company and their competitors were at satisfying these characteristics. The following table of information was the result of their efforts:

	Importance	Our Company	Company A	Company B
Accept Major Credit Cards	4	4.5	4.1	4.7
800# for Ordering	5	4.3	4.5	4.2
Clear Order Form	2	4.0	2.0	3.2
Uncomplicated Shipping Rates	5	2.1	4.2	4.5
Clear Exchange Terms	4	2.3	3.5	4.0
Clear Warranty Terms	4	2.6	4.8	4.5
Easy to Use Size Tables	5	2.1	1.8	2.0
Accurate Descriptions	3	2.3	2.2	1.8
Complete Descriptions	3	2.5	1.9	2.1
Not too Large	4	1.5	4.1	2.1
Good Index	1	2.1	2.3	2.0
Environmentally Sound	5	2.1	2.5	2.0
Clear Colors	3	3.5	3.1	4.0
True Colors	5	3.5	3.2	4.2
Lots of Pictures	4	4.2	4.0	4.8

Now that the team had a good idea of what their customers wanted in a clothing catalog, they needed to develop a list of design requirements for the catalog. The team brainstormed the following list of design requirements:

Number of Pages
Number of Products Offered
Number of Colors
2 Tier Shipping Rates
Paper Type
Warranty on Services
Hidden Tabs
Standard Description Format

Paper Weight
Type Style (Font)
Single 800# for both sales and
   service
Warranty on Goods
Recyclable Document
800# on Each Page

Next the team had to perform some initial analysis on this descriptive design data that they had collected. Using the above list, they created the following Affinity Diagram that outlined the design characteristics for the new catalog:

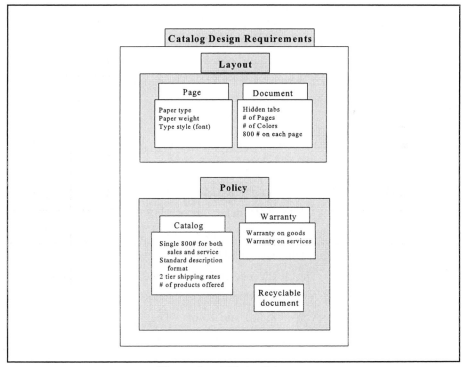

Figure 2. Affinity Diagram

After again consulting with knowledgeable people outside the team to obtain a "reality check" on their design requirements, the team began to complete the "House of Quality." The customer requirements for the catalog (from Figure 1) were transcribed into the far left column of the "house." The design requirements (from Figure 2) were transcribed across the top row of the "house." Finally, the degree of importance and company ratings for each customer requirement were transcribed on the right side of the "house."

The team then assessed, for each customer requirement/design requirement combination, if there was a relationship between the customer requirement and the design requirement. If there was a relationship, they also determined how strong the relationship was and assigned a weight to it. Then, using sales point information obtained from a marketing analysis, they completed the rest of the body of the "house." (See step-by-step calculation directions in Chapter 11 for importance weights and relative weights for customer and design requirements.)

Once the body of the "house" was completed, the team then turned their attention to the roof. They assessed all of the possible correlations, both positive and negative, between the design requirements. They also identified the basic form of the target values for each design requirement. Their initial "house" is shown in Figure 3.

After analyzing their initial "house," the team found the following:

1.   Their highest priority challenge would be to develop an environmentally sound catalog (made from recycled materials as well as being recyclable) that would also present clear and accurate photographs of their clothing products. In fact, the roof of the "house" clearly showed that the recyclable requirement was highly correlated with several other design requirements. The team decided to conduct a series of designed experiments to determine the optimum combination of paper weight, paper type (various recycled fibers) and color. They also decided to use the survey results from several customer focus groups as a measured response variable.

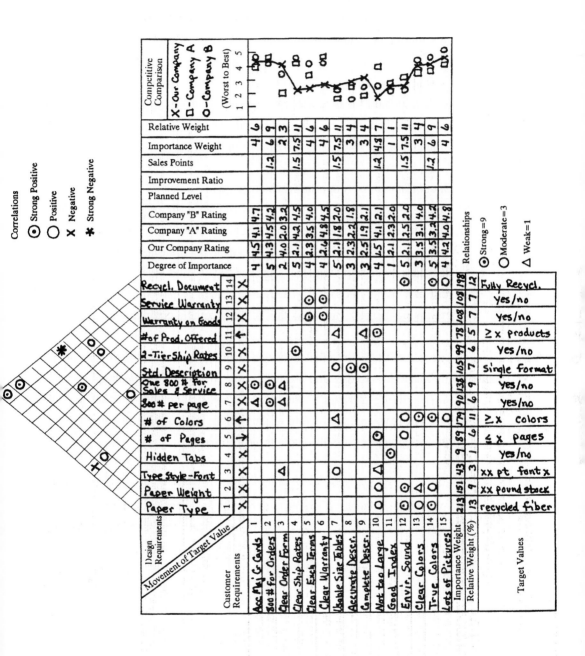

Figure 3. House of Quality

2.   Developing a single 800 number for all sales and service information was very important. They commissioned a side team to work with the customer support and marketing organizations within the company to change the existing company policies and procedures so that this new "one stop service" would be supported throughout both organizations. They also recognized that this concept would need to be deployed through the whole company and not solely in the catalog operations.

3.   One of the most important customer requirements, "easy to use size tables," was not strongly related to any of their design requirements. They also noticed that neither they nor their competitors were doing a good job meeting this requirement. (Note: This analysis could have been built into the chart by completing the "planned level" and "improvement ratio" portions of the "house.") The team decided to pay special attention to improving their size tables as this could become a competitive advantage.

The team then presented a status briefing to the company president, outlining their activities to date and planned actions. They also emphasized how the "house" was a living document that not only assisted in the decision making process, but also gave sound structure to the design. It could easily be built upon for future development efforts throughout the company.

Case Study 28: Developing a Risk Management Model[1]

In 1994, the Air Force's senior leadership became concerned about the threat of Information Warfare (IW) attacks through the Internet and other sources. They tasked the Electronics Systems Center (ESC), located at Hanscom Air Force Base in Massachusetts, to develop a comprehensive "fix plan" that would address all major weapon systems' vulnerabilities to IW attacks. Where existing technologies failed to meet the need, ESC would recommend investment in new technologies. But first, how great is the IW risk? How would different IW attacks affect mission accomplishment? Where are the best places to put countermeasures for the least cost? To answer these questions, ESC's Information Warfare Product Group developed a powerful analytical method to isolate specific IW risks and then develop tailored solutions. The method combines Quality Function Deployment (QFD) with Failure Modes, Effects, and Criticality Analysis (FMECA) to link mission objectives, tasks, and information assets together with recognized vulnerabilities and projected threats. Extending QFD a step further, the same linkage is used to predict the reduction in mission risk that would result from using various countermeasures. Based on the predictions, the most cost-effective solution is easy to determine.

The QFD Chain of Matrices

The chain of matrices set up for this problem is shown in Figure 1. The first two matrices reflect the weapon system users' rank-ordering of their mission objectives and the wartime tasks needed to achieve them. ESC filled in these two matrices based on interviews with Air Combat Command experts on wartime operations. In similar fashion, ESC used interviews with weapon system developers from Air Force Materiel Command to identify all of the information assets (data bases, networks, etc.) used to accomplish each task. The assets, rank-ordered by their importance to various tasks, appear in the third matrix. Once the assets were

---

[1]Our special thanks to Mr. Jim Watters, a support contractor assigned to Electronic Systems Center (ESC), for providing an excellent study on how QFD and FMEA can be used outside of a manufacturing setting.

determined, IW specialists at ESC and the MITRE Corporation paired known vulnerabilities in each asset with potential threats identified by the Air Force intelligence community. The resulting risks facing each asset, shown rank-ordered in the fourth matrix and its related Pareto chart, could then be translated into risks threatening each mission objective. This translation appears in the fifth matrix and its Pareto. The completed chain of matrices provides an analytical tool as IW specialists apply the effect of various countermeasures to the risk numbers in the fourth matrix. The end result shows up immediately in the fifth matrix as reduced risk to mission objectives. Differences are easily spotted using the Paretos (see Figures 8a and 8b). By weighing the cost and the effect of different combinations of countermeasures, analysts can find the best mix.

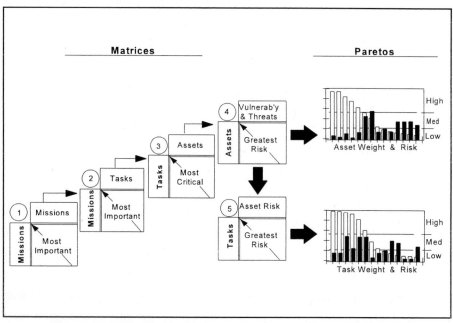

Figure 1. QFD Chain of Matrices and Associated Pareto Charts

The Matrices in More Detail

Figure 2 shows the first matrix in more detail. The values in the blocks show the relative importance among the mission objectives. The relative weight is obtained by adding the row values and dividing by the total of all the rows. The values for this hypothetical weapon system would come from interviews with Air Combat Command experts. In this case, the most important mission objective is Air-to-Ground Attack, as it carries about 70% of the weight. Any IW threat to this mission objective would be among the first to be remedied. Afterward, to the extent funds are available, risks to the other, less important mission objectives would be addressed.

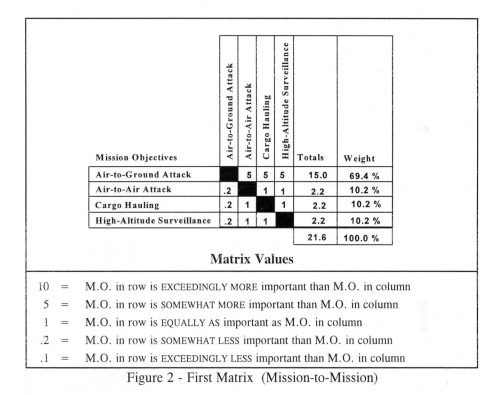

Mission Objectives	Air-to-Ground Attack	Air-to-Air Attack	Cargo Hauling	High-Altitude Surveillance	Totals	Weight
Air-to-Ground Attack		5	5	5	15.0	69.4 %
Air-to-Air Attack	.2		1	1	2.2	10.2 %
Cargo Hauling	.2	1		1	2.2	10.2 %
High-Altitude Surveillance	.2	1	1		2.2	10.2 %
					21.6	100.0 %

Matrix Values

10	=	M.O. in row is EXCEEDINGLY MORE important than M.O. in column
5	=	M.O. in row is SOMEWHAT MORE important than M.O. in column
1	=	M.O. in row is EQUALLY AS important as M.O. in column
.2	=	M.O. in row is SOMEWHAT LESS important than M.O. in column
.1	=	M.O. in row is EXCEEDINGLY LESS important than M.O. in column

Figure 2 - First Matrix (Mission-to-Mission)

The second matrix, shown in Figure 3, matches operational tasks to each mission objective. It answers the question: What specific tasks are needed to achieve each Mission Objective? As with the first matrix, the values are obtained by interviewing experts in the use of the weapon system, but this time using the 1, 3, 9 scale typical of QFD exercises. Relative weights are computed for each task as the sum of the products of each Mission Objective (M.O.) weight and the task's importance value for that M.O. For example, the task "Identify Mobile SAM Sites" is scored as a 9 (required) for the Air-to-Ground Attack mission objective; a 3 (important) for Air-to-Air Attack; and a 9 for Cargo Hauling and High-Altitude Surveillance. The sum of the products (0.694 * 9) + (0.102 * 3) + (0.102 * 9) + (0.102 * 9) = 8.39 for the task relative weight. Notably absent are the "roof," the "basement," and the "right wing" typically seen in a QFD "house of quality." The "roof," showing the cross-correlation between the columns, could certainly be added, although for IW risk assessments done to date its omission has not been a problem. Each column entry has been treated as independent of the others — not necessarily a bad assumption if the entries are chosen properly. The "basement," which contains target values for the column entries, does not lend itself to this particular case since there are not really target values for any of the relative weights or relative risks. The "right wing," however, is used in matrix #4, although in a way somewhat different than discussed earlier in this book. The details will be discussed later.

		Update System Data Base While On Station (9.00)	Maintain Station-Keeping Off Target (9.00)	Disseminate Mission Critical Data (9.00)	Carry Out Air-To-Air Refueling (9.00)	Receive and React to Threat Warnings (9.00)	Power-Off Preflight and Launch (9.00)	Initialize Avionics with Mission Data Base (9.00)	Discriminate Between Friendly and Enemy Forces (8.39)	Identify Mobile SAM Sites (8.39)	Maintain Unbroken Comm with Ground Forces (7.78)	Maintain Unbroken Comm with Airborne Forces (7.78)	Record In-Flight Data for Post-Mission Analysis (7.78)
Mission Objectives	**Rel Wt**	Task Importance to M.O.											
Air-to-Ground Attack	0.694	9	9	9	9	9	9	9	9	9	9	9	9
Air-to-Air Attack	0.102	9	9	9	9	9	9	9	3	3	3	3	3
Cargo Hauling	0.102	9	9	9	9	9	9	9	9	9	3	3	3
High-Altitude Surveillance	0.102	9	9	9	9	9	9	9	9	9	9	9	9

9 = Task required for mission accomplishment.

3 = Task important to mission accomplishment.

1 = Task aids some in mission accomplishment.

0 = Task n/a.

Operational Tasks

Figure 3. Second Matrix (Mission-to-Task)

Once the tasks have been weighted the third matrix, relating information assets to each of the tasks, can be constructed[1]. The third matrix, a portion of which is shown in Figure 4, uses a scale that is somewhat different from the previous one. Each asset's criticality to a task is judged by the impact its loss would have on the accomplishment of the task. If the asset's loss would have no effect on task accomplishment, that would rate a zero entry. If the asset's loss would cause the system operator to resort to some acceptable work-around, that would be scored as a two. If the loss would degrade task accomplishment even with the use of a work-around, a four is entered. If the asset's loss prevents task accomplishment, the entry is a six. As with matrix #2, the relative weights are calculated as the sum of the products of the task relative weights and the asset criticality values.

	Rel Wt	615	572	420	411	327	303	281	272	264
6 = Loss of asset prevents task accomplishment. 4 = Loss of asset degrades task accomplishment. 2 = Loss of asset requires a work-around to do task. 0 = Asset is n/a.	**Information Assets**	Executive Subsystem	Executive Software	Data Link Software	Data Link Subsystem	Radar Software	Radar Subsystem	Communications Software	Communications Subsystem	Maintenance MIS
Operational Tasks	Rel Wt	Asset Criticality to Task								
Update System Data Base While On Station	9.00	6	6					2	2	
Maintain Station-Keeping Off Target	9.00	6	6	6	4	6	6	2	2	
Disseminate Mission Critical Data	9.00	6	6	6	6			2	2	
Carry Out Air-To-Air Refueling	9.00	4	4	6	6			4	4	
Receive and React to Threat Warnings	9.00	6	6	6	4			2	2	
Power-Off Preflight and Launch	9.00									6
Initialize Avionics with Mission Data Base	9.00	6								
Discriminate Between Friendly and Enemy Forces	8.39	4	6		4	6	6			
Identify Mobile SAM Sites	8.39	4	6		4	6	6			
Maintain Unbroken Comm with Ground Forces	7.78	6	6	6	4	6	6	2	2	
Maintain Unbroken Comm with Airborne Forces	7.78	6	6	4	4			4	4	
Record In-Flight Data for Post-Mission Analysis	7.78	6							2	

Figure 4. Portion of Third Matrix (Task-to-Asset)

[1]For this analysis, "information assets" include a range of items that create, store, manipulate, or transport information. Computers, networks, databases, software and their creators all fall under this definition, although they are at different levels of abstraction.

Now that the system's information-oriented parts and their functions are well defined, the risks facing them can be determined. Each asset — a network node, a data base, etc. — must be examined for its vulnerability to various types of IW attack. In ESC's analysis, the impact of an attack on a given asset was predicted in terms of the time needed to get the asset back "on line." While this was being done, threat analysts estimated the probability that such an attack would occur.

The product of impact and likelihood is defined as risk to the asset, or Asset Risk. Going a step further, the product of Asset Risk and the asset's criticality to task is called Task Risk. Figure 5 shows how Asset Risk is developed using the fourth matrix. For each asset, the values in the impact (I) and Likelihood (L) columns are used to compute the risk facing the asset from each type of attack. The far right-hand column holds the maximum risk value (I * L) for each asset from all of the threats.

For example, in Figure 5, the Maintenance Management Information System (MIS) faces a risk of 4 * 7 = 28 to attacks using malicious code and 4 * 3 = 12 to attacks using data manipulation.[2] In each case, the impact value of 4 means that the asset will only be partially repairable after the attack. However, the likelihood values (7 and 3) mean that there could be nearly a 90% chance of a malicious code attack but less than a 50% chance of an attack using data manipulation. Clearly, the greater risk to the Maintenance MIS is the malicious code.

In Figure 5, the System Risk numbers at the top represent the overall, relative risk to the different types of IW attack. They are computed as the sum of the products of each asset's relative weight times the impact and likelihood values for each threat. These numbers give a good indication of where the biggest overall problem lies, but the more useful result comes from the maximum risk values in the right-hand column: The right countermeasure(s) to protect the asset(s) most vital to the primary mission objective(s). This result is found using the Pareto diagrams in Figures 7 and 8. But first, the risk to each task is computed using the fifth and final matrix.

---

[2]The details of the different IW attacks are available from ESC/ICW. Those interested should call (617) 271-8220 and ask to speak to the assessment team.

Information Assets	Sys Risk / Rel Wt	Malicious Code I	L	Chipping I	L	Spoofing I	L	Data Manipulation I	L	Max Asset Risk
Sys Risk		26,592		8,284		3,885		3,168		
Executive Subsystem	615			4	1					4
Executive Software	572	4	3							12
Data Link Software	420	4	3							12
Data Link Subsystem	411			4	1	3	1			4
Radar Software	327	4	3							12
Radar Subsystem	303			4	1	3	1			4
Communications Software	281	4	3							12
Communications Subsystem	272			4	1	3	1			4
Maintenance MIS	264	4	7	4	1			4	3	28
Comm SW Developer	206			4	1	3	1			4
Radar Integration Lab	201									
Navigation Subsystem	103					3	1			3

Impact Scale	Likelihood Scale
1 = Asset not significantly affected	
2 = Asset recoverable in a short time	1 = 1% - 19% chance of attack
3 = Asset recoverable in a long time	3 = 20% - 49% chance of attack
4 = Asset only partially recoverable	7 = 50% - 89% chance of attack
5 = Asset not recoverable at all	9 = 90% - 99% chance of attack
6 = Damage migrates to other assets	

Figure 5. Portion of Fourth Matrix (Asset-to-Threat)

Shown in Figure 6, the fifth matrix uses the same lists of tasks and assets, plus the same asset criticality values, as found in the third matrix. Here, however, they are used to compute the maximum risk to each of the tasks. Found in the far right-hand column of Figure 6, each Task Risk value is computed as the maximum of the products of each asset's risk (from the top row) and its criticality to the task. For example, the greatest risk facing the task called "Update System Data Base While On Station" is the maximum of the products 4 * 6 = 24, 12 * 6 = 72, 12 * 2 = 24, and 4 * 2 = 8. In this case, the maximum is 72, meaning that the Executive Software, scored as a must-have for this task, is itself facing a relatively high risk of attack. From this, the Executive Software would be a prime candidate for the application of a countermeasure.

Max Risk	4	12	12	4	12	4	12	4	28	
6 = Loss of asset prevents task accomplishment. 4 = Loss of asset degrades task accomplishment. 2 = Loss of asset requires a work-around to do task. 0 = Asset is n/a.	Information Assets	Executive Subsystem	Executive Software	Data Link Software	Data Link Subsystem	Radar Software	Radar Subsystem	Communications Software	Communications Subsystem	Maintenance MIS
Operational Tasks	Rel Wt	Asset Criticality to Task								Max Task Risk
Update System Data Base While On Station	9.00	6	6					2	2		72
Maintain Station-Keeping Off Target	9.00	6	6	6	4	6	6	2	2		72
Disseminate Mission Critical Data	9.00	6	6	6	6			2	2		72
Carry Out Air-To-Air Refueling	9.00	4	4	6	6			4	4		72
Receive and React to Threat Warnings	9.00	6	6	6	4			2	2		72
Power-Off Preflight and Launch	9.00									6	168
Initialize Avionics with Mission Data Base	9.00	6									24
Discriminate Between Friendly and Enemy Forces	8.39	4	6		4	6	6				72
Identify Mobile SAM Sites	8.39	4	6		4	6	6				72
Maintain Unbroken Comm with Ground Forces	7.78	6	6	6	4	6	6	2	2		72
Maintain Unbroken Comm with Airborne Forces	7.78	6	6	4	4			4	4		72
Record In-Flight Data for Post-Mission Analysis	7.78	6							2		24

Figure 6. Portion of Fifth Matrix (Task Risk)

Incorporating countermeasure analysis into this process is a simple matter. Once the five matrices are done and the maximum Asset Risks and Task Risks are identified, the Task Risks that are unacceptably high are chosen. The assets that drive those risks are identified from the fifth matrix, and the threats that must be countered are found from the fourth matrix. In each case, the effects of a countermeasure are considered and the values for Impact, Likelihood, or Asset Criticality adjusted to reflect the effect of the countermeasure. For instance, a countermeasure against data manipulation attacks might be data partitioning, which limits the damage an attacker can create, improving recovery from a partial one to a quick and complete one. The Impact value would drop from 4 to 2, reducing the risk. Similarly, adding a backup system to a critical asset might reduce the asset's criticality to a task due to redundancy. That would not affect the Asset Risk value, but it would reduce Task Risk — the important bottom line. Finally, changing to a highly complicated and uncommon operating system might make an attack infeasible for all but the best-equipped and most talented attacker. That would reduce the Likelihood value, again dropping the Asset Risk. The effect of these measures can best be seen using Pareto charts, as discussed in the next paragraph. The same charts simplify the choice of tasks and assets to protect as well.

Asset Risk and Task Risk - Using the Pareto Charts

Figure 7a shows the maximum IW risk facing each of the assets in the weapon system. The assets are ordered according to their relative weight, which is based on the tasks they support and their criticality to those tasks. The risk scale is defined as follows: Low risk is presumed to be acceptable, requiring no mitigating action; medium risk should be mitigated but may be overlooked on a case by case basis; high risk demands mitigating action. Clearly, any asset with a high relative weight and a medium or greater risk is a candidate for some type of countermeasure. The decision to apply a countermeasure to an asset depends on the task risk that results from the asset being threatened. Task risk is determined using the linkage between assets and tasks, via the QFD matrices. These enable us to plot task risk as shown in Figure 7b.

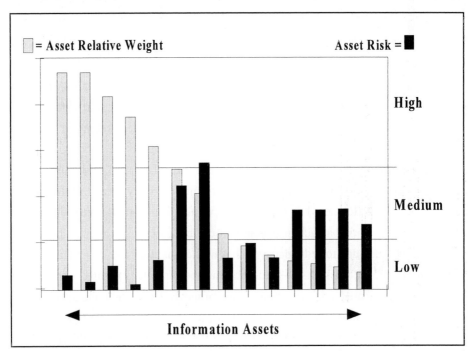

Figure 7a. Asset Risk vs. Asset Weight

In an unconstrained environment, all risk could be mitigated without concern over costs. However, in these times of severe fiscal constraints, program managers need to know how to reach their goals with the minimum outlay. The Pareto diagram in Figure 7b allows the program manager to quickly identify the weapon system tasks that must be protected from IW impact. Using the task-to-asset relationships contained in the QFD matrices and the nature of the risks facing the most critical assets, quick and accurate selection of countermeasures can be made. In the case shown in Figure 7b, the first six tasks score high on the relative weight scale, meaning they play a large part in achieving mission objectives. They are the tasks to be protected. The 3rd, 5th and 6th tasks display significant risk, and the 4th shows a risk level just barely in the medium category. From the QFD matrices it is seen that assets number six and seven are used in performing the tasks at hand.

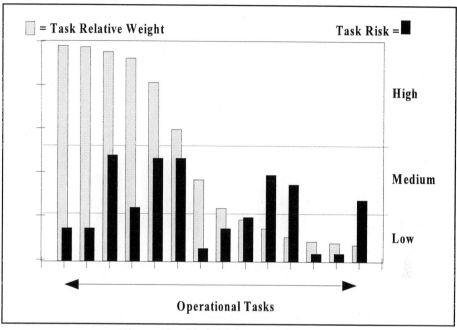

Figure 7b. Task Risk vs. Task Weight

Suppose mitigating actions are taken to reduce the risk facing the 6th and 7th assets in Figure 7a. These may include procedural changes, installation of new equipment or software, or some other measure. The estimated effects of these actions are incorporated into the QFD matrices and the analysis repeated. The resulting drop in asset risk is automatically carried through to the various tasks, depending on the assets' criticality to each task and the degree of risk mitigation to the asset.

The results of this hypothetical case are shown in Figure 8a, where the dotted lines indicate the original asset risk before countermeasure application. The real payoff is shown in Figure 8b, where the dotted lines show the task risk before countermeasure application. This is the view the program manager can use to judge whether more or less mitigating action is needed.

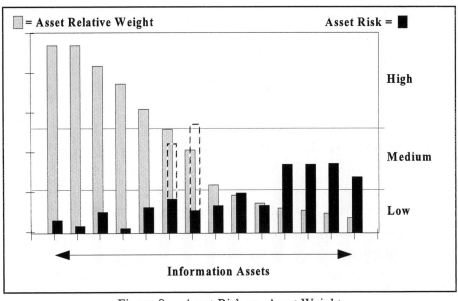

Figure 8a. Asset Risk vs. Asset Weight
(After Applying Countermeasures)

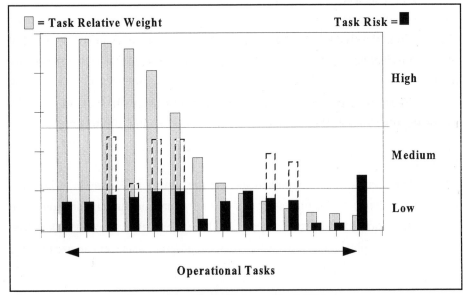

Figure 8b. Task Risk vs. Task Weight
(After Applying Countermeasures)

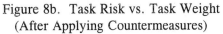

Case Study 29: Lemon/Lime Juicer Project[1]

A design team decided to use QFD to help develop and produce an improved kitchen tool for obtaining the juice of a lemon or lime. Many people cook with these juices, but they typically use a fork or some other blunt object to extract the juice from the fruit. This usually leads to inefficient extraction of the juice as well as seeds and pulp winding up in the prepared dish. People tend to avoid the juicers they use for larger citrus fruit since this creates more clean up activity as well as a perception of extra effort.

The team developed the House of Quality (HOQ) shown in Figure 1 for the project. For benchmark information, the company's current orange juicer product was compared to their competitors' products.

From this matrix, it was clear the "easy to use" was a critical customer requirement. As a result, minimizing the angular force and the pressure required to obtain the juice were key technical requirements.

To better involve technical people from outside the team, the team completed the matrix shown in Figure 2. The functions, sometimes referred to as the "voice of the engineer," were compared against the Design Requirements obtained from the HOQ. After the primary function of "squeeze juice," it was clear the new product needed to also strain the juice and facilitate the hand action of the user.

Next, the team had to develop a concept for the new juicer. They decided to use Pugh's Concept Selection Matrix, which is shown in Figure 3.

---

[1]Contributed by Peter Hofmann from MCI Telecommunications Corp.

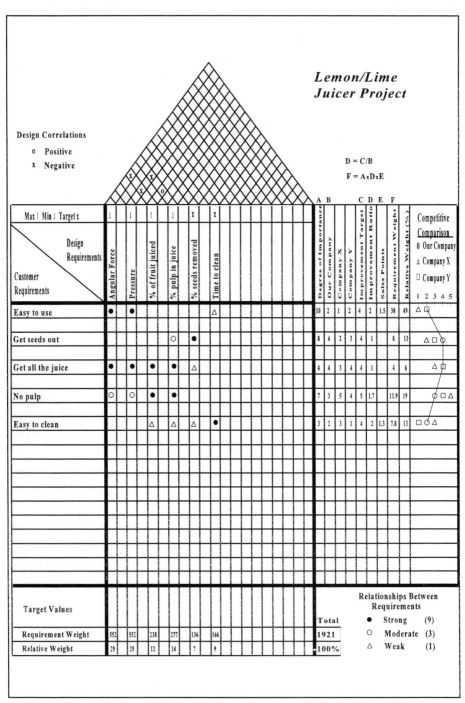

Figure 1. House of Quality

Functions

Design Requirements

Relationships Between Requirements		
●	Strong	(9)
○	Moderate	(3)
▵	Weak	(1)

Design Requirements		Squeeze Juice	Strain Juice	Facilitate Hand Action	Separate Seeds	Work Quickly					
Angular Force	29	●		○							
Pressure	29	●		○							
% of fruit juiced	12	●	○								
% pulp in juice	14		●								
% seeds removed	7				●						
Time to clean	9					○					
										Total
Requirement Weight		630	162	174	63	27					1056
Relative Weight		60	15	16	6	3					100 %

Figure 2. Functions vs Design Requirements Matrix

Design Options

Key Requirements	❶	❷	❸	❹	❺	❻
Easy to use		+	-	-	+	+
Get seeds out		s	-	-	-	s
Get all the juice	D	s	-	s	s	+
No pulp		s	-	-	-	s
Easy to clean		s	+	+	-	s
Angular force		s	+	+	+	+
Pressure	A	s	-	s	+	+
% of fruit juiced		s	-	+	s	+
% of pulp in juice		s	-	-	-	+
% seeds removed		s	+	+	-	s
Time to clean	T	s	-	-	-	s
Squeeze juice		s	-	+	-	+
Strain juice		s	-	-	-	s
Facilitate hand action		s	-	-	+	+
Separate seeds	U	s	-	-	-	s
Work quickly		s	s	s	+	+
Cost		+	+	+	+	-
Reliability		+	+	+	+	+
Safety	M	+	-	-	-	+
Producibility		+	+	+	+	-
+'s		5	6	8	8	11
-'s		0	13	9	10	2

Figure 3. Pugh's Concept Selection Matrix

The concepts evaluated included:

1. Traditional glass reamer with strainer
2. Plastic reamer with strainer
3. Fork
4. Wooden seafood reamer without strainer
5. Kitchen mallet
6. New prototype (see Figure 4).

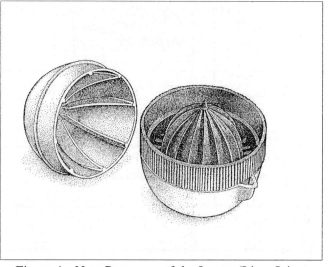

Figure 4. New Prototype of the Lemon/Lime Juicer

The team first compared concepts 2-5 against concept 1. After reviewing the strengths and weaknesses of the existing concepts, concept 6 was developed to take advantage of the strengths of the other concepts and better satisfy the key requirements identified. The team selected this new concept to develop.

The new concept had three parts. The "base" resembled a standard plastic reamer with strainer. To this base, two new parts, a "cover" and a "collector," were added. The following matrix (Figure 5) was used to identify which of these parts were critical to the success of the final product. Clearly, the base was the part that needed the most focus during development.

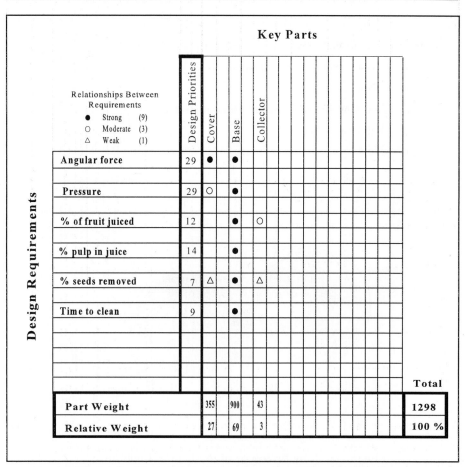

Figure 5.   Design Requirements vs Parts/Service Details

Given that the design of the base (i.e., shape of the reamer and dimensions of the strainer holes) were critical to the success of the design, the team ran some designed experiments to optimize these parameters. The results are summarized in the matrix shown in Figure 6.

Experiment	Factors	Treatments	Method	Optimal Combination
1. Optimize shape of juicer	Cover Shape Base Shape	Cover$_1$ Cover$_2$ Base$_1$ Base$_2$	Classical (with % fruit juiced as response)	Cover$_2$ Base$_1$
2. Optimize hole size/ shape in strainer	hole size hole shape	Size$_1$ Size$_2$ Shape$_1$ Shape$_2$	Taguchi (with % of pulp in juice as response)	Size$_1$ Shape$_2$

Figure 6. Design of Experiments Scorecard

After firming up the final design, the team wanted to check their design against the requirements they identified in the HOQ. Specifically, they wanted to identify those potential failure modes of their design that might impact key design requirements before they began production. They used the matrix in Figure 7 to perform their analysis.

After the analysis, the team found that they had to ensure the overall integrity of the base since "juicer collapses" and "point breaks off" were critical failure modes. Also, maintaining tolerances between the cover and the base were also important. They also found that some requirements may have been missed. Since the "collector cracks" failure mode is not related to any design requirement, a customer requirement to "collect juice" with a corresponding design requirement of "% juice collected" should have been captured.

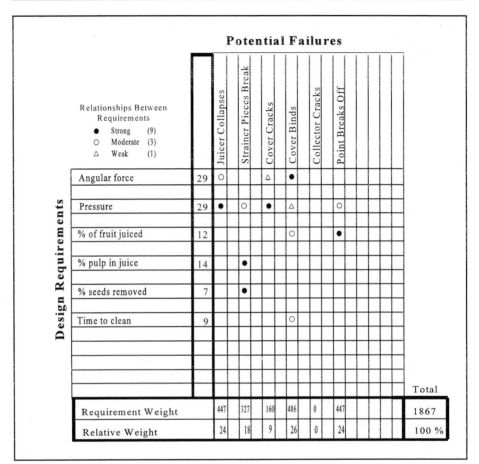

Figure 7. Potential Failures vs Key Design Requirements

Finally, the team used the matrix in Figure 8 to help ensure that the base was produced as designed. They also hoped that this extra emphasis on the base would prevent the most damaging potential failure modes from occurring.

Key Part/ Service Details	Process Flow	Process Controls					Special Training Requirements
		What	Where	When	Method	Target	
Base	Load Hopper →	Amt of Material	Mold Machine	As req'd	visual	> ¼ full	
	Preheat Mat →	Temp	"	15 min	x̄-R	200 deg	SPC
	Mold Part →	Pressure	"	15 min	x̄-R	70 psi	SPC
	Eject →						
	Cool →	Temp	Holding Area			?	
	Measure	Critical Dim		Random sample/15min	x̄-R		SPC

Figure 8. Delivered Quality Matrix

Case Study 30:   Statistical Analysis of a Randomization Algorithm[1]

In theory, it is difficult to define a hash function which is capable of creating random data from non-random data. This case study presents the results of a statistical analysis of the randomization properties of the MD4[2] Algorithm [7][3]. The MD4 message digest algorithm is a fast, compact hash function which maps an arbitrarily-long string of bits onto a 128-bit quantity. For a complete description of the algorithm, the reader is referred to Rivest [7]. The investigation of MD4 consisted of a series of six statistical tests in which a large number of 128-bit outputs was generated and then examined for randomness, or the lack thereof. The results of these tests are as follows.

The first test conducted was a byte parity test. The appropriate hypotheses for this Chi-Square test are presented as follows:

H_0: Odd/Even parity of bytes are equally likely.

H_1: Odd/Even parity of bytes are not equally likely.

From a total of one million iterated applications of MD4, each of the 16 byte positions of the one million outputs was examined for parity. The results of this test are shown in Table 1. It is apparent that the null hypothesis cannot be rejected, indicating that each byte is equally likely to be odd or even. In fact, the extremely high P-value (.9813) might lend statistical credence to the algorithm's "purposeful smashing of bytes."

---

[1]This case study is provided by Philip L. Mayfield of Digital Computations and Mark J. Kiemele, one of the co-authors.

[2]MD4 is the product of Ron Rivest, MIT Laboratory for Computer Science, 1990.

[3]References for this case study are on the last page of this case study.

Byte Position	Actual	Expected	χ^2 Contribution
1	500979	500000	1.916882
2	499412	500000	.691488
3	499985	500000	.000450
4	500510	500000	.520200
5	500513	500000	.526338
6	499849	500000	.045602
7	499780	500000	.096800
8	499808	500000	.073728
9	500624	500000	.778752
10	499776	500000	.100352
11	499787	500000	.090738
12	499922	500000	.012168
13	500242	500000	.117128
14	499414	500000	.686792
15	499939	500000	.007442
16	500347	500000	.240818
		Total	5.905678
		P-value	.9813

Table 1. Byte Parity Test

A second test conducted is a check for uniformity in the bivariate distribution of byte position versus byte value. The hypotheses tested are as follows:

H_0: Bivariate distribution of byte position vs byte value is uniform.

H_1: Bivariate distribution is not uniform.

Three million iterated applications of MD4 were performed, and the results of examining the decimal integer value of each byte in each of the three million outputs are shown in Table 2.

		Byte Position			
		1	**2**	....	**16**
	0	11645	11591	....	11678
V	**1**	11722	11658	....	11780
A	.	.	.		.
L	.	.	.		.
U	.	.	.		.
E	.	.	.		.
	.	.	.		.
	255	11832	11823	....	11483

Bivariate Frequency Distribution

E(X) = 11718.75		χ^2_0 = 3895.87	
Min = 11354		df = 4095	
Max = 12099		p = .54	

Table 2. Uniformity of Byte Position vs Byte Value

The results indicate that the distribution is indeed uniform. One can also conclude independence between position and value. That is, given a particular byte position, the byte value is equally likely to be any of the 256 possible values. Similarly, given a particular value, it is equally likely to occur in any of the 16 byte positions.

A third frequency test was then conducted, this time at the bit level. The hypotheses for this test are expressed as follows:

H_0: The distribution of 1's across all 128 bit positions is uniform.

H_1: This distribution is not uniform.

Another three million outputs from MD4 were generated and each of the bit positions examined to determine the frequency of 1's in each position. Table 3 provides the results of this test. These results again indicate uniformity across the 128 bit positions, i.e., each bit is equally likely to be a 0 or 1.

Bit Position	Actual	Expected	χ^2 Contribution
1	1499496	1500000	.169
2	1500769	1500000	.394
3	1501119	1500000	.835
4	1500256	1500000	.037
5	1500256	1500000	.044
.	.	.	.
.	.	.	.
.	.	.	.
.	.	.	.
.	.	.	.
126	1499975	1500000	.000
127	1499466	1500000	.190
128	1500624	1500000	.260

$$\chi_0^2 \quad = \quad 70.657$$

Min = 1,498,191   df = 127

Max = 1,502,403   p = .85

Table 3. Frequency Test for Bit Positions

The fourth test, a gap test, examined another set of one million outputs from MD4. Each output was scanned for the number of 0's between successive 1's. For example, the string "10010110001" has gaps of 2, 1, 0, and 3, respectively. Table 4 shows the total number of observed gaps for gaps of size 24 or less. No gaps of size 25 or larger were encountered. The successive halving of the number of observed gaps for an incremental gap size of 1 is what we would expect to see if the probability of a 1 or 0 in each bit position is .5.

Gap Size	Observed #	Gap Size	Observed #
0	32263081	13	3583
1	16247658	14	1793
2	8063542	15	881
3	4001042	16	421
4	1984905	17	209
5	983004	18	121
6	488068	19	59
7	241362	20	27
8	120025	21	18
9	59219	22	5
10	29843	23	3
11	14471	24	2
12	7337		

Table 4. Gap Test

The fifth test conducted was one in which the difference (in absolute value) between the number of 1's and the number of 0's occurring in each of one million outputs was noted.   The observed (actual) frequencies, as well as expected frequencies (under the assumption that a 0 or 1 in any position is equally likely), are shown in Table 5. These results clearly support the assumption.

Difference	Actual	Expected	χ^2
0	70331	70386	.043
2	138539	138606	.032
4	132566	132306	.511
6	122122	122433	.790
8	109900	109829	.046
10	94974	95504	2.941
12	80948	80496	2.538
14	65734	65757	.008
16	52053	52058	.000
18	39905	39935	.023
20	29959	29681	2.604
22	21426	21370	.147
24	14928	14903	.042
26	10117	10064	.219
28	6508	6581	.810
30	4205	4165	.384
32	2532	2551	.142
34	1480	1512	.677
36	870	866	.018
38	490	480	.208
40	277	257	1.556
42	115	133	2.436
44	67	67	.000
46	32	32	.000
48	14	15	.600
50	4	7	1.286
52	5	5.833	.119
54	0	2.436	2.436
56	1	.980	.000
χ_0^2 = 18.603	df = 27		P = .884

Table 5. Differences between # of 1's and # of 0's

A final test was conducted to examine the avalanche effect of MD4. A series of 30,000 comparisons was made, where each comparison compared two outputs of MD4. The two outputs compared were the outputs corresponding to two "almost identical" inputs to MD4. That is, if A is a 32,000-byte (256,000-bit) string and A' is also a 32,000-byte string which differs from A in only 1 bit position, we then compared MD4(A) and MD4(A'), each being a 128-bit string. We looked at the Hamming distance between MD4(A) and MD4(A'), i.e., the number of bit position changes that occurred between MD4(A) and MD4(A'). The hypotheses tested can be described as follows:

H_0:   The distribution of Hamming distances is binomial
with n = 128 and p = .5

H_1:   This distribution is not binomial with n = 128 and p = .5

The frequency distribution of Hamming distances that occurred among the 30,000 comparisons is shown in the histogram of Figure 1.

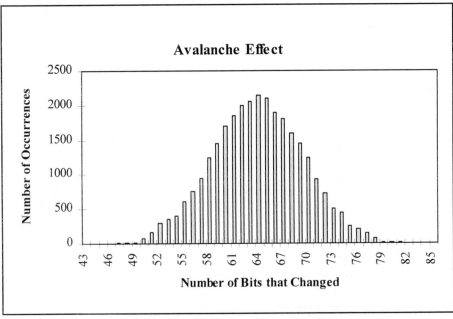

Figure 1. Avalanche Effect: Hamming Distance

This figure suggests that, on the average, about half (64) of the bits will change. While this figure gives us an indication of how many bits will change, Table 6 shows us that, of the bits that do change, each of the bit positions tends to contribute equally to the number of changes. Clearly, the avalanche effect demonstrated is one in which the outputs for two "almost identical" inputs appear to be as random as any other two randomly chosen 128-bit strings.

The results shown here indicate that MD4 is a byte smasher extraordinaire. These random properties of MD4, together with its speed and compactness, make it a potentially valuable tool for a variety of applications, including virus detection, gaining access to secure files, and compressing large files prior to signing them with a public-key algorithm such as RSA.

Bit Position	Actual	Expected	χ^2 Contribution
1	14995	15000	.002
2	14997	15000	.001
3	15123	15000	1.009
4	14956	15000	.129
5	14915	15000	.482
.	.	.	.
.	.	.	.
.	.	.	.
.	.	.	.
.	.	.	.
126	14855	15000	1.402
127	15029	15000	.056
128	15092	15000	.564

$$\chi^2_0 \;=\; 65.148$$

Min = 14811   df = 127

Max = 15294   p = .89

Table 6. Avalanche Effect: How often each of the 128 bit positions
changed in the 30,000 comparisons

References

1.   Damgård, Ivan Bjerre.   Design Principles for Hash Functions. *Proceedings CRYPTO '89*, pp 395-405.

2.   Denning, Dorothy E.R. *Cryptography and Data Security*.   Addison-Wesley, 1982.

3.   Knuth, Donald E. *The Art of Computer Programming (2nd ed), Vol 2/Seminumerical Algorithms*.   Addison-Wesley, 1981.

4.   Merkle, Ralph C.   A Fast Software One-Way Hash Function. *Journal of Cryptology*, Vol. 3, No. 1 (1990), pp 43-58.

5.   Merkle, Ralph C.   One-Way Hash Functions and DES. *Proceedings CRYPTO '89*, pp 407-419.

6.   Patterson, Wayne. *Mathematical Cryptology for Computer Scientists and Mathematicians*.   Rowman & Littlefield, 1987.

7.   Rivest, Ronald L.   The MD4 Message Digest Algorithm. *Proceedings CRYPTO '90*, pp 281-291.

Bibliography

[Afrm 89]   *Air Force R & M 2000 Variability Reduction Process Guidebook*
              (Draft). HQ USAF/LE-RD, Pentagon, Washington D.C., 1989.

[Augu 86]   Augustine, N.R. *Augustine's Laws*. Viking Penguin, Inc. NY,
              1986.

[Bain 78]   Bain, Lee J. *Statistical Analysis of Reliability and Life-Testing
              Models*. Marcel Dekker, Inc., New York, 1978.

[Barl 75]   Barlow, R.E. and Proschan, F. *Statistical Theory of Reliability
              and Life-Testing*. Holt, Rinehart, and Winston, New York, 1975.

[Barl 65]   Barlow, R.E. and Proschan, F. *Mathematical Theory of
              Reliability*. John Wiley & Sons, New York, 1965.

[Benn 84]   Bennis, Warren. "The Four Competencies of Leadership."
              Training & Development Journal 1984, 38 (8), pp 14-19.

[Berw 91]   Berwick, D.M., Godfrey, A.B., and Roessner, J. *Curing Health
              Care*. Jossey-Bass Publishers, San Francisco, CA, 1991.

[Bort 98]   Bortkiewicz, L. *Das Gesetz der Kleinen Zahlen*. Leipzig,
              Teubner, 1898.

[Box 88]   Box, G.E.P., Bisgaard, S. and Fung, C. "An Explanation and
              Critique of Taguchi's Contributions to Quality Engineering."
              Center for Quality Improvement, University of Wisconsin,
              Madison, WI, 1988.

[Box 78]   Box, G.E.P., Hunter, W.G. and Hunter, J.S. *Statistics for
              Experimenters*. John Wiley & Sons, New York, 1978.

[Burs 88]   Burstein, C. and Sedlak, K. "The Federal Productivity
              Improvement Effort: Current Status and Future Agenda."
              National Productivity Review, Spring 1988.

[Busi 87]   "The Push for Quality." *Business Week*. June 9, 1987.

[Camp 94]   Campbell and Wallace. *Statistics Packet* (3rd edition). National
               Clearinghouse for the Defense of Battered Women, Philadelphia,
               PA, February, 1994.

[Clau 89]   Clausing, Don. "Quality Function Deployment." *Taguchi
               Methods and QFD; How's and Why's for Management*, Nancy
               Ryan, Editor, ASI Press, Dearborn, MI, 1989.

[Comm 93]   The Commonwealth Fund. *First Comprehensive National Health
               Survey of American Women Finds Them at Significant Risk*.
               News Release, The Commonwealth Fund, New York, NY, 14
               July 1993.

[Crim 92]   *Crime in the U.S.: Uniform Crime Reports*. U.S. Department of
               Justice, FBI, p. 17, 1992-1995.

[Degr 86]   DeGroot, M.H., Fienberg, S.E. and Kadane, J.B. *Statistics and
               the Law*. John Wiley & Sons, Inc., New York, 1986.

[Demi 86]   Deming, W. Edwards, *Out of the Crisis*. MIT center for
               Advanced Engineering Study, 1986.

[Demi 75]   Deming, W. Edwards. "On Some Statistical Aids Toward
               Economic Production," *Interfaces*, Vol. 5, No. 4, August 1975,
               p 5. The Institute for Management Sciences, Providence, RI.

[Demi 67]   Deming, Dr. W.E. "What Happened in Japan?" *Industrial
               Quality Control*, Vol. 24, No. 3, August 1967, pp 89-93.

[Ders 95]   Dershowitz, Alan. *Spousal Abuse Doesn't Predict Murder*. The
               Times Union, Albany, NY, January 16, 1995, p. A6.

[Dert 89]   Dertoregos, Michael L., Lester, Richard K. and Solow, Robert M.
               Made in America, Regaining the Productive Edge. MIT Press,
               Cambridge, MA, 1989.

[Devo 87]   Devore, Jay L. *Probability and Statistics for Engineering and the
               Sciences* (Second Edition). Brooks/Cole Publishing Company,
               Monterey, CA, 1987.

[DOEK 97]   DOE KISS Version 1.1.   Digital Computations, Inc., Albuquerque, NM, 1997.

[Focu 84]   "Focus on Statistics."   Woodrow Wilson National Fellowship Foundation, Princeton, NJ, 1984.

[Ford 85]   Ford Motor Company. *Continuing Process Control and Process Capability Improvement*. Statistical Methods Office, Ford Motor Company, September, 1985.

[Free 80]   Freedman, D., Pisani, R. and Purves, R.   *Statistics*.   W.W. Norton & Company, New York, 1980.

[Freu 88]   Freund, John E. *Modern Elementary Statistics* (Seventh Edition). Prentice-Hall, Inc., Englewood Cliffs, NJ, 1988.

[Galv 92]   Galvin, Robert.   Presentation, PACE Conference, Philadelphia, PA, May 92.

[Goal 88]   Goal/QPC. *The Memory Jogger: A Pocket Guide of Tools for Continuous Improvement*. Goal/QPC, Methuen, MA 1988.

[Graz 89]   Grazier, Peter B. "Before It's Too Late, Employee Involvement... An Idea Whose Time Has Come." Teambuilding, Inc., Chadds Ford, PA, 1989.

[Gros 89]   Grosh, Doris Lloyd. *A Primer of Reliability Theory*. John Wiley & Sons, New York, 1989.

[Harr 88]   Harry, Mikel J. *The Nature of Six Sigma Quality*. Motorola, Inc., Government Electronics Group, 1988.

[Haus 88]   Hauser, John R. and Clausing, Don. "The House of Quality." *The Harvard Business Review*, May-June 1988.

[Hine 80]   Hines, William W. and Montgomery, Douglas C. *Probability and Statistics in Engineering and Management Science* (Second Edition). John Wiley & Sons, New York, 1980.

[Hogg 87]   Hogg, Robert V. and Ledolter, Johannes. *Engineering Statistics*. Macmillan Publishing Company, New York, 1987.

4 Bibliography

[Hopk 93]   Hopkins, Bert. Information from personal interview with Mr. Hopkins. Bedford, MA, Jan 1993.

[Ishi 87]   Ishikawa, K. *Guide to Quality Control*. Kraus International Publications, White Plains, NY, 1987.

[John 80]   Johnson, Robert. *Elementary Statistics* (Third Edition). Wadsworth, Inc., Belmont, CA 1980.

[John 92]   Johnson and Dumas. "How to Improve Quality if You're Not In Manufacturing." *Training*, Nov 1992.

[Kapu 77]   Kapur, K.C. and Lamberson, L.R. *Reliability in Engineering Design*. John Wiley & Sons, New York, 1977.

[Kell 90]   Kelly, Michael J., Chairman of the Board, KELCO Industries, 1990. Permission to include the information on KELCO granted by Mr. Kelly.

[King 87]   King, Robert. "Better Designs in Half the Time — Implementing QFD in the USA." Presented at the Growth Opportunity Alliance of Lawrence, MA, 1987.

[Lars 85]   Larson, Richard J. and Marx, Morris L. *An Introduction to Probability and Its Applications*. Prentice-Hall, Inc., Englewood Cliffs, NJ 1985.

[Lead 90]   *Leading Edge*, Vol. xxxii, No. 6. Air Force Systems Command, Andrews AFB, MD, 1990.

[Made 89]   *Made In America*, MIT Press, 1989.

[Malc 90]   "Malcolm Baldrige National Quality Award." *Application Guidelines*. U.S. Department of Commerce, Gaithersburg, MD, 1990.

[Mann 74]   Mann, N.R., Schafer, R.E. and Singpurwalla, N.D. *Methods for Statistical Analysis of Reliability and Life Data*. John Wiley & Sons, New York, 1974.

[McCl 86]   McClave, James T. and Dietrich, Frank H. II. *A First Course in Statistics* (Second Edition). Dellen Publishing Company, San Francisco, CA, 1986.

[McDe 96]   McDermott, Robin E., Mikulak, Raymond J., and Beauregard, Michael R. *The Basics of FMEA.* Quality Resources, New York, NY, 1996.

[McGh 85]   McGhee, John W. *Introductory Statistics.* West Publishing Company, St. Paul, MN. 1985.

[Mend 88]   Mendenhall, William and Sincich, Terry. *Statistics for the Engineering and Computer Sciences* (Second Edition). Dellen Publishing Company, San Francisco, CA, 1988.

[Metc 79]   Metcalf & Eddy, Inc., *Wastewater Engineering: Treatment, Disposal, Reuse* (Second Edition). McGraw-Hill, New York, 1979.

[Mont 91]   Montgomery, D.C. *Introduction to Statistical Quality Control* (Second Edition). John Wiley & Sons, New York, 1991.

[Mont 90]   Montgomery, D.C. *Design and Analysis of Experiments* (Third Edition). John Wiley & Sons, New York, 1990.

[Moro 82]   Moroney, M.H. *Facts from Figures.* Penguin Books, Ltd, New York, 1982.

[Moto 88]   Motorola, Inc. "1988 Annual Report." 1988.

[Myer 90]   Myers, Raymond H. *Classical and Modern Regression with Applications* (Second Edition). PWS-KENT Publishing Company, Boston, MA, 1990.

[Nais 84]   Naisbett, John. *Megatrends.* Warner Books, New York, 1984.

[Nete 88]   Neter, J., Wasserman, W. and Whitmore, G.A. *Applied Statistics* (Third Edition). Allyn and Bacon, Inc., Newton, MA, 1988.

[Nora 86]   Nora, John J., Rogers, C. Raymond and Strainy, Robert E. *Transforming the Workplace.* Princeton Research Press, 1986.

[Ostl 75]   Ostle, Bernard and Mensing, Richard W. *Statistics in Research* (Third Edition). Iowa State University Press, Ames, Iowa, 1975.

[Para 91]   Vos Savant, Marilyn. "Ask Marilyn." *Parade Magazine*, February 17, 1991, p 12.

[Park 96]   Parker, Deborah C. "She's Got the Scoop on Statistical Process Control." *Engineering News*, U. of Massachusetts, Vol 4, No. 1, Fall 1996.

[Peav 85]   Peavy, Howard S., Rowe, Donald R. and Tchobanoglous, George. *Environmental Engineering*, McGraw-Hill, New York, 1985.

[Piat 91]   Piatteli-Palmarini, Massimo. "Probability Blindness: Neither Rational nor Capricious." *Bostonia*, March/April 1991.

[Pote 95]   *Potential Failure Mode and Effects Analysis (FMEA) Reference Manual.* Chrysler, Ford, and General Motors Corporations, 1995.

[Rice 88]   Rice, John A. *Mathematical Statistics and Data Analysis*. Wadsworth, Inc., Belmont, CA, 1988.

[Ryan 89]   Ryan, Thomas P. *Statistical Methods for Quality Improvement*. John Wiley & Sons, New York, 1989.

[Schm 96]   Schmidt, S. R., Kiemele, M. J., and Berdine, R. J. *Knowledge Based Management*. Air Academy Press, Colorado Springs, CO, 1996.

[Schm 94]   Schmidt, Stephen R. and Launsby, Robert G. *Understanding Industrial Designed Experiments* (Fourth Edition). Air Academy Press, Colorado Springs, CO, 1992.

[SPCK 97]   *SPC KISS* Version 2.0. Digital Computations, Inc., Albuquerque, NM, 1997.

[Stam 95]   Stamatis, D.H. *Failure Mode and Effect Analysis: FMEA from Theory to Execution*. ASQC Quality Press, Milwaukee, WI, 1995.

[Stat 95]   Stata, R. and d'Arbeloff, A. "Teaching Top Dogs New Tricks."
              The Boston Globe, August 22, 1995.

[Sull 89]   Sullivan, L.P.   Quality Function Deployment: The Latent
              Potential of Phases III & IV."   Presented at the
              AIAA/ADPA/NSIA First Total Quality Management Symposium,
              Denver, CO, 1-3 November 1989.

[Tagu 87]   Taguchi, Genichi. *System of Experimental Design, Vol I and II*.
              Kraus International Publications, White Plains, NY, 1987.

[Triv 82]   Trivedi, K.S. *Probability and Statistics with Reliability, Queuing,
              and Computer Science Applications*.   Prentice-Hall, Inc.,
              Englewood Cliffs, NJ, 1982.

[USA 88]   "USA Faces New Trade Challenge." *U.S.A. Today*, April 1988.

[Walp 85]   Walpole, Ronald E. and Myers, Raymond H. *Probability and
              Statistics for Engineers and Scientists* (Third Edition). Macmillan
              Publishing Company, New York, NY, 1985.

[Walt 86]   Walton, Mary. *The Deming Method*. Dodd, Meade & Company,
              New York, 1986.

[Wats 90]   Watson, F.D. and Schmidt, S.R. *Quality Through Leadership:
              TQM in Action*. Air Academy Press, Colorado Springs, CO,
              1990.

[Wood 83]   Woodgate, R.W. *Handbook of Machine Soldering*. John Wiley
              & Sons, Inc., New York, 1983.

Row	Column 1–5	6–10	11–15	16–20	21–25	26–30	31–35
1	50527	38063	89093	22648	96867	34685	62753
2	40483	25946	46848	63507	86984	05724	73793
3	80462	82084	21308	94040	77165	29466	52079
4	90545	19515	78767	05365	88790	57080	41826
5	19169	99595	73714	19238	72251	05922	87552
6	52665	26336	40764	26694	19988	67596	36786
7	90799	50016	47988	52706	16346	31800	00971
8	45285	63834	42650	90117	15536	72206	44917
9	19441	42344	28192	99346	23158	19639	48645
10	37222	08805	23852	40256	85211	32672	64952
11	37065	26209	66719	27944	16944	81393	35310
12	48899	17553	47175	31199	26672	19188	38099
13	44622	87968	72342	43257	54973	60454	12025
14	73638	18241	52230	89038	67734	44774	55787
15	42221	90634	01822	75261	08195	60196	28252
16	89921	11815	52797	74868	34720	20249	95911
17	38555	79182	87003	42410	09189	29731	44611
18	26213	44234	84459	21895	01186	21979	70360
19	68455	81564	04625	69299	58598	41757	58267
20	91866	70002	70036	66806	00693	89570	22355
21	99347	68962	57338	30529	02209	82909	71087
22	82928	44796	04134	13362	69602	24334	86744
23	16242	87679	08615	30668	50641	63391	00152
24	71385	29319	42870	79158	67826	32664	21638
25	51420	85638	02017	31406	16928	66618	74314
26	07700	56663	07642	72029	15147	43861	48231
27	38514	82498	84649	38830	02403	07300	42212
28	39756	23408	97266	33946	58018	78874	25099
29	12796	42649	09878	25483	30689	39905	99100
30	93589	01789	63970	76873	33782	38460	87506
31	11213	10296	36181	92910	14471	21381	04219
32	61233	66409	94223	11695	36021	92241	85479
33	98421	75296	41506	71427	57443	64683	94253
34	04977	06872	34764	00911	95711	00664	08983
35	59298	13735	95985	56378	94783	54725	05701
36	15017	53836	00873	37602	85937	05872	74263
37	29311	95124	47110	83427	44799	68966	49907
38	82609	13408	68318	84181	36411	20749	83699
39	39952	93967	91401	50899	75843	47751	55027
40	09731	99395	53883	53855	88758	14556	83042
41	45821	84519	07493	12282	80953	13641	67947
42	84309	87119	70436	73462	98107	82617	35553
43	02929	44362	01024	67928	43415	79721	67653
44	42989	49536	64420	38132	58927	32576	51847
45	63063	49898	06019	98905	37156	84088	64254
46	57588	27694	01911	79714	87368	96074	90567
47	26093	88111	84279	85312	51340	23309	98691
48	73300	95180	55945	73516	31579	81353	15724
49	54085	91495	91575	46931	54054	18622	64911
50	76838	77175	69384	09011	69902	88684	81569

Binomial Probabilities calculated from: $P(x) = \binom{n}{x} p^x(1-p)^{n-x}$

		p										
n	x	.01	.02	.03	.04	.05	.06	.07	.08	.09	x	n
2	0	.9801	.9604	.9409	.9216	.9025	.8836	.8649	.8464	.8281	2	
	1	.0198	.0392	.0582	.0768	.0950	.1128	.1302	.1472	.1630	1	
	2	.0001	.0004	.0009	.0016	.0025	.0036	.0049	.0064	.0081	0	2
3	0	.9703	.9412	.9127	.8847	.8574	.8306	.8044	.7787	.7536	3	
	1	.0294	.0576	.0847	.1106	.1354	.1590	.1816	.2031	.2236	2	
	2	.0003	.0012	.0026	.0046	.0071	.0102	.0137	.0177	.0221	1	
	3	.0000	.0000	.0000	.0001	.0001	.0002	.0003	.0005	.0007	0	3
4	0	.9606	.9224	.8853	.8493	.8145	.7807	.7481	.7164	.6857	4	
	1	.0388	.0753	.1095	.1416	.1715	.1993	.2252	.2492	.2713	3	
	2	.0006	.0023	.0051	.0088	.0135	.0191	.0254	.0325	.0402	2	
	3	.0000	.0000	.0001	.0002	.0005	.0008	.0013	.0019	.0027	1	
	4	.0000	.0000	.0000	.0000	.0000	.0000	.0000	.0000	.0001	0	4
5	0	.9510	.9039	.8587	.8154	.7738	.7339	.6957	.6591	.6240	5	
	1	.0480	.0922	.1328	.1699	.2036	.2342	.2618	.2866	.3086	4	
	2	.0010	.0038	.0082	.0142	.0214	.0299	.0394	.0498	.0610	3	
	3	.0000	.0001	.0003	.0006	.0011	.0019	.0030	.0043	.0060	2	
	4	.0000	.0000	.0000	.0000	.0000	.0001	.0001	.0002	.0003	1	
	5	.0000	.0000	.0000	.0000	.0000	.0000	.0000	.0000	.0000	0	5
6	0	.9415	.8858	.8330	.7828	.7351	.6899	.6470	.6064	.5679	6	
	1	.0571	.1085	.1546	.1957	.2321	.2642	.2922	.3164	.3370	5	
	2	.0014	.0055	.0120	.0204	.0305	.0422	.0550	.0688	.0833	4	
	3	.0000	.0002	.0005	.0011	.0021	.0036	.0055	.0080	.0110	3	
	4	.0000	.0000	.0000	.0000	.0001	.0002	.0003	.0005	.0008	2	
	5	.0000	.0000	.0000	.0000	.0000	.0000	.0000	.0000	.0000	1	
	6	.0000	.0000	.0000	.0000	.0000	.0000	.0000	.0000	.0000	0	6
7	0	.9321	.8681	.8080	.7514	.6983	.6485	.6017	.5578	.5168	7	
	1	.0659	.1240	.1749	.2192	.2573	.2897	.3170	.3396	.3578	6	
	2	.0020	.0076	.0162	.0274	.0406	.0555	.0716	.0886	.1061	5	
	3	.0000	.0003	.0008	.0019	.0036	.0059	.0090	.0128	.0175	4	
	4	.0000	.0000	.0000	.0001	.0002	.0004	.0007	.0011	.0017	3	
	5	.0000	.0000	.0000	.0000	.0000	.0000	.0000	.0001	.0001	2	
	6	.0000	.0000	.0000	.0000	.0000	.0000	.0000	.0000	.0000	1	
	7	.0000	.0000	.0000	.0000	.0000	.0000	.0000	.0000	.0000	0	7
8	0	.9227	.8508	.7837	.7214	.6634	.6096	.5596	.5132	.4703	8	
	1	.0746	.1389	.1939	.2405	.2793	.3113	.3370	.3570	.3721	7	
	2	.0026	.0099	.0210	.0351	.0515	.0695	.0888	.1087	.1288	6	
	3	.0001	.0004	.0013	.0029	.0054	.0089	.0134	.0189	.0255	5	
	4	.0000	.0000	.0001	.0002	.0004	.0007	.0013	.0021	.0031	4	
	5	.0000	.0000	.0000	.0000	.0000	.0000	.0001	.0001	.0002	3	
	6	.0000	.0000	.0000	.0000	.0000	.0000	.0000	.0000	.0000	2	
	7	.0000	.0000	.0000	.0000	.0000	.0000	.0000	.0000	.0000	1	
	8	.0000	.0000	.0000	.0000	.0000	.0000	.0000	.0000	.0000	0	8
9	0	.9135	.8337	.7602	.6925	.6302	.5730	.5204	.4722	.4279	9	
	1	.0830	.1531	.2116	.2597	.2985	.3292	.3525	.3695	.3809	8	
	2	.0034	.0125	.0262	.0433	.0629	.0840	.1061	.1285	.1507	7	
	3	.0001	.0006	.0019	.0042	.0077	.0125	.0186	.0261	.0348	6	
	4	.0000	.0000	.0001	.0003	.0006	.0012	.0021	.0034	.0052	5	
	5	.0000	.0000	.0000	.0000	.0000	.0001	.0002	.0003	.0005	4	
	6	.0000	.0000	.0000	.0000	.0000	.0000	.0000	.0000	.0000	3	
	7	.0000	.0000	.0000	.0000	.0000	.0000	.0000	.0000	.0000	2	
	8	.0000	.0000	.0000	.0000	.0000	.0000	.0000	.0000	.0000	1	
	9	.0000	.0000	.0000	.0000	.0000	.0000	.0000	.0000	.0000	0	9
		.99	.98	.97	.96	.95	.94	.93	.92	.91	x	n
		p										

n	x	.10	.15	.20	.25	.30	.35	.40	.45	.50	x	n
2	0	.8100	.7225	.6400	.5625	.4900	.4225	.3600	.3025	.2500	2	
	1	.1800	.2550	.3200	.3750	.4200	.4550	.4800	.4950	.5000	1	
	2	.0100	.0225	.0400	.0625	.0900	.1225	.1600	.2025	.2500	0	2
3	0	.7290	.6141	.5120	.4219	.3430	.2746	.2160	.1664	.1250	3	
	1	.2430	.3251	.3840	.4219	.4410	.4436	.4320	.4084	.3750	2	
	2	.0270	.0574	.0960	.1406	.1890	.2389	.2880	.3341	.3750	1	
	3	.0010	.0034	.0080	.0156	.0270	.0429	.0640	.0911	.1250	0	3
4	0	.6561	.5220	.4096	.3164	.2401	.1785	.1296	.0915	.0625	4	
	1	.2916	.3685	.4096	.4219	.4116	.3845	.3456	.2995	.2500	3	
	2	.0486	.0975	.1536	.2109	.2646	.3105	.3456	.3675	.3750	2	
	3	.0036	.0115	.0256	.0469	.0756	.1115	.1536	.2005	.2500	1	
	4	.0001	.0005	.0016	.0039	.0081	.0150	.0256	.0410	.0625	0	4
5	0	.5905	.4437	.3277	.2373	.1681	.1160	.0778	.0503	.0312	5	
	1	.3280	.3915	.4096	.3955	.3601	.3124	.2592	.2059	.1562	4	
	2	.0729	.1382	.2048	.2637	.3087	.3364	.3456	.3369	.3125	3	
	3	.0081	.0244	.0512	.0879	.1323	.1811	.2304	.2757	.3125	2	
	4	.0004	.0022	.0064	.0146	.0283	.0488	.0768	.1128	.1562	1	
	5	.0000	.0001	.0003	.0010	.0024	.0053	.0102	.0185	.0312	0	5
6	0	.5314	.3771	.2621	.1780	.1176	.0754	.0467	.0277	.0156	6	
	1	.3543	.3993	.3932	.3560	.3025	.2437	.1866	.1359	.0938	5	
	2	.0984	.1762	.2458	.2966	.3241	.3280	.3110	.2780	.2344	4	
	3	.0146	.0415	.0819	.1318	.1852	.2355	.2765	.3032	.3125	3	
	4	.0012	.0055	.0154	.0330	.0595	.0951	.1382	.1861	.2344	2	
	5	.0001	.0004	.0015	.0044	.0102	.0205	.0369	.0609	.0938	1	
	6	.0000	.0000	.0001	.0002	.0007	.0018	.0041	.0083	.0156	0	6
7	0	.4783	.3206	.2097	.1335	.0824	.0490	.0280	.0152	.0078	7	
	1	.3720	.3960	.3670	.3115	.2471	.1848	.1306	.0872	.0547	6	
	2	.1240	.2097	.2753	.3115	.3177	.2985	.2613	.2140	.1641	5	
	3	.0230	.0617	.1147	.1730	.2269	.2679	.2903	.2918	.2734	4	
	4	.0026	.0109	.0287	.0577	.0972	.1442	.1935	.2388	.2734	3	
	5	.0002	.0012	.0043	.0115	.0250	.0466	.0774	.1172	.1641	2	
	6	.0000	.0001	.0004	.0013	.0036	.0084	.0172	.0320	.0547	1	
	7	.0000	.0000	.0000	.0001	.0002	.0006	.0016	.0037	.0078	0	7
8	0	.4305	.2725	.1678	.1001	.0576	.0319	.0168	.0084	.0039	8	
	1	.3826	.3847	.3355	.2670	.1977	.1373	.0896	.0548	.0312	7	
	2	.1488	.2376	.2936	.3115	.2965	.2587	.2090	.1569	.1094	6	
	3	.0331	.0839	.1468	.2076	.2541	.2786	.2787	.2568	.2188	5	
	4	.0046	.0185	.0459	.0865	.1361	.1875	.2322	.2627	.2734	4	
	5	.0004	.0026	.0092	.0231	.0467	.0808	.1239	.1719	.2188	3	
	6	.0000	.0002	.0011	.0038	.0100	.0217	.0413	.0703	.1094	2	
	7	.0000	.0000	.0001	.0004	.0012	.0033	.0079	.0164	.0312	1	
	8	.0000	.0000	.0000	.0000	.0001	.0002	.0007	.0017	.0039	0	8
9	0	.3874	.2316	.1342	.0751	.0404	.0207	.0101	.0046	.0020	9	
	1	.3874	.3679	.3020	.2253	.1556	.1004	.0605	.0339	.0176	8	
	2	.1722	.2597	.3020	.3003	.2668	.2162	.1612	.1110	.0703	7	
	3	.0446	.1069	.1762	.2336	.2668	.2716	.2508	.2119	.1641	6	
	4	.0074	.0283	.0661	.1168	.1715	.2194	.2508	.2600	.2461	5	
	5	.0008	.0050	.0165	.0389	.0735	.1181	.1672	.2128	.2461	4	
	6	.0001	.0006	.0028	.0087	.0210	.0424	.0743	.1160	.1641	3	
	7	.0000	.0000	.0003	.0012	.0039	.0098	.0212	.0407	.0703	2	
	8	.0000	.0000	.0000	.0001	.0004	.0013	.0035	.0083	.0176	1	
	9	.0000	.0000	.0000	.0000	.0000	.0001	.0003	.0008	.0020	0	9
		.90	.85	.80	.75	.70	.65	.60	.55	.50	x	n

p

n	x	.01	.02	.03	.04	.05	.06	.07	.08	.09		
10	0	.9044	.8171	.7374	.6648	.5987	.5386	.4840	.4344	.3894	10	
	1	.0914	.1667	.2281	.2770	.3151	.3438	.3643	.3777	.3851	9	
	2	.0042	.0153	.0317	.0519	.0746	.0988	.1234	.1478	.1714	8	
	3	.0001	.0008	.0026	.0058	.0105	.0168	.0248	.0343	.0452	7	
	4	.0000	.0000	.0001	.0004	.0010	.0019	.0033	.0052	.0078	6	
	5	.0000	.0000	.0000	.0000	.0001	.0001	.0003	.0005	.0009	5	
	6	.0000	.0000	.0000	.0000	.0000	.0000	.0000	.0000	.0001	4	
	7	.0000	.0000	.0000	.0000	.0000	.0000	.0000	.0000	.0000	3	
	8	.0000	.0000	.0000	.0000	.0000	.0000	.0000	.0000	.0000	2	
	9	.0000	.0000	.0000	.0000	.0000	.0000	.0000	.0000	.0000	1	
	10	.0000	.0000	.0000	.0000	.0000	.0000	.0000	.0000	.0000	0	10
12	0	.8864	.7847	.6938	.6127	.5404	.4759	.4186	.3677	.3225	12	
	1	.1074	.1922	.2575	.3064	.3413	.3645	.3781	.3837	.3827	11	
	2	.0060	.0216	.0438	.0702	.0988	.1280	.1565	.1835	.2082	10	
	3	.0002	.0015	.0045	.0098	.0173	.0272	.0393	.0532	.0686	9	
	4	.0000	.0001	.0003	.0009	.0021	.0035	.0067	.0104	.0153	8	
	5	.0000	.0000	.0000	.0001	.0002	.0004	.0008	.0014	.0024	7	
	6	.0000	.0000	.0000	.0000	.0000	.0000	.0001	.0001	.0003	6	
	7	.0000	.0000	.0000	.0000	.0000	.0000	.0000	.0000	.0000	5	
	8	.0000	.0000	.0000	.0000	.0000	.0000	.0000	.0000	.0000	4	
	9	.0000	.0000	.0000	.0000	.0000	.0000	.0000	.0000	.0000	3	
	10	.0000	.0000	.0000	.0000	.0000	.0000	.0000	.0000	.0000	2	
	11	.0000	.0000	.0000	.0000	.0000	.0000	.0000	.0000	.0000	1	
	12	.0000	.0000	.0000	.0000	.0000	.0000	.0000	.0000	.0000	0	12
15	0	.8601	.7386	.6333	.5421	.4633	.3953	.3367	.2863	.2430	15	
	1	.1303	.2261	.2938	.3388	.3658	.3785	.3801	.3734	.3605	14	
	2	.0092	.0323	.0636	.0988	.1348	.1691	.2003	.2273	.2496	13	
	3	.0004	.0029	.0085	.0178	.0307	.0468	.0653	.0857	.1070	12	
	4	.0000	.0002	.0008	.0022	.0049	.0090	.0148	.0223	.0317	11	
	5	.0000	.0000	.0001	.0002	.0006	.0013	.0024	.0043	.0069	10	
	6	.0000	.0000	.0000	.0000	.0000	.0001	.0003	.0006	.0011	9	
	7	.0000	.0000	.0000	.0000	.0000	.0000	.0000	.0000	.0001	8	
	8	.0000	.0000	.0000	.0000	.0000	.0000	.0000	.0000	.0000	7	
	9	.0000	.0000	.0000	.0000	.0000	.0000	.0000	.0000	.0000	6	
	10	.0000	.0000	.0000	.0000	.0000	.0000	.0000	.0000	.0000	5	
	11	.0000	.0000	.0000	.0000	.0000	.0000	.0000	.0000	.0000	4	
	12	.0000	.0000	.0000	.0000	.0000	.0000	.0000	.0000	.0000	3	
	13	.0000	.0000	.0000	.0000	.0000	.0000	.0000	.0000	.0000	2	
	14	.0000	.0000	.0000	.0000	.0000	.0000	.0000	.0000	.0000	1	
	15	.0000	.0000	.0000	.0000	.0000	.0000	.0000	.0000	.0000	0	15
20	0	.8179	.6676	.5438	.4420	.3585	.2901	.2342	.1887	.1516	20	
	1	.1652	.2725	.3364	.3683	.3774	.3703	.3526	.3282	.3000	19	
	2	.0159	.0528	.0988	.1458	.1887	.2246	.2521	.2711	.2818	18	
	3	.0010	.0065	.0183	.0364	.0596	.0860	.1139	.1414	.1672	17	
	4	.0000	.0006	.0024	.0065	.0133	.0233	.0364	.0523	.0703	16	
	5	.0000	.0000	.0002	.0009	.0022	.0048	.0088	.0145	.0222	15	
	6	.0000	.0000	.0000	.0001	.0003	.0008	.0017	.0032	.0055	14	
	7	.0000	.0000	.0000	.0000	.0000	.0001	.0002	.0005	.0011	13	
	8	.0000	.0000	.0000	.0000	.0000	.0000	.0000	.0000	.0002	12	
	9	.0000	.0000	.0000	.0000	.0000	.0000	.0000	.0000	.0000	11	
	10	.0000	.0000	.0000	.0000	.0000	.0000	.0000	.0000	.0000	10	
	11	.0000	.0000	.0000	.0000	.0000	.0000	.0000	.0000	.0000	9	
	12	.0000	.0000	.0000	.0000	.0000	.0000	.0000	.0000	.0000	8	
	13	.0000	.0000	.0000	.0000	.0000	.0000	.0000	.0000	.0000	7	
	14	.0000	.0000	.0000	.0000	.0000	.0000	.0000	.0000	.0000	6	
	15	.0000	.0000	.0000	.0000	.0000	.0000	.0000	.0000	.0000	5	
	16	.0000	.0000	.0000	.0000	.0000	.0000	.0000	.0000	.0000	4	
	17	.0000	.0000	.0000	.0000	.0000	.0000	.0000	.0000	.0000	3	
	18	.0000	.0000	.0000	.0000	.0000	.0000	.0000	.0000	.0000	2	
	19	.0000	.0000	.0000	.0000	.0000	.0000	.0000	.0000	.0000	1	
	20	.0000	.0000	.0000	.0000	.0000	.0000	.0000	.0000	.0000	0	20
		.99	.98	.97	.96	.95	.94	.93	.92	.91	x	n

n	x	.10	.15	.20	.25	.30	.35	.40	.45	.50		
10	0	.3487	.1969	.1074	.0563	.0282	.0135	.0060	.0025	.0010	10	
	1	.3874	.3474	.2684	.1877	.1211	.0725	.0403	.0207	.0098	9	
	2	.1937	.2759	.3020	.2816	.2335	.1757	.1209	.0763	.0439	8	
	3	.0574	.1298	.2013	.2503	.2668	.2522	.2150	.1665	.1172	7	
	4	.0112	.0401	.0881	.1460	.2001	.2377	.2508	.2384	.2051	6	
	5	.0015	.0085	.0264	.0584	.1029	.1536	.2007	.2340	.2461	5	
	6	.0001	.0012	.0055	.0162	.0368	.0689	.1115	.1596	.2051	4	
	7	.0000	.0001	.0008	.0031	.0090	.0212	.0425	.0746	.1172	3	
	8	.0000	.0000	.0001	.0004	.0014	.0043	.0106	.0229	.0439	2	
	9	.0000	.0000	.0000	.0000	.0001	.0005	.0016	.0042	.0098	1	
	10	.0000	.0000	.0000	.0000	.0000	.0000	.0001	.0003	.0010	0	10
12	0	.2824	.1422	.0687	.0317	.0138	.0057	.0022	.0008	.0002	12	
	1	.3766	.3012	.2062	.1267	.0712	.0368	.0174	.0075	.0029	11	
	2	.2301	.2924	.2835	.2323	.1678	.1088	.0639	.0339	.0161	10	
	3	.0852	.1720	.2362	.2581	.2397	.1954	.1419	.0923	.0537	9	
	4	.0213	.0683	.1329	.1936	.2311	.2367	.2128	.1700	.1208	8	
	5	.0038	.0193	.0532	.1032	.1585	.2039	.2270	.2225	.1934	7	
	6	.0005	.0040	.0155	.0401	.0792	.1281	.1766	.2124	.2256	6	
	7	.0000	.0006	.0033	.0115	.0291	.0591	.1009	.1489	.1934	5	
	8	.0000	.0001	.0005	.0024	.0078	.0199	.0420	.0762	.1208	4	
	9	.0000	.0000	.0001	.0004	.0015	.0048	.0125	.0277	.0537	3	
	10	.0000	.0000	.0000	.0000	.0002	.0008	.0025	.0068	.0161	2	
	11	.0000	.0000	.0000	.0000	.0000	.0001	.0003	.0010	.0029	1	
	12	.0000	.0000	.0000	.0000	.0000	.0000	.0000	.0001	.0002	0	12
15	0	.2059	.0874	.0352	.0134	.0047	.0016	.0005	.0001	.0000	15	
	1	.3432	.2312	.1319	.0668	.0305	.0126	.0047	.0016	.0005	14	
	2	.2669	.2856	.2309	.1559	.0916	.0476	.0219	.0090	.0032	13	
	3	.1285	.2184	.2501	.2252	.1700	.1110	.0634	.0318	.0139	12	
	4	.0428	.1156	.1876	.2252	.2186	.1792	.1268	.0780	.0417	11	
	5	.0105	.0449	.1032	.1651	.2061	.2123	.1859	.1404	.0916	10	
	6	.0019	.0132	.0430	.0917	.1472	.1906	.2066	.1914	.1527	9	
	7	.0003	.0030	.0138	.0393	.0811	.1319	.1771	.2013	.1964	8	
	8	.0000	.0005	.0035	.0131	.0348	.0710	.1181	.1647	.1964	7	
	9	.0000	.0001	.0007	.0034	.0116	.0298	.0612	.1048	.1527	6	
	10	.0000	.0000	.0001	.0007	.0030	.0096	.0245	.0515	.0916	5	
	11	.0000	.0000	.0000	.0001	.0006	.0024	.0074	.0191	.0417	4	
	12	.0000	.0000	.0000	.0000	.0001	.0004	.0016	.0052	.0139	3	
	13	.0000	.0000	.0000	.0000	.0000	.0001	.0003	.0010	.0032	2	
	14	.0000	.0000	.0000	.0000	.0000	.0000	.0000	.0001	.0005	1	
	15	.0000	.0000	.0000	.0000	.0000	.0000	.0000	.0000	.0000	0	15
20	0	.1216	.0388	.0115	.0032	.0008	.0002	.0000	.0000	.0000	20	
	1	.2702	.1368	.0576	.0211	.0068	.0020	.0005	.0001	.0000	19	
	2	.2852	.2293	.1369	.0669	.0278	.0100	.0031	.0008	.0002	18	
	3	.1901	.2428	.2054	.1339	.0716	.0323	.0123	.0040	.0011	17	
	4	.0898	.1821	.2182	.1897	.1304	.0738	.0350	.0139	.0046	16	
	5	.0319	.1028	.1746	.2023	.1789	.1272	.0746	.0365	.0148	15	
	6	.0089	.0454	.1091	.1686	.1916	.1712	.1244	.0746	.0370	14	
	7	.0020	.0160	.0545	.1124	.1643	.1844	.1659	.1221	.0739	13	
	8	.0004	.0046	.0222	.0609	.1144	.1614	.1797	.1623	.1201	12	
	9	.0001	.0011	.0074	.0271	.0654	.1158	.1597	.1771	.1602	11	
	10	.0000	.0002	.0020	.0099	.0308	.0686	.1171	.1593	.1762	10	
	11	.0000	.0000	.0005	.0030	.0120	.0336	.0710	.1185	.1602	9	
	12	.0000	.0000	.0001	.0008	.0039	.0136	.0355	.0727	.1201	8	
	13	.0000	.0000	.0000	.0002	.0010	.0045	.0146	.0366	.0739	7	
	14	.0000	.0000	.0000	.0000	.0002	.0012	.0049	.0150	.0370	6	
	15	.0000	.0000	.0000	.0000	.0000	.0003	.0013	.0049	.0148	5	
	16	.0000	.0000	.0000	.0000	.0000	.0000	.0003	.0013	.0046	4	
	17	.0000	.0000	.0000	.0000	.0000	.0000	.0000	.0002	.0011	3	
	18	.0000	.0000	.0000	.0000	.0000	.0000	.0000	.0000	.0002	2	
	19	.0000	.0000	.0000	.0000	.0000	.0000	.0000	.0000	.0000	1	
	20	.0000	.0000	.0000	.0000	.0000	.0000	.0000	.0000	.0000	0	20
		.90	.85	.80	.75	.70	.65	.60	.55	.50	x	n

p

Poisson Probabilities calculated from: $P(x) = \dfrac{\lambda^x e^{-\lambda}}{x!}$

x	.1	.2	.3	.4	.5	.6	.7	.8	.9
0	.9048	.8187	.7408	.6703	.6065	.5488	.4966	.4493	.4066
1	.0905	.1637	.2222	.2681	.3033	.3293	.3476	.3595	.3659
2	.0045	.0164	.0333	.0536	.0758	.0988	.1217	.1438	.1647
3	.0002	.0011	.0033	.0072	.0126	.0198	.0284	.0383	.0494
4	.0000	.0001	.0003	.0007	.0016	.0030	.0050	.0077	.0111
5	.0000	.0000	.0000	.0001	.0002	.0004	.0007	.0012	.0020
6	.0000	.0000	.0000	.0000	.0000	.0000	.0001	.0002	.0003

x	1.0	1.5	2.0	2.5	3.0	3.5	4.0	4.5	5.0
0	.3679	.2231	.1353	.0821	.0498	.0302	.0183	.0111	.0067
1	.3679	.3347	.2707	.2052	.1494	.1057	.0733	.0500	.0337
2	.1839	.2510	.2707	.2565	.2240	.1850	.1465	.1125	.0842
3	.0613	.1255	.1804	.2138	.2240	.2158	.1954	.1687	.1404
4	.0153	.0471	.0902	.1336	.1680	.1888	.1954	.1898	.1755
5	.0031	.0141	.0361	.0668	.1008	.1322	.1563	.1708	.1755
6	.0005	.0035	.0120	.0278	.0504	.0771	.1042	.1281	.1462
7	.0001	.0008	.0034	.0099	.0216	.0385	.0595	.0824	.1044
8	.0000	.0001	.0009	.0031	.0081	.0169	.0298	.0463	.0653
9	.0000	.0000	.0002	.0009	.0027	.0066	.0132	.0232	.0363
10	.0000	.0000	.0000	.0002	.0008	.0023	.0053	.0104	.0181
11	.0000	.0000	.0000	.0000	.0002	.0007	.0019	.0043	.0082
12	.0000	.0000	.0000	.0000	.0001	.0002	.0006	.0016	.0034
13	.0000	.0000	.0000	.0000	.0000	.0001	.0002	.0006	.0013
14	.0000	.0000	.0000	.0000	.0000	.0000	.0001	.0002	.0005
15	.0000	.0000	.0000	.0000	.0000	.0000	.0000	.0001	.0002

x	5.5	6.0	6.5	7.0	7.5	8.0	9.0	10.0	11.0
0	.0041	.0025	.0015	.0009	.0006	.0003	.0001	.0000	.0000
1	.0225	.0149	.0098	.0064	.0041	.0027	.0011	.0005	.0002
2	.0618	.0446	.0318	.0223	.0156	.0107	.0050	.0023	.0010
3	.1133	.0892	.0688	.0521	.0389	.0286	.0150	.0076	.0037
4	.1558	.1339	.1118	.0912	.0729	.0573	.0337	.0189	.0102
5	.1714	.1606	.1454	.1277	.1094	.0916	.0607	.0378	.0224
6	.1571	.1606	.1575	.1490	.1367	.1221	.0911	.0631	.0411
7	.1234	.1377	.1462	.1490	.1465	.1396	.1171	.0901	.0646
8	.0849	.1033	.1188	.1304	.1373	.1396	.1318	.1126	.0888
9	.0519	.0688	.0858	.1014	.1144	.1241	.1318	.1251	.1085
10	.0285	.0413	.0558	.0710	.0858	.0993	.1186	.1251	.1194
11	.0143	.0225	.0330	.0452	.0585	.0722	.0970	.1137	.1194
12	.0065	.0113	.0179	.0263	.0366	.0481	.0728	.0948	.1094
13	.0028	.0052	.0089	.0142	.0211	.0296	.0504	.0729	.0926
14	.0011	.0022	.0041	.0071	.0113	.0169	.0324	.0521	.0728
15	.0004	.0009	.0018	.0033	.0057	.0090	.0194	.0347	.0534
16	.0001	.0003	.0007	.0014	.0026	.0045	.0109	.0217	.0367
17	.0000	.0001	.0003	.0006	.0012	.0021	.0058	.0128	.0237
18	.0000	.0000	.0001	.0002	.0005	.0009	.0029	.0071	.0145
19	.0000	.0000	.0000	.0001	.0002	.0004	.0014	.0037	.0084
20	.0000	.0000	.0000	.0000	.0001	.0002	.0006	.0019	.0046
21	.0000	.0000	.0000	.0000	.0000	.0001	.0003	.0009	.0024
22	.0000	.0000	.0000	.0000	.0000	.0000	.0001	.0004	.0012
23	.0000	.0000	.0000	.0000	.0000	.0000	.0000	.0002	.0006
24	.0000	.0000	.0000	.0000	.0000	.0000	.0000	.0001	.0003
25	.0000	.0000	.0000	.0000	.0000	.0000	.0000	.0000	.0001

x	12	13	14	15	16	17	18	19	20
0	.0000	.0000	.0000	.0000	.0000	.0000	.0000	.0000	.0000
1	.0001	.0000	.0000	.0000	.0000	.0000	.0000	.0000	.0000
2	.0004	.0002	.0001	.0000	.0000	.0000	.0000	.0000	.0000
3	.0018	.0008	.0004	.0002	.0001	.0000	.0000	.0000	.0000
4	.0053	.0027	.0013	.0006	.0003	.0001	.0001	.0000	.0000
5	.0127	.0070	.0037	.0019	.0010	.0005	.0002	.0001	.0001
6	.0255	.0152	.0087	.0048	.0026	.0014	.0007	.0004	.0002
7	.0437	.0281	.0174	.0104	.0060	.0034	.0019	.0010	.0005
8	.0655	.0457	.0304	.0194	.0120	.0072	.0042	.0024	.0013
9	.0874	.0661	.0473	.0324	.0213	.0135	.0083	.0050	.0029
10	.1048	.0859	.0663	.0486	.0341	.0230	.0150	.0095	.0058
11	.1144	.1015	.0844	.0663	.0496	.0355	.0245	.0164	.0106
12	.1144	.1099	.0984	.0829	.0661	.0504	.0368	.0259	.0176
13	.1056	.1099	.1060	.0956	.0814	.0658	.0509	.0378	.0271
14	.0905	.1021	.1060	.1024	.0930	.0800	.0655	.0514	.0387
15	.0724	.0885	.0989	.1024	.0992	.0906	.0786	.0650	.0516
16	.0543	.0719	.0866	.0960	.0992	.0963	.0884	.0772	.0646
17	.0383	.0550	.0713	.0847	.0934	.0963	.0936	.0863	.0760
18	.0255	.0397	.0554	.0706	.0830	.0909	.0936	.0911	.0844
19	.0161	.0272	.0409	.0557	.0699	.0814	.0887	.0911	.0888
20	.0097	.0177	.0286	.0418	.0559	.0692	.0798	.0866	.0888
21	.0055	.0109	.0191	.0299	.0426	.0560	.0684	.0783	.0846
22	.0030	.0065	.0121	.0204	.0310	.0433	.0560	.0676	.0769
23	.0016	.0037	.0074	.0133	.0216	.0320	.0438	.0559	.0669
24	.0008	.0020	.0043	.0083	.0144	.0226	.0328	.0442	.0557
25	.0004	.0010	.0024	.0050	.0092	.0154	.0237	.0336	.0446
26	.0002	.0005	.0013	.0029	.0057	.0101	.0164	.0246	.0343
27	.0001	.0002	.0007	.0016	.0034	.0063	.0109	.0173	.0254
28	.0000	.0001	.0003	.0009	.0019	.0038	.0070	.0117	.0181
29	.0000	.0001	.0002	.0004	.0011	.0023	.0044	.0077	.0125
30	.0000	.0000	.0001	.0002	.0006	.0013	.0026	.0049	.0083
31	.0000	.0000	.0000	.0001	.0003	.0007	.0015	.0030	.0054
32	.0000	.0000	.0000	.0001	.0001	.0004	.0009	.0018	.0034
33	.0000	.0000	.0000	.0000	.0001	.0002	.0005	.0010	.0020
34	.0000	.0000	.0000	.0000	.0000	.0001	.0002	.0006	.0012
35	.0000	.0000	.0000	.0000	.0000	.0001	.0001	.0003	.0007
36	.0000	.0000	.0000	.0000	.0000	.0000	.0001	.0002	.0004
37	.0000	.0000	.0000	.0000	.0000	.0000	.0000	.0001	.0002
38	.0000	.0000	.0000	.0000	.0000	.0000	.0000	.0000	.0001
39	.0000	.0000	.0000	.0000	.0000	.0000	.0000	.0000	.0001

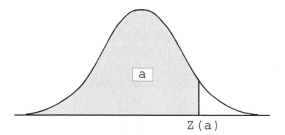

Cumulative Probabilities

Each 4-digit entry is the area "a" under the standard normal curve
from -∞ to Z(a). Reference the sketch above.

Z	.00	.01	.02	.03	.04	.05	.06	.07	.08	.09
.0	.5000	.5040	.5080	.5120	.5160	.5199	.5239	.5279	.5319	.5359
.1	.5398	.5438	.5478	.5517	.5557	.5596	.5636	.5675	.5714	.5753
.2	.5793	.5832	.5871	.5910	.5948	.5987	.6026	.6064	.6103	.6141
.3	.6179	.6217	.6255	.6293	.6331	.6368	.6406	.6443	.6480	.6517
.4	.6554	.6591	.6628	.6664	.6700	.6736	.6772	.6808	.6844	.6879
.5	.6915	.6950	.6985	.7019	.7054	.7088	.7123	.7157	.7190	.7224
.6	.7257	.7291	.7324	.7357	.7389	.7422	.7454	.7486	.7517	.7549
.7	.7580	.7611	.7642	.7673	.7704	.7734	.7764	.7794	.7823	.7852
.8	.7881	.7910	.7939	.7967	.7995	.8023	.8051	.8078	.8106	.8133
.9	.8159	.8186	.8212	.8238	.8264	.8289	.8315	.8340	.8365	.8389
1.0	.8413	.8438	.8461	.8485	.8508	.8531	.8554	.8577	.8599	.8621
1.1	.8643	.8665	.8686	.8708	.8729	.8749	.8770	.8790	.8810	.8830
1.2	.8849	.8869	.8888	.8907	.8925	.8944	.8962	.8980	.8997	.9015
1.3	.9032	.9049	.9066	.9082	.9099	.9115	.9131	.9147	.9162	.9177
1.4	.9192	.9207	.9222	.9236	.9251	.9265	.9279	.9292	.9306	.9319
1.5	.9332	.9345	.9357	.9370	.9382	.9394	.9406	.9418	.9429	.9441
1.6	.9452	.9463	.9474	.9484	.9495	.9505	.9515	.9525	.9535	.9545
1.7	.9554	.9564	.9573	.9582	.9591	.9599	.9608	.9616	.9625	.9633
1.8	.9641	.9649	.9656	.9664	.9671	.9678	.9686	.9693	.9699	.9706
1.9	.9713	.9719	.9726	.9732	.9738	.9744	.9750	.9756	.9761	.9767
2.0	.9772	.9778	.9783	.9788	.9793	.9798	.9803	.9808	.9812	.9817
2.1	.9821	.9826	.9830	.9834	.9838	.9842	.9846	.9850	.9854	.9857
2.2	.9861	.9864	.9868	.9871	.9875	.9878	.9881	.9884	.9887	.9890
2.3	.9893	.9896	.9898	.9901	.9904	.9906	.9909	.9911	.9913	.9916
2.4	.9918	.9920	.9922	.9925	.9927	.9929	.9931	.9932	.9934	.9936

z	.00	.01	.02	.03	.04	.05	.06	.07	.08	.09
2.5	.9938	.9940	.9941	.9943	.9945	.9946	.9948	.9949	.9951	.9952
2.6	.9953	.9955	.9956	.9957	.9959	.9960	.9961	.9962	.9963	.9964
2.7	.9965	.9966	.9967	.9968	.9969	.9970	.9971	.9972	.9973	.9974
2.8	.9974	.9975	.9976	.9977	.9977	.9978	.9979	.9979	.9980	.9981
2.9	.9981	.9982	.9982	.9983	.9984	.9984	.9985	.9985	.9986	.9986
3.0	.9987	.9987	.9987	.9988	.9988	.9989	.9989	.9989	.9990	.9990
3.1	.9990	.9991	.9991	.9991	.9992	.9992	.9992	.9992	.9993	.9993
3.2	.9993	.9993	.9994	.9994	.9994	.9994	.9994	.9995	.9995	.9995
3.3	.9995	.9995	.9995	.9996	.9996	.9996	.9996	.9996	.9996	.9997
3.4	.9997	.9997	.9997	.9997	.9997	.9997	.9997	.9997	.9997	.9998

Selected Percentiles

Each entry is $Z(a)$ where $P[\ Z \leq Z(a)\ ] = a$. For example,
$P(Z \leq 1.645) = .95$ so $Z(.95) = 1.645$.

a	.10	.05	.025	.02	.01	.005	.001
Z(a)	−1.282	−1.645	−1.960	−2.054	−2.326	−2.576	−3.090

a	.90	.95	.975	.98	.99	.995	.999
Z(a)	1.282	1.645	1.960	2.054	2.326	2.576	3.090

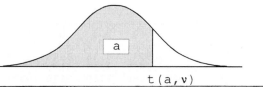

$$t(a,\nu)$$

df				a			
ν	.75	.90	.95	.975	.99	.995	.9995
1	1.000	3.078	6.314	12.706	31.821	63.657	636.619
2	0.816	1.886	2.920	4.303	6.965	9.925	31.599
3	0.765	1.638	2.353	3.182	4.541	5.841	12.924
4	0.741	1.533	2.132	2.776	3.747	4.604	8.610
5	0.727	1.476	2.015	2.571	3.365	4.032	6.869
6	0.718	1.440	1.943	2.447	3.143	3.707	5.959
7	0.711	1.415	1.895	2.365	2.998	3.499	5.408
8	0.706	1.397	1.860	2.306	2.896	3.355	5.041
9	0.703	1.383	1.833	2.262	2.821	3.250	4.781
10	0.700	1.372	1.812	2.228	2.764	3.169	4.587
11	0.697	1.363	1.796	2.201	2.718	3.106	4.437
12	0.695	1.356	1.782	2.179	2.681	3.055	4.318
13	0.694	1.350	1.771	2.160	2.650	3.012	4.221
14	0.692	1.345	1.761	2.145	2.624	2.977	4.140
15	0.691	1.341	1.753	2.131	2.602	2.947	4.073
16	0.690	1.337	1.746	2.120	2.583	2.921	4.015
17	0.689	1.333	1.740	2.110	2.567	2.898	3.965
18	0.688	1.330	1.734	2.101	2.552	2.878	3.922
19	0.688	1.328	1.729	2.093	2.539	2.861	3.883
20	0.687	1.325	1.725	2.086	2.528	2.845	3.850
21	0.686	1.323	1.721	2.080	2.518	2.831	3.819
22	0.686	1.321	1.717	2.074	2.508	2.819	3.792
23	0.685	1.319	1.714	2.069	2.500	2.807	3.768
24	0.685	1.318	1.711	2.064	2.492	2.797	3.745
25	0.684	1.316	1.708	2.060	2.485	2.787	3.725
26	0.684	1.315	1.706	2.056	2.479	2.779	3.707
27	0.684	1.314	1.703	2.052	2.473	2.771	3.690
28	0.683	1.313	1.701	2.048	2.467	2.763	3.674
29	0.683	1.311	1.699	2.045	2.462	2.756	3.659
30	0.683	1.310	1.697	2.042	2.457	2.750	3.646
40	0.681	1.303	1.684	2.021	2.423	2.704	3.551
60	0.679	1.296	1.671	2.000	2.390	2.660	3.460
120	0.677	1.289	1.658	1.980	2.358	2.617	3.373
∞	0.674	1.282	1.645	1.960	2.326	2.576	3.291

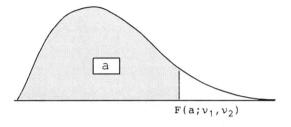

$$F(a;\nu_1,\nu_2)$$

The F distribution is skewed to the right and has two parameters:

    ν_1 = numerator degrees of freedom, and

    ν_2 = denominator degrees of freedom.

The random variable F is denoted by $F(\nu_1,\nu_2)$. Entries in the tables denote the positive values $F(a;\nu_1,\nu_2)$ such that

$$P[F(\nu_1,\nu_2) \le F(a;\nu_1,\nu_2)] = a \quad \text{(reference the above sketch)}.$$

The superscripted entries, like 4053^2, should be read as 405300.

Because the F distribution has two parameters, any set of tables containing even a few percentiles will be extensive. We have included the following percentiles in the pages that follow: 50th, 75th, 90th, 95th, 97.5^{th}, 99th, and 99.9^{th}. This set of seven percentiles is given for each of 20 different values that ν_1 and ν_2 can assume (i.e., 400 possible combinations of ν_1 and ν_2).

Percentiles less than the 50th can be found by using the following relationship:

$$F(a;\nu_1,\nu_2) = \frac{1}{F(1-a;\nu_2,\nu_1)}$$

Please note that the order of the parameters is reversed in the two F expressions. For example, if a = .05, ν_1 = 8, and ν_2 = 30, then

$$F(.05;8,30) = \frac{1}{F(.95;30,8)} = \frac{1}{3.08} = .325$$

ν_2	a	ν_1 1	2	3	4	5	6	7	8	9	10
1	.50	1.00	1.50	1.71	1.82	1.89	1.94	1.98	2.00	2.03	2.04
	.75	5.83	7.50	8.20	8.58	8.82	8.98	9.10	9.19	9.26	9.32
	.90	39.9	49.5	53.6	55.8	57.2	58.2	58.9	59.4	59.9	60.2
	.95	161	200	216	225	230	234	237	239	241	242
	.975	648	800	864	900	922	937	948	957	963	969
	.99	4052	5000	5403	5625	5764	5859	5928	5982	6022	6056
	.999	4053^2	5000^2	5404^2	5625^2	5764^2	5859^2	5929^2	5981^2	6023^2	6056^2
2	.50	.667	1.00	1.13	1.21	1.25	1.28	1.30	1.32	1.33	1.34
	.75	2.57	3.00	3.15	3.23	3.28	3.31	3.34	3.35	3.37	3.38
	.90	8.53	9.00	9.16	9.24	9.29	9.33	9.35	9.37	9.38	9.39
	.95	18.5	19.0	19.2	19.2	19.3	19.3	19.4	19.4	19.4	19.4
	.975	38.5	39.0	39.2	39.2	39.3	39.3	39.4	39.4	39.4	39.4
	.99	98.5	99.0	99.2	99.2	99.3	99.3	99.4	99.4	99.4	99.4
	.999	998	999	999	999	999	999	999	999	999	999
3	.50	.585	.881	1.00	1.06	1.10	1.13	1.15	1.16	1.17	1.18
	.75	2.02	2.28	2.36	2.39	2.41	2.42	2.43	2.44	2.44	2.44
	.90	5.54	5.46	5.39	5.34	5.31	5.28	5.27	5.25	5.24	5.23
	.95	10.1	9.55	9.28	9.12	9.01	8.94	8.89	8.85	8.81	8.79
	.975	17.4	16.0	15.4	15.1	14.9	14.7	14.6	14.5	14.5	14.4
	.99	34.1	30.8	29.5	28.7	28.2	27.9	27.7	27.5	27.3	27.2
	.999	167	149	141	137	135	133	132	131	130	129
4	.50	.549	.828	.941	1.00	1.04	1.06	1.08	1.09	1.10	1.11
	.75	1.81	2.00	2.05	2.06	2.07	2.08	2.08	2.08	2.08	2.08
	.90	4.54	4.32	4.19	4.11	4.05	4.01	3.98	3.95	3.94	3.92
	.95	7.71	6.94	6.59	6.39	6.26	6.16	6.09	6.04	6.00	5.96
	.975	12.2	10.6	9.98	9.60	9.36	9.20	9.07	8.98	8.90	8.84
	.99	21.2	18.0	16.7	16.0	15.5	15.2	15.0	14.8	14.7	14.5
	.999	74.1	61.2	56.2	53.4	51.7	50.5	49.7	49.0	48.5	48.0
5	.50	.528	.799	.907	.965	1.00	1.02	1.04	1.05	1.06	1.07
	.75	1.69	1.85	1.88	1.89	1.89	1.89	1.89	1.89	1.89	1.89
	.90	4.06	3.78	3.62	3.52	3.45	3.40	3.37	3.34	3.32	3.30
	.95	6.61	5.79	5.41	5.19	5.05	4.95	4.88	4.82	4.77	4.74
	.975	10.0	8.43	7.76	7.39	7.15	6.98	6.85	6.76	6.68	6.62
	.99	16.3	13.3	12.1	11.4	11.0	10.7	10.5	10.3	10.2	10.1
	.999	47.2	37.1	33.2	31.1	29.7	28.8	28.2	27.6	27.2	26.9
6	.50	.515	.780	.886	.942	.977	1.00	1.02	1.03	1.04	1.05
	.75	1.62	1.76	1.78	1.79	1.79	1.78	1.78	1.78	1.77	1.77
	.90	3.78	3.46	3.29	3.18	3.11	3.05	3.01	2.98	2.96	2.94
	.95	5.99	5.14	4.76	4.53	4.39	4.28	4.21	4.15	4.10	4.06
	.975	8.81	7.26	6.60	6.23	5.99	5.82	5.70	5.60	5.52	5.46
	.99	13.7	10.9	9.78	9.15	8.75	8.47	8.26	8.10	7.98	7.87
	.999	35.5	27.0	23.7	21.9	20.8	20.0	19.5	19.0	18.7	18.4
7	.50	.506	.767	.871	.926	.960	.983	1.00	1.01	1.02	1.03
	.75	1.57	1.70	1.72	1.72	1.71	1.71	1.70	1.70	1.69	1.69
	.90	3.59	3.26	3.07	2.96	2.88	2.83	2.78	2.75	2.72	2.70
	.95	5.59	4.74	4.35	4.12	3.97	3.87	3.79	3.73	3.68	3.64
	.975	8.07	6.54	5.89	5.52	5.29	5.12	4.99	4.90	4.82	4.76
	.99	12.2	9.55	8.45	7.85	7.46	7.19	6.99	6.84	6.72	6.62
	.999	29.2	21.7	18.8	17.2	16.2	15.5	15.0	14.6	14.3	14.1

ν_2	a	ν_1 11	12	15	20	24	30	40	60	120	∞
1	.50	2.05	2.07	2.09	2.12	2.13	2.15	2.16	2.17	2.18	2.20
	.75	9.36	9.41	9.49	9.58	9.63	9.67	9.71	9.76	9.80	9.85
	.90	60.5	60.7	61.2	61.7	62.0	62.3	62.5	62.8	63.1	63.3
	.95	243	244	246	248	249	250	251	252	253	254
	.975	973	977	985	993	997	1000	1010	1010	1010	1020
	.99	6080	6110	6160	6210	6230	6260	6290	6310	6340	6370
	.999	6090^2	6110^2	6160^2	6210^2	6230^2	6260^2	6290^2	6310^2	6340^2	6370^2
2	.50	1.35	1.36	1.38	1.39	1.40	1.41	1.42	1.43	1.43	1.44
	.75	3.39	3.39	3.41	3.43	3.43	3.44	3.45	3.46	3.47	3.48
	.90	9.40	9.41	9.42	9.44	9.45	9.46	9.47	9.47	9.48	9.49
	.95	19.4	19.4	19.4	19.4	19.5	19.5	19.5	19.5	19.5	19.5
	.975	39.4	39.4	39.4	39.4	39.5	39.5	39.5	39.5	39.5	39.5
	.99	99.4	99.4	99.4	99.4	99.5	99.5	99.5	99.5	99.5	99.5
	.999	999	999	999	999	999	999	999	999	999	999
3	.50	1.19	1.20	1.21	1.23	1.23	1.24	1.25	1.25	1.26	1.27
	.75	2.45	2.45	2.46	2.46	2.46	2.47	2.47	2.47	2.47	2.47
	.90	5.22	5.22	5.20	5.18	5.18	5.17	5.16	5.15	5.14	5.13
	.95	8.76	8.74	8.70	8.66	8.63	8.62	8.59	8.57	8.55	8.53
	.975	14.4	14.3	14.3	14.2	14.1	14.1	14.0	14.0	13.9	13.9
	.99	27.1	27.1	26.9	26.7	26.6	26.5	26.4	26.3	26.2	26.1
	.999	129	128	127	126	126	125	125	124	124	123
4	.50	1.12	1.13	1.14	1.15	1.16	1.16	1.17	1.18	1.18	1.19
	.75	2.08	2.08	2.08	2.08	2.08	2.08	2.08	2.08	2.08	2.08
	.90	3.91	3.90	3.87	3.84	3.83	3.82	3.80	3.79	3.78	3.76
	.95	5.94	5.91	5.86	5.80	5.77	5.75	5.72	5.69	5.66	5.63
	.975	8.79	8.75	8.66	8.56	8.51	8.46	8.41	8.36	8.31	8.26
	.99	14.4	14.4	14.2	14.0	13.9	13.8	13.7	13.7	13.6	13.5
	.999	47.7	47.4	46.8	46.1	45.8	45.4	45.1	44.7	44.4	44.0
5	.50	1.08	1.09	1.10	1.11	1.12	1.12	1.13	1.14	1.14	1.15
	.75	1.89	1.89	1.89	1.88	1.88	1.88	1.88	1.87	1.87	1.87
	.90	3.28	3.27	3.24	3.21	3.19	3.17	3.16	3.14	3.12	3.10
	.95	4.71	4.68	4.62	4.56	4.53	4.50	4.46	4.43	4.40	4.36
	.975	6.57	6.52	6.43	6.33	6.28	6.23	6.18	6.12	6.07	6.02
	.99	9.96	9.89	9.72	9.55	9.47	9.38	9.29	9.20	9.11	9.02
	.999	26.6	26.4	25.9	25.4	25.1	24.9	24.6	24.3	24.1	23.8
6	.50	1.05	1.06	1.07	1.08	1.09	1.10	1.10	1.11	1.12	1.12
	.75	1.77	1.77	1.76	1.76	1.75	1.75	1.75	1.74	1.74	1.74
	.90	2.92	2.90	2.87	2.84	2.82	2.80	2.78	2.76	2.74	2.72
	.95	4.03	4.00	3.94	3.87	3.84	3.81	3.77	3.74	3.70	3.67
	.975	5.41	5.37	5.27	5.17	5.12	5.07	5.01	4.96	4.90	4.85
	.99	7.79	7.72	7.56	7.40	7.31	7.23	7.14	7.06	6.97	6.88
	.999	18.2	18.0	17.6	17.1	16.9	16.7	16.4	16.2	16.0	15.7
7	.50	1.04	1.04	1.05	1.07	1.07	1.08	1.08	1.09	1.10	1.10
	.75	1.69	1.68	1.68	1.67	1.67	1.66	1.66	1.65	1.65	1.65
	.90	2.68	2.67	2.63	2.59	2.58	2.56	2.54	2.51	2.49	2.47
	.95	3.60	3.57	3.51	3.44	3.41	3.38	3.34	3.30	3.27	3.23
	.975	4.71	4.67	4.57	4.47	4.42	4.36	4.31	4.25	4.20	4.14
	.99	6.54	6.47	6.31	6.16	6.07	5.99	5.91	5.82	5.74	5.65
	.999	13.9	13.7	13.3	12.9	12.7	12.5	12.3	12.1	11.9	11.7

						v_1					
v_2	a	1	2	3	4	5	6	7	8	9	10
8	.50	.499	.757	.860	.915	.948	.971	.988	1.00	1.01	1.02
	.75	1.54	1.66	1.67	1.66	1.66	1.65	1.64	1.64	1.64	1.63
	.90	3.46	3.11	2.92	2.81	2.73	2.67	2.62	2.59	2.56	2.54
	.95	5.32	4.46	4.07	3.84	3.69	3.58	3.50	3.44	3.39	3.35
	.975	7.57	6.06	5.42	5.05	4.82	4.65	4.53	4.43	4.36	4.30
	.99	11.3	8.65	7.59	7.01	6.63	6.37	6.18	6.03	5.91	5.81
	.999	25.4	18.5	15.8	14.4	13.5	12.9	12.4	12.0	11.8	11.5
9	.50	.494	.749	.852	.906	.939	.962	.978	.990	1.00	1.01
	.75	1.51	1.62	1.63	1.63	1.62	1.61	1.60	1.60	1.59	1.59
	.90	3.36	3.01	2.81	2.69	2.61	2.55	2.51	2.47	2.44	2.42
	.95	5.12	4.26	3.86	3.63	3.48	3.37	3.29	3.23	3.18	3.14
	.975	7.21	5.71	5.08	4.72	4.48	4.32	4.20	4.10	4.03	3.96
	.99	10.6	8.02	6.99	6.42	6.06	5.80	5.61	5.47	5.35	5.26
	.999	22.9	16.4	13.9	12.6	11.7	11.1	10.7	10.4	10.1	9.89
10	.50	.490	.743	.845	.899	.932	.954	.971	.983	.992	1.00
	.75	1.49	1.60	1.60	1.59	1.59	1.58	1.57	1.56	1.56	1.55
	.90	3.28	2.92	2.73	2.61	2.52	2.46	2.41	2.38	2.35	2.32
	.95	4.96	4.10	3.71	3.48	3.33	3.22	3.14	3.07	3.02	2.98
	.975	6.94	5.46	4.83	4.47	4.24	4.07	3.95	3.85	3.78	3.72
	.99	10.0	7.56	6.55	5.99	5.64	5.39	5.20	5.06	4.94	4.85
	.999	21.0	14.9	12.6	11.3	10.5	9.92	9.52	9.20	8.96	8.75
11	.50	.486	.739	.840	.893	.926	.948	.964	.977	.986	.994
	.75	1.47	1.58	1.58	1.57	1.56	1.55	1.54	1.53	1.53	1.52
	.90	3.23	2.86	2.66	2.54	2.45	2.39	2.34	2.39	2.27	2.25
	.95	4.84	3.98	3.59	3.36	3.20	3.09	3.01	2.95	2.90	2.85
	.975	6.72	5.26	4.63	4.28	4.04	3.88	3.76	3.66	3.59	3.53
	.99	9.65	7.21	6.22	5.67	5.32	5.07	4.89	4.74	4.63	4.54
	.999	19.7	13.8	11.6	10.3	9.58	9.05	8.66	8.35	8.12	7.92
12	.50	.484	.735	.835	.888	.921	.943	.959	.972	.981	.989
	.75	1.46	1.56	1.56	1.55	1.54	1.53	1.52	1.51	1.51	1.50
	.90	3.18	2.81	2.61	2.48	2.39	2.33	2.28	2.24	2.21	2.19
	.95	4.75	3.89	3.49	3.26	3.11	3.00	2.91	2.85	2.80	2.75
	.975	6.55	5.10	4.47	4.12	3.89	3.73	3.61	3.51	3.44	3.37
	.99	9.33	6.93	5.95	5.41	5.06	4.82	4.64	4.50	4.39	4.30
	.999	18.6	13.0	10.8	9.63	8.89	8.38	8.00	7.71	7.48	7.29
15	.50	.478	.726	.826	.878	.911	.933	.948	.960	.970	.977
	.75	1.43	1.52	1.52	1.51	1.49	1.48	1.47	1.46	1.46	1.45
	.90	3.07	2.70	2.49	2.36	2.27	2.21	2.16	2.12	2.09	2.06
	.95	4.54	3.68	3.29	3.06	2.90	2.79	2.71	2.64	2.59	2.54
	.975	6.20	4.76	4.15	3.80	3.58	3.41	3.29	3.20	3.12	3.06
	.99	8.68	6.36	5.42	4.89	4.56	4.32	4.14	4.00	3.89	3.80
	.999	16.6	11.3	9.34	8.25	7.57	7.09	6.74	6.47	6.26	6.08
20	.50	.472	.718	.816	.868	.900	.922	.938	.950	.959	.966
	.75	1.40	1.49	1.48	1.47	1.45	1.44	1.43	1.42	1.41	1.40
	.90	2.97	2.59	2.38	2.25	2.16	2.09	2.04	2.00	1.96	1.94
	.95	4.35	3.49	3.10	2.87	2.71	2.60	2.51	2.45	2.39	2.35
	.975	5.87	4.46	3.86	3.51	3.29	3.13	3.01	2.91	2.84	2.77
	.99	8.10	5.85	4.94	4.43	4.10	3.87	3.70	3.56	3.46	3.37
	.999	14.8	9.95	8.10	7.10	6.46	6.02	5.69	5.44	5.24	5.08

ν_2	a	ν_1									
		11	12	15	20	24	30	40	60	120	∞
8	.50	1.02	1.03	1.04	1.05	1.06	1.07	1.07	1.08	1.08	1.09
	.75	1.63	1.62	1.62	1.61	1.60	1.60	1.59	1.59	1.58	1.58
	.90	2.52	2.50	2.46	2.42	2.40	2.38	2.36	2.34	2.32	2.29
	.95	3.31	3.28	3.22	3.15	3.12	3.08	3.04	3.01	2.97	2.93
	.975	4.24	4.20	4.10	4.00	3.95	3.89	3.84	3.78	3.73	3.67
	.99	5.73	5.67	5.52	5.36	5.28	5.20	5.12	5.03	4.95	4.86
	.999	11.4	11.2	10.8	10.5	10.3	10.1	9.92	9.73	9.54	9.34
9	.50	1.01	1.02	1.03	1.04	1.05	1.05	1.06	1.07	1.07	1.08
	.75	1.58	1.58	1.57	1.56	1.56	1.55	1.55	1.54	1.53	1.53
	.90	2.40	2.38	2.34	2.30	2.28	2.25	2.23	2.21	2.18	2.16
	.95	3.10	3.07	3.01	2.94	2.90	2.86	2.83	2.79	2.75	2.71
	.975	3.91	3.87	3.77	3.67	3.61	3.56	3.51	3.45	3.39	3.33
	.99	5.18	5.11	4.96	4.81	4.73	4.65	4.57	4.48	4.40	4.31
	.999	9.71	9.57	9.24	8.90	8.72	8.55	8.37	8.19	8.00	7.81
10	.50	1.01	1.01	1.02	1.03	1.04	1.05	1.05	1.06	1.06	1.07
	.75	1.55	1.54	1.53	1.52	1.52	1.51	1.51	1.50	1.49	1.48
	.90	2.30	2.28	2.24	2.20	2.18	2.16	2.13	2.11	2.08	2.06
	.95	2.94	2.91	2.85	2.77	2.74	2.70	2.66	2.62	2.58	2.54
	.975	3.66	3.62	3.52	3.42	3.37	3.31	3.26	3.20	3.14	3.08
	.99	4.77	4.71	4.56	4.41	4.33	4.25	4.17	4.08	4.00	3.91
	.999	8.58	8.44	8.13	7.80	7.64	7.47	7.30	7.12	6.94	6.76
11	.50	1.00	1.01	1.02	1.03	1.03	1.04	1.05	1.05	1.06	1.06
	.75	1.52	1.51	1.50	1.49	1.49	1.48	1.47	1.47	1.46	1.45
	.90	2.23	2.21	2.17	2.12	2.10	2.08	2.05	2.03	2.00	1.97
	.95	2.82	2.79	2.72	2.65	2.61	2.57	2.53	2.49	2.45	2.40
	.975	3.47	3.43	3.33	3.23	3.17	3.12	3.06	3.00	2.94	2.88
	.99	4.46	4.40	4.25	4.10	4.02	3.94	3.86	3.78	3.69	3.60
	.999	7.76	7.62	7.32	7.01	6.85	6.68	6.52	6.35	6.17	6.00
12	.50	.995	1.00	1.01	1.02	1.03	1.03	1.04	1.05	1.05	1.06
	.75	1.50	1.49	1.48	1.47	1.46	1.45	1.45	1.44	1.43	1.42
	.90	2.17	2.15	2.11	2.06	2.04	2.01	1.99	1.96	1.93	1.90
	.95	2.72	2.69	2.62	2.54	2.51	2.47	2.43	2.38	2.34	2.30
	.975	3.32	3.28	3.18	3.07	3.02	2.96	2.91	2.85	2.79	2.72
	.99	4.22	4.16	4.01	3.86	3.78	3.70	3.62	3.54	3.45	3.36
	.999	7.14	7.01	6.71	6.40	6.25	6.09	5.93	5.76	5.59	5.42
15	.50	.984	.989	1.00	1.01	1.02	1.02	1.03	1.03	1.04	1.05
	.75	1.44	1.44	1.43	1.41	1.41	1.40	1.39	1.38	1.37	1.36
	.90	2.04	2.02	1.97	1.92	1.90	1.87	1.85	1.82	1.79	1.76
	.95	2.51	2.48	2.40	2.33	2.39	2.25	2.20	2.16	2.11	2.07
	.975	3.01	2.96	2.86	2.76	2.70	2.64	2.59	2.52	2.46	2.40
	.99	3.73	3.67	3.52	3.37	3.29	3.21	3.13	3.05	2.96	2.87
	.999	5.93	5.81	5.54	5.25	5.10	4.95	4.80	4.64	4.47	4.31
20	.50	.972	.977	.989	1.00	1.01	1.01	1.02	1.02	1.03	1.03
	.75	1.39	1.39	1.37	1.36	1.35	1.34	1.33	1.32	1.31	1.29
	.90	1.91	1.89	1.84	1.79	1.77	1.74	1.71	1.68	1.64	1.61
	.95	2.31	2.28	2.20	2.12	2.08	2.04	1.99	1.95	1.90	1.84
	.975	2.72	2.68	2.57	2.46	2.41	2.35	2.29	2.22	2.16	2.09
	.99	3.29	3.23	3.09	2.94	2.86	2.78	2.69	2.61	2.52	2.42
	.999	4.94	4.82	4.56	4.29	4.15	4.01	3.86	3.70	3.54	3.38

		ν_1									
ν_2	a	1	2	3	4	5	6	7	8	9	10
24	.50	.469	.714	.812	.863	.895	.917	.932	.944	.953	.961
	.75	1.39	1.47	1.46	1.44	1.43	1.41	1.40	1.39	1.38	1.38
	.90	2.93	2.54	2.33	2.19	2.10	2.04	1.98	1.94	1.91	1.88
	.95	4.26	3.40	3.01	2.78	2.62	2.51	2.42	2.36	2.30	2.25
	.975	5.72	4.32	3.72	3.38	3.15	2.99	2.87	2.78	2.70	2.64
	.99	7.82	5.61	4.72	4.22	3.90	3.67	3.50	3.36	3.26	3.17
	.999	14.0	9.34	7.55	6.59	5.98	5.55	5.23	4.99	4.80	4.64
30	.50	.466	.709	.807	.858	.890	.912	.927	.939	.948	.955
	.75	1.38	1.45	1.44	1.42	1.41	1.39	1.38	1.37	1.36	1.35
	.90	2.88	2.49	2.28	2.14	2.05	1.98	1.93	1.88	1.85	1.82
	.95	4.17	3.32	2.92	2.69	2.53	2.42	2.33	2.27	2.21	2.16
	.975	5.57	4.18	3.59	3.25	3.03	2.87	2.75	2.65	2.57	2.51
	.99	7.56	5.39	4.51	4.02	3.70	3.47	3.30	3.17	3.07	2.98
	.999	13.3	8.77	7.05	6.12	5.53	5.12	4.82	4.58	4.39	4.24
40	.50	.463	.705	.802	.854	.885	.907	.922	.934	.943	.950
	.75	1.36	1.44	1.42	1.40	1.39	1.37	1.36	1.35	1.34	1.33
	.90	2.84	2.44	2.23	2.09	2.00	1.93	1.87	1.83	1.79	1.76
	.95	4.08	3.23	2.84	2.61	2.45	2.34	2.25	2.18	2.12	2.08
	.975	5.42	4.05	3.46	3.13	2.90	2.74	2.62	2.53	2.45	2.39
	.99	7.31	5.18	4.31	3.83	3.51	3.29	3.12	2.99	2.89	2.80
	.999	12.6	8.25	6.60	5.70	5.13	4.73	4.44	4.21	4.02	3.87
60	.50	.461	.701	.798	.849	.880	.901	.917	.928	.937	.945
	.75	1.35	1.42	1.41	1.38	1.37	1.35	1.33	1.32	1.31	1.30
	.90	2.79	2.39	2.18	2.04	1.95	1.87	1.82	1.77	1.74	1.71
	.95	4.00	3.15	2.76	2.53	2.37	2.25	2.17	2.10	2.04	1.99
	.975	5.29	3.93	3.34	3.01	2.79	2.63	2.51	2.41	2.33	2.27
	.99	7.08	4.98	4.13	3.65	3.34	3.12	2.95	2.82	2.72	2.63
	.999	12.0	7.76	6.17	5.31	4.76	4.37	4.09	3.87	3.69	3.54
120	.50	.458	.697	.793	.844	.875	.896	.912	.923	.932	.939
	.75	1.34	1.40	1.39	1.37	1.35	1.33	1.31	1.30	1.29	1.28
	.90	2.75	2.35	2.13	1.99	1.90	1.82	1.77	1.72	1.68	1.65
	.95	3.92	3.07	2.68	2.45	2.29	2.18	2.09	2.02	1.96	1.91
	.975	5.15	3.80	3.23	2.89	2.67	2.52	2.39	2.30	2.22	2.16
	.99	6.85	4.79	3.95	3.48	3.17	2.96	2.79	2.66	2.56	2.47
	.999	11.4	7.32	5.79	4.95	4.42	4.04	3.77	3.55	3.38	3.24
∞	.50	.455	.693	.789	.839	.870	.891	.907	.918	.927	.934
	.75	1.32	1.39	1.37	1.35	1.33	1.31	1.29	1.28	1.27	1.25
	.90	2.71	2.30	2.08	1.94	1.85	1.77	1.72	1.67	1.63	1.60
	.95	3.84	3.00	2.60	2.37	2.21	2.10	2.01	1.94	1.88	1.83
	.975	5.02	3.69	3.12	2.79	2.57	2.41	2.29	2.19	2.11	2.05
	.99	6.63	4.61	3.78	3.32	3.02	2.80	2.64	2.51	2.41	2.32
	.999	10.8	6.91	5.42	4.62	4.10	3.74	3.47	3.27	3.10	2.96

ν_2	a	11	12	15	20	24	30	40	60	120	∞
24	.50	.967	.972	.983	.994	1.00	1.01	1.01	1.02	1.02	1.03
	.75	1.37	1.36	1.35	1.33	1.32	1.31	1.30	1.29	1.28	1.26
	.90	1.85	1.83	1.78	1.73	1.70	1.67	1.64	1.61	1.57	1.53
	.95	2.21	2.18	2.11	2.03	1.98	1.94	1.89	1.84	1.79	1.73
	.975	2.59	2.54	2.44	2.33	2.27	2.21	2.15	2.08	2.01	1.94
	.99	3.09	3.03	2.89	2.74	2.66	2.58	2.49	2.40	2.31	2.21
	.999	4.50	4.39	4.14	3.87	3.74	3.59	3.45	3.29	3.14	2.97
30	.50	.961	.966	.978	.989	.994	1.00	1.01	1.01	1.02	1.02
	.75	1.35	1.34	1.32	1.30	1.29	1.28	1.27	1.26	1.24	1.23
	.90	1.79	1.77	1.72	1.67	1.64	1.61	1.57	1.54	1.50	1.46
	.95	2.13	2.09	2.01	1.93	1.89	1.84	1.79	1.74	1.68	1.62
	.975	2.46	2.41	2.31	2.20	2.14	2.07	2.01	1.94	1.87	1.79
	.99	2.91	2.84	2.70	2.55	2.47	2.39	2.30	2.21	2.11	2.01
	.999	4.11	4.00	3.75	3.49	3.36	3.22	3.07	2.92	2.76	2.59
40	.50	.956	.961	.972	.983	.989	.994	1.00	1.01	1.01	1.02
	.75	1.32	1.31	1.30	1.28	1.26	1.25	1.24	1.22	1.21	1.19
	.90	1.73	1.71	1.66	1.61	1.57	1.54	1.51	1.47	1.42	1.38
	.95	2.04	2.00	1.92	1.84	1.79	1.74	1.69	1.64	1.58	1.51
	.975	2.33	2.29	2.18	2.07	2.01	1.94	1.88	1.80	1.72	1.64
	.99	2.73	2.66	2.52	2.37	2.29	2.20	2.11	2.02	1.92	1.80
	.999	3.75	3.64	3.40	3.15	3.01	2.87	2.73	2.57	2.41	2.23
60	.50	.951	.956	.967	.978	.983	.989	.994	1.00	1.01	1.01
	.75	1.29	1.29	1.27	1.25	1.24	1.22	1.21	1.19	1.17	1.15
	.90	1.68	1.66	1.60	1.54	1.51	1.48	1.44	1.40	1.35	1.29
	.95	1.95	1.92	1.84	1.75	1.70	1.65	1.59	1.53	1.47	1.39
	.975	2.22	2.17	2.06	1.94	1.88	1.82	1.74	1.67	1.58	1.48
	.99	2.56	2.50	2.35	2.20	2.12	2.03	1.94	1.84	1.73	1.60
	.999	3.43	3.31	3.08	2.83	2.69	2.56	2.41	2.25	2.09	1.89
120	.50	.945	.950	.961	.972	.978	.983	.989	.994	1.00	1.01
	.75	1.27	1.26	1.24	1.22	1.21	1.19	1.18	1.16	1.13	1.10
	.90	1.62	1.60	1.55	1.48	1.45	1.41	1.37	1.32	1.26	1.19
	.95	1.87	1.83	1.75	1.66	1.61	1.55	1.50	1.43	1.35	1.25
	.975	2.10	2.05	1.95	1.82	1.76	1.69	1.61	1.53	1.43	1.31
	.99	2.40	2.34	2.19	2.03	1.95	1.86	1.76	1.66	1.53	1.38
	.999	3.12	3.02	2.78	2.53	2.40	2.26	2.11	1.95	1.76	1.54
∞	.50	.939	.945	.956	.967	.972	.978	.983	.989	.994	1.00
	.75	1.24	1.24	1.22	1.19	1.18	1.16	1.14	1.12	1.08	1.00
	.90	1.57	1.55	1.49	1.42	1.38	1.34	1.30	1.24	1.17	1.00
	.95	1.79	1.75	1.67	1.57	1.52	1.46	1.39	1.32	1.22	1.00
	.975	1.99	1.94	1.83	1.71	1.64	1.57	1.48	1.39	1.27	1.00
	.99	2.25	2.18	2.04	1.88	1.79	1.70	1.59	1.47	1.32	1.00
	.999	2.84	2.74	2.51	2.27	2.13	1.99	1.84	1.66	1.45	1.00

ν_1

Entry is $\chi^2(a,\nu)$, where

$P[\chi^2(\nu) \le \chi^2(a,v)] = a.$

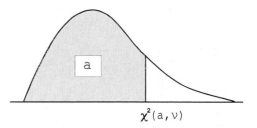

$\chi^2(a,\nu)$

df	a									
ν	.005	.010	.025	.050	.100	.900	.950	.975	.990	.995
1	.0⁴39	.0³16	.0³98	.0²39	.0158	2.71	3.84	5.02	6.63	7.88
2	.0100	.0201	.0506	.103	.211	4.61	5.99	7.38	9.21	10.60
3	.072	.115	.216	.352	.584	6.25	7.81	9.35	11.34	12.84
4	.207	.0297	.484	.711	1.064	7.78	9.49	11.14	13.28	14.86
5	.412	.554	.831	1.145	1.61	9.24	11.07	12.83	15.09	16.75
6	.676	.872	1.24	1.64	2.20	10.64	12.59	14.45	16.81	18.55
7	.989	1.24	1.69	2.17	2.83	12.02	14.07	16.01	18.48	20.28
8	1.34	1.65	2.18	2.73	3.49	13.36	15.51	17.53	20.09	21.96
9	1.73	2.09	2.70	3.33	4.17	14.68	16.92	19.02	21.67	23.59
10	2.16	2.56	3.25	3.94	4.87	15.99	18.31	20.48	23.21	25.19
11	2.60	3.05	3.82	4.57	5.58	17.28	19.68	21.92	24.73	26.76
12	3.07	3.57	4.40	5.23	6.30	18.55	21.03	23.34	26.22	28.30
13	3.57	4.11	5.01	5.89	7.04	19.81	22.36	24.74	27.69	29.82
14	4.07	4.66	5.63	6.57	7.79	21.06	23.68	26.12	29.14	31.32
15	4.60	5.23	6.26	7.26	8.55	22.31	25.00	27.49	30.58	32.80
16	5.14	5.81	6.91	7.96	9.31	23.54	26.30	28.85	32.00	34.27
17	5.70	6.41	7.56	8.67	10.09	24.77	27.59	30.19	33.41	35.72
18	6.26	7.01	8.23	9.39	10.86	25.99	28.87	31.53	34.81	37.16
19	6.84	7.63	8.91	10.12	11.65	27.20	30.14	32.85	36.19	38.58
20	7.43	8.26	9.59	10.85	12.44	28.41	31.41	34.17	37.57	40.00
21	8.03	8.90	10.28	11.59	13.24	29.62	32.67	35.48	38.93	41.40
22	8.64	9.54	10.98	12.34	14.04	30.81	33.92	36.78	40.29	42.80
23	9.26	10.20	11.69	13.09	14.85	32.01	35.17	38.08	41.64	44.18
24	9.89	10.86	12.40	13.85	15.66	33.20	36.42	39.36	42.98	45.56
25	10.52	11.52	13.12	14.61	16.46	34.38	37.65	40.65	44.31	46.93
26	11.16	12.20	13.84	15.38	17.29	35.56	38.89	41.92	45.64	48.29
27	11.81	12.88	14.57	16.15	18.11	36.74	40.11	43.19	46.96	49.64
28	12.46	13.56	15.31	16.93	18.94	37.92	41.34	44.46	48.28	50.99
29	13.12	14.26	16.05	17.71	19.77	39.09	42.56	45.72	49.59	52.34
30	13.79	14.95	16.79	18.49	20.60	40.26	43.77	46.98	50.89	53.67
40	20.71	22.16	24.43	26.51	29.05	51.81	55.76	59.34	63.69	66.77
50	27.99	29.71	32.36	34.76	37.69	63.17	67.50	71.42	76.15	79.49
60	35.53	37.48	40.48	43.19	46.46	74.40	79.08	83.30	88.38	91.95
80	51.17	53.54	57.15	60.39	64.28	96.58	101.9	106.6	112.3	116.3
100	67.33	70.06	74.22	77.93	82.36	118.5	124.3	129.6	135.8	140.2

Discrete Distributions

Name	Parameters	Probability Mass Function (PMF)	Mean	Variance
Uniform	a = smallest b = biggest (Let n = b - a + 1)	$P(x) = \dfrac{1}{n}$ for x = a, a + 1, ..., b	$\dfrac{a+b}{2}$	$\dfrac{n^2 - 1}{12}$
Binomial	n = number of trials p = probability of success on any trial (Let q = 1 - p)	$P(x) = \dbinom{n}{x} p^x q^{n-x}$ for x = 0, 1, ..., n	np	npq
Geometric	p = probability of success on any trial (Let q = 1 - p)	$P(x) = pq^{x-1}$ for x = 1, 2, ...	$\dfrac{1}{p}$	$\dfrac{q}{p^2}$
Negative Binomial	p = probability of success on any trial r = number of successes (Let q = 1 - p)	$P(x) = \dbinom{x-1}{r-1} p^r q^{x-r}$ for x = r, r + 1, ...	$\dfrac{r}{p}$	$\dfrac{rq}{p^2}$
Hyper-geometric	N = size of population n = size of sample selected D = size of subpopulation of interest (Let $p = \dfrac{D}{N}$ and q = 1 - p)	$P(x) = \dfrac{\dbinom{D}{x}\dbinom{N-D}{n-x}}{\dbinom{N}{n}}$ for x = 0, 1, ..., min (n, D)	np	$\left(\dfrac{N-n}{N-1}\right)npq$
Poisson	λ = average number of occurrences per interval	$P(x) = \dfrac{\lambda^x e^{-\lambda}}{x!}$ for x = 0, 1, . . .	λ	λ

Continuous Distributions

Name	Parameters	Probability Density Function (PDF)	Mean	Variance
Uniform	a = smallest b = biggest	$f(x) = \dfrac{1}{b - a}$ for $a \leq x \leq b$	$\dfrac{a+b}{2}$	$\dfrac{(b - a)^2}{12}$
Normal	μ = mean (center) σ = standard deviation (dispersion)	$f(x) = \dfrac{1}{\sigma\sqrt{2\pi}} e^{-\frac{1}{2}\left(\frac{x-\mu}{\sigma}\right)^2}$ for $-\infty < x < \infty$	μ	σ^2
Exponential	λ = average number of occurrences per interval	$f(x) = \lambda e^{-\lambda x}$ for $x \geq 0$	$\dfrac{1}{\lambda}$	$\dfrac{1}{\lambda^2}$

NOTATION:

CL	= center line	n	=	sample size
UCL	= upper control limit	$\bar{n}$	=	average sample size
LCL	= lower control limit	$\bar{p}$	=	average proportion of defectives
R	= range of sample	$\bar{c}$	=	average count of defects
$\bar{R}$	= average of ranges	$\bar{u}$	=	average count of defects per unit area of observation
$\bar{x}$	= average of readings	$\bar{a}$	=	average area of observation
$\bar{\bar{x}}$	= average of averages	$\hat{\sigma}$	=	estimated overall process standard deviation
$\bar{s}$	= average of sample standard deviations			

CONSTANTS:

n	A_2	A_3	B_3	B_4	d_2	D_3	D_4	E_2
2	1.88	2.66	.00	3.27	1.13	.00	3.27	2.66
3	1.02	1.95	.00	2.57	1.69	.00	2.57	1.77
4	.73	1.63	.00	2.27	2.06	.00	2.28	1.46
5	.58	1.43	.00	2.09	2.33	.00	2.11	1.29
6	.48	1.29	.03	1.97	2.53	.00	2.00	1.18
7	.42	1.18	.12	1.88	2.70	.08	1.92	1.11
8	.37	1.10	.19	1.82	2.85	.14	1.86	1.05
9	.34	1.03	.24	1.76	2.97	.18	1.82	1.01
10	.31	.98	.28	1.72	3.08	.22	1.78	.98
11	.29	.93	.32	1.68	3.17	.26	1.74	
12	.27	.89	.35	1.65	3.26	.28	1.72	
13	.25	.85	.38	1.62	3.34	.31	1.69	
14	.24	.82	.41	1.59	3.41	.33	1.67	
15	.22	.79	.43	1.57	3.47	.35	1.65	
16	.21	.76	.45	1.55	3.53	.36	1.64	
17	.20	.74	.47	1.53	3.59	.38	1.62	
18	.19	.72	.48	1.52	3.64	.39	1.61	
19	.19	.70	.50	1.50	3.69	.40	1.60	
20	.18	.68	.51	1.49	3.74	.42	1.59	

SOURCE: A_2, A_3, B_3, B_4, d_2, D_3, D_4, E_2 reprinted with permission from *ASTM Manual on the Presentation of Data and Control Chart Analysis* (Philadelphia, PA:ASTM 1976), pp.134-36. Copyright ASTM.

FORMULAS:

Chart	CL	UCL	LCL	Comments
x̄ - R	$\bar{\bar{x}}$	$\bar{\bar{x}} + A_2\bar{R}$	$\bar{\bar{x}} - A_2\bar{R}$	$\hat{\sigma} = \dfrac{\bar{R}}{d_2}$
	$\bar{R}$	$D_4\bar{R}$	$D_3\bar{R}$	use when $n < 10$
individuals with moving range	$\bar{x}$	$\bar{x} + E_2\bar{R}$	$\bar{x} - E_2\bar{R}$	$\hat{\sigma} = \dfrac{\bar{R}}{d_2}$
	$\bar{R}$	$D_4\bar{R}$	$D_3\bar{R}$	use $n = 2$
x̄ - s	$\bar{\bar{x}}$	$\bar{\bar{x}} + A_3\bar{s}$	$\bar{\bar{x}} - A_3\bar{s}$	$\hat{\sigma} = \bar{s}$
	$\bar{s}$	$B_4\bar{s}$	$B_3\bar{s}$	use when $n \geq 10$ or when n varies
np	$n\bar{p}$	$n\bar{p} + 3\sqrt{n\bar{p}(1-\bar{p})}$	$n\bar{p} - 3\sqrt{n\bar{p}(1-\bar{p})}$	n is fixed size
p	$\bar{p}$	$\bar{p} + 3\sqrt{\dfrac{\bar{p}(1-\bar{p})}{\bar{n}}}$	$\bar{p} - 3\sqrt{\dfrac{\bar{p}(1-\bar{p})}{\bar{n}}}$	use n_i instead of $\bar{n}$ if n_i's vary widely
c	$\bar{c}$	$\bar{c} + 3\sqrt{\bar{c}}$	$\bar{c} - 3\sqrt{\bar{c}}$	fixed area of observation
u	$\bar{u}$	$\bar{u} + 3\sqrt{\dfrac{\bar{u}}{\bar{a}}}$	$\bar{u} - 3\sqrt{\dfrac{\bar{u}}{\bar{a}}}$	use a_i instead of $\bar{a}$ if a_i's vary widely

BACKGROUND:

What follows is a brief description of how the control limits for the $\bar{x}$ - R Chart are derived in terms of Shewhart's constants. First, we define a new random variable Y as $Y = \dfrac{R}{\sigma}$. Y is called the ***relative range variable*** which relates the range random variable R to the standard deviation σ, a constant, in a normal distribution. Simulation studies and/or extensive theory can be used to show the following results:

$$E(Y) = d_2 \qquad \text{and} \qquad \sigma(Y) = \sigma_Y = d_3$$

That is, the expected value (E operator) or mean value of Y is a constant called d_2 in the literature and the standard deviation of Y is another constant called d_3. Both d_2 and d_3 are dependent on the sample or subgroup size n.

Since $Y = \dfrac{R}{\sigma}$, $E(Y) = \dfrac{E(R)}{\sigma} = d_2$. We then can use $\bar{R}$ from the R-chart to estimate E(R) and solve for $\hat{\sigma}$ as follows:

$$\frac{\bar{R}}{\hat{\sigma}} = d_2 \qquad \text{or} \qquad \hat{\sigma} = \frac{\bar{R}}{d_2}$$

Thus, we can use $\bar{R}$ from the R-chart along with the constant d_2 to estimate the process standard deviation, σ.

We can now go one step farther to compute the UCL and LCL of the $\bar{x}$-chart. Since we are dealing with the $\bar{x}$ (or child) distribution,

$$\begin{matrix} \text{UCL} \\ \text{LCL} \end{matrix} = \bar{\bar{x}} \pm \frac{3\hat{\sigma}}{\sqrt{n}} = \bar{\bar{x}} \pm \frac{3\bar{R}}{d_2 \sqrt{n}}$$

Since d_2 depends on n, practical considerations suggest using the traditional constant called $A_2 = \dfrac{3}{d_2 \sqrt{n}}$ and rewriting the formulas for UCL and LCL as

$$\begin{aligned} \text{UCL} \\ \text{LCL} \end{aligned} = \overline{\overline{x}} \pm A_2 \overline{R}$$

The upper and lower control limits for the R-chart are written as

$$\begin{aligned} \text{UCL} \\ \text{LCL} \end{aligned} = \overline{R} \pm 3\hat{\sigma}_R,$$

where $\hat{\sigma}_R$ is an estimated standard deviation of the range (R) distribution.

Since $Y = \dfrac{R}{\sigma}$, $R = Y\sigma$ and $\hat{\sigma}_R = \hat{\sigma}_Y \hat{\sigma}$. But we already know that $\hat{\sigma}_Y = d_3$ and

$\hat{\sigma} = \dfrac{\overline{R}}{d_2}$. Thus, $\hat{\sigma}_R = d_3\left(\dfrac{\overline{R}}{d_2}\right)$ and $\begin{aligned} \text{UCL} \\ \text{LCL} \end{aligned} = \overline{R} \pm 3\hat{\sigma}_R = \overline{R} \pm 3 d_3 \dfrac{\overline{R}}{d_2}$ or

$$\begin{aligned} \text{UCL} \\ \text{LCL} \end{aligned} = \overline{R}\left(1 \pm \frac{3 d_3}{d_2}\right)$$

If we let $D_4 = 1 + \dfrac{3 d_3}{d_2}$ and $D_3 = 1 - \dfrac{3 d_3}{d_2}$, we obtain the traditional forms for the

UCL and LCL of the R-chart, namely

$$\textbf{UCL} = \textbf{D}_4 \overline{R}$$

$$\textbf{LCL} = \textbf{D}_3 \overline{R}$$

Glossary of Terms

α **(alpha) risk** the probability of concluding the alternative hypothesis (H_1) when the null hypothesis (H_0) is true.

aliasing when two factors or interaction terms are set at identical levels throughout the entire experiment (i.e., the two columns are 100% correlated).

alternative hypothesis the hypothesis to be accepted if the null hypothesis is rejected. It is denoted by H_1.

analysis of variance (ANOVA) a procedure for partitioning the total variation. It is often used to compare more than two population means.

assignable cause (of variation) significant, identifiable change in a response which is caused by some specific variable from the cause and effect diagram.

attribute data (quality) data coming basically from GO/NO-GO, pass/fail determinations of whether units conform to standards. Also includes noting presence or absence of a quality characteristic.

average (of a statistical sample) ($\bar{x}$) also called the sample mean, it is the arithmetic average value of all of the sample values. It is calculated by adding all of the sample values together and dividing by the number of elements (n) in the sample.

β **(Beta) risk** the probability of concluding the null hypothesis (H_0) when the alternative (H_1) is true.

balanced design a 2-level experimental design is balanced if each factor is run the same number of times at the high and low levels.

bar chart a graphical method which depicts how data fall into different categories.

bell-shaped curve a curve or distribution showing a central peak and tapering off smoothly and symmetrically to "tails" on either side. A normal (Gaussian) curve is bell-shaped.

bias (in measurement) systematic error which leads to a difference between the average result of a population of measurements and the true, accepted value of the quantity being measured.

bimodal distribution a frequency distribution which has two peaks. Usually an indication of samples from two processes incorrectly analyzed as a single process.

binomial distribution (probability distribution) given that a trial can have only two possible outcomes (yes/no, pass/fail, heads/tail), of which one outcome has probability p and the other probability $q(q = 1 - p)$, the probability that the outcome represented by p occurs x times in n trials is given by the binomial distribution.

binomial experiment an experiment involving a sequence of n identical and independent trials, where there are only two possible outcomes (e.g., success or failure) for each trial.

binomial random variable a discrete random variable which represents the number of successes out of n identical and independent trials.

c-chart for attribute data: a control chart of the number of defects found in areas of equal opportunity for finding defects. The c-chart is used where each unit typically has a number of defects. It is based on the Poisson distribution.

calibration (of instrument) adjusting an instrument using a reference standard to reduce the difference between the average reading of the instrument and the "true" value of the standard being measured, i.e., to reduce measurement bias.

capability (of process) a measure of quality for a process usually expressed as sigma capability, C_{pk}, or defects per million (dpm). It is obtained by comparing the actual process with the specification limit(s).

cause and effect diagram a pictorial diagram showing possible causes (process inputs) for a given effect (process output).

Central Limit Theorem (CLT) if samples of size n are drawn from a population, and the values of $\bar{x}$ are calculated for each sample, the shape of the distribution is found to approach a normal distribution for sufficiently large n. This theorem allows one to use the assumption of a normal distribution when dealing with $\bar{x}$. "Sufficiently large" depends on the population's distribution and what range of $\bar{x}$ is being considered; for practical purposes, the easiest approach may be to take a number of samples of a desired size and see if their means are normally distributed. If not, the sample size should be increased. This theorem is one of the most important results in all of statistics and is the heart of inferential statistics.

central tendency a measure of the point about which a group of values is clustered; some measures of central tendency are mean, mode, and median.

characteristic a process output which can be measured and monitored for control and capability.

chi-square the test statistic used when testing the null hypothesis of independence in a contingency table or when testing the null hypothesis of a set of data following a prescribed distribution.

chi-square distribution the distribution of chi-square statistics.

class boundary the juncture or dividing point between two adjacent cells in a frequency distribution or histogram.

classes cells or intervals in a frequency distribution or histogram.

class frequency the number of observations occurring in a given class.

class interval each cell or subinterval of a histogram.

class interval width the length (or width) of each class interval in a histogram.

coefficient of determination the square of the sample correlation coefficient.

coefficient of variation the ratio of the standard deviation to the mean. It is a standardized method of looking at variation.

common causes of variation those sources of variability in a process which are truly random, i.e., inherent in the process itself.

complement the complement of event A, denoted A^* or $\bar{A}$, is everything in the sample space outside of A.

conditional probability (A given B) the probability that A will occur given the fact that B has already occurred.

confidence coefficient the probability that a confidence interval will capture or enclose an unknown population parameter.

confidence interval range within which a parameter of a population (e.g., mean, standard deviation, etc.) may be expected to fall, on the basis of measurement, with some specified confidence level or confidence coefficient.

confidence limits the upper and lower boundaries of a confidence interval.

contingency table a two-dimensional table constructed for classifying count data, the purpose of which is to determine if two variables are dependent (or contingent) on each other.

continuous random variable a variable that is based on measurement data (inches, pounds, liters, etc.). It supposedly has an uncountable number of possible outcomes.

control (of process) a process is said to be in a state of statistical control if the process exhibits only random variation (as opposed to systematic variation and/or variation with known sources). When monitoring control with control charts, a state of control is exhibited when all points remain between set control limits without any abnormal (non-random) patterns.

control chart the basic tool of statistical process control. It consists of a run chart, together with statistically determined upper and lower control limits and a centerline.

control limits upper and lower bounds in a control chart that are determined by the process itself. They can be used to detect special causes of variation. They are usually set at $\pm$ 3 standard deviations from the centerline.

correlation coefficient a measure of the linear relationship between two random variables.

C_p during process capability studies, C_p is a capability index which shows the process capability potential but does not consider how centered the process is. C_p may range in value from 0 to infinity, with a large value indicating greater potential capability. A value of 1.33 or greater is usually desired.

C_{pk} during process capability studies, C_{pk} is an index used to compare the natural tolerance of a process with the specification limits. C_{pk} has a value equal to C_p if the process is centered on the nominal; if C_{pk} is negative, the process mean is outside the specification limits; if C_{pk} is between 0 and 1 then the natural tolerances of the process fall outside the spec limits. If C_{pk} is larger than 1, the natural tolerances fall completely within the spec limits. A value of 1.33 or greater is usually desired.

cumulative sum chart (CuSum) a cumulative sum chart plots the cumulative deviation of each subgroup's average from the nominal value. If the process consistently produces parts near the nominal, the CuSum chart shows a line which is essentially horizontal. If the process begins to shift, the line will show an upward or downward trend. The CuSum chart is sensitive to small shifts in process average.

defect departure of a quality characteristic from its acceptable level or state, i.e., the measured value of the characteristic is outside of specification. Also referred to as non-conformance to requirements.

defective unit a sample (part) which contains one or more defects, making the sample unacceptable for its intended, normal usage.

degrees of freedom a parameter in the t, F, and χ^2 distributions. It is a measure of the amount of information available for estimating the population variance, σ^2. It is the number of independent observations minus the number of parameters estimated.

deviation the difference between an observed value and the mean or average of all observed values.

discrete random variable a variable that is based on count data (number of defects, number of births, number of deaths, etc.). It supposedly has a countable number of possible outcomes.

dispersion (of a statistical sample) the tendency of the values of the elements in a sample to differ from each other. Dispersion is commonly expressed in terms of the range of the sample (difference between the lowest and highest values) or by the standard deviation.

estimation an approach to making inferences about population parameters. This includes both point estimates and interval estimates (confidence intervals).

event any possible subset of the totality of basic outcomes from an experiment.

exhaustive a term that usually refers to the classes of a frequency distribution or histogram. It means that, for every data value in the data set, there is a class into which that data value falls.

experiment any activity or process that is capable of generating two or more possible outcomes.

experimental design purposeful changes to the inputs (factors) of a process in order to observe corresponding changes in the outputs (responses).

exponential distribution a probability distribution mathematically described by an exponential function. Used to describe the probability that a product survives a length of time t in service, under the assumption that the probability of a product failing in any small time interval is independent of time.

F-distribution distribution of F-statistics.

F-ratio a ratio of two independent estimates of experimental error. If there isn't an effect, the ratio should be close to 1.

F-statistic a test statistic used to compare the variances from two normal populations.

factor an input to a process which can be manipulated during experimentation.

failure rate the average number of failures per unit time. Used for assessing reliability of a product in service.

fault tree analysis (FTA) a technique for evaluating the possible causes which might lead to the failure of a product. For each possible failure, the possible causes of the failure are determined; then the situations leading to those causes are determined; and so forth, until all paths leading to possible failures have been traced. The result is an inverted tree whose branches lead to the root causes of the failure.

feedback using the results of a process to control it. The feedback principle has wide application. An example would be using control charts to keep production personnel informed on the results of a process. This allows them to make suitable adjustments to the process. Some form of feedback on the results of a process is essential in order to keep the process under control.

fishbone diagram a wiring diagram which a group can use to organize and document its thoughts during a brainstorming session. The backbone of the fish represents the response being measured. The "ribs" represent the types of factors that affect the response. Also known as cause and effect diagrams or Ishikawa diagrams.

flow chart or diagram (for programs, decision making, process development) a pictorial representation of a process indicating the main steps, branches, and eventual outcomes of the process.

fractional factorial an orthogonal subset of a full factorial. It can be used if one is willing to assume that some interactions will not occur and then assign a factor to that interaction column.

freehand regression line the best-fit line drawn by "eyeballing."

frequency the number of observations falling into a given class.

frequency distribution for a sample drawn from a statistical population, the number of times each outcome or class of outcomes is observed.

frequency polygon a graphical representation of the frequency of data that can be obtained by connecting adjacent midpoints of the tops of the bars in a histogram. It is easier to compare two polygons than it is to compare two histograms.

full factorial all possible combinations of the factors and levels. Given k factors, all with two levels, there will be 2^k runs. If all factors have 3 levels, then there will be 3^k runs.

goodness-of-fit any measure of how well a set of data matches a proposed distribution. Chi-square is the most common measure for frequency distributions. Simple visual inspection of a histogram is a less quantitative, but equally valid, way to determine goodness-of-fit.

grand average overall average of data.

histogram a bar chart that depicts the frequencies of numerical or measurement data.

hypothesis test a procedure whereby one of two mutually exclusive and exhaustive statements about a population parameter is concluded. Information from a sample is used to infer something about a population from which the sample was drawn.

independent events events such that the occurrence of one does not affect nor is affected by the occurrence of the other.

individuals chart with moving range a control chart used when sample or subgroup size is 1. The individual values are plotted on the chart. The individuals chart is always accompanied by a moving range chart, usually using two consecutive individual readings to calculate the moving range points.

interaction two factors (input variables) are said to interact if one factor's effect on the response is dependent upon the level of the other factor.

interval estimate an interval (usually called a confidence interval) which consists of a lower limit and an upper limit that are computed from sample data. The interval enclosed by these two numbers should capture an unknown population parameter with a certain level of confidence.

Ishikawa diagram see cause and effect diagram or fishbone diagram.

just-in-time (JIT) manufacturing a strategy that coordinates scheduling, inventory, and production to move away from the batch mode of production in order to improve quality and reduce inventories.

LCL (lower control limit) for control charts: the limit above which the process subgroup statistics ($\bar{x}$, R, sigma) must remain when the process is in control. Typically 3 standard deviations below the center line.

least squares a method of curve-fitting that defines the "best" fit as the one that minimizes the sum of the squared deviations of the data points from the fitted curve.

level a setting or value of a factor.

level of significance a measure of the outcome of a hypothesis test. It is the P-value or probability of making a Type I error.

loss function a technique for quantifying loss due to product deviations from target values.

lower confidence limit the smaller of the two numbers that form a confidence interval.

LSL (lower specification limit)   the lowest value of a product dimension or measurement which is acceptable.

mean   the average of a set of values. We usually use $\bar{x}$ or $\bar{y}$ to denote a sample mean, whereby we use the Greek letter μ to denote a population mean.

measures of central tendency   numerical measures that depict the center of a data set. The most commonly used measures are the mean and the median.

median   the middle value of a data set when the values are arranged in either ascending or descending order.

MTBF (mean time between failures)   mean time between successive failures of a repairable product. This is a measure of product reliability.

multicollinearity   the existence of strong correlations between input factors or independent variables.

multiple regression   a model where several independent variables are used to predict one dependent variable.

mutually exclusive events   events that cannot both happen simultaneously.

natural tolerances (of a process)   3 standard deviations on either side of the center point (mean value). In a normally distributed process, the natural tolerances encompass 99.73% of all measurements.

nominal   for a product whose size is of concern: the desired mean value for the particular dimension, the target value.

nonconforming unit   a sample (part) which has one or more nonconformities, making the sample unacceptable for its intended use.

normal distribution   the distribution characterized by the smooth, bell-shaped curve.

np-chart   for attribute data: a control chart of the number of defective units in a subgroup. Assumes a constant subgroup size. Based on the binomial distribution.

null hypothesis (H_0)   the conclusion that typically includes equality, i.e., H_0: $\mu_1 = \mu_2$ or H_0: $\sigma_1 = \sigma_2$.

one-at-a-time approach   a popular, but inefficient way to conduct a designed experiment.

outcome   a possible result from an experiment. It may also be referred to as a basic outcome.

out of control (of a process)   a process is said to be out of control if it exhibits variations larger than its control limits, or shows a systematic pattern of variation.

p-chart (percent defective)   for attribute data: a control chart of the proportion of defective units (or fraction defective) in a subgroup.   Based on the binomial distribution.

P-value   the probability of making a Type I error.   This value comes from the data itself.   It also provides the exact level of significance of a hypothesis test.

parameter   a numerical measure of some aspect of a population.

Pareto diagram   a bar chart for attribute (or categorical) data that is presented in descending order of frequency.

percent defective   for acceptance sampling: the percentage of units in a lot which are defective, i.e., of unacceptable quality.

percentiles   a method used to describe the location of values within a data set.   It divides the entire range of a data set into 100 equal parts called percentiles.   This is typically the finest resolution used (as opposed to deciles, which use 10 equal parts; or quartiles, which use 4 equal parts; or the median, which uses two equal parts).

point estimate   the use of a sample statistic (or single value) such as $\bar{x}$ to estimate a population parameter μ.

Poisson distribution   a probability distribution for the number of occurrences per unit interval (time or space); λ = average number of occurrences per interval is the only parameter. The Poisson distribution is a good approximation of the binomial distribution for the case where n is large and p is small. $\lambda = np$.

population   a set or collection of objects or individuals.   It can also be the corresponding set of values which measure a certain characteristic of a set of objects or individuals.

pre-control a method of controlling a process based on the specification limits. It is used to prevent the manufacture of defective units, but does not work toward minimizing variation of the process. The area between the specifications are split into zones (green, yellow and red) and adjustments made when a specified number of points fall in the yellow or red zones.

predicted value of y a value of the response (dependent) variable y that is computed from the prediction equation for some particular value of the input (independent) variable x. It is labeled $\hat{y}$ and is computed from $\hat{y} = b_0 + b_1 x$.

primary reference standard for measurements: a standard maintained by the National Bureau of Standards for a particular measuring unit. The primary reference standard duplicates as nearly as possible the international standard and is used to calibrate other (transfer) standards, which in turn are used to calibrate measuring instruments for industrial use.

probability a measure of the likelihood of a given event occurring. It is a measure that takes on values between 0 and 1 inclusive, with 1 being the certain event and 0 meaning that there is relatively no chance at all of the event occurring. How probabilities are assigned is another matter. The relative frequency approach to assigning probabilities is one of the most common.

probability distribution the assignment of probabilities to all of the possible outcomes from an experiment. This assignment is usually portrayed by way of a table, graph, or formula.

process capability comparing actual process performance with process specification limits. There are various measures of process capability, such as C_{pk}, σ_{level}, and dpm (defects per million).

process control a process is said to be in control or it is a stable, predictable process if all special causes of variation have been removed. Only common causes or natural variation remains in the process.

quality assurance the function of assuring that a product or service will satisfy given needs. The function includes necessary verification, audits, and evaluations of quality factors affecting the intended usage and customer satisfaction. This function is normally the responsibility of one or more upper management individuals overseeing the quality assurance program.

quality characteristic a particular aspect of a product which relates to its ability to perform its intended function.

quality control the process of maintaining an acceptable level of product quality.

quality function the function of maintaining product quality levels; i.e., the execution of quality control.

quality specifications particular specifications of the limits within which each quality characteristic of a product is to be maintained in order to meet the minimum functional requirements of the customer.

$\overline{R}$ average range value displayed as the centerline on a range control chart. Value is set at the time control limit(s) are calculated.

R-chart a control chart of the range of variation among the individual elements of a sample — i.e., the difference between the largest and smallest elements — as a function of time, or lot number, or similar chronological variable.

random varying with no discernable pattern.

random sample a sample selected from a population in such a way that every element of the population had an equally likely chance of being selected.

random variable a definition of the possible outcomes of interest from a given experiment.

range a measure of the variability in a data set. It is a value, namely the difference between the largest and smallest values in a data set.

regression analysis a statistical technique for determining the mathematical relation between a measured quantity and the variables it depends on.

regression line the line that is fit to a set of data points by using the method of least squares.

reliability the probability that a product will function properly for some specified period of time, under specified conditions.

repeatability (of a measurement) the extent to which repeated measurements of a particular object with a particular instrument produce the same value.

reproducibility   the variation between individual people taking the same measurement and using the same gaging.

residual   the difference between an observed value and a predicted value: residual $= y - \hat{y}$.

resolution (of a measuring instrument)   the smallest unit of measure which an instrument is capable of indicating.

Rule of Thumb (ROT)   a simplified, practical procedure that can be used in place of a formal statistical test that will produce approximately the same result.

run   in SPC, a set of consecutive units, i.e., sequential in time;   in DOE, an experimental combination of factor settings

run chart   a basic graphical tool that charts a process over time, recording either individual readings or averages over time.

sample   a set of values or items selected from some population.

sample size   the number of elements, or units, in a sample.

sampling   the process of selecting a sample from a population and determining the properties of the sample. The sample is chosen in such a way that its properties are representative of the population.

sampling distribution   the probability distribution for some sample statistic.

sampling distribution of the mean   the distribution of all means (or averages) from all possible samples of a fixed size.

sampling variation   the variation of a sample's properties from the properties of the population from which it was drawn.

scatterplot   a two-dimensional plot for displaying bivariate data.

short-run SPC   a set of techniques used for SPC in low-volume, short duration manufacturing.

sigma (σ)   the standard deviation of a statistical population.

sigma level a commonly used measure of process capability that represents the number of standard deviations between the center of a process and the closest specification limit.

sigma limits for histograms: lines marked on the histogram showing the points n standard deviations above and below the mean.

significance level see level of significance.

simple linear regression a model where one independent variable is used to predict one dependent variable.

simulation (modeling) using a mathematical model of a system or process to predict the performance of the real system. The model consists of a set of equations or logic rules which operate on numerical values representing the operating parameters of the system. The result of the equation is a prediction of the system's output.

skewed distribution a distribution that (graphically) has a longer tail on the right than it does on the left (or vice versa), i.e., it is not symmetric about its center point. Numerically, a distribution is skewed if the mean and the median are not the same.

slope the term b_1 in the prediction equation $\hat{y} = b_0 + b_1 x$.

special causes of variation those nonrandom causes of variation that can be detected by the use of control charts and good process documentation.

specification limits the bounds of acceptable values for a given product or process. They should be customer driven.

SSE sum of squares due to error, or sum of squared residuals.

SSR sum of squares due to regression, or the sum of squared deviations of the predicted values from the mean.

SST total sum of squares, or the sum of squared deviations of the observed values from the mean.

stability (of a process) a process is said to be stable if it shows no recognizable pattern of change.

standard (measurement) a reference item providing a known value of a quantity to be measured. Standards may be primary — i.e., the standard essentially defines the unit of measure — or secondary (transfer) standards, which have been compared to the primary standard (directly or by way of an intermediate transfer standard). Standards are used to calibrate instruments which are then employed to make routine measurements.

standard deviation one of the most common measures of variability in a data set or in a population.

standardized normal distribution a normal distribution or random variable having a mean and standard deviation of 0 and 1, respectively. It is denoted by the symbol Z and is also called the Z distribution.

statistic a value calculated from a random sample which is used to estimate a population parameter.

statistical control (of a process) a process is said to be in a state of statistical control when it exhibits only random variation.

statistical inference the process of drawing conclusions about a population on the basis of statistics.

statistical process control (SPC) the use of basic graphical and statistical methods for analyzing and controlling the variation of a process, and thus continuously improving the process.

statistical quality control (SQC) the application of statistical methods for measuring and improving the quality of processes. SPC is one method included in SQC.

statistics numerical measures obtained from a sample (as opposed to parameters, which are numerical measures of a population).

stem-and-leaf plot a graphical means of displaying data. It is similar to a histogram but provides more specific information about the elements within a class (or stem).

subgroup for control charts: a sample of units from a given process, all taken at or near the same time.

systematic variation (of a process) variation which exhibits a predictable pattern. The pattern may be cyclic (i.e., a recurring pattern) or may progress linearly (i.e., a trend).

t-distribution a symmetric, bell-shaped distribution that resembles the standardized normal (or Z) distribution, but it typically has more area in its tails than does the Z distribution. That is, it has greater variability than the Z distribution.

t-test a hypothesis test of population means when small samples are involved.

test statistic a single value which combines the evidence obtained from sample data. The P-value in a hypothesis test is directly related to this value.

tolerance the permissible range of variation in a particular dimension of a product. Tolerances are often set by engineering requirements to ensure that components will function together properly. In DOE, a measure (from 0 to 1) of the independence among independent variables.

total quality management (TQM) a management philosophy of integrated controls, including engineering, purchasing, financial administration, marketing and manufacturing, to ensure customer satisfaction and economical cost of quality.

trend a gradual, systematic change with time or some other variable.

two-level design an experiment where all factors are set at one of two levels, denoted as low and high (-1 and +1).

two-tailed test also known as a two-sided test, it is a hypothesis test with a two-sided alternative hypothesis. That is, one could possibly err on either side of the center.

Type I error concluding H_1 (or rejecting H_0) when H_0 is really true.

Type II error concluding H_0 when H_1 is really true.

u-chart for attribute data: a control chart of the average number of defects per part in a subgroup.

UCL (upper control limit) for control charts: the upper limit below which a process statistic ($\bar{x}$, R, etc.) must remain to be in control. Typically this value is 3 standard deviations above the center line.

uniform distribution a distribution in which all outcomes are equally likely.

upper confidence limit the larger of the two numbers that form a confidence interval.

USL (upper specification limit) the highest value of a product dimension or measurement which is acceptable.

variability the property of exhibiting variation, i.e., changes or differences, in key measurements of a process.

variables quantities which are subject to change or variability.

variables data concerning the values of a variable, as opposed to attribute data. A dimensional value can be recorded and is only limited in value by the resolution of the measurement system.

variance a measure of variability in a data set or population. It is the square of the standard deviation.

x̄ and R charts for variables data: control charts for the average and range of subgroups of data.

x̄ and s charts for variables data: control charts for the average and standard deviation (sigma) of subgroups of data.

y-intercept the term b_0 in the prediction equation $\hat{y} = b_0 + b_1 x$.

z-value a standardized value formed by subtracting the mean and then dividing this difference by the standard deviation.

Table of Symbols

α	probability of a Type I error
β	probability of a Type II error
Δ	effect
$\Delta/2$	half-effect
Σ	(capital sigma) summation
σ	(lower case sigma) population standard deviation
σ^2	population variance
$\hat{\sigma}$	estimated standard deviation
s	sample standard deviation
s^2	sample variance
s_p	pooled sample standard deviation
s_{xy}	sample covariance
μ	population mean
n	sample size
N	population size
$\bar{x}$	sample mean (for data set x)
$\tilde{x}$	sample median (for data set x)
E(X)	expected value of X (a population mean)
V(X)	variance of X (a population variance)
f(x)	notation for a probability density function
F(x)	notation for a cumulative distribution function
λ	average rate of occurrence
π	population proportion
ν	degrees of freedom
p	probability or sample proportion
q	probability or sample proportion equal to $1 - p$
P_{80}	80th percentile
$\approx$	approximately equal to
$\lceil x \rceil$	the smallest integer greater than or equal to x
e	base of natural logarithm (≈ 2.718)
b_0	intercept of the simple linear regression line

b_1	slope of the simple linear regression line
$r = R$	sample linear correlation coefficient
R^2	sample coefficient of determination
$\mathbb{R}^2$	population coefficient of determination
T	target or nominal specification value
χ_0^2	calculated χ^2 test statistic
F_0	calculated F test statistic
t_0	calculated t test statistic
Z_0	calculated standardized normal Z test statistic
H_0	null hypothesis
H_1	alternative hypothesis
S	sample space
$\cap$	intersect ("and")
$\cup$	union ("or")
$A^* = \bar{A}$	complement of event A
$\varnothing$	empty set
▮	denotes the end of an example
P(A)	probability of event A (a marginal probability)
P(A∩B)	probability of event A *and* event B (a joint probability)
P(A|B)	probability of event A *given* that event B has already occurred (a conditional probability)
P(A∪B)	probability of event A *or* event B
n!	n factorial
$\binom{n}{k}$	number of combinations of n things taken k at a time
$\hat{y}$	predicted y (response) value
df	degrees of freedom
e_i	error of prediction for the ith data point or i^{th} residual
$e_{i,-i}$	press residual
CDF	cumulative distribution function
CE	cause and effect
CNX	constant, noise, experimental

dpm	defects per million
C_p	process potential index (assumes a process is centered on target)
C_{pk}	process capability index (does not assume a centered process)
L	loss
ANOVA	analysis of variance
COPQ	cost of poor quality
DET	probability of escaped detection
DOE	design of experiments
LCL	lower control limit (for control charts)
LSL	lower specification limit
MR	multiple regression
MSB	mean square between
MSE	mean square error
OCC	probability of cause occurring
pdf	probability density function
PF	process flow
pmf	probability mass function
QFD	quality function deployment
ROT	rule of thumb
RPN	risk priority number
RV	random variable
S/N	signal-to-noise ratio
SEV	severity of failure effect
SLR	simple linear regression
SOP	standard operating procedure
SPC	statistical process control
SSB	sum of squares due to between group difference
SSE	sum of squares due to error
SSR	sum of squares due to regression
SST	sum of squares total
TQM	total quality management
UCL	upper control limit (for control charts)
USL	upper specification limit

Student Version
User's Guide

Keep It
Simple Statistically

Air
Academy
Associates LLC
(719) 531-0777

License Agreement

This book cannot be returned for credit or refund if the seal on the disk envelope has been broken or tampered with in any way. If you do not accept the terms of this license, you must return the book with the disk seal unbroken immediately to the party from whom you received it.

License Grant

This license is granted to the Licensee to possess, use, and make limited copies of the software licensed hereunder only on the terms and conditions specifically set out in this document. The Licensee evidences its acceptance of these terms and conditions by opening the product package or by any use of the licensed product.

The Licensee may:

• Use one copy of the Software on a single computer ("Dedicated Computer");

• Make one copy of the Software for archival purposes, or copy the Software onto the hard disk of your computer and retain the original for archival purposes;

• Transfer the entire Product on a permanent basis to another person or entity, provided you retain no copies of the Product and the transferee agrees to the terms of this agreement;

• Access the Software from a hard disk, over a network, or any other method you choose, so long as you otherwise comply with this agreement;

• Use on a network, however, only one computer may be using the program for each copy of the program owned.

Disclaimer of Warranty

This Software and Manual are sold "AS IS" and without warranties as to performance or merchantability. The seller's salespersons may have made statements about this software. Any such statements do not constitute warranties and shall not be relied on by the buyer in deciding whether to purchase this Program.

The Program is sold without any express or implied warranties whatsoever Because of the diversity of conditions and hardware under which this Program may be used, no warranty of fitness for a particular purpose is offered. The user must assume the entire risk of using the Program. Any liability of seller or manufacturer will be limited exclusively to Product replacement or refund of the purchase price.

Limited Warranty

This Product is warranted to be free of defects in materials and workmanship for a period of 90 days from your receipt of this Product. Workmanship is defined as the media, documentation, and associated packaging. If the Product fails to comply with the warranty set forth herein, the entire liability and your exclusive remedy will be replacement of the disk or, at our option, reasonable effort to make the Product meet the warranty set forth. This limited warranty applies only if you return all copies of the Product, along with a copy of your paid invoice, to the seller, shipping prepaid.

Limit of Liability

In no event shall Digital Computations, Inc. (DC) or Air Academy Associates,LLC (AAA), or its suppliers be liable for any damages whatsoever (including, but not limited to, damages for loss of profits, business interruption, loss of information, or other pecuniary loss) arising out of the use of or inability to use this Product, even if DC has been advised of the possibility of such damages.

Digital Computations, Inc. and Air Academy Associates, LLC, disclaim all other warranties, either express or implied, including but not limited to implied warranties of merchantability and fitness for a particular purpose, with respect to the Product.

General

This agreement constitutes the entire agreement between you and DC and AAA and supersedes any prior agreement concerning the contents of this package. It shall not be modified except by written agreement dated subsequent to the date of this agreement and signed by an authorized DC representative. DC is not bound by any provision of any purchase order, receipt, acceptance, confirmation, correspondence, or otherwise, unless DC specifically agrees to the provision in writing. This agreement is governed by the Federal laws of the United States of America.

Save This License
For Future Reference.

Installation

INSTALLING **SPC KISS**

System Requirements

To use SPC KISS Student Version for Windows or Macintosh, you need a system capable of running Microsoft® Excel 5.0 or higher.

Windows™

Installing SPC KISS

To install SPC KISS Student Version, insert the floppy into your floppy drive and double-click on the Setup.exe icon (for Windows 3.1x, run A:\Setup.exe.). Follow the instructions on the screen.

Running SPC KISS

When the installation is complete, you will see the window that contains the SPC Student Version icons. To run SPC KISS, double-click on the icon labeled SPC KISS Student Version.

You can also run SPC KISS from the Start menu in Windows 95 or Windows NT. The default folder is SPC KISS Student Version.

Either of these methods will launch Microsoft Excel and SPC KISS Student Version. See the Program Description Overview (following) for more information on how SPC KISS works.

Uninstalling SPC KISS

If you need to remove SPC KISS from your hard drive, use the Uninstall SPC KISS Student Version icon. The icon is located in the window that contains all the SPC Student Version icons. You can also uninstall SPC KISS by going to the Start menu in Windows 95 or Windows NT. The default folder is SPC KISS Student Version.

Macintosh®

Installing SPC KISS

To install SPC KISS, open Setup using these instructions.
1.   Insert the SPC KISS installation disk into the floppy drive.
2.   From the installation disk, launch the "Setup" program. This is done by double clicking on the "Setup.exe" icon.
3.   Follow the installation instructions. You can accept the default directory destination, or change the name of the destination folder.

Running SPC KISS

After SPC KISS is installed, you will be able to run it by selecting the SPC KISS menu item from Excel.

Note: You will not see the SPC KISS menu option unless you have a worksheet open in Excel.

TROUBLESHOOTING

Excel 97 Users

When opening SPC KISS Student Version you may see the following dialog box.

If you want SPC KISS Student Version to run you must click 'Enable Macros'.

Windows Users

If SPC KISS and Excel do not start when you launch SPC KISS (using either method described in the section "Running SPC KISS"), you will need to add-in SPC KISS to Excel. To do this, simply open the file "ADDIN.XLA", located in the same directory in which you installed SPC KISS. The file "ADDIN.XLA" will add SPC KISS to the list of Excel Add-Ins and you will always have access to SPC KISS, regardless of how you start Excel.

Note: You should not start Excel from the SPC KISS icon after you add in SPC KISS using "ADDIN.XLA".

All Excel Users

1. *Problem* - Error message stating "Runtime Error 1005: Unable to set zoom properties of the Windows Class".

    Solution - From Excel, select "Tools" - "Options" - "Chart" from the menu bar. The chart type must be set to "Built-In", not "MS Excel 4.0".

2. *Problem* - Error message stating "Cannot find SPCDEF.XLS"

    Solution - Copy SPCDEF.XLS from the Excel directory to the XLSTART directory (probably C:\EXCEL\XLSTART).

3. *Problem* - Error message stating "File format not supported", "File Format No Longer Available", or "Cannot access file SPCKISS".

    Solution - Uninstall and reinstall Excel.

Program Description

OVERVIEW

SPC KISS allows you to analyze your data using several different functions. You may create diagrams, control charts, or use analysis tools (such as multiple regression, correlation matrix, independence test matrix) to analyze your data. You may also access a variety of probability distributions, as well as conduct a Measurement System Analysis.

Note: The SPC KISS Student Version is limited in functionality to a data area of 30x5 (or 5x30). Other limitations are listed in the on-line help system under Student Version Limits. The full version of SPC KISS can be obtained from Air Academy Associates, LLC, at (800) 748-1277.

Before You Begin

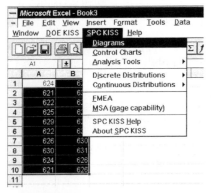

For all diagrams, charts, and analysis functions, you may begin with selecting your data range by clicking on the Excel cell in the upper left corner of your data and, while holding the left mouse button down, drag the mouse down or down and to the right until you have highlighted the desired range of data (release the mouse button). You may also highlight non-contiguous data by selecting the first group of data, then holding down the Control (Ctrl) key while you select the next group of data. Then go to the "SPC KISS" pull-down menu item and select "Diagrams", "Control Charts", or "Analysis Tools".

DIAGRAMS AND CONTROL CHARTS

For the creation of diagrams and control charts, you will be prompted to follow the four step process described here. For any questions, see the on-line help system for information.

Step 1 - Selecting Your Diagram or Control Chart

You may choose from six diagrams or seven control charts. Simply click on the diagram or chart of your choice to proceed to the next step.

Step 2 - Selecting Your Data

If you have already chosen the correct data range (see *Before You Begin*), select Next > to proceed.

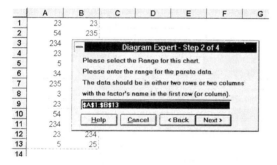

If you need to change the range, simply select the new range of data using the mouse. Then select Next > to continue. (You can drag the Step 2 dialog box out of the way so you can see all of your data if necessary.)

You may also type in the correct range in the edit box of the Step 2 window; be sure that you input the range in the proper format.

Example: $A$1:$B$13 denotes numbers in cells A1 through and including B13 on your Excel worksheet. You can type A1:B13 to specify that range.

Remember: Each diagram and control chart accepts data in a different format. See the on-line help system for information on specific data formats.

Step 3 - Options

Diagrams

You have several options for viewing your data in the completed diagram. Each diagram has its own set of options. See the on-line help system for information on a specific diagram's options.

Control Charts

There are three different methods available to establish control limits: Shewhart control limits, control limits based upon standard deviations, and manual control limits. See the on-line help system for more information.

Step 4 - Diagram Chart Titles

With SPC KISS, you have the ability to edit the titles on your diagram. Each diagram has its own default titles to begin with, and you can change those defaults by saving your chart titles with the Save Options button.

ANALYSIS TOOLS

For any questions, see the on-line help system for information, including the data format for each analysis tool.

Analysis Tools consist of the following: multiple regression, correlation matrix, F test matrix, t test matrix, and independence test matrix.

For all of the analysis tools, select the data on the spreadsheet, then select the analysis tool that you want to use. For multiple regression, the data must be in columns. For all other tools, the data may be in rows or columns.

Below is an example data set for a Correlation Matrix. If you label your columns or rows, those labels will be used in the output to identify the factors. If you do not label your factors, SPC KISS will identify them as A, B, C, etc., in the output table.

Question A	Question B	Question C
1	4	3
2	2	2
4	5	3
3	3	4
5	1	2
2	4	1
3	2	4

In the output table below, the highlighted number shows the correlation coefficient between Question B and Question C.

Correlation Matrix			
	Question A	Question B	Question C
Question A	1.000		
Question B	0.563	1.000	
Question C	0.431	-0.070	1.000

DISCRETE AND CONTINUOUS DISTRIBUTIONS

For any questions, see the on-line help system for information.

You can access the Binomial, Poisson, Discrete Uniform, Continuous Uniform, Exponential and Normal distributions. (The full version of SPC KISS has six discrete distributions and fifteen continuous distributions.) Select the distribution you want from the menu bar. SPC KISS will add a spreadsheet to the current workbook with the distribution ready for use.

MEASUREMENT SYSTEM ANALYSIS (MSA)

OR GAGE CAPABILITY

For any questions, see the on-line help system for information.

For MSA, the data must be in columns (not rows) and each data set much contain at least two columns. One data set might consist of the same operator measuring 20 different items (i.e., 20 rows) two times each (i.e., 2 columns). A second data set might be another 20x2 matrix in the spreadsheet which corresponds to a second operator's measurements on the same 20 items as the first operator measured. Data sets normally correspond to different operators, different measuring devices (with same operator), or possibly different operator/measuring device combinations.

Answers to Selected Problems

CHAPTER 1

1.1   Politics, medicine, law, sports, and advertising (there are a number of other possible answers).

1.2   "Continuous improvement" means using statistical tools to reduce waste and improve quality by continually refining processes. It is important in keeping a global competitive advantage. Consumers constantly demand higher quality at lower prices, calling for more improvement in goods and services.

1.3   Statistics is the medium through which knowledge is gained to enhance our decision making process.

1.4   **THE DECLARATION OF INDEPENDENCE**
This phrase was encoded by taking each letter in this phrase and moving three letters beyond it in the alphabet to produce the corresponding symbol in the cryptogram. This particular technique of encoding is known as the Caesar Cipher, because Julius Caesar is known to have used this method more than 20 centuries ago.

1.5   The new philosophy of quality focuses on deviations or variability, whereas the old philosophy dealt with meeting specifications.

1.7   KBM is both a philosophy and a strategy. It has three primary ingredients: questions managers need to answer, questions managers need to ask, and the tools required to answer the questions and improve the scorecard.

1.9   Continuous process improvement and the use of statistical tools to gain the knowledge needed to drive the improvement.

1.12   A valid metric is an objective indicator or measure used to facilitate process improvement.

CHAPTER 2

2.6   One possible histogram is shown here.

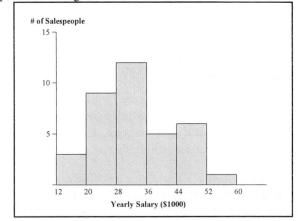

2.16   Age: continuous
Eye color: discrete
Height: continuous

2.18   (a) 79.25 inches   (b) 80.50 inches   c) skewed left

2.19   (a) 14 inches   (b) 24.20 in^2 (c) 4.92 inches   (d) inches

2.20   (d) $288   (e) $296.77   (f) $296.77   (g) $409.09
(h) $172.76   (i) $16,117.17

2.21   (a) S_{xy} = 1288.13, indicating a positive linear relationship between
the variables
(b)   r = .99892
(c)   an extremely strong positive linear relationship between CPU time
and I/O operations

2.22   (a) 78.485   (b) 15.606   (c) 478.788   (d) .908
(e) Yes, a fairly strong positive linear relationship

CHAPTER 3

3.1   b

3.2   a

3.3   (a) $S = \{1,2,3,4,5,6\}$ (b) $B = \{1,2,3\}$ (c) $C = \{3,6\}$
        (d) $B \cup C = \{1,2,3,6\}$ (e) $B \cap C = \{3\}$ (f) $C^* = \{1,2,4,5\}$

3.4   (a) $1/2$ (b) $1/3$ (c) $2/3$
        (d) $1/6$ (e) $2/3$ (f) 1

3.5   (a) $P(B) \cdot P(C) = (1/2)(1/3) = 1/6 = P(B \cap C)$
        (b) Yes, because $P(B) \cdot P(C^*) = (1/2)(2/3) = 1/3 = P(B \cap C^*)$
        (c) Yes, because $P(B^*) \cdot P(C) = (1/2)(1/3) = 1/6 = P(B^* \cap C)$
        (d) Yes, because $P(B^*) \cdot P(C^*) = (1/2)(2/3) = 1/3 = P(B^* \cap C^*)$

3.6   Since A and B are mutually exclusive, $P(A \cap B) = 0$
        Since A and B are independent, $P(A \cap B) = P(A) \cdot P(B)$
        Therefore, $P(A) \cdot P(B) = 0$ implies $P(A) = 0$ or $P(B) = 0$

3.7   (a)   Employee Class B and Overtime Pay are independent.
        (b)   No, because not all joint probabilities are equal to the products of
              their respective marginal probabilities.

3.8

	SNOW	NO SNOW	TOTAL
ON TIME	.16	.18	.34
LATE	.64	.02	.66
TOTAL	.80	.20	1.00

3.9   (a) .57   (b) .76   (c) .78

3.10   (a) .08   (b) .16

3.11   .9

3.12   (a) .49   (b) .7   (c) .91

3.14   (a)

	C	C*	Total
P	.0014	.0086	.01
N	.0105	.1395	.15
S	.0420	.7980	.84
Total	.0539	.9461	1.00

(b)   .026

3.15   .1045

3.16   .2579

3.17   11,232,000

3.18 7.4×10^{11}

3.19   9,261,000

3.20 $1.0910866882 \times 10^{37}$

3.21 $1.2154866003 \times 10^{14}$

3.22   (a) 18!   (b) (2)(16!)

3.23   2,598,960

3.24   1,166,803,110

CHAPTER 4

4.1   (a) .5   (b) .45 (assuming "0" is an even integer)
         (c) .7   (d) .25

4.2   (a) .6   (b) 1.1   (c) $880   (d) .4

4.3   (a) 1.9   (b) 1.19   (c) .2025

4.4   (a) 2/3   (b) 3.5   (c) 1.7078

4.5   (a)

t	0	1	2	3	4	5	6	7	8	9	10
P(t)	.075	.1375	.1725	.17	.1575	.1275	.08	.045	.0225	.01	.0025

　　　　(b) E(T) = 3.35   V(T) = 4.4275
　　　　(c) E(D) = .15   V(D) = 4.4275
　　　　(d) .43

4.6   .1209, .2272

4.7   .3487

4.8   .6331

4.9   (a) 6   (b) 3.6   (c) .2131

4.10   .7361

4.11   (a) .7385   (b) 1   (c) .818

4.12   .0154

4.13   (a) .2019   (b) .3230   (c) .4751

4.14   (a) Binomial with n=12,000,000 and p=.000001
　　　　(b) Using λ = np = 12, P(X ≥ 5) = .9924

4.15   (a) .4405   (b) .3840   (c) because P(X=5) is not included in either

4.16   .0116

4.17   (a) .1024   (b) .0027

4.18   (a) .5889   (b) 240.77 sec   (c) 10th percentile of X

4.19   .0124

4.20   .0456

4.21   .0228

4.22   (a) 0.13%   (b) 1068
      (c) 68.27% within $\pm$ 1σ
         95.45% within $\pm$ 2σ
         99.73% within $\pm$ 3σ

4.23   .3679

4.24   (a) .0498   (b) .0498

4.25   (a) .2532   (b) .7468

4.26   for Treatment A: .1587
      for Treatment B: .2466;   choose B

4.27   1.0000

4.28   .3918

4.29 μ = 1926, σ = 800

4.30   (a) 22.12%   (b) $1389.40

4.31   .5719

4.32   .4869

4.33   (a) 0.62%   (b) $25,000

CHAPTER 5

5.2

$\bar{x}$	1.00	1.25	1.50	1.75	2.00	2.25	2.50	2.75	3.00
$P(\bar{x})$	1/81	4/81	10/81	16/81	19/81	16/81	10/81	4/81	1/81

$$E(\bar{X}) = 2 \qquad \sigma^2(\bar{X}) = 1/6$$

5.4   .3594

5.5 $3.031 \leq \mu$

5.6   1, .5; must quadruple the sample size in order to cut the standard error of the mean in half.

5.7   (a) n = 1   (b) .9676

5.8 $222.952 \leq \mu \leq 257.048$

5.9   (a) 2   (b) 1/3
          (c) approximately normal with $E(\bar{Y})=2$ and $\sigma(\bar{Y})=1/3$
          (d) approximately Z-distributed

5.10   (a) 2   (b) .8944
          (c) unknown because n=5 is small
          (d) unknown because the form of $\bar{Y}$ is unknown

5.11 $51,744 \leq \mu \leq 68,256$; the assumption of normality is necessary in order to use the t distribution because n is small.

5.12   135

5.13 $-33.12 \leq \mu \leq 38.12$

5.14   35-12 = 23 more households

5.15 $7.58 \leq \mu$

5.18   (a) 68.27%   (b) .975   (c) .0000
          (d)   The process is not in control because the likelihood of a stable, predictable process producing a sample mean of .3 is highly unlikely.

5.19 α, σ, and n

5.20 $2.52 \leq \mu \leq 5.68$

5.21   (a) 278   (b) $42 \leq \mu \leq 46$ (c) 97.74% confident

5.22   (a) Z-distributed   (b) t-distributed with n−1 df
(c) the sample size must be large enough (n ≥ 30)

5.23 $222.74 \leq \mu$

5.24 $\pi \leq .112$

5.26   (a) $933.82 \leq \mu \leq 952.18$ (b) $933.82 \leq \mu$
(c) the same, because a 90% two-sided confidence interval distributes the
area α equally on both sides (i.e., .05 on each side, thereby producing the
same lower limit as a 95% lower confidence interval).

5.30   The young pilot's interpretation is not correct. The proper interpretation
is that the <u>mean</u> (or <u>average</u>) noise level of <u>all</u> T-37s is at least 85 db. The
pilot is interpreting a confidence interval (or limit) more as a percentile,
which it is not.

CHAPTER 6

6.1   Conclude H_1: $\mu \neq 6$ with 99.74% confidence

6.2   (a) $P = .097$, so fail to reject H_0: $\mu = 0$
(b) $P = .005$, so reject H_0: $\mu = 0$ with 99.5% confidence

6.3 $P = .2033$ so could conclude H_1: $\mu > 3000$ with only 79.67% confidence,
far from overwhelming evidence.

6.4   (a) H_0: $\mu \geq 2000$ (b) H_0: $\mu \leq 2000$
             H_1: $\mu < 2000$ H_1: $\mu > 2000$

6.5   (a) $Z_0 = -3.13$ and $P = .0009$, so yes with 99.91% confidence
(b)   e.g., cost of living may be lower in that area

6.6   (a) $Z_0 = 3.44$ and $P = .0003$, so can conclude H_1: $\mu > 2.80$ with
99.97% confidence
(b)   34% (if obtained from raw data alone)
31.21% (if assuming $\mu = \bar{x} = 2.85$ and $\sigma = s = .1028$)

6.7 (a) Normality must be assumed in order to use a t test

(b) $t_0 = 4.024$. Since $P < .001 < \alpha = .005$, we can conclude $H_1: \mu \neq 2.0$ with at least 99.9% confidence

(c) cannot conclude they are different

(d) same conclusion as in (b)

6.8 $t_0 = 3.118$ and $P = .0024$ so conclude $H_1: \mu > 12$ with 99.76% confidence

6.9 $Z_0 = .57$ and $P = .2843$ so fail to reject $H_0: \pi \leq .5$

6.10 $Z_0 = -3.18$ and $P = .0007$ so conclude $H_1: \mu_2 - \mu_1 < 0$ with 99.93% confidence

6.11 (a) No   (b) Yes, with more than 99.95% confidence

6.13 $\chi^2_0 = 25.5$ and $P = .8529$, so not significantly better

6.14 (a) No ($P = .3340$ from $Z_0 = .966$)

(b) The West Coast market share appears to exceed the 15% level with a significance of $P = .0147$ (from $Z_0 = 2.178$).

6.16 (d) $\chi^2_0 = 5.213$ and $P > .10$, so fail to reject hypothesis that the distribution is normal.

6.17 (b) fail to reject $H_0: \mu_1 = \mu_2$

(c) fail to reject H_0: the two makes have the same mileage distribution ($P >> .10$)

6.18 (a) $\chi^2_0 = 39.08$ and $P < .005$, so conclude that degree and sex are not independent with more than 99.5% confidence

6.19 fail to reject H_0: Binomial with $n = 12$, $p = .5$

CHAPTER 7

7.1   (c) $\Sigma x = 841$ $\Sigma y = 1885$ $\Sigma x^2 = 58985$
                  $\Sigma y^2 = 300153$ $\Sigma xy = 132506$
                  $\bar{x} = 70.0833$ $\bar{y} = 157.0833$

      (d) $\hat{y} = -465.345 + 8.881x$

      (e)   0. Yes, this will always result in 0.

      (f)   4050.91667

      (g)   3542.88327

      (h)   508.0334

      (i)   Yes

      (j)   .935

      (k) $R^2 = .875$ or 87.5% of the variability in y can be accounted for by y's linear relationship with x.

      (l)

Source	SS	df	MS	F_0
Regression	3542.88327	1	3542.88327	69.7372
Error	508.03340	10	50.80334	
Total	4050.91667	11		

      (m) $F(.99; 1,10) = 10.0$. Since $F_0 = 69.7372$, $P < .001$. We conclude H_1: $\mathbb{R}^2 \neq 0$ with at least 99.9% confidence.

      (n)   [132.81, 179.84]

7.2 $\hat{y} = 63.507 - 1.521x$ $F_0 = 90.849$
Since $P < .001$, conclude H_1: $\mathbb{R}^2 \neq 0$ with at least 99.9% confidence. A very useful model.

7.3   (a)   y as a function of x_1 alone yields
                 $\hat{y} = 8.167 + .005x_1$ with $F_0 = 1.064$ and $P = .318$

      (b)   y as a function of x_2 alone yields
                 $\hat{y} = 5.667 + .217x_2$ with $F_0 = 30.844$ and $P = .000$

      (c)   y as a function of both x_1 and x_2 yields
                 $\hat{y} = 1.667 + .005x_1 + .217x_2$ with $F_0 = 19.36$ and $P = .000$

Although the model in (c) gives an R^2 of .721 while the model in (b) yields an R^2 of .658, the model in (b) may be more useful because of a single predictor variable (x_2). The launch range variable (x_1) contributes very little to the miss distance predictions.

7.4   (b) $\hat{y} = 4.675 - .061x$
    (c) $\hat{y}|_{x=31} = 2.784$
    (d)

Source	SS	df	MS	F_0
Regression	4.441	1	4.441	10.402
Error	9.393	22	0.427	
Total	13.834	23		

    (e) $F_0 = 10.402$; conclude H_1: $\beta_1 \neq 0$ with 99.6% confidence
    (f) $P = .004$
    (g)   significant linear relationship between temperature and o-ring failures
    (h)   [.718, 4.906]
    (i)   fairly likely there would be at least one o-ring failure at 31° F

7.5   (a) $\hat{y} = -3.447 + .751x_1$; $F_0 = 97.852$; $P = .000$
    (b) $\hat{y} = -20.975 + 1.157x_2$; $F_0 = 49.267$; $P = .000$
    (c) $\hat{y} = -7.991 + .663x_1 + .160x_2$; $F_0 = 47.083$; $P = .000$
    (d) x_2 turns out to be quite insignificant when both x_1 and x_2 are in the model; model (a) also has a lower standard error; thus, model (a) is a good, simple model to use.

CHAPTER 8

8.2   (a)

Factor:	A	B	AB	C	AC	BC	ABC
F-Value:	2589.6	.831	.713	.092	6.76	.713	.001

    $\hat{y} = 3.282 + .942\,A + .048\,AC + .006\,C$

    (b)   None of the factors are significant for minimizing the variance. Thus, to determine the optimal settings, set $\hat{y} = 3$, $C = -1$ (which assumes the "-1" vendor is less expensive), and solve for A in the above equation. This results in $A = -.31$, which means:
    Set Mold Temperature to 3.69° C;
    Set Pour Time to whatever is most convenient/least expensive;
    Use Vendor A, the one associated with the "-1" setting.

CHAPTER 9

9.1   (a)   The process appears to be in statistical control. That is, there are no points outside the control limits and no long runs.

   (b) $\hat{\sigma} = \bar{R}/d_2 = .027/2.06 = .013$

   $Z_{USL} = (1.08 - 1.044)/.013 = 2.769$

   $Z_{LSL} = (1.00 - 1.044)/.013 = -3.385$

   $Z_{min} = 2.769$

   $C_{pk} = 2.769/3 = .923$

   (c)   There appears to be somewhat of an inverse relationship between the process location and the process variation. This is something the analyst should investigate because such a pattern would not result from only common causes of variation. In light of further analysis, the answer to (a) might turn out to be "no."

9.2 $C_{pk} = .107$; the process is capable at the $0.321 \, \sigma$ level.

CHAPTER 10

10.1   .9512, .6065, .3679, .2231, .1353, .0821, .0498

10.2   (a) $R_1(20) = .8078$ (b) $R_1(20|10) = .8091$

   $R_2(20) = .3859$ $R_2(20|10) = .4118$

10.3   (a)   .80, .63, .42, .24, .14, .10, .02

   (c) $\hat{\lambda} = 1/\bar{x} = 1/.425 = 2.35 = h(t)$

10.4   .3679

10.5   (a) .1003   (b) .6056   (c) .5361   (d) .7402   (e) .7402

10.7   .88407

10.8   (a) Parallel System   (b) .9975

10.9   (a)   for Components 1, 2, and 3, MTBF $= 20$ hours

   for Component 4, MTBF $= 22$ hours

   (b)   .9451

10.11   Using $p = R(9) = e^{-.05(9)} = .6376$, $n = 13$ helicopters

Index